MECHANICS OF SOLIDS

It is a massy wheel
Fixed on the summit of the highest mount,
To whose huge spokes ten thousand lesser things
are mortis'd and adjoin'd; which when it falls,
Each small annexement, petty consequence,
Attends the boist'rous ruin.

Shakespeare: *Hamlet* II, iii

CARLISLE THOMAS FRANCIS ROSS, DSc

Carlisle Thomas Francis Ross was born in Kharagpur, India and educated in Bangalore at St Joseph's European High School during the closing years of the British Raj. Coming to England, he attended the Chatham Technical School for Boys (1948-51), followed by part time education at the Royal Dockyard Technical College (1951-56) at Chatham, Kent where he served a five-year shipwright apprenticeship. He proceeded to King's College, Newcastle-upon-Tyne (University of Durham), obtaining a B.Sc. (Hons) degree in naval architecture (1956-59); also working as a part-time draughtsman at HM Dockyard during university vacations.

For the next two years he laid the foundations of his powerful industrial experience as Designer in the Project Design Office at Vickers-Armstrongs (Shipbuilders), Barrow-in-Furness (1959-61). His outstanding work was acknowledged by promotion to the position of Deputy Chief of the Project Design Office. He next worked as a research graduate in the Department of Engineering, University of Manchester (1961-62), where in 1963 he gained his Ph.D. for research in stress analysis of pressure vessels.

He brought his industrial experience from Vickers-Armstrongs into teaching, first as Lecturer in Civil and Structural Engineering at Constantine College of Technology, Middlesborough (now University of Teesside) (1964-66), and later to Portsmouth Polytechnic (now University of Portsmouth) (1966-71) as Senior Lecturer in Mechanical Engineering, where he remains as Professor of Structural Mechanics.

His research research is based on computational methods, tested experimentally with colleagues. He has made important discoveries on the buckling of ring-stiffened cylinders and cones under external pressure, and has also developed the application of microcomputers on finite element analysis. In 1971 he co-invented the tube-stiffened pressure hull, making for greater submarine strength. In 1987 followed his structurally more efficient "bean can" pressure hull invention, and also the cheaper satisfactory dome-cup end for submarines. In 1995, with colleague David Jordan, came the invention of an important method for improving stability of roll-on roll-off car ferries, reducing risk of vessels capsizing in the event of flooding on the car deck. Currently he is developing a large drilling rig and a dredger, to enable oil and gas exploration in water to a depth of 5000 metres. This important application stems from his earlier research on pressure vessels when in Manchester University.

His contributions to engineering science were recognised in 1992 by the award of a Doctor of Science degree from the Council of National Academic Awards, London.

Mechanics of Solids

C.T.F. ROSS, BSc., PhD, DSc, CEng, FRINA
Professor of Structural Dynamics
University of Portsmouth
Portsmouth

WP
WOODHEAD
PUBLISHING

Oxford Cambridge Philadelphia New Delhi

Published by Woodhead Publishing Limited,
80 High Street, Sawston, Cambridge CB22 3HJ, UK
www.woodheadpublishing.com; woodheadpublishingonline.com

Woodhead Publishing, 1518 Walnut Street, Suite 1100, Philadelphia,
PA 19102-3406, USA

Woodhead Publishing India Private Limited, G-2, Vardaan House, 7/28 Ansari Road,
Daryaganj, New Delhi – 110002, India
www.woodheadpublishingindia.com

First published by Horwood Publishing Limited, 1999
Reprinted by Woodhead Publishing Limited, 2012

British Library Cataloguing in Publication Data
A catalogue record for this book is available from the British Library

ISBN 978-1-898563-67-9

To my children
Nicolette and Jonathan

Contents

Preface

Failure of most components, be they electrical or mechanical devices, can be attributed to stress values. These components vary from small electrical switches to supertankers, or from automobile components to aircraft structures or bridges. Thus, some knowledge of stress analysis is of great importance to all engineers, be they electrical or mechanical engineers, civil engineers or naval architects, or others.

Although, today, there is much emphasis on designing structures by methods dependent on computers, it is very necessary for the engineer to be capable of interpreting computer output, and also to know how to use the computer program correctly.

The importance of this book, therefore, is that it introduces the fundamental concepts and principles of statics and stress analysis, and then applies these concepts and principles to a large number of practical problems which do not require computers. The author considers that it is essential for all engineers who are involved in structural design, whether they use computer methods or not, to be at least familiar with the fundamental concepts and principles that are discussed and demonstrated in this book.

The book should appeal to undergraduates in all branches of engineering, construction and architecture and also to other students on BTEC and similar courses in engineering and construction. It contains a large number of worked examples, which are presented in great detail, so that many readers will be able to grasp the concepts by working under their own initiative. Most of the chapters contain a section on examples for practice, where readers can test their newly acquired skills.

Chapter 1 is on statics, and after an introduction of the principles of statics, the method is applied to plane pin-jointed trusses, and to the calculation of bending moments and shearing forces in beams. The stress analysis of cables, supporting distributed and point loads, is also investigated. Chapter 2 is on simple stress and strain, and after some fundamental definitions are given, application is made to a number of practical problems, including compound bars and problems involving stresses induced by temperature change. Chapter 3 is on geometrical properties, and shows the reader how to calculate second moments of area and the positions of centroids for a number of different two-dimensional shapes. These principles are then extended to 'built-up' sections, such as the cross-sections of 'I' beams, tees, etc. Chapter 4 is on stresses due to the bending of symmetrical sections, and after proving the well-known formulae for bending stresses, the method is applied to a number of practical problems involving bending stresses, and also the problem on combined and direct stress. Composite beams are also considered.

Chapter 5 is on beam deflections due to bending, and after deriving the differential equation relating deflection and bending moment, based on small-deflection elastic theory, this equation is then applied to a number of statically determinate and statically indeterminate beams. The area–moment theorem is also derived, and then applied to the deflection of a cantilever with a varying cross-section.

Chapter 6 is on torsion. After deriving the well-known torsion formulae, these are then applied to a number of circular-section shafts, including compound shafts. The method is also applied to the stress analysis of close-coiled helical springs, and then extended to the stress analysis of thin-walled open and closed non-circular sections. The elastic–plastic stress analysis of circular-section shafts is also considered.

Chapter 7 is on complex stress and strain, and commences with the derivation of the equation for direct stress and shear stress at any angle to the co-ordinate stresses. These relationships are later extended to determine the principal stresses in terms of the co-ordinate stresses. The method is also applied to two-dimensional strains, and the equations for both two-dimensional stress and two-dimensional strain are applied to a number of practical problems, including the analysis of strains recorded from shear pairs and strain rosettes. Mohr's circle for stress and strain is also introduced. Compliance and stiffness matrices are derived for the application to composite structures.

Chapter 8 is on membrane theory for thin-walled circular cylinders and spheres, and commences with a derivation of the elementary formulae for hoop and longitudinal stress in a thin-walled circular cylinder under uniform internal pressure. A similar process is also used for determining the membrane stress in a thin-walled spherical shell under uniform internal pressure. These formulae are then applied to a number of practical examples.

Chapter 9 is on energy methods, and commences by stating the principles of the most popular energy theorems in stress analysis. Expressions are then derived for the strain energy of rods, beams and torque bars, and these expressions are then applied to a number of problems involving thin curved bars and rigid-jointed frames. The method is also used for investigating stresses and bending moments in rods and beams under impact, and an application to a beam supported by a wire is also considered. The unit load method is also introduced. The elastic–plastic bending of beams is considered, together with the plastic design of beams.

Chapter 10 is on the theories of elastic failure. The five major theories of elastic failure are introduced, and through the use of two worked examples, some of the differences between the five major theories are demonstrated. Chapter 11 is on thick cylinders and spheres, and commences by determining the equations for the hoop and radial stresses of thick cylinders under pressure. The Lamé line is introduced and applied to a number of cases, including thick compound tubes with interference fits. The plastic theory of thick tubes is discussed, as are the theories for thick spherical shells and for rotating rings and discs.

Chapter 12 is on the buckling of struts, and commences by discussing the Euler theory for axially loaded struts. The theory is extended to investigate the inelastic instability of axially loaded struts through the use of the Rankine–Gordon formula, together with BS 449. The chapter also considers eccentrically loaded and initially curved struts, and derives the Perry–Robertson formula.

Chapter 13 is on the unsymmetrical bending of beams. This chapter covers the unsymmetrical bending of symmetrical section beams and the bending and deflections of beams of unsymmetrical cross-sections. It also shows how to calculate the second moments of area of unsymmetrical cross-sections.

Chapter 14 is on shear stresses in bending and shear deflections. This chapter shows how both vertical and horizontal shearing stresses can occur owing to bending in the vertical plane. The theory is extended to determine shear stresses in open and closed thin-walled curved tubes and to calculate shear centre positions for these structures. Shear deflections due to bending are also discussed.

Chapter 15 is on the matrix displacement method, and commences by introducing the finite element method. A stiffness matrix is obtained for a rod element and applied to a statically indeterminate plane pin-jointed truss, with the aid of a worked example.

Chapter 16 is on experimental strain analysis, and discusses a number of different methods in experimental strain analysis, in particular those in electrical resistance strain gauges and photoelasticity.

Notation

Unless otherwise stated, the following symbols are used:

E	Young's modulus of elasticity
F	shearing force (SF)
G	shear of rigidity modulus
g	acceleration due to gravity
I	second moment of area
J	polar second moment of area
K	bulk modulus
l	length
M	bending moment (BM)
P	load or pressure
R, r	radius
T	torque or temperature change
t	thickness
W	concentrated load
w	load/unit length
$\hat{x}$	maximum value of x
α	coefficient of linear expansion
γ	shear strain
ε	direct or normal strain
θ	angle of twist
λ	load factor
ν	Poisson's ratio
ρ	density
σ	direct or normal stress
τ	shear stress
P_e	Euler buckling load
P_R	Rankine buckling load

σ_{yp}	yield stress
W_c	plastic collapse load
τ_{yp}	yield stress in shear
$[k]$	= elemental stiffness matrix in local co-ordinates
$[k^0]$	= elemental stiffness matrix in global co-ordinates
$\{p_i\}$	= a vector of internal nodal forces
$\{q^0\}$	= a vector of external nodal forces in global co-ordinates
$\{u_i\}$	= a vector of nodal displacements in local co-ordinates
$\{u_i^0\}$	= a vector of nodal displacements in global co-ordinates
$[K_{11}]$	= that part of the system stiffness matrix that corresponds to the 'free' displacements
$[\Xi]$	= a matrix of directional cosines
$[I]$	= identity matrix
[ ]	= a square or rectangular matrix
{ }	= a column vector
⌊ ⌋	= a row vector
[0]	= a null matrix
NA	neutral axis
KE	kinetic energy
PE	potential energy
UDL	uniformly distributed load
WD	work done
2E11	2×10^{11}
3.2E-3	3.2×10^{-3}
*	multiplier
⇒	vector defining the direction of rotation, according to the *right-hand screw rule.* The direction of rotation, according to the right-hand screw rule, can be obtained by pointing the right hand in the direction of the open arrow, and rotating it *clockwise.*

Some SI units in stress analysis

s	second (time)
m	metre
kg	kilogram (mass)
N	newton (force)
Pa	pascal (pressure) = 1 N/m^2
MPa	megapascal (10^6 pascals)
bar	(pressure), where 1 bar = 10^5 N/m^2 = 14.5 lbf/in^2
kg/m^3	kilograms/cubic metre (density)
W	watt (power), where 1 watt = 1 ampere * 1 volt = 1 N m/s = 1 joule/s
hp	horse-power (power), where 1 hp = 745.7 W

Author's note on the SI system

Is it not interesting to note that an *apple* weighs approximately 1 *newton*?

Parts of the Greek alphabet commonly used in mathematics

α	alpha
β	beta
γ	gamma
δ	delta
Δ	delta (capital)
ε	epsilon
ζ	zeta
η	eta
θ	theta
κ	kappa
λ	lambda
μ	mu
ν	nu
ξ	xi
Ξ	xi (capital)
π	pi
σ	sigma
Σ	sigma (capital)
τ	tau
ϕ	phi
χ	chi
ψ	psi
ω	omega
Ω	omega (capital)

1

Statics

1.1 Introduction

All the structures that are analyzed in the present chapter are assumed to be in *equilibrium*. This is a fundamental assumption that is made in the structural design of most structural components, some of which can then be designed quite satisfactorily from statical considerations alone. Such structures are said to be *statically determinate*.

Most modern structures, however, cannot be solved by considerations of statics alone, as there are more unknown 'forces' than there are simultaneous equations obtained from observations of statical equilibrium. Such structures are said to be *statically indeterminate*, and analysis of this class of structure will not be carried out here.

Examples of statically determinate and statically indeterminate pin-jointed trusses are given in Figures 1.1 and 1.2, where the applied concentrated loads and support reactions (R and H) are shown by arrows.

1.2 Plane pin-jointed trusses

The structures of Figures 1.1 and 1.2 are called pin-jointed trusses because the joints are assumed to be held together by smooth frictionless pins. The reason for this assumption is that the solution of trusses with pin joints is considerably simpler than if the joints were assumed welded (i.e. rigid joints). It should be noted that the external loads in Figures 1.1 and 1.2 are assumed to be applied at the joints, and providing this assumption is closely adhered to, the differences between the internal member forces determined from a pin-joined truss calculation and those obtained from a rigid-jointed 'truss' calculation will be small as shown in Example 1.5.

The members of a pin-jointed truss are called *rods*, and these elements are assumed to withstand loads axially, so that they are in tension or in compression or in a state of zero load. When a rod is subjected to tension it is called a *tie* and when it is in compression, it is called a *strut*, as shown in Figure 1.3. The internal resisting

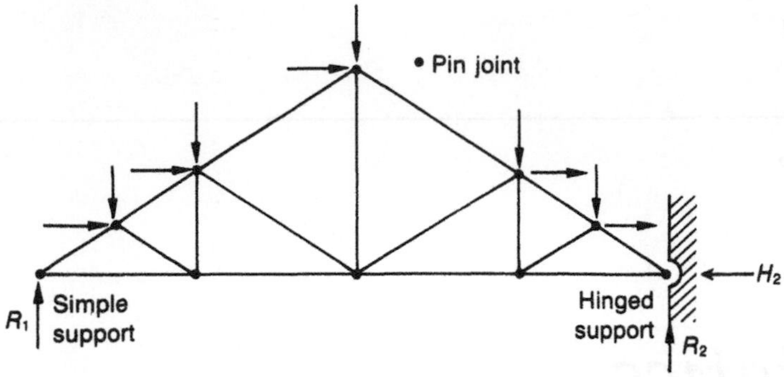

Figure 1.1 Statically determinate pin-jointed truss

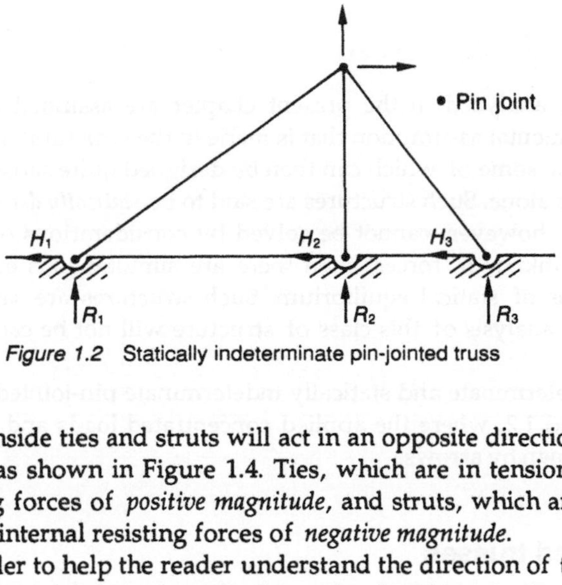

Figure 1.2 Statically indeterminate pin-jointed truss

forces inside ties and struts will act in an opposite direction to the externally applied forces, as shown in Figure 1.4. Ties, which are in tension, are said to have internal resisting forces of *positive magnitude*, and struts, which are in compression, are said to have internal resisting forces of *negative magnitude*.

In order to help the reader understand the direction of the internal forces acting in a rod, the reader can imagine his or her hands and arms acting together as a single rod element in a truss. If the external forces acting on the hands are trying to pull the reader apart, so that he or she will be in tension, then it will be necessary for the reader to pull inwards to achieve equilibrium. That is, if the reader is in tension, then he or she will have to pull inwards as shown by the diagram for the tie in Figure 1.4. Alternatively, if the reader is in compression, then he or she will have to push outwards to achieve equilibrium, as shown by the strut in Figure 1.4.

The structure of Figure 1.2 is said to be statically indeterminate to the first degree, or that it has one *redundant member*; this means that for the structure to be classified as a structure, one rod may be removed. If two rods were removed from Figure 1.2 or one from Figure 1.1, the structure (or part of it) would become a *mechanism*, and collapse. It should be noted, however, that if a rod were removed from Figure 1.2,

Tie (and externally applied loads)
Strut (and externally applied loads)

Figure 1.3 External loads acting on ties and struts

Tie, with internal resisting forces acting inwards (positive)
Strut, with internal resisting forces acting outwards (negative)

Figure 1.4 Sign conventions for ties and struts

although the structure can still be classified as a structure and not a mechanism, it may become too weak and collapse in any case.

1.3 Criterion for sufficiency of bracing

A simple formula, which will be given without proof, to test whether or not a pin-jointed truss is statically determinate, is as follows:

$$2j = r + R \tag{1.1}$$

where

j = number of pin joints

r = number of rods

R = minimum number of reacting forces

To test equation (1.1), consider the pin-jointed truss of Figure 1.1:

$$j = 10 \quad r = 17 \quad R = 3$$

Therefore

$$2 \times 10 = 17 + 3$$

i.e. statically determinate.

Now consider the truss of Figure 1.2:

$$j = 4 \quad r = 3 \quad R = 6$$

or

$$2 \times 4 = 3 + 6 \tag{1.2}$$

(i.e. there is one redundancy).

NB If $2j > r + R$, then the structure (or part of it) is a mechanism and will collapse under load. Similarly, from equation (1.2), it can be seen that if $2j < r + R$, the structure may be statically indeterminate.

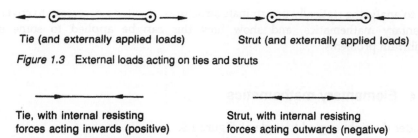

Prior to analyzing statically determinate structures, it will be necessary to cover some elementary mathematics, and show how this can be applied to some simple equilibrium problems.

1.4 Elementary mathematics

Consider the right-angled triangle of Figure 1.5.
From elementary trigonometry

$$\sin\theta = \frac{bc}{ac}$$

$$\cos\theta = \frac{ab}{ac} \qquad (1.3)$$

$$\tan\theta = \frac{bc}{ab}$$

It will now be shown how these simple trigonometrical formulae can be used for elementary statics.

A **scalar** is a quantity which only has magnitude. Typical scalar quantities are length, time, mass and temperature; that is, they have magnitude, but there is no direction of these quantities.

A **vector** is a quantity which has both magnitude and direction. Typical vector quantities are force, weight, velocity, acceleration, etc. A vector quantity, say **F**, can be displayed by a scaled drawing, as shown by Figure 1.6.

The same vector quantity F can be represented by the two vectors F_x and F_y shown in Figure 1.6, where F_x is the horizontal component of **F** and F_y is the vertical component of **F**. Both F_x and F_y can be related to **F** and the angle that **F** acts with respect to the horizontal, namely θ, as follows.

From equation (1.3),

$$\sin\theta = \frac{F_y}{F}$$

or

$$F_y = F\sin\theta \qquad (1.4)$$

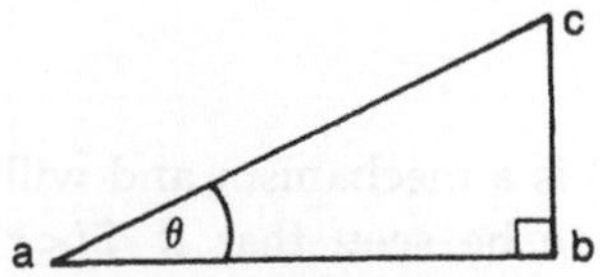

Figure 1.5 Right-angled triangle

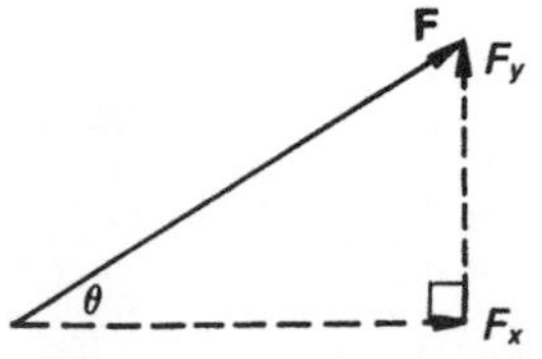

Figure 1.6 A vector quantity

From equation (1.3),

$$\cos\theta = \frac{F_x}{\mathbf{F}}$$

or

$$F_x = \mathbf{F}\cos\theta \qquad (1.5)$$

That is, the **horizontal** component of **F** is $F\cos\theta$ and the **vertical** component of **F** is $F\sin\theta$.

It will now be shown how expressions such as equations (1.4) and (1.5) can be used for elementary statics.

1.5 Equilibrium considerations

In *two dimensions*, the following three equilibrium considerations apply:

(a) Vertical equilibrium must be satisfied, i.e.

Upward forces = downward forces

(b) Horizontal equilibrium must be satisfied, i.e.

Forces to the left = forces to the right

(c) Rotational equilibrium must be satisfied at all points, i.e.

Clockwise couples = counter-clockwise couples

Use of the above vertical and horizontal equilibrium considerations is self-explanatory, but use of the rotational equilibrium consideration is not quite as self-evident; this deficiency will now be clarified through the use of a worked example.

EXAMPLE 1.1

Determine the reactions R_A and R_B acting on the ends of the horizontal beams of Figures 1.7 and 1.8. Neglect the self-weight of the beam.

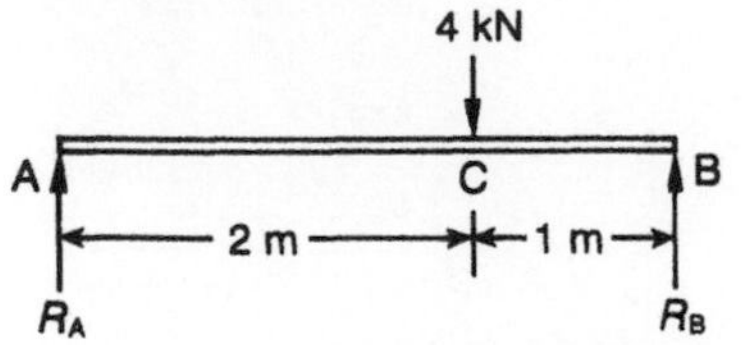

Figure 1.7 Beam with a point load at C

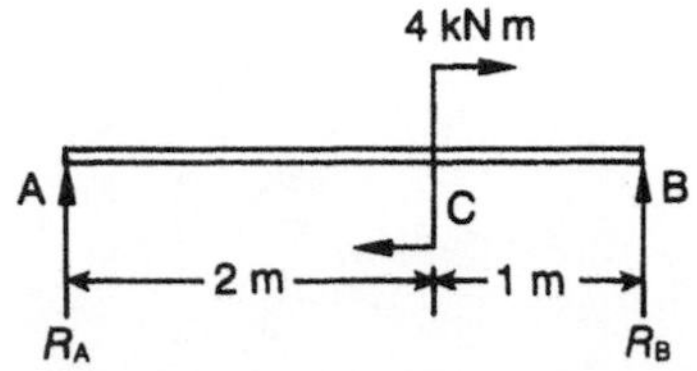

Figure 1.8 Beam with a clockwise couple at C

SOLUTION

(a) Beam with a point load at C

Take moments about the point B in Figure 1.7. In practice, moments could be taken anywhere, but by taking moments about B, the reaction R_B would not exert a couple about B, and this would leave only one unknown, namely R_A. It would have been just as convenient to have taken moments about A.

By taking moments about B, it is meant that

Clockwise couples about B = counter-clockwise couples about B

That is, for rotational equilibrium about B,

$$R_A \times 3 = 4 \times 1$$

or

$$R_A = \frac{4}{3} = 1.333 \text{ kN}$$

Resolving forces vertically (i.e. seeking vertical equilibrium),

Upward forces = downward forces

or

$$R_A + R_B = 4 \text{ kN}$$

but

$$R_A = 1.333 \text{ kN}$$

Therefore

$$R_B = 2.667 \text{ kN}$$

(b) Beam with a clockwise couple about C

Taking moments about B (i.e. seeking rotational equilibrium about B),

Clockwise couples about B = counter-clockwise couples about B

$$R_A \times 3 + 4 = 0$$

or

$$R_A = -1.333 \text{ kN}$$

That is, R_A is acting downwards.

Resolving forces vertically (i.e. seeking vertical equilibrium),

Upward forces = downward forces

$$R_A + R_B = 0$$

$$R_B = -R_A = 1.333 \text{ kN}$$

To demonstrate the static analysis of pin-jointed trusses, the following two examples will be considered.

EXAMPLE 1.2

Determine the internal forces in the members of the pin-jointed truss of Figure 1.9.

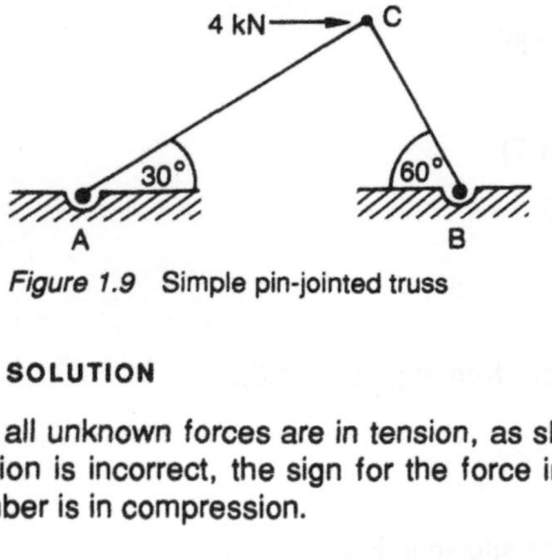

Figure 1.9 Simple pin-jointed truss

SOLUTION

Assume all unknown forces are in tension, as shown by the arrows in Figure 1.10. If this assumption is incorrect, the sign for the force in the rod will be negative, indicating that the member is in compression.

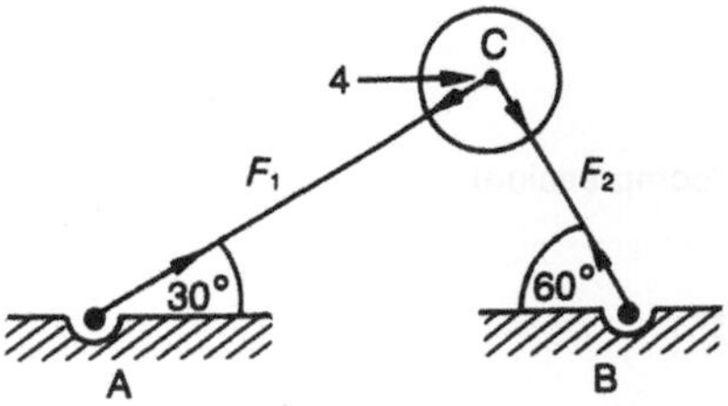

Figure 1.10 Assumed internal forces in simple truss

By drawing an imaginary circle around joint C, equilibrium can be sought around the joint with the aid of the free-body diagram of Figure 1.11.

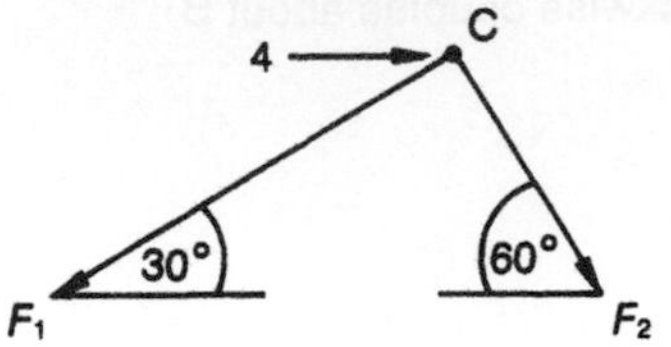

Figure 1.11 Equilibrium around joint C (free-body diagram)

Resolving forces vertically

Upward forces = downward forces

$$0 = F_1 \sin 30° + F_2 \sin 60°$$

Therefore

$$F_1 = \frac{-F_2 \sin 60°}{\sin 30°}$$

$$F_1 = -1.732 F_2 \qquad (1.6)$$

Resolving forces horizontally

Forces to the left = forces to the right

$$F_1 \cos 30° = 4 + F_2 \cos 60° \qquad (1.7)$$

Substituting equation (1.6) into equation (1.7),

$$-1.732 F_2 \cos 30° = 4 + F_2 \cos 60°$$

$$-1.5 F_2 = 4 + 0.5 F_2$$

$$F_2 = -2 \text{ kN}$$

(i.e. this member is in compression). Hence, from equation (1.6),

$$F_1 = -1.732 \times -2 \text{ kN}$$

$$F_1 = 3.464 \text{ kN}$$

The forces in the members of this truss are shown in Figure 1.12.

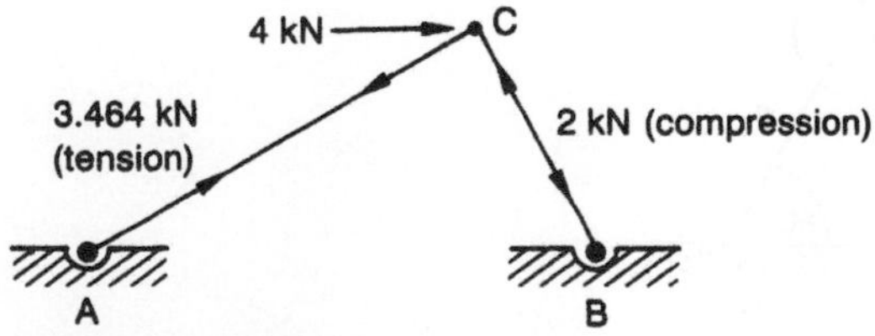

Figure 1.12 Internal forces in truss

EXAMPLE 1.3

Determine the internal forces in the members of the statically determinate pin-jointed truss shown in Figure 1.13.

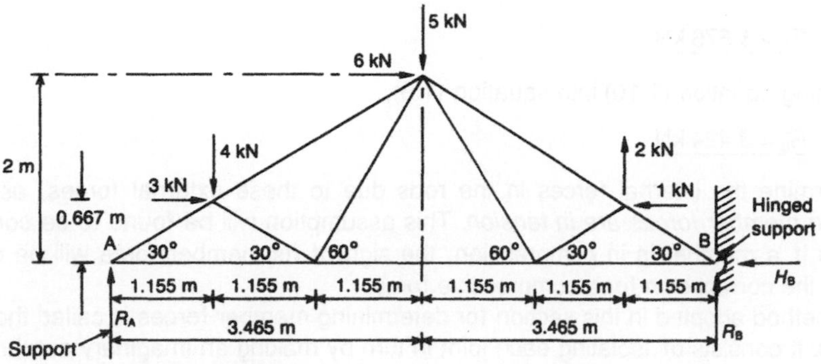

Figure 1.13 Pin-jointed truss

SOLUTION

From Figure 1.13 it can be seen that to achieve equilibrium, it will be necessary for the three reactions R_A, R_B and H_B to act. These reactions balance vertical and horizontal forces and rotational couples.

To determine the unknown reactions, it will be necessary to obtain three simultaneous equations from the three equilibrium considerations described in Section 1.5.

Consider *horizontal equilibrium*:

Forces to the left = forces to the right

$$H_B = 3 + 6 - 1$$

$$H_B = 8 \text{ kN} \quad (1.8)$$

Consider *vertical equilibrium*:

Upward forces = downward forces

$$R_A + R_B + 2 = 4 + 5$$

or

$$\underline{R_B = 7 - R_A} \quad (1.9)$$

Consider *rotational equilibrium*. Now rotational equilibrium can be considered at any point in the plane of the truss, but to simplify arithmetic, it is better to take moments about a point through which an unknown force acts, so that this unknown force has no moment about this point. In this case, it will be convenient to *take moments* about A or B.

Taking moments about B means that

Clockwise couples about B = counter-clockwise couples about B

$$R_A \times 6.93 + 3 \times 0.667 + 6 \times 2 + 2 \times 1.155 = 4 \times 5.775 + 5 \times 3.465 + 1 \times 0.667$$

i.e.

$$\underline{R_A = 3.576 \text{ kN}} \tag{1.10}$$

Substituting equation (1.10) into equation (1.9),

$$\underline{R_B = 3.424 \text{ kN}} \tag{1.11}$$

To determine the internal forces in the rods due to these external forces, assume all *unknown member forces are in tension.* This assumption will be found to be convenient, because if a member is in compression, the sign of its member force will be negative, which is the correct sign for a compressive force.

The method adopted in this section for determining member forces is called the *method of joints.* It consists of isolating each joint in turn by making an imaginary cut through the members around that joint and then considering vertical and horizontal equilibrium between the internal member forces and the external forces at that joint (i.e. a free-body diagram of the joint). As only vertical and horizontal equilibrium is considered at each joint, it is necessary to start the analysis at a joint where there are only *two unknown member forces.* In this case, consideration must first be made at either joint 1 or joint 7 (see Figure 1.14).

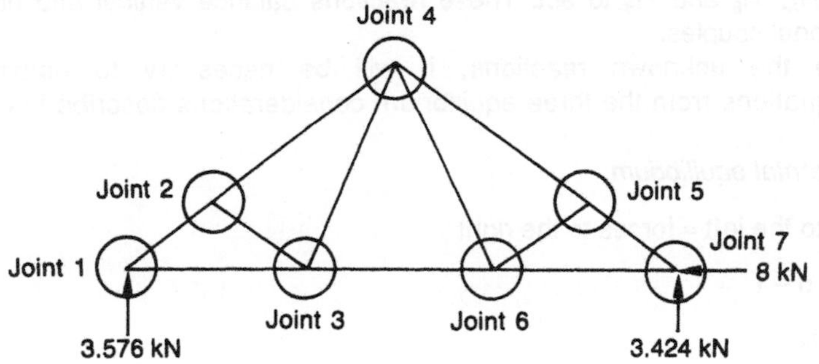

Figure 1.14 Joint numbers for pin-jointed trust

Joint 1 (see the free-body diagram of Figure 1.15)

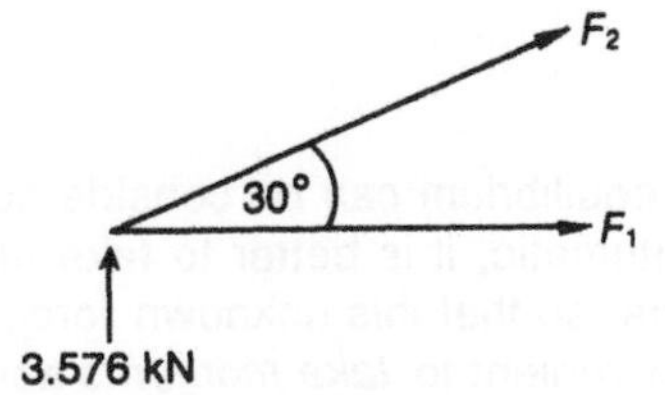

Figure 1.15 Joint 1 (free-body diagram)

Resolving forces vertically

Upward forces = downward forces

$3.576 + F_2 \sin 30° = 0$

Therefore

$F_2 = -7.152$ kN (compression) (1.12)

Resolving forces horizontally

Forces to the left = forces to the right

$0 = F_2 \cos 30° + F_1$ or $F_1 = -F_2 \cos 30°$

Therefore

$F_1 = 6.194$ kN (tension) (1.13)

Joint 2 (see the free-body diagram of Figure 1.16)

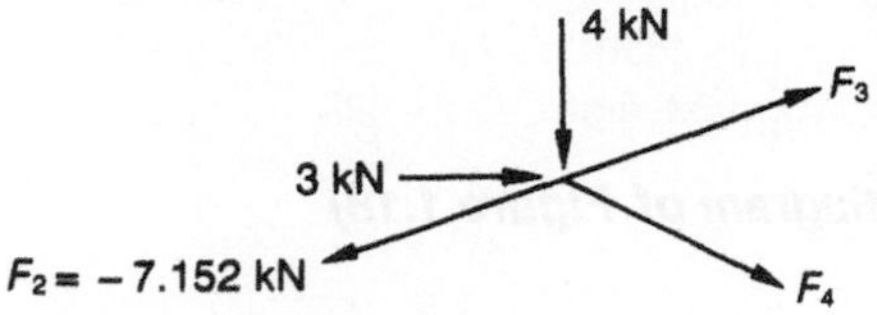

Figure 1.16 Joint 2 (free-body diagram)

Resolving horizontally

$F_3 \cos 30° + F_4 \cos 30° + 3 = F_2 \cos 30°$

Therefore

$F_3 = -10.616 - F_4$ (1.14)

Resolving vertically

$F_3 \sin 30° = F_4 \sin 30° + F_2 \sin 30° + 4$

or

$F_4 = F_3 - 0.848$ (1.15)

Substituting equation (1.14) into (1.15),

$F_4 = -5.732$ kN (compression) (1.16)

From equation (1.14)

$F_3 = -4.884$ kN (compression) (1.17)

Joint 3 (see the free-body diagram of Figure 1.17)

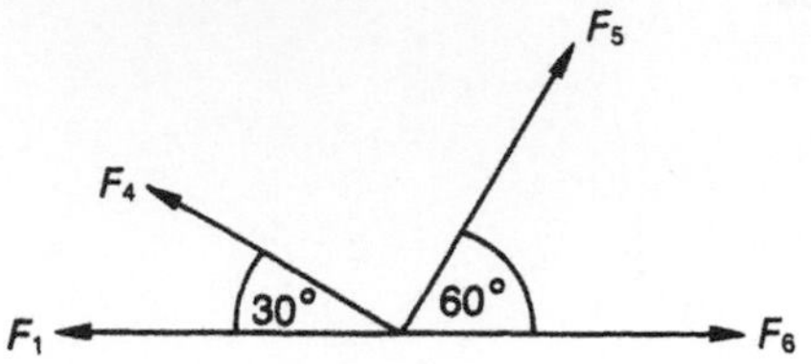

Figure 1.17 Joint 3 (free-body diagram)

Resolving vertically

$F_5 \sin 60° + F_4 \sin 30° = 0$

$\underline{F_5 = 3.309 \text{ kN (tensile)}}$ (1.18)

Resolving horizontally

$F_1 + F_4 \cos 30° = F_6 + F_5 \cos 60°$

$\underline{F_6 = -0.424 \text{ kN (compression)}}$ (1.19)

Joint 4 (see the free-body diagram of Figure 1.18)

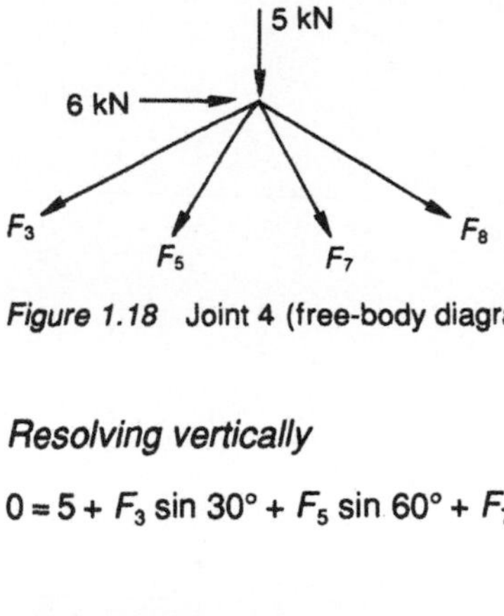

Figure 1.18 Joint 4 (free-body diagram)

Resolving vertically

$0 = 5 + F_3 \sin 30° + F_5 \sin 60° + F_7 \sin 60° + F_8 \sin 30°$

or

$$0.866 F_7 + 0.5 F_8 = -5.425 \tag{1.20}$$

Resolving horizontally

$6 + F_8 \cos 30° + F_7 \cos 60° = F_3 \cos 30° + F_5 \cos 60°$

or

$$F_8 = -0.577 F_7 - 9.902 \tag{1.21}$$

Substituting equation (1.21) into (1.20),

$$\underline{F_7 = -0.82 \text{ kN (compression)}} \tag{1.22}$$

$$\underline{F_8 = -9.43 \text{ kN (compression)}} \tag{1.23}$$

Joint 5 (see the free-body diagram of Figure 1.19)

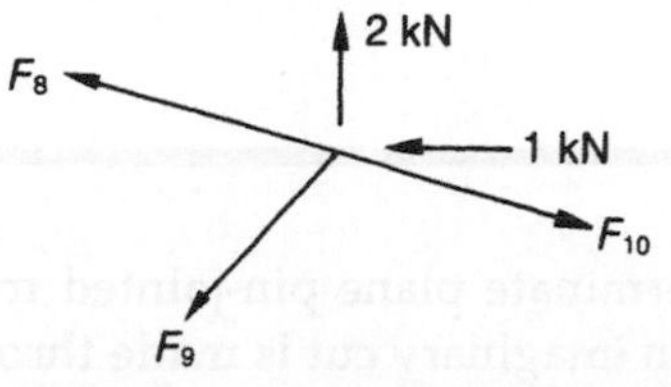

Figure 1.19 Joint 5 (free-body diagram)

Resolving vertically

$2 + F_8 \sin 30° = F_{10} \sin 30° + F_9 \sin 30°$

or

$$F_9 + F_{10} = -5.43 \tag{1.24}$$

Resolving horizontally

$F_8 \cos 30° + F_9 \cos 30° + 1 = F_{10} \cos 30°$

$-9.43 + F_9 + 1.155 = F_{10}$

or

$$F_{10} - F_9 = -8.28 \tag{1.25}$$

Adding (1.24) and (1.25),

$2F_{10} = -13.7$

Therefore

$$\underline{F_{10} = -6.85 \text{ kN (compression)}} \tag{1.26}$$

Substituting (1.26) into (1.25),

$$\underline{F_9 = 1.43 \text{ kN (tensile)}} \tag{1.27}$$

Joint 7 (see the free-body diagram of Figure 1.20)

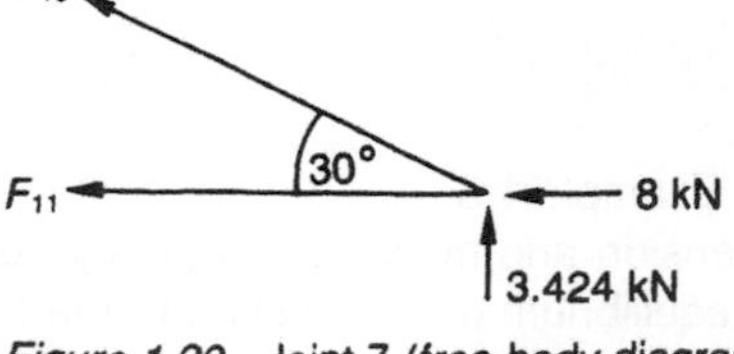

Figure 1.20 Joint 7 (free-body diagram)

Only one unknown force is required, namely F_{11}; therefore it will be easier to consider joint 7, rather than joint 6.

Resolving horizontally

$$F_{11} + 8 + F_{10} \cos 30° = 0$$

Therefore

$$\underline{F_{11} = -2.07 \text{ kN (compression)}} \quad (1.28)$$

Another method for analyzing statically determinate plane pin-jointed trusses is called the *method of sections*. In this method, an imaginary cut is made through the truss, so that there are no more than three unknown internal forces to be determined, across that section. By considering equilibrium, these unknown internal forces can be calculated. Prior to applying the method, however, it is usually necessary first to determine the support reactions acting on the framework. To demonstrate the method, the following example will be considered.

EXAMPLE 1.4

Using the method of sections, determine the forces in the three members, namely a, b and c, of Figure 1.21.

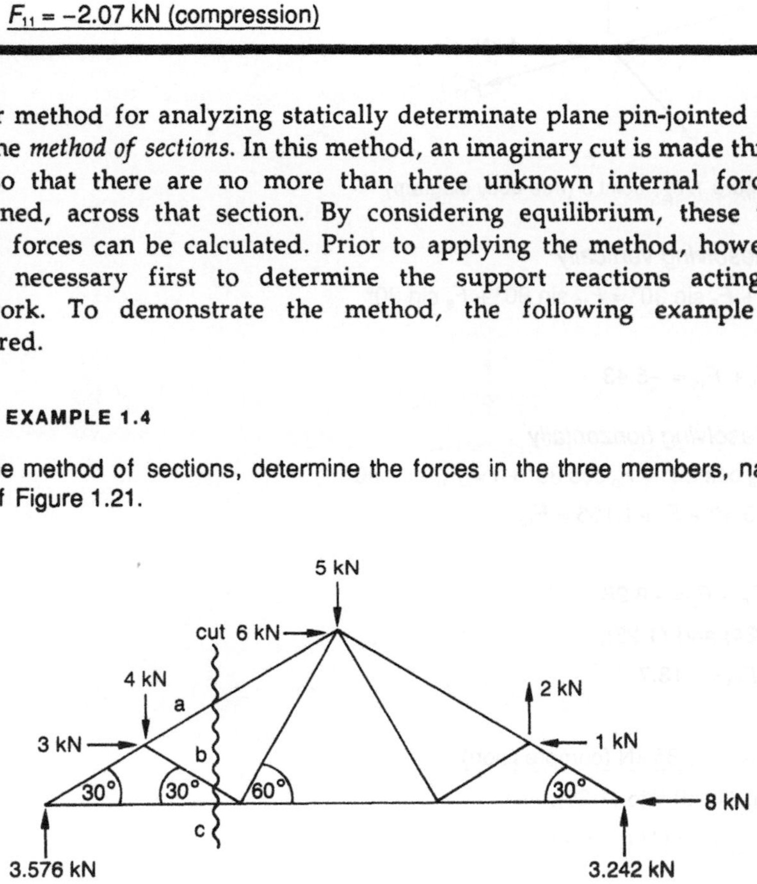

Figure 1.21 Pin-jointed truss

SOLUTION

It can be seen that this problem is the same as Example 1.3.

Assume all unknown internal forces are in tension and make an imaginary cut through the frame, as shown in Figure 1.21. Consider equilibrium of the section to the left of this cut, as shown by the free-body diagram of Figure 1.22.

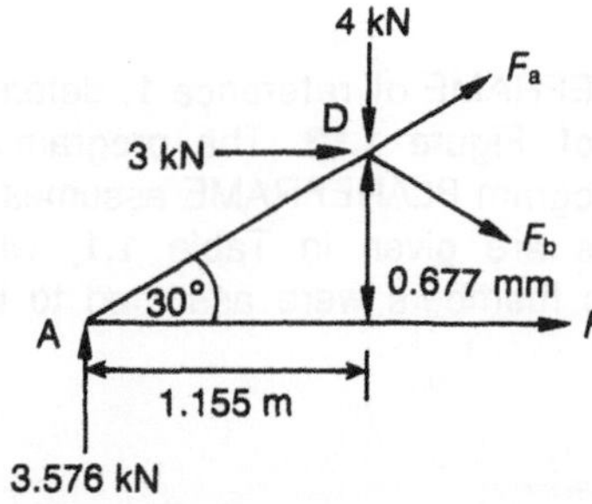

Figure 1.22 Section of truss (free-body diagram)

Taking moments about D

Clockwise couples about D = counter-clockwise couples about D

$$3.576 \times 1.155 = F_c \times 0.667$$

$$\underline{F_c = 6.192 \text{ kN}}$$

Resolving forces vertically

Upward forces = downward forces

$$3.576 + F_a \sin 30° = 4 + F_b \sin 30°$$

$$(F_a - F_b)\sin 30° = 0.424$$

$$F_a = 0.848 + F_b \tag{1.29}$$

Resolving forces horizontally

Forces to the left = forces to the right

$$3 + F_c + F_a \cos 30° + F_b \cos 30° = 0$$

or

$$(F_a + F_b)\cos 30° = -9.192 \tag{1.30}$$

Substituting equation (1.29) into (1.30),

$$(0.848 + 2F_b)0.866 = -9.192$$

$$F_b = \frac{-11.462}{2} = -5.731 \text{ kN} \tag{1.31}$$

Substituting equation (1.31) into (1.29),

$$\underline{F_a = -4.883 \text{ kN}}$$

Thus

$$F_a = -4.883 \text{ kN} \quad F_b = -5.731 \text{ kN} \quad F_c = 6.192 \text{ kN}$$

as before.

EXAMPLE 1.5

Using the computer programs TRUSS and PLANEFRAME of reference 1, determine the member forces in the plane pin-jointed truss of Figure 1.13. The program TRUSS assumes that the frame has pin joints, and the program PLANEFRAME assumes that the frame has rigid (or welded) joints; the results are given in Table 1.1, where for PLANEFRAME the cross-sectional areas of the members were assumed to be 1000 times greater than their second moments of area.

Table 1.1 **Member forces in framework**

Force	Pin-jointed truss (kN)	Rigid-jointed truss (kN)
F_1	6.194	6.043
F_2	-7.152	-7.021
F_3	-4.884	-4.843
F_4	-5.730	-5.601
F_5	3.309	3.258
F_6	-0.424	-0.417
F_7	-0.823	-0.868
F_8	-9.426	-9.384
F_9	1.426	1.421
F_{10}	-6.847	-6.829
F_{11}	-2.071	-2.086

SOLUTION

As can be seen from Table 1.1, the assumption that the joints are pinned gives similar results to the much more difficult problem of assuming that the joints are rigid (or welded), providing, of course, that the externally applied loads are at the joints. If the applied loads are applied between the joints, then bending can occur, which can be catered for by PLANEFRAME, but not by TRUSS; thus TRUSS and PLANEFRAME will only agree if the applied loads are placed at the joints.

1.6 Bending moment and shearing force

These are very important in the analysis of beams and rigid-jointed frameworks, but the latter structures will not be considered in the present text, because they are much more suitable for computer analysis [1].

Once again, as in Section 1.2, all the beams will be assumed to be statically determinate and in equilibrium.

1.6.1 Definition of bending moment (M)

A *bending moment, M,* acting at any particular section on a beam, in equilibrium, can be defined as the resultant of all the couples acting on one side of the beam at that particular section. The resultant of the couples acting on either side of the appropriate section can be considered, as the beam is in equilibrium, so that the beam will be either in a *sagging* condition or in a *hogging* one, at that section, as shown in Figure 1.23.

The sign convention for bending moments will be that a sagging moment is assumed to be *positive* and a hogging one is assumed to be *negative*. The units for bending moment are N m, kN m, etc.

1.6.2 Definition of shearing force (F)

A *shearing force, F,* acting at any particular section on a *horizontal beam,* in equilibrium, can be defined as the resultant of the *vertical forces* acting on one side of the beam at that particular section. The resultant of the vertical forces acting on either side of the appropriate section can be considered, as the beam is in equilibrium, as shown in Figure 1.24 which also shows the sign conventions for positive and negative shearing forces. The units for shearing force are N, kN, MN, etc.

Prior to analyzing beams it will be necessary to describe the symbols used for loads and supports and the various types of beam.

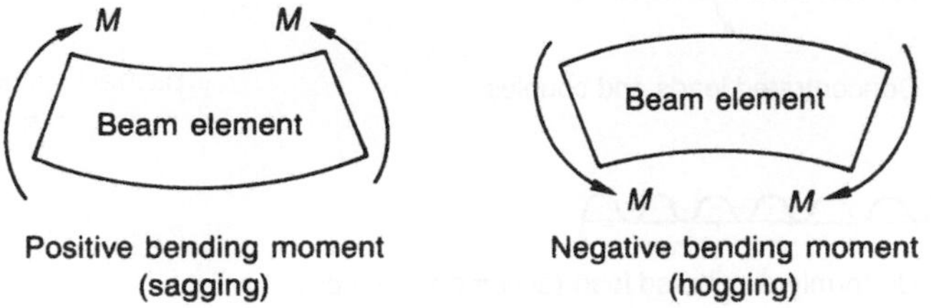

Figure 1.23 Sagging and hogging bending action

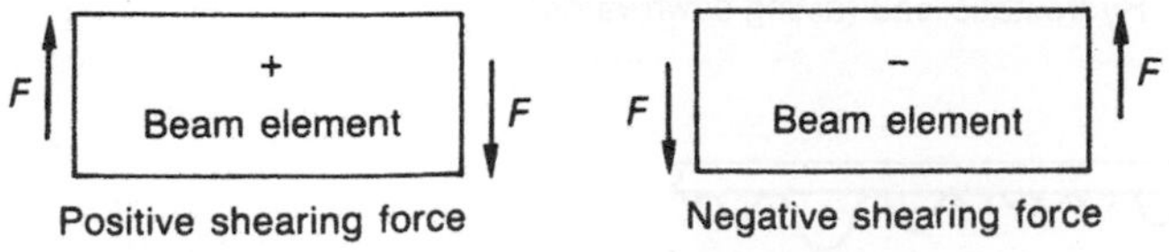

Figure 1.24 Positive and negative shearing forces

1.7 Loads

These can take various forms, including concentrated loads and couples, which are shown as arrows in Figure 1.25, and distributed loads, as shown in Figures 1.26 to 1.28.

Concentrated loads are assumed to act at points. This is, in general, a pessimistic assumption. Typical concentrated loads appear in the form of loads transmitted through wheels, hanging weights, etc. The units for concentrated loads are N, kN, MN, etc.

Uniformly distributed loads are assumed to be distributed uniformly over the length or part of the length of beam. Typical uniformly distributed loads are due to wind load, snow load, self-weight, etc. The units for uniformly distributed loads are N/m, kN/m, etc.

Hydrostatic (or trapezoidal) loads are assumed to increase or decrease linearly with length, as shown in Figure 1.27. They usually appear in containers carrying liquids or in marine structures, etc., which are attempting to contain the water, etc. The units of hydrostatic loads are N/m, kN/m, etc.

Varying distributed loads do not have the simpler shapes of uniformly distributed and hydrostatic loads described earlier. Typical cases of varying distributed loads are those due to the self-weight of a ship, together with the buoyant forces acting on its hull.

Wind loads can be calculated from momentum considerations, as follows:

$$\text{Wind pressure} = \rho v^2 = P$$

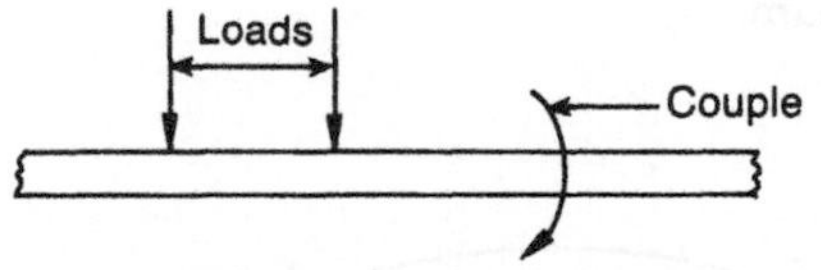

Figure 1.25 Concentrated loads and couples

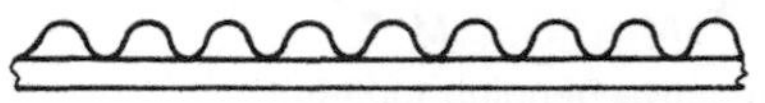

Figure 1.26 Uniformly distributed load (acting downwards)

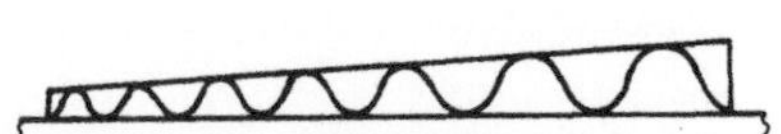

Figure 1.27 Hydrostatic load (acting downwards)

Figure 1.28 A varying distributed load (acting upwards)

where

ρ = density of air (kg/m^3)

v = velocity of air (m/s)

P = pressure in pascals (N/m^2)

This simple expression assumes that the wind acts on a flat surface and that the wind is turned through 90° from its original direction, so that the calculation for P is, in general, overestimated, and on the so-called 'safe' side.

For *air* at standard temperature and pressure (*s.t.p.*),

$\rho \fallingdotseq 1.293\ kg/m^3$

so that, when

$v = 44.7$ m/s (100 miles/h)

then

$P = 2584$ Pa (0.375 psi)

For an *atomic blast*,

$P = 0.103$ MPa (15 psi)

so that

$v = 283$ m/s (633 miles/h)

NB For a standard living room of dimensions, say, 5 m × 5 m × 2.5 m, the air contained in the room will weigh about 80.8 kg (178.1 lbf), and for a large hall of dimensions, say, 30 m × 30 m × 15 m, the air contained in the hall will weigh about 17 456 kg (17.1 tons)!

Structural engineers usually calculate the pressure due to wind acting on buildings as $\rho v^2/2$.

In general, the pressure due to water increases linearly with the depth of the water, according to the expression

$$P = \rho g h$$

where

P = pressure in pascals (N/m^2)

h = depth of water (m)

g = acceleration due to gravity (m/s^2)

ρ = density of water (kg/m^3), which for pure water, at normal temperature and pressure, is 1000 kg/m^3, and for sea water is 1020 kg/m^3

At a *depth* of 1000 m (in sea water)

$P \doteqdot 10$ MPa (1451 psi)

so that, for a submarine of average diameter 10 m and of length 100 m, the total load due to water pressure will be about 37 700 MN (3.78 E6 tons)!

NB 1 MN ≒ 100 tons.

1.8 Types of beam

The simplest types of beam are those shown in Figures 1.29 and 1.30.

The beam of Figure 1.29, which is statically determinate, is supported on two knife edges or *simple supports*. In practice, however, a true knife-edge support is not possible, as such edges will not have zero area.

The beam of Figure 1.30, which is also statically determinate, is rigidly fixed (encastré) at its right end and is called a *cantilever*.

It can be seen from Figures 1.29 and 1.30 that in both cases, there are two reacting 'forces', which in the case of Figure 1.29 are R_1 and R_2 and in the case of Figure 1.30 are the vertical reaction R_1 and the restraining couple M_1. These beams are said to be statically determinate, because the two unknown reactions can be found from two simultaneous equations, which can be obtained by resolving forces vertically and taking moments.

Statically indeterminate beams cannot be analyzed through simple statical considerations alone, because for such beams, the number of unknown 'reactions' is more than the equations that can be derived from statical observations. For example, the propped cantilever of Figure 1.31 has one redundant reacting 'force' (i.e. either R_1 or M_2), and the beam of Figure 1.32, which is encastré at both ends, has two

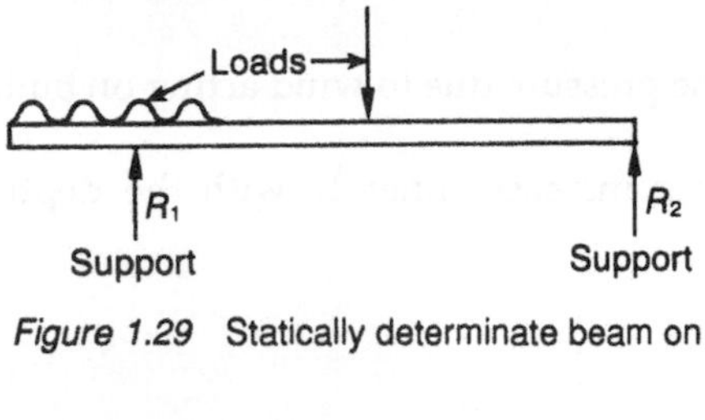

Figure 1.29 Statically determinate beam on simple supports

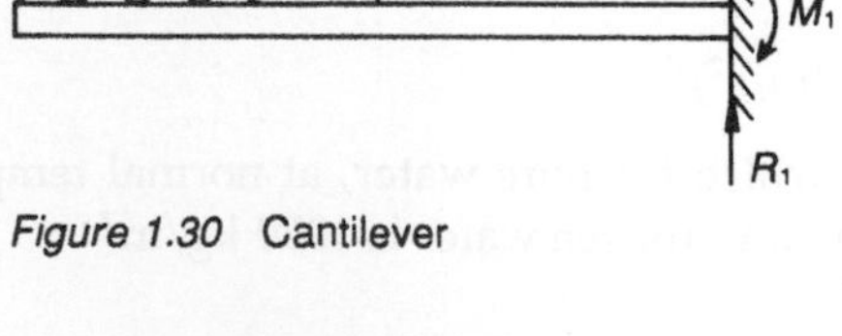

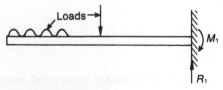

Figure 1.30 Cantilever

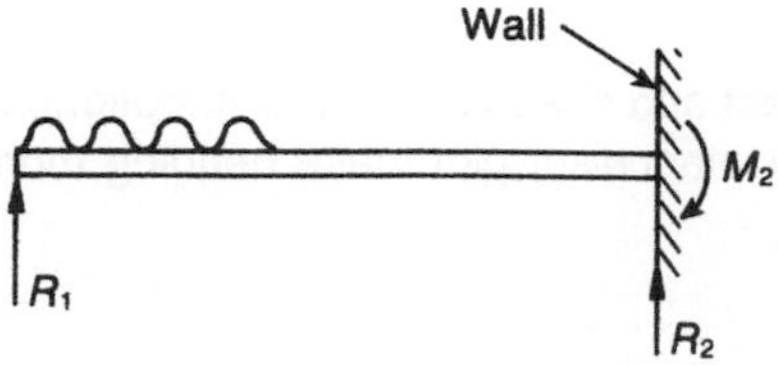

Figure 1.31 Propped cantilever

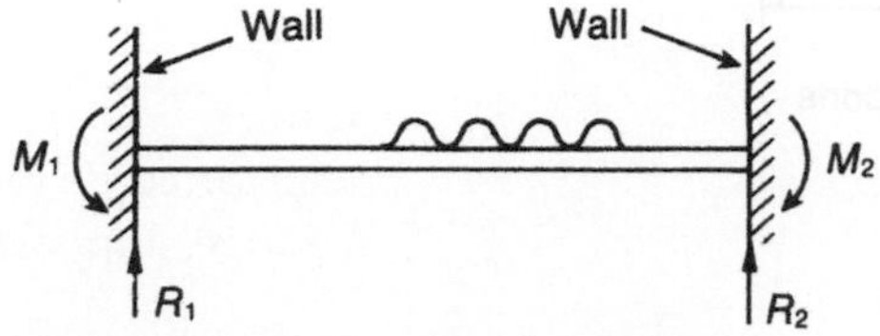

Figure 1.32 Beam with encastré ends

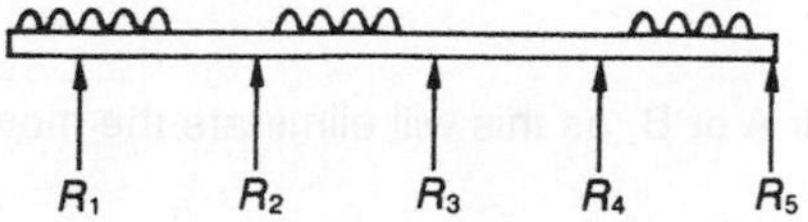

Figure 1.33 Continuous beam with several supports

redundant reacting 'forces' (i.e. either M_1 and M_2 or R_1 and M_1 or R_2 and M_2). These 'forces' are said to be redundant, because if they did not exist, the structure could still be classified as a structure, as distinct from a mechanism. Other statically indeterminate beams are like the continuous beam of Figure 1.33, which has three redundant reacting forces (i.e. it is statically indeterminate to the third degree).

NB By an 'encastré end', it is meant that all movement, including rotation, is prevented at that end.

1.9 Bending moment and shearing force diagrams

To demonstrate bending moment and shearing force action, Example 1.6 will first be considered, but on this occasion, the shearing forces and bending moments acting on both sides of any particular section will be calculated. The reason for doing this is to demonstrate the nature of bending moment and shearing force.

EXAMPLE 1.6

Determine expressions for the bending moment and shearing force distributions along the length of the beam in Figure 1.34. Hence, or otherwise, plot these bending moment and shearing force distributions.

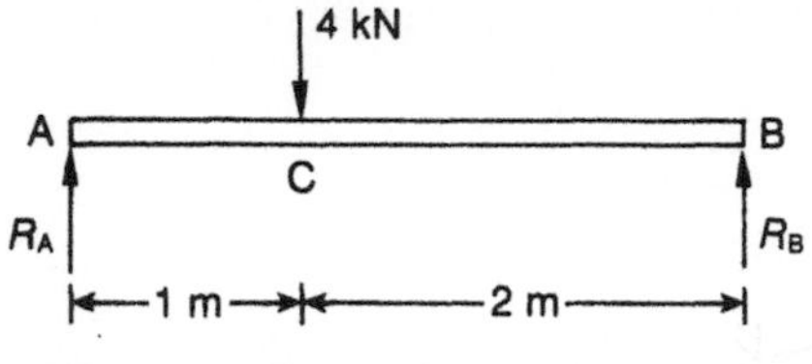

Figure 1.34 Beam with end supports

SOLUTION

First, it will be necessary to calculate the reactions R_A and R_B, and this can be done by taking moments about a suitable point and resolving forces vertically.

Take moments about B

It is convenient to take moments about either A or B, as this will eliminate the moment due to either R_A or R_B, respectively.

Clockwise moments about B = counter-clockwise moments about B

$$R_A \times 3\text{ m} = 4\text{ kN} \times 2\text{ m}$$

Therefore

$$\underline{R_A = 2.667\text{ kN}}$$

Resolve forces vertically

Upward forces = downward forces

$$R_A + R_B = 4\text{ kN}$$

Therefore

$$\underline{R_B = 1.333\text{ kN}}$$

To determine the bending moment between A and C

Consider any distance x between A and C, as shown in Figure 1.35. From the figure, it can be seen that at any distance x, between A and C, the reaction R_A causes a bending moment equal to $R_A x$, which is sagging (i.e. positive). It can also be seen from Figure 1.35, that the forces to the right of the beam cause an equal and opposite moment, so that the beam bends at this point in the manner shown in Figure 1.35(b). Therefore

$$\text{Bending moment} = M = 2.667x \tag{1.32}$$

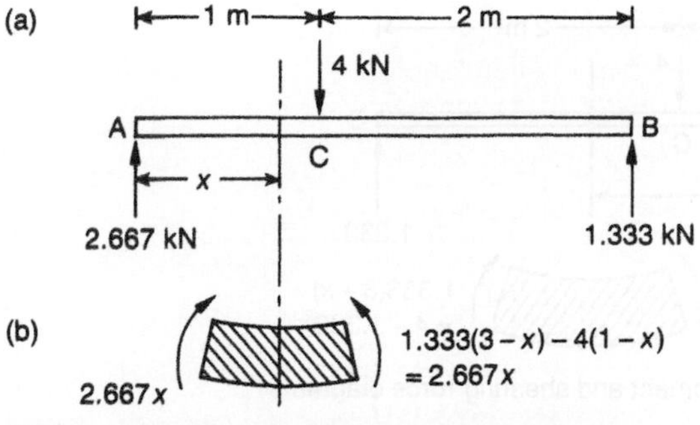

Figure 1.35 Bending moment between A and C

Equation (1.32) can be seen to be a straight line, which increases linearly from zero at A to a maximum value of 2.667 kN m at C.

To determine the shearing force between A and C

Consider any distance x between A and C as shown in Figure 1.36. From Figure 1.36(b), it can be seen that the vertical forces on the left of the beam tend to cause the left part of the beam at x to 'slide' upwards, whilst the vertical forces on the right of the beam tend to cause the right part of the beam at x to 'slide' downwards, i.e.

$$\text{Shearing force at } x = F = 2.667 \text{ kN} \tag{1.33}$$

Equation (1.33) shows the shearing force to be constant between A and C, and is said to be positive, because the right side tends to move downwards, as shown in Figure 1.36(b).

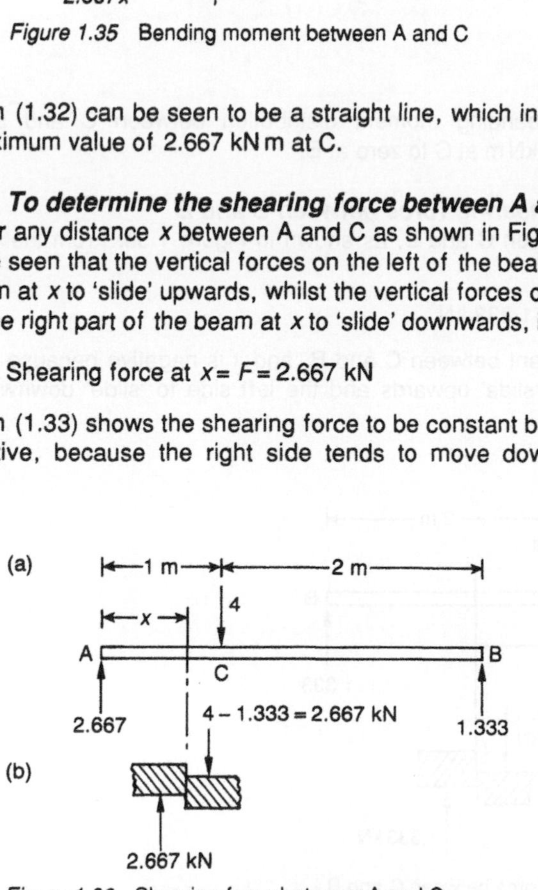

Figure 1.36 Shearing force between A and C

To determine the bending moment between C and B

Consider any distance x between C and B, as shown in Figure 1.37. From the figure, it can be seen that at any distance x

$$\text{Bending moment} = M = 4 - 1.333x \text{ (sagging)} \tag{1.34}$$

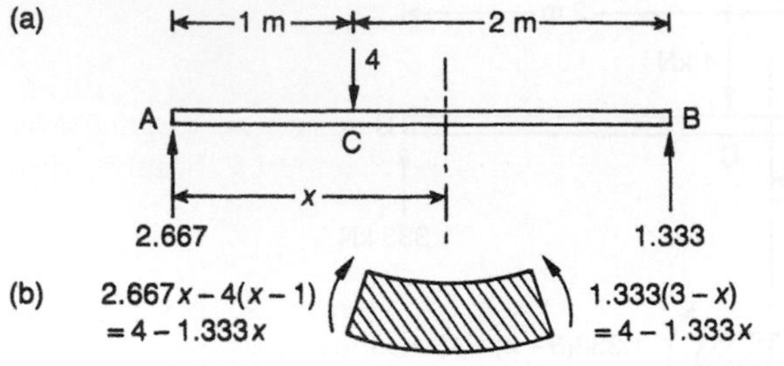

Figure 1.37 Bending moment and shearing force diagrams

Equation (1.34) shows the bending moment distribution between C and B to be decreasing linearly from 2.667 kN m at C to zero at B.

To determine the shearing force between C and B

Consider any distance x between C and B, as shown in Figure 1.38. From the figure, it can be seen that

$$\text{Shearing force} = F = -1.333 \text{ kN} \tag{1.35}$$

The shearing force F is constant between C and B, and it is negative because the right side of any section tends to 'slide' upwards and the left side to 'slide' downwards, as shown in Figure 1.38(b).

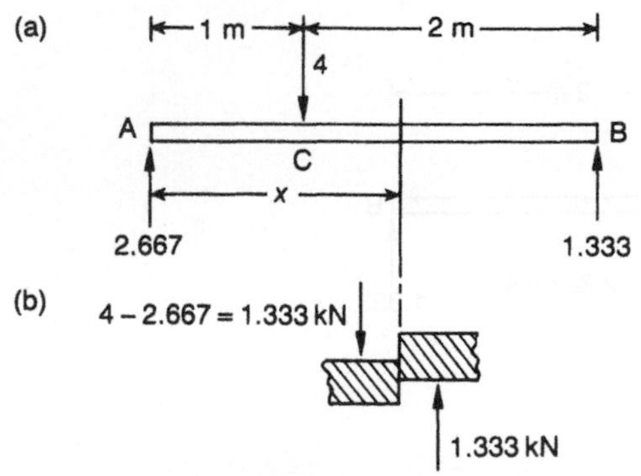

Figure 1.38 Shearing force between C and B

Bending moment and shearing force diagrams

To obtain the bending moment diagram, it is necessary to plot equations (1.32) and (1.34) in the manner shown in Figure 1.39(b), and to obtain the shearing force diagram, it is necessary to plot equations (1.33) and (1.35) in the manner shown in Figure 1.39(c).

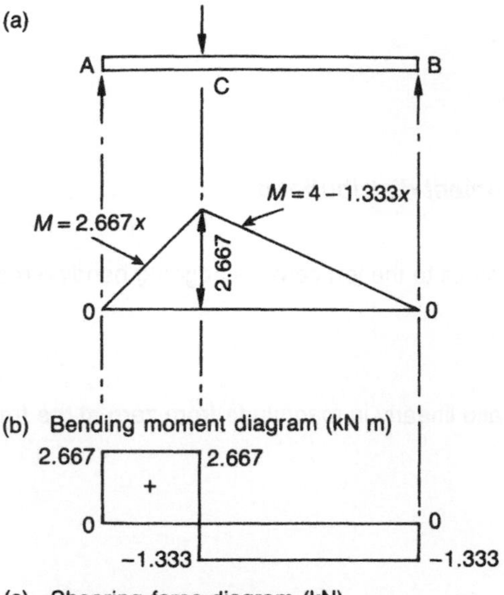

Figure 1.39 Bending moment and shearing force diagrams

From the calculations carried out above, it can be seen that to determine the bending moment or shearing force at any particular section, it is only necessary to consider the resultant of the forces on *one side of the section*. Either side of the appropriate section can be considered, as the beam is in equilibrium.

EXAMPLE 1.7

Determine the bending moment and shearing force distributions for the cantilever of Figure 1.40. Hence, or otherwise, plot the bending moment and shearing force diagrams.

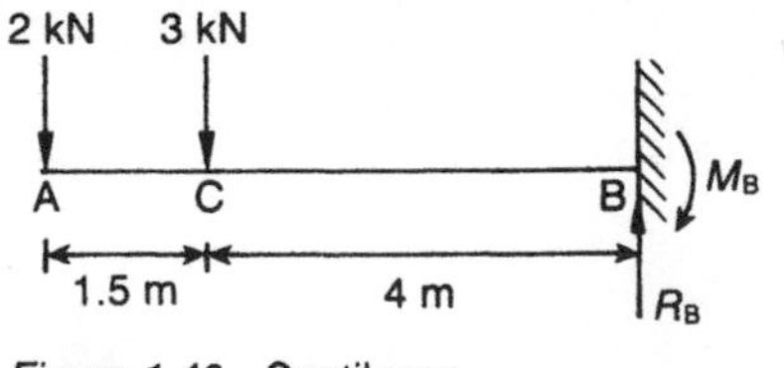

Figure 1.40 Cantilever

SOLUTION

To determine R_B and M_B

Resolving forces vertically

$\underline{R_B = 2 + 3 = 5 \text{ kN}}$

Taking moments about B

$M_B = 2 \times 5.5 + 3 \times 4$

$\underline{M_B = 23 \text{ kN m}}$

To determine bending moment distributions

Consider span AC

At any given distance x from A, the forces to the left cause a hogging bending moment of $2x$, as shown in Figure 1.41, i.e.

$$M = -2x \tag{1.36}$$

Equation (1.36) can be seen to increase linearly in magnitude from zero at the free end to 3 kN m at C.

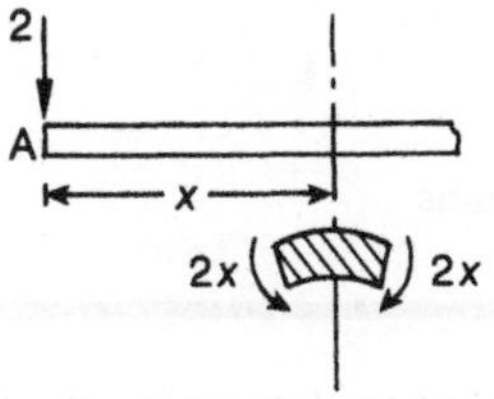

Figure 1.41 Bending moment between A and C

Consider span CB

At any distance x from A, the forces to the left cause a hogging bending moment of $2x + 3(x - 1.5)$, as shown in Figure 1.42, i.e.

$$M = -2x - 3(x - 1.5)$$

$$M = -5x + 4.5 \tag{1.37}$$

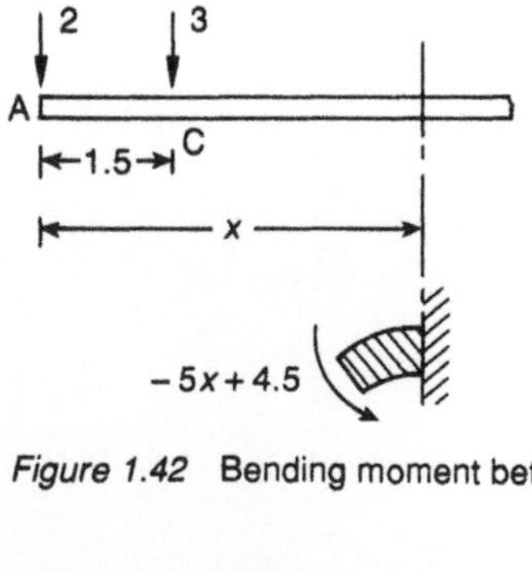

Figure 1.42 Bending moment between C and B

Equation (1.37) can be seen to increase linearly in magnitude from 3 kN m at C to 23 kN m at B.

To determine the shearing force distributions

Consider span AC

At any distance x, the resultant of the vertical forces to the left of this section causes a shearing force equal to

$$F = -2 \text{ kN} \qquad (1.38)$$

as shown in Figure 1.43.

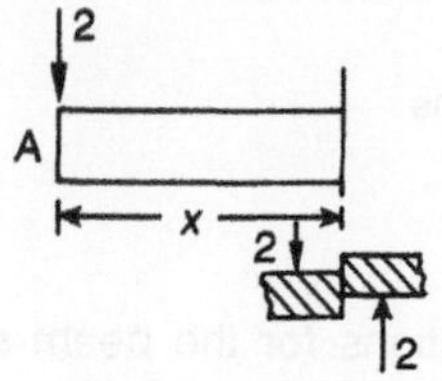

Figure 1.43 Shearing force between A and C

Consider span CB

At any given distance x, the resultant of the vertical forces to the left of this section causes a shearing force equal to

$$\underline{F = -2 - 3 = -5 \text{ kN}} \qquad (1.39)$$

as shown in Figure 1.44.

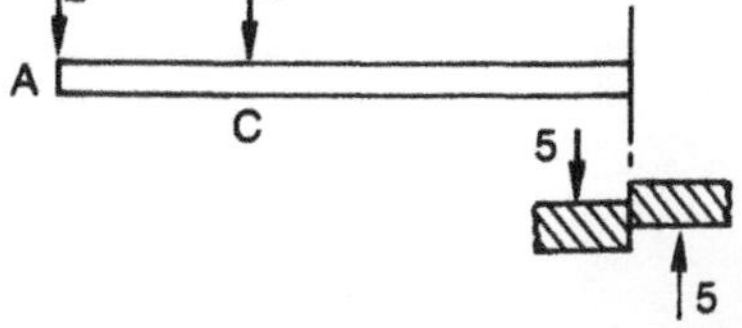

Figure 1.44 Shearing force between C and B

Bending moment and shearing force diagrams

Plots of the bending moment and shearing force distributions, along the length of the cantilever, are shown in Figure 1.45.

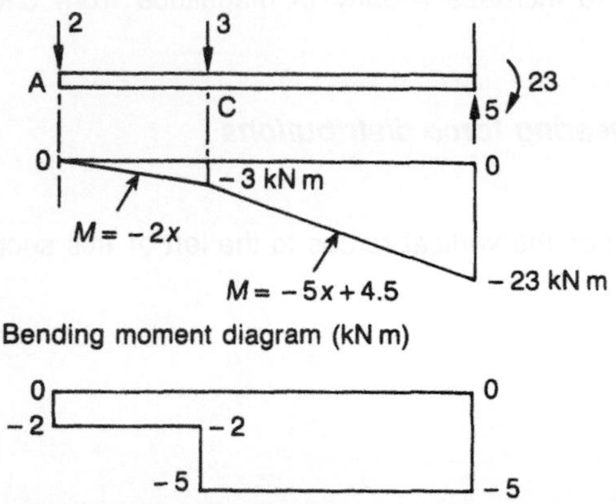

(b) Bending moment diagram (kN m)

(c) Shearing force diagram (kN)

Figure 1.45 Bending moment and shearing force diagrams

EXAMPLE 1.8

Determine the bending moment and shearing force distributions for the beam shown in Figure 1.46 and plot the diagrams. Determine also the position of the point of contraflexure. It is convenient to divide the beam into spans AB, BC and CD, because there are discontinuities at B and C.

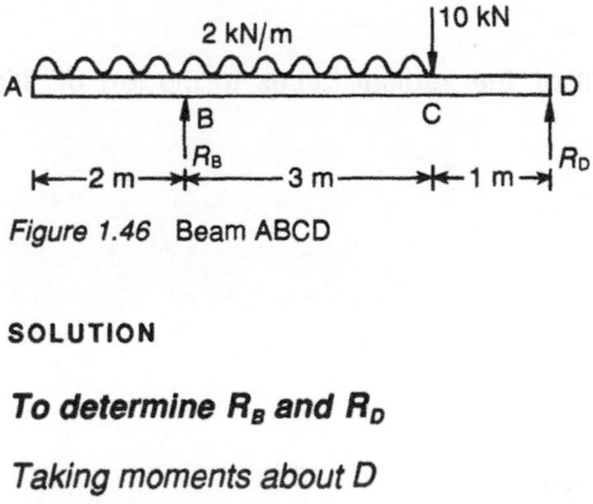

Figure 1.46 Beam ABCD

SOLUTION

To determine R_B and R_D

Taking moments about D

$R_B \times 4 = 10 \times 1 + 5 \times 2 \times 3.5$

$\underline{R_B = 11.25 \text{ kN}}$

Resolving forces vertically

$R_B + R_D = 10 + 5 \times 2$

$\underline{R_D = 8.75 \text{ kN}}$

Consider span AB

From Figure 1.47, it can be seen that at any distance x from the left end, the bending

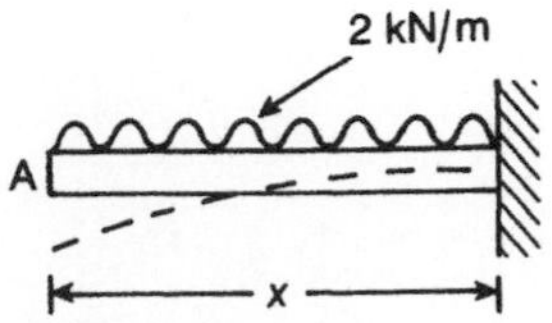

Figure 1.47

moment *M*, which is hogging, is given by

$$M = -2 \times x \times \frac{x}{2} = -x^2 \qquad (1.40)$$

Equation (1.40) can be seen to be parabolic, which increases in magnitude from zero at A to −4 kN m at B. The equation is obtained by multiplying the weight of the load, which is $2x$, by the lever, which is $x/2$. It is negative, because the beam is hogging at this section.

Furthermore, from Figure 1.47, it can be seen that the resultant of the vertical forces to the left of the section is $2x$, causing the left to 'slide' downwards, i.e.

$$\text{Shearing force} = F = -2x \qquad (1.41)$$

Equation (1.41) is linear, increasing in magnitude from zero at A to 4 kN at B.

Consider span BC

At any distance x in Figure 1.48,

$$M = -2 \times x \times \frac{x}{2} + 11.25(x - 2)$$

or

$$\underline{M = -x^2 + 11.25x - 22.5 \text{ (parabolic)}} \qquad (1.42)$$

$$@ \; \underline{x = 2 \quad M_B = -4 \text{ kN m}}$$

$$@ \; \underline{x = 5 \quad M_C = 8.75 \text{ kN m}}$$

At any distance x,

$$F = 11.25 - 2 \times x$$

$$\underline{F = 11.25 - 2x \text{ (linear)}} \qquad (1.43)$$

$$@ \; \underline{x = 2 \quad F_B = 7.25 \text{ kN}}$$

$$@ \; \underline{x = 5 \quad F_C = 1.25 \text{ kN}}$$

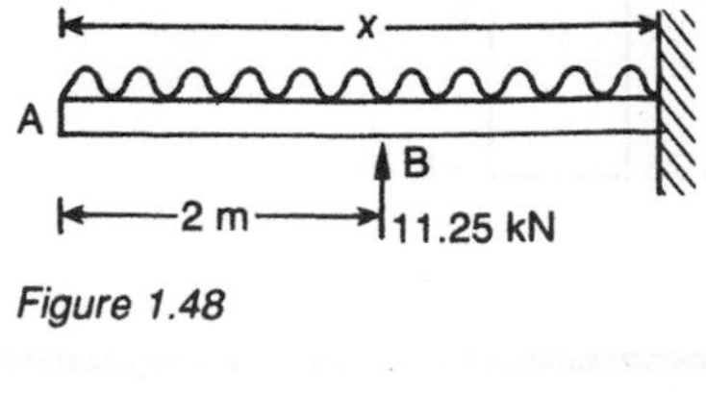

Figure 1.48

Consider span CD

At any distance x in Figure 1.49

$$M = 11.25(x-2) - 2 * 5 * (x-2.5) - 10(x-5)$$

$$\underline{M = 52.5 - 8.75x \text{ (linear)}} \qquad (1.44)$$

$$@\ \underline{x = 5 \quad M_C = 8.75 \text{ kN m}}$$

$$@\ \underline{x = 6 \quad M_D = 0 \text{ (as required)}}$$

At any distance x in Figure 1.49,

$$F = 11.25 - 2 \times 5 - 10$$

$$\underline{F = -8.75 \text{ kN (constant)}} \qquad (1.45)$$

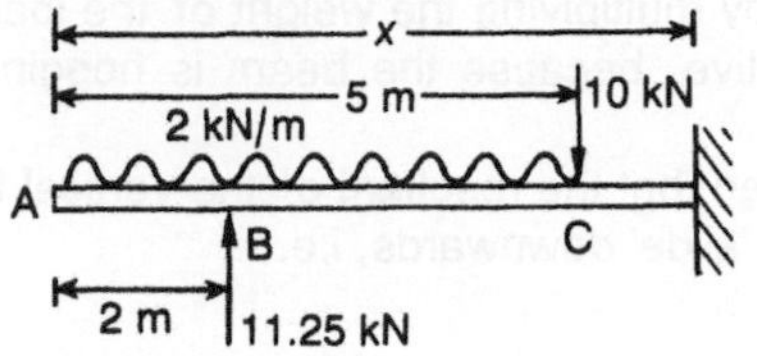

Figure 1.49

Bending moment and shearing force diagrams

From equations (1.40) to (1.45), the bending moment and shearing force diagrams can be plotted, as shown in Figure 1.50.

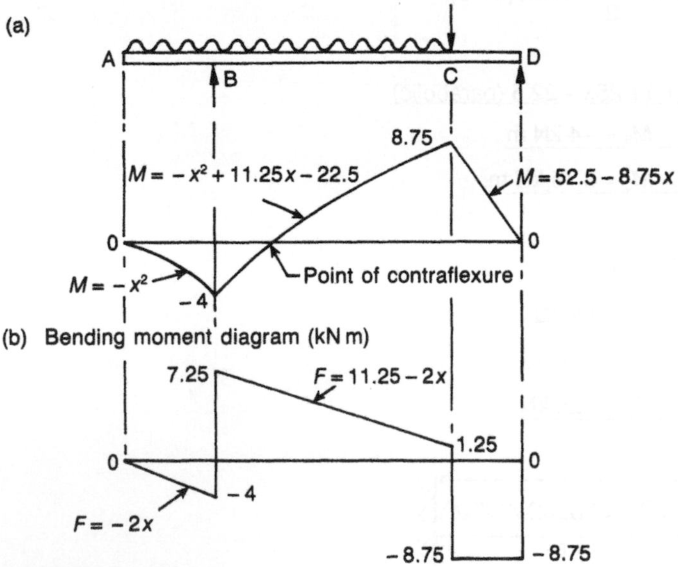

Figure 1.50

1.10 Point of contraflexure

The *point of contraflexure* is the point on a beam where the bending moment changes sign from a positive value to a negative one, or vice versa, as shown in Figure 1.50(b).

For this case, M must be zero between B and C, i.e. equation (1.42) must be used, so that

$$-x^2 + 11.25x - 22.5 = 0$$

or

$$x = \frac{-11.25 + \sqrt{11.25^2 - 4 \times 22.5}}{-2}$$

$$\underline{x = 2.6\text{m}}$$

i.e. the point of contraflexure is 2.6 m from A or 0.6 m to the right of B.

EXAMPLE 1.9

Determine the bending moment and shearing force diagrams for the simply supported beam of Figure 1.51, which is acted upon by a clockwise couple of 3 kN m at B and a counter-clockwise couple of 5 kN m at C.

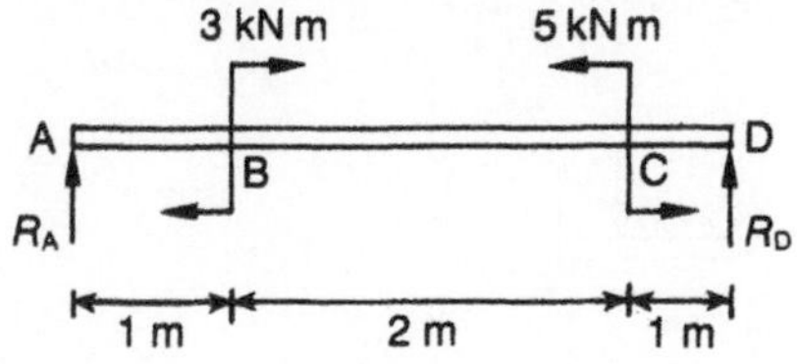

Figure 1.51 Beam with couples

SOLUTION

To determine R_A and R_D

Take moments about D

$$R_A \times 4 + 3 = 5$$

$$\underline{R_A = 0.5 \text{ kN}}$$

Resolve vertically

$$R_A + R_D = 0$$

Therefore

$$\underline{R_D = -0.5 \text{ kN}}$$

i.e. R_D acts vertically downwards.

To determine the bending moment and shearing force distributions

Consider span AB

At any distance x in Figure 1.52,

$$M = 0.5x \text{ (sagging)} \tag{1.46}$$

$$@ \text{ A} \quad M_A = 0$$

$$@ \text{ B} \quad M_B = 0.5 \text{ kN m}$$

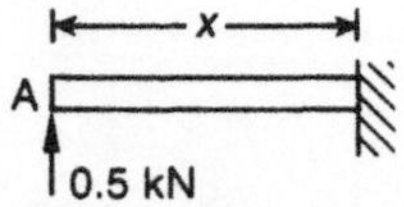

Figure 1.52

Similarly, considering vertical forces only,

$$F = 0.5 \text{ kN (constant)} \tag{1.47}$$

Consider span BC

At any distance x in Figure 1.53,

$$M = 0.5x + 3 \text{ (sagging)} \tag{1.48}$$

$$@ \text{ B} \quad M_B = 3.5 \text{ kN m}$$

$$@ \text{ C} \quad M_C = 4.5 \text{ kN m}$$

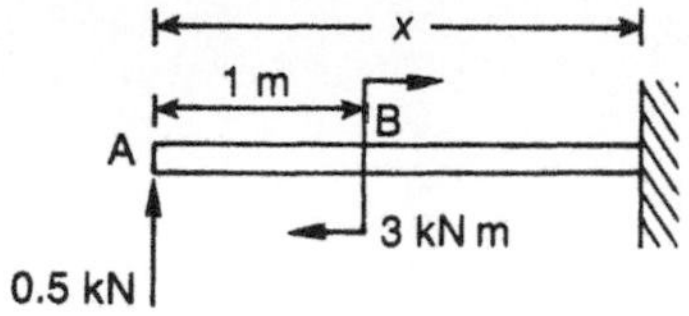

Figure 1.53

Similarly, considering vertical forces only,

$$F = 0.5 \text{ kN (constant)} \tag{1.49}$$

Consider span CD

At any distance x in Figure 1.54,

$$M = 0.5x + 3 - 5$$

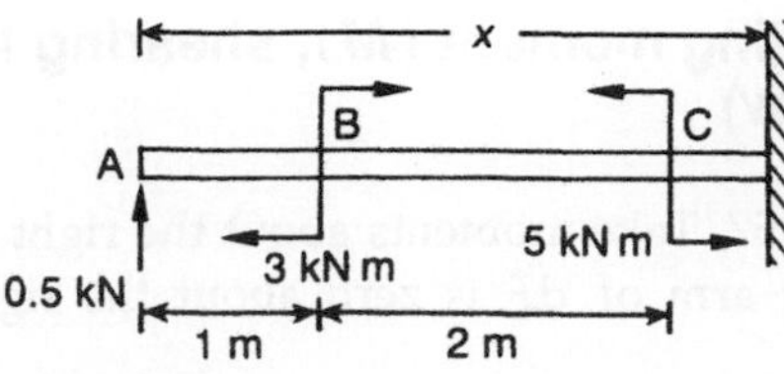

Figure 1.54

or

$$M = 0.5x - 2 \text{ (hogging)} \tag{1.50}$$

@ C $M_C = -0.5$ kN m

@ D $M_D = 0$

Similarly, considering vertical forces only,

$$F = 0.5 \text{ kN (constant)} \tag{1.51}$$

The bending moment and shearing force distributions can be obtained by plotting equations (1.46) to (1.51), as shown in Figure 1.55.

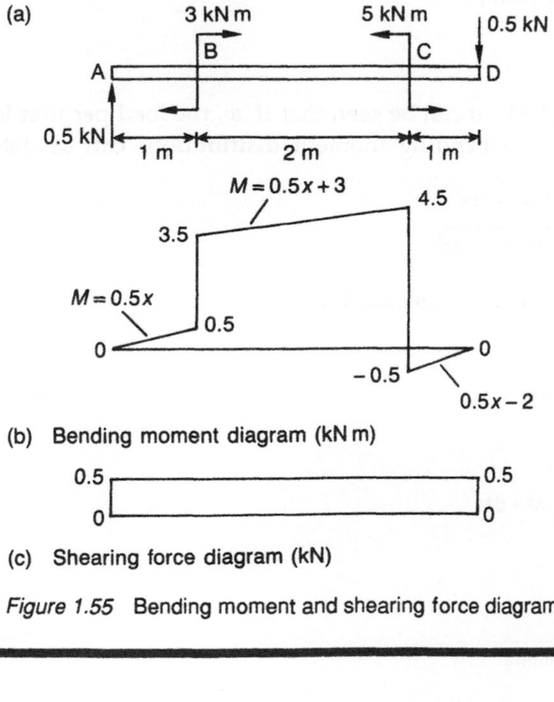

Figure 1.55 Bending moment and shearing force diagrams

1.11 Relationship between bending moment (M), shearing force (F) and intensity of load (w)

Consider the beam in Figures 1.56 and 1.57. Take moments about the right edge, as this will eliminate $\mathrm{d}F$ because the lever arm of $\mathrm{d}F$ is zero about the right edge; additionally, neglect the term $w\,\mathrm{d}x\,\mathrm{d}x/2$:

$$M + F\,\mathrm{d}x = M + \mathrm{d}M$$

or

$$\frac{\mathrm{d}M}{\mathrm{d}x} = F \tag{1.52}$$

i.e. the derivative of the bending moment with respect to x is equal to the shearing force at x.

Resolving forces vertically,

$$F + w\,\mathrm{d}x = F + \mathrm{d}F$$

Therefore

$$\frac{\mathrm{d}F}{\mathrm{d}x} = w \tag{1.53}$$

i.e. the derivative of the shearing force with respect to x is equal to w, the load per unit length, at x.

From equations (1.52) and (1.53)

$$\frac{\mathrm{d}^2M}{\mathrm{d}x^2} = w$$

From equations (1.52) and (1.53), it can be seen that if w, the load per unit length, is known, the shearing force and bending moment distributions can be determined

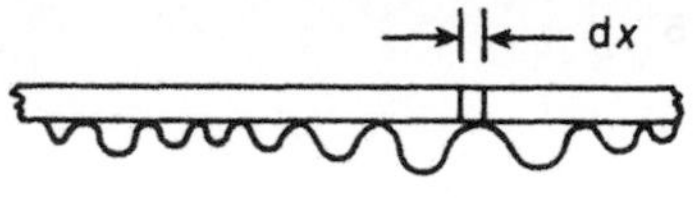

Figure 1.56 Beam with (positive) distributed load

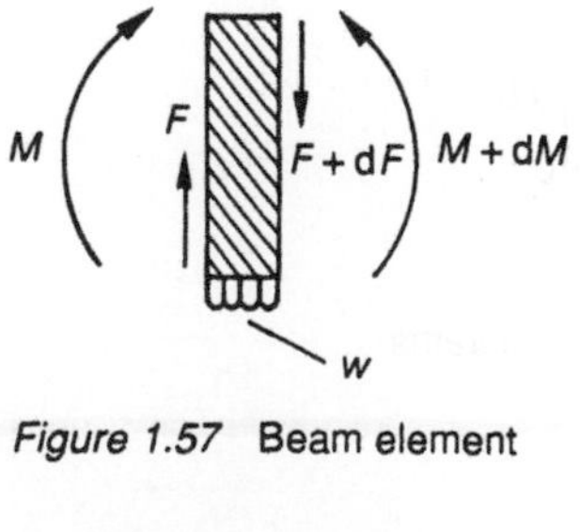

Figure 1.57 Beam element

through repeated integration and the appropriate substitution of boundary conditions. Concentrated loads, however, present a problem, but, in general, these can be approximated by either rectangles or trapeziums or triangles.

EXAMPLE 1.10

Determine the shearing force and bending moment distributions for the hydrostatically loaded beam of Figure 1.58, which is simply supported at its ends. Find, also, the position and value of the maximum value of bending moment, and plot the bending moment and shearing force diagrams.

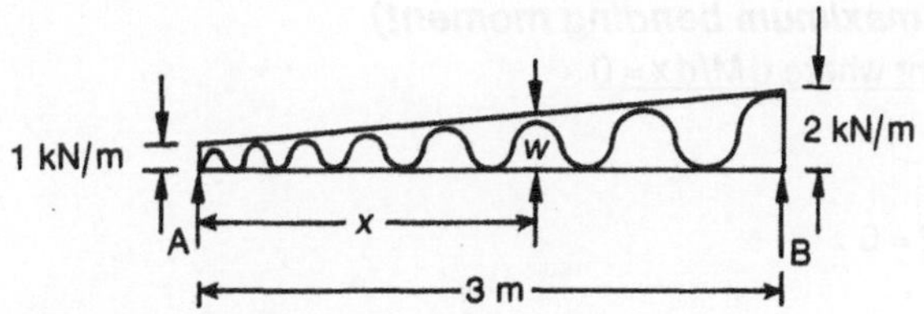

Figure 1.58 Hydrostatically loaded beam

SOLUTION

At any distance x, the load per unit length is

$$w = -1 - 0.333x \tag{1.54}$$

Now,

$$\frac{dF}{dx} = w$$

Therefore

$$\underline{F = -x - 0.166\,67x^2 + A} \tag{1.55}$$

Now,

$$\frac{dM}{dx} = F$$

Therefore

$$\underline{M = -0.5x^2 - 0.055\,56x^3 + Ax + B} \tag{1.56}$$

As there are two unknowns, two *boundary conditions* will be required to obtain the two simultaneous equations:

$$@\ x = 0 \quad M = 0$$

Therefore from equation (1.56),

$$\underline{B = 0}$$

$$@\ x = 3 \quad M = 0$$

Therefore from equation (1.56),

$$\underline{A = 2}$$

i.e.

$$F = -x - 0.1667x^2 + 2 \tag{1.57}$$

and

$$M = -0.5x^2 - 0.05666x^3 + 2x \tag{1.58}$$

To obtain $\hat{M}$ (the maximum bending moment)

$\hat{M}$ occurs at the point where $dM/dx = 0$

i.e.

$$-0.16667x^2 - x + 2 = 0$$

or

$$\underline{x = 1.583 \text{ m (to the right of A)}} \tag{1.59}$$

Substituting equation (1.59) into (1.58):

$$\underline{\hat{M} = 1.693 \text{ kN m}}$$

Bending moment and shearing force diagrams

The bending moment and shearing force distributions can be obtained from equations (1.57) and (1.58), as shown in Figure 1.59.

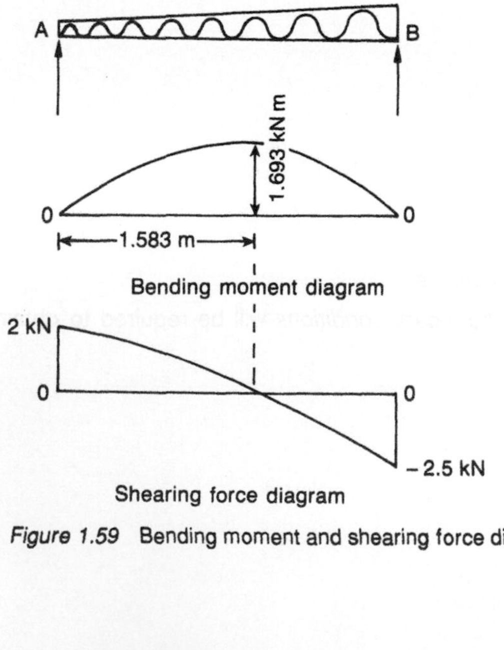

Figure 1.59 Bending moment and shearing force diagrams

EXAMPLE 1.11

A barge of uniform width 10 m and length 100 m can be assumed to be of weight 40 000 kN, which is uniformly distributed over its entire length.

Assuming that the barge is horizontal and is in equilibrium, determine the bending moment and shearing force distributions when the barge is subjected to upward buoyant forces from a wave, which is of sinusoidal shape, as shown in Figure 1.60. The wave, whose height between peaks and trough is 3 m, may be assumed to have its peaks at the ends of the barge and its trough at the mid-length of the barge (amidships ⊕).

ρ = density of water = 1020 kg/m^3

g = 9.81 m/s^2

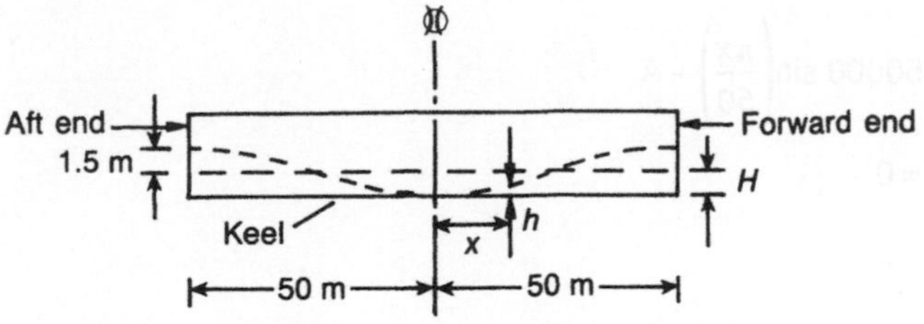

Figure 1.60 Barge subjected to a sinusoidal wave

SOLUTION

To determine H, the depth of the still waterline

$$40\,000\,000 = 1020 \times 9.81 \times 10 \times H \times 100$$

Therefore

$$H = 3.998 \text{ m or, say, } 4 \text{ m}$$

At any distance x from amidships, the height of the water above the keel is

$$h = 4 - 1.5\cos\left(\frac{\pi x}{50}\right) \tag{1.60}$$

The *upward load* per unit length acting on the barge will be due to the buoyancy, i.e. the buoyant load/unit length at x is

$$w_b = 1200 \times 9.81 \times 10 \times \left[4 - 1.5\cos\left(\frac{\pi x}{50}\right)\right]$$

$$w_b = 400000 - 150000\cos\left(\frac{\pi x}{50}\right) \tag{1.61}$$

Now *downward load* per unit length is due to the self-weight of the barge, or 400 000 N/m. Therefore

$$w = w_b - 400\,000$$

$$\underline{w = -150000\cos\left(\frac{\pi x}{50}\right)}$$

Now

$$\frac{dF}{dx} = w = -150000\cos\left(\frac{\pi x}{50}\right)$$

Therefore

$$F = -\frac{50}{\pi} \times 150000\sin\left(\frac{\pi x}{50}\right) + A$$

$$@\ x = 0 \quad F = 0$$

Therefore

$$\underline{A = 0}$$

Now,

$$\frac{dM}{dx} = F = -\frac{50}{\pi} \times 150000\sin\left(\frac{\pi x}{50}\right)$$

Therefore

$$M = \left(\frac{50}{\pi}\right)^2 \times 150000\cos\left(\frac{\pi x}{50}\right) + B$$

$$@\ x = 50 \quad M = 0$$

Therefore

$$\underline{B = 3.8 \times 10^7}$$

and

$$\underline{M = 3.8 \times 10^7\left[1 + \cos\left(\frac{\pi x}{50}\right)\right]} \tag{1.62}$$

and so

$$\underline{F = -2387320\sin\left(\frac{\pi x}{50}\right)} \tag{1.63}$$

$\hat{M}$ occurs at amidships and is equal to <u>76 MN m</u>.

Bending moment and shearing force diagrams

The bending moment and shearing force diagrams are shown in Figure 1.61.

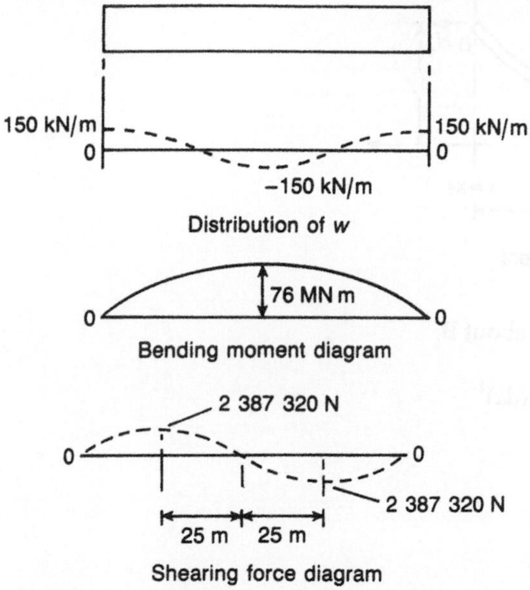

Figure 1.61 Bending moment and shearing force diagrams

1.12 Cables

Cables, when acting as load-carrying members, appear in a number of different forms varying from power lines to cables used in suspension bridges, and from rods used in pre-stressed concrete to cables used in air-supported structures. When cables are used for pre-stressed concrete and for air-supported structures, the cables or rods are initially placed under tension, where, providing their stress values are within the elastic limit, their bending stiffness will increase with tension. These problems, however, which are non-linear, are beyond the scope of this book and will not be discussed further.

Prior to analysis, it will be necessary to obtain the appropriate differential equation that governs the deflection of cables.

Consider an element of cable, loaded with a distributed load, w, which is uniform with respect to the x axis, as shown in Figure 1.62.

Taking moments about A,

$$V_2 \times dx = H \times dy + \frac{w}{2} \times (dx)^2 \tag{1.64}$$

Neglecting higher-order terms, equation (1.64) becomes

$$V_2 = H \times \left.\frac{dy}{dx}\right|_{x=x_B} \tag{1.65}$$

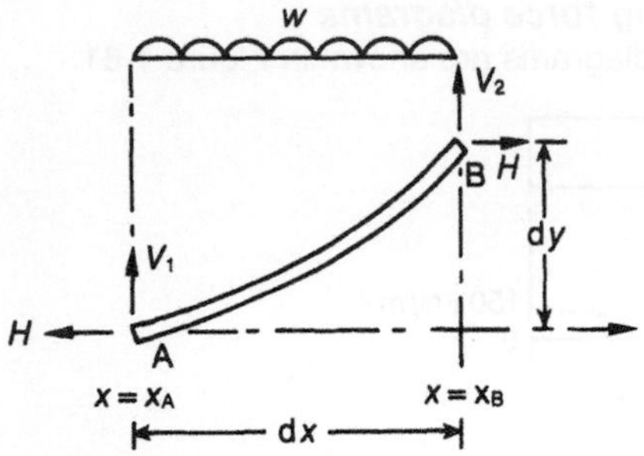

Figure 1.62 Cable element

Similarly, by taking moments about B,

$$H \times dy + V_1 \times dx = \frac{w}{2} \times (dx)^2$$

or

$$V_1 = -H \times \left.\frac{dy}{dx}\right|_{x=x_A} \tag{1.66}$$

Resolving vertically,

$$V_1 + V_2 = w \times dx$$

or

$$w = \frac{V_1 + V_2}{dx} \tag{1.67}$$

Substituting equations (1.65) and (1.66) into (1.67),

$$w = \frac{H[(dy/dx)_{x=x_B} - (dy/dx)_{x=x_A}]}{dx}$$

$$= H \times \frac{d^2y}{dx^2}$$

i.e.

$$\frac{d^2y}{dx^2} = \frac{w}{H} \tag{1.68}$$

In general, when w varies with x, equation (1.68) becomes

$$\frac{d^2y}{dx^2} = \frac{w(x)}{H} \tag{1.69}$$

where $w(x)$ is the value of the load/unit length at any distance x.

1.12.1 Cable under self-weight

Consider an infinitesimally small length of cable, under its own weight, as shown in Figure 1.63. As the element is infinitesimal,

$$(ds)^2 = (dx)^2 + (dy)^2$$

or

$$ds = dx\left[1 + \left(\frac{dy}{dx}\right)^2\right]^{1/2} \tag{1.70}$$

Let

w_s = weight/unit length of cable in the s direction

$w(x)$ = weight/unit length of cable in the x direction, at any distance x

Then

$$w(x) = w_s\left[1 + \left(\frac{dy}{dx}\right)^2\right]^{1/2} \tag{1.71}$$

Substituting equation (1.71) into equation (1.69),

$$\frac{d^2y}{dx^2} = \frac{w_s}{H}\left[1 + \left(\frac{dy}{dx}\right)^2\right]^{1/2} \tag{1.72}$$

Solution of equation (1.72) can be achieved by letting

$$\frac{dy}{dx} = Y \tag{1.73}$$

and

$$\frac{d^2y}{dx^2} = \frac{dY}{dx} \tag{1.74}$$

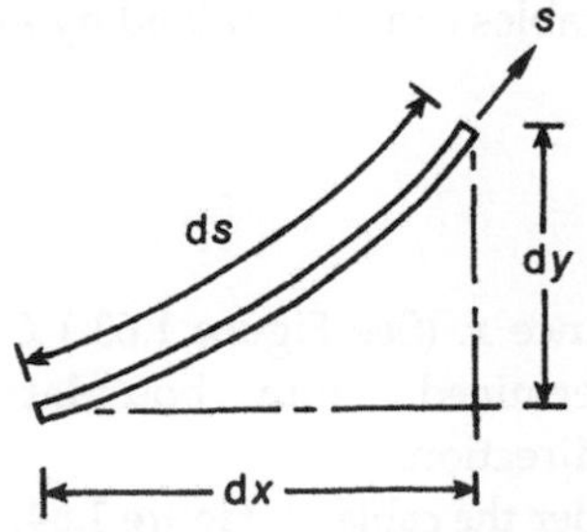

Figure 1.63 Infinitesimal length of cable under self-weight

Substituting equation (1.73) into (1.72),

$$\frac{dY}{dx} = \frac{w_s}{H}(1 + Y^2)^{1/2}$$

or

$$\frac{dY}{(1 + Y^2)^{1/2}} = \frac{w_s}{H} \times dx$$

which by inspection, yields the following solution:

$$\sinh^{-1}(Y) = w_s H \times x + C_1$$

or

$$Y = \sinh\left(\frac{w_s}{H} x + C_1\right) \tag{1.75}$$

Substituting equation (1.73) into (1.75),

$$\frac{dy}{dx} = \sinh\left(\frac{w_s}{H} x + C_1\right)$$

or

$$dy = \sinh\left(\frac{w_s}{H} x + C_1\right) dx$$

Therefore

$$y = \frac{H}{w_s} \cosh\left(\frac{w_s}{H} x + C_1\right) + C_2 \tag{1.76}$$

where C_1 and C_2 are arbitrary constants which can be obtained from boundary value considerations.

Equation (1.76) can be seen to be the equation for a *catenary*, which is how a cable deforms naturally under its own weight.

The solution of equation (1.76) for practical cases is very difficult, but a good approximation for the small-deflection theory of cables can be obtained by assuming a parabolic variation for y, as in equation (1.77):

$$y = \frac{w}{2H} x^2 + C_1 x + C_2 \tag{1.77}$$

where y is the deflection of the cable at any distance x. (See Figure 1.63.) C_1 and C_2 are arbitrary constants which can be determined from boundary value considerations, and w is load/unit length in the x direction.

To illustrate the solution of equation (1.77), consider the cable of Figure 1.64, which is supported at the same level at its ends. Let T be the tension in the cable at any distance x.

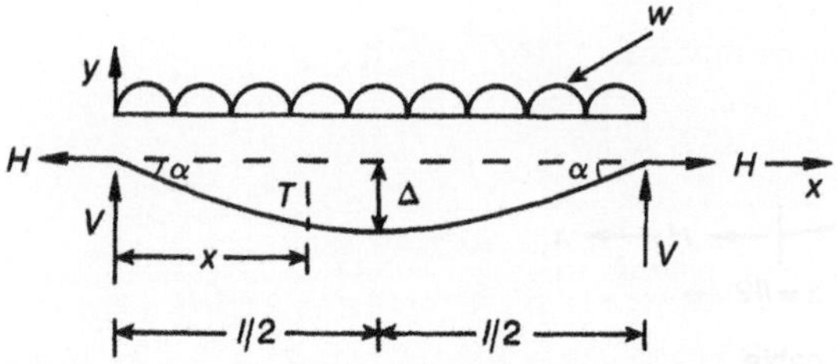

Figure 1.64 Cable with a uniformly distributed load

1.12.2 To determine C_1 and C_2

At $x = 0$, $y = 0$. Therefore

$$\underline{C_2 = 0} \tag{1.78}$$

Now,

$$\frac{dy}{dx} = \frac{w}{H}x + C_1$$

At $x = l/2$,

$$\frac{dy}{dx} = 0$$

Therefore

$$0 = \frac{wl}{2H} + C_1$$

or

$$C_1 = -\frac{wl}{2H} \tag{1.79}$$

Substituting equations (1.78) and (1.79) into (1.77),

$$\underline{y = \frac{wx^2}{2H} - \frac{wlx}{2H} - \frac{w}{2H}(x^2 - lx)} \tag{1.80}$$

To determine the sag Δ, substitute $x = l/2$ in equation (1.80), i.e.

$$\Delta = \frac{w}{2H}\left(\frac{l^2}{4} - \frac{l^2}{2}\right)$$

or

$$\underline{\text{Sag} = \Delta = -\frac{wl^2}{8H}} \tag{1.81}$$

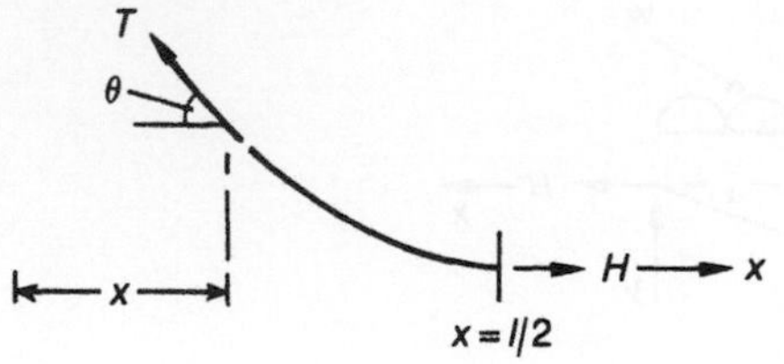

Figure 1.65 Element of cable

From equation (1.81), it can be seen that if the sag is known, H can be found, and, hence, from equilibrium considerations, V can be calculated.

1.12.3 *To determine* T

Let

T = tension in the cable at any distance x, as shown by Figures 1.64 and 1.65

θ = angle of cable with the x axis, at x

Resolving horizontally,

$$T \cos \theta = H \tag{1.82}$$

From equation (1.82), it can be seen that the maximum tension in the cable, namely $\hat{T}$, will occur at its steepest gradient, so that in this case

$$\hat{T} = \frac{H}{\cos \alpha} = H \sec \alpha \tag{1.83}$$

where

$\hat{T}$ = maximum tension in the cable, which is at the ends of the cable in Figure 1.64

α = slope of this cable at its ends

EXAMPLE 1.12

Determine the maximum tension in the cable of Figure 1.66, where

w = 120 N/m and H = 20 kN

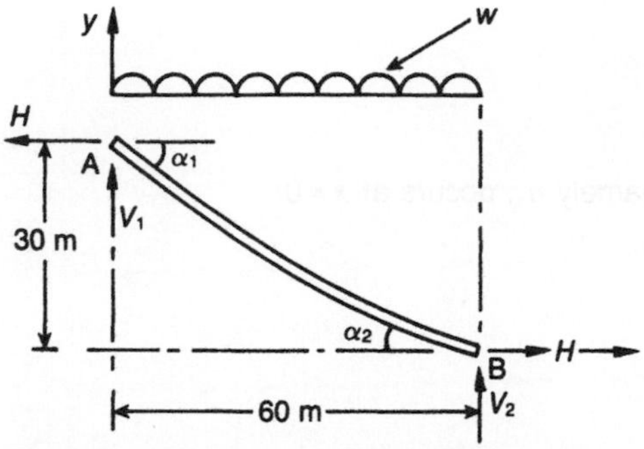

Figure 1.66 Cable supported at different levels

SOLUTION

From equation (1.77),

$$y = \frac{wx^2}{2H} + C_1x + C_2 \tag{1.84}$$

$$@\ x = 0 \quad y = 30$$

Therefore

$$\underline{C_2 = 30} \tag{1.85}$$

$$@\ x = 60 \quad y = 0$$

Therefore

$$0 = \frac{60}{H}\,60^2 + 60C_1 + 30$$

$$C_1 = -\frac{3600}{H} - 0.5 \tag{1.86}$$

But

$$H = 20\,000$$

Therefore

$$\underline{C_1 = -0.68} \tag{1.87}$$

therefore

$$y = \frac{120x^2}{2 \times 20000} - 0.68x + 30$$

or

$$y = 3\text{E} - 3x^2 - 0.68x + 30$$

and

$$\frac{dy}{dx} = 6E - 3x - 0.68 \qquad (1.88)$$

By inspection, the maximum slope, namely α_1, occurs at $x = 0$:

$$\alpha_1 = \tan^{-1}(-0.68) = -34.22°$$

From equation (1.75),

$$\hat{T} = \frac{H}{\cos\alpha_1} = \frac{20 \text{ kN}}{0.827}$$

$$\hat{T} = 24.19 \text{ kN}$$

1.12.4 Cables under concentrated loads

Cables in this category are assumed to deform, as shown in Figures 1.67 to 1.69.

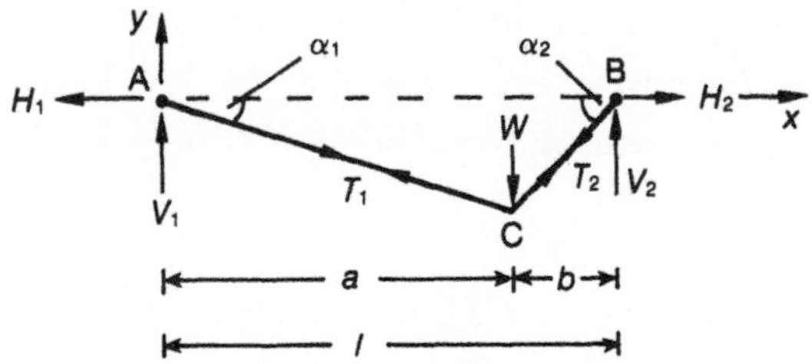

Figure 1.67 Cable with a single concentrated load

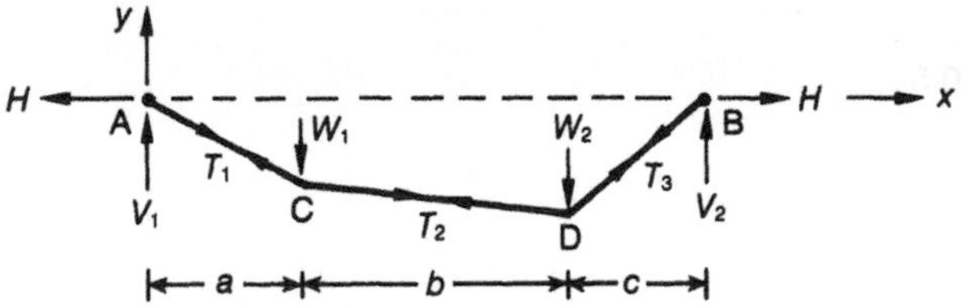

Figure 1.68 Cable with two concentrated loads, but with the end supports at the same level

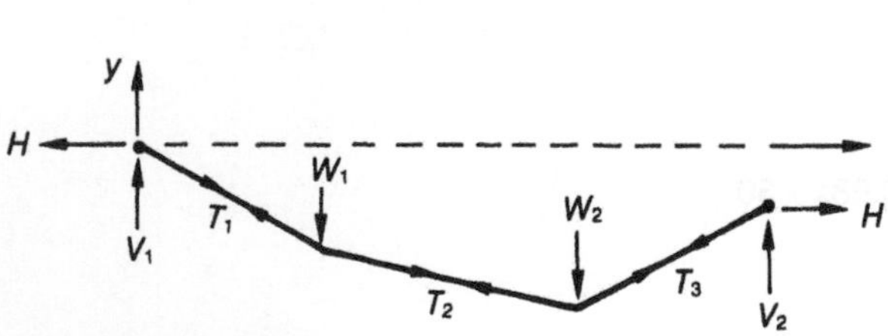

Figure 1.69 Cable with two concentrated loads, but with the end supports at different levels

EXAMPLE 1.13

Determine expressions for T_1, T_2, H, V_1 and V_2, in terms of α_1 and α_2, for the cable of Figure 1.67.

SOLUTION

Resolving vertically

$$V_1 + V_2 = W \tag{1.89}$$

Taking moments about A

$$V_2 * l = W * a$$

Therefore

$$\underline{V_2 = Wa/l}$$

and from equation (1.88)

$$\underline{V_1 = Wb/l}$$

Consider the point C

Resolving horizontally at the point C

$$T_1 \cos \alpha_1 = T_2 \cos \alpha_2$$

$$T_1 = T_2 \cos \alpha_2 / \cos \alpha_1 \tag{1.90}$$

Resolving vertically at the point C

$$T_1 \sin \alpha_1 + T_2 \sin \alpha_2 = W$$

or

$$T_1 = \frac{W}{\sin \alpha_1} - T_2 \frac{\sin \alpha_2}{\sin \alpha_1} \tag{1.91}$$

Equating (1.90) and (1.91),

$$\frac{W}{\sin \alpha_1} - T_2 \frac{\sin \alpha_2}{\sin \alpha_1} = \frac{T_2 \cos \alpha_2}{\cos \alpha_1}$$

or

$$T_2\left(\frac{\cos \alpha_2}{\cos \alpha_1} + \frac{\sin \alpha_2}{\sin \alpha_1}\right) + \frac{W}{\sin \alpha_1}$$

Therefore

$$\underline{T_2 = \frac{W}{\sin \alpha_1} \Big/ \left(\frac{\cos \alpha_2}{\cos \alpha_1} + \frac{\sin \alpha_2}{\sin \alpha_1}\right)} \tag{1.92}$$

Substituting equation (1.92) into (1.91), T_1 can be determined.

H_1 and H_2 can be determined by resolution, as follows.

Resolving horizontally

$$H_1 = H_2 = T_1 \cos \alpha_1 = T_2 \cos \alpha_2 \quad (1.93)$$

EXAMPLE 1.14

Determine the tensions T_1, T_2 and T_3 that act in the cable of Figure 1.70. Hence, or otherwise, determine the end forces H_1, V_1, H_2 and V_2.

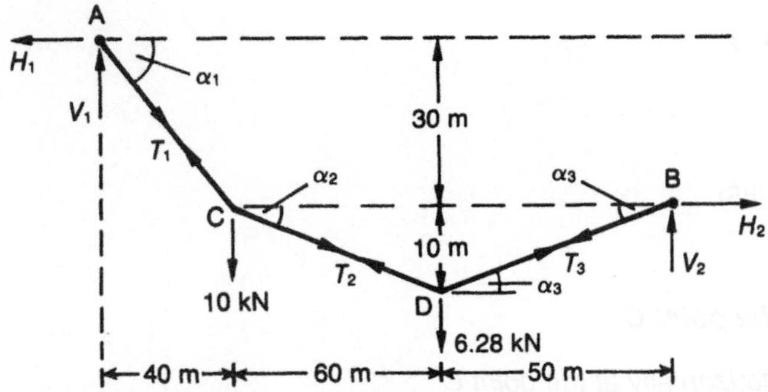

Figure 1.70 Cable with concentrated loads

SOLUTION

Resolving vertically

$$V_1 + V_2 = 30 \quad (1.94)$$

$\alpha_1 = \tan^{-1}(0.75) = 36.87°$

$\alpha_2 = \tan^{-1}(10/60) = 9.46°$

$\alpha_3 = \tan^{-1}(10/50) = 11.31°$

To determine T_1, T_2 and T_3

Resolving horizontally at C

$$T_1 \cos \alpha_1 = T_2 \cos \alpha_2$$

or

$$0.8\,T_1 = 0.986\,T_2$$

Therefore

$$T_1 = 1.233\,T_2 \quad (1.95)$$

Resolving vertically at C

$$T_1 \sin \alpha_1 = 10 + T_2 \sin \alpha_2$$

$$0.6\,T_1 = 10 + 0.164\,T_2$$

or

$$\underline{T_1 = 16.67 + 0.273\,T_2} \qquad (1.96)$$

Equating (1.95) and (1.96),

$$1.233\,T_2 = 16.67 + 0.273\,T_2$$

Therefore

$$\underline{T_2 = 17.37 \text{ kN}} \qquad (1.97)$$

Substituting equation (1.97) into (1.96),

$$\underline{T_1 = 21.42 \text{ kN}} \qquad (1.98)$$

Resolving horizontally at D

$$T_3 \cos \alpha_3 = T_2 \cos \alpha_2$$

$$\underline{T_3 = 17.47 \text{ kN}} \qquad (1.99)$$

To determine H_1, V_1, H_2 and V_2

Resolving vertically at A

$$\underline{V_1 = T_1 \sin \alpha_1 = 12.85 \text{ kN}}$$

Resolving horizontally at A

$$\underline{H_1 = T_1 \cos \alpha_1 = 17.14 \text{ kN}}$$

Resolving vertically at B

$$\underline{V_2 = T_3 \sin \alpha_3 = 3.43 \text{ kN}}$$

Resolving horizontally at B

$$\underline{H_2 = T_3 \cos \alpha_3 = 17.13 \text{ kN}}$$

Note that

$$\underline{H_1 = H_2 \text{ (as required)}}$$

1.13 Suspension bridges

The use of cables to improve the structural efficiency of bridges is widely adopted throughout the world, especially for suspension bridges. In this case, the cables are

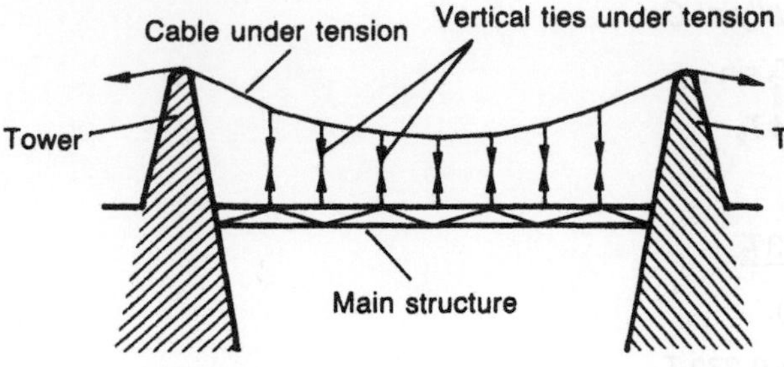

Figure 1.71 Suspension bridge

placed under tension between towers, so that the cables exert upward forces to support the main structure of the bridge, via vertical tie-bars, as shown in Figure 1.71. Most of the world's longest bridges are, in fact, suspension bridges.

EXAMPLES FOR PRACTICE 1

For all problems, neglect **self-weight**.

1. Determine the reactions R_A and R_B for the simply supported beams of Figures 1.72(a) and (b).

(a)

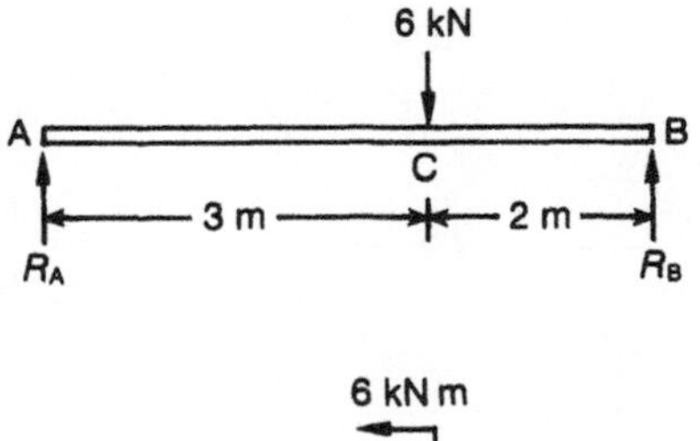

(b)

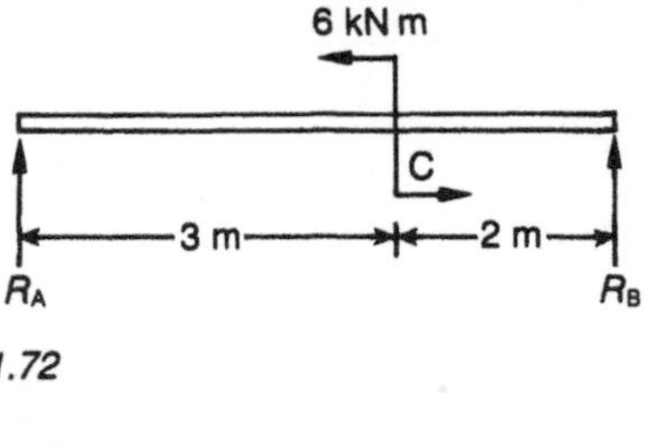

Figure 1.72

{(a) $R_A = 2.4$ kN, $R_B = 3.6$ kN; (b) $R_A = 1.2$ kN, $R_B = -1.2$ kN}

2. Determine the reactions R_A and R_B for the simply supported beams of Figures 1.73(a) and (b).

{(a) $R_A = 2.67$ kN, $R_B = 5.33$ kN;
(b) $R_A = R_B = 0$}

(a)

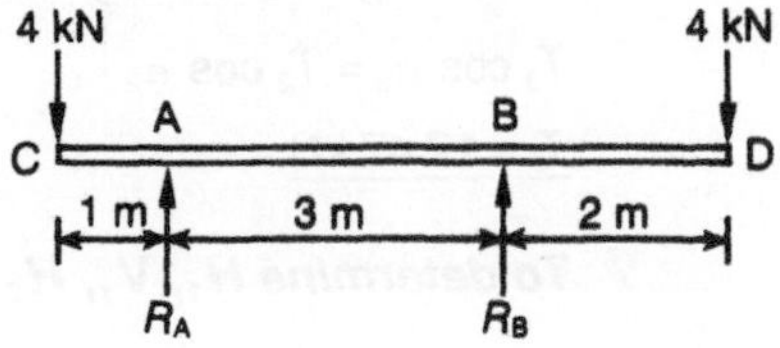

(b)

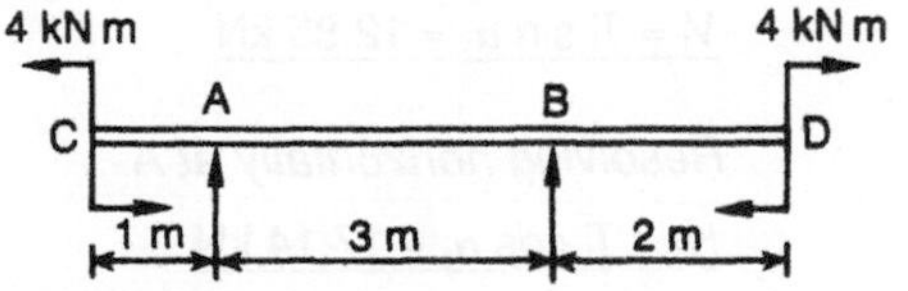

Figure 1.73

3. Determine the internal forces in the members of the plane pin-jointed truss of Figures 1.74(a) and (b).

{(a) $F_{ac} = 8$ kN, $F_{bc} = -6.928$ kN;
(b) $F_{ab} = 1.732$ kN, $F_{ac} = -2$ kN, $F_{bc} = -3.464$ kN}

(a)

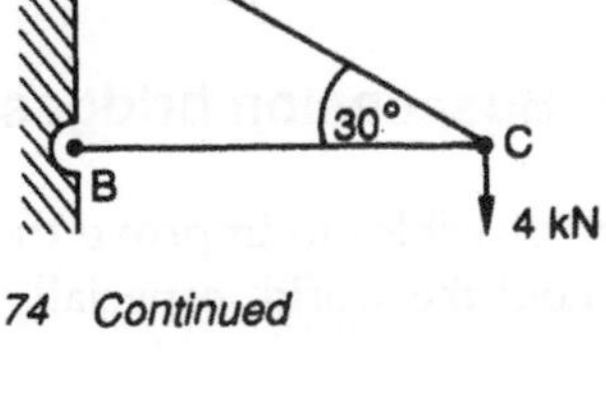

Figure 1.74 Continued

(b)

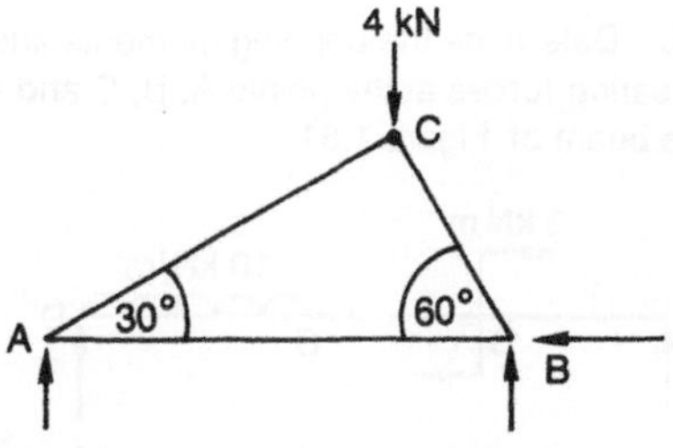

Figure 1.74

4. A plane pin-jointed truss is firmly pinned at its base, as shown in Figure 1.75. Determine the forces in the members of this truss, stating whether they are in tension or compression.

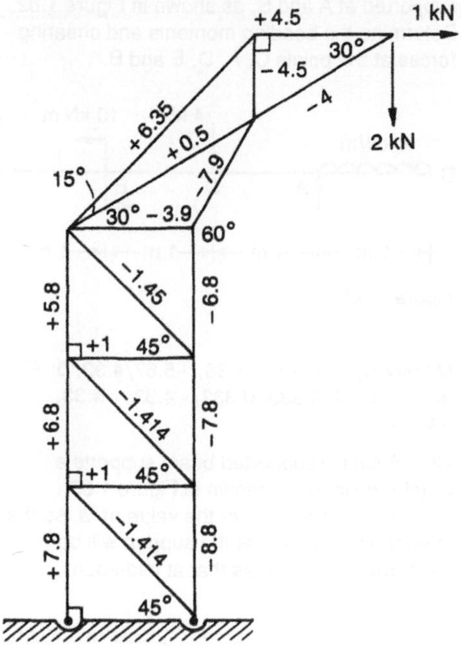

Figure 1.75

{Forces in kN; + tension, − compression}

5. The plane pin-jointed truss of Figure 1.76 is firmly pinned at A and B and subjected to two point loads at point F.

Determine the forces in the members, stating whether they are tensile or compressive.

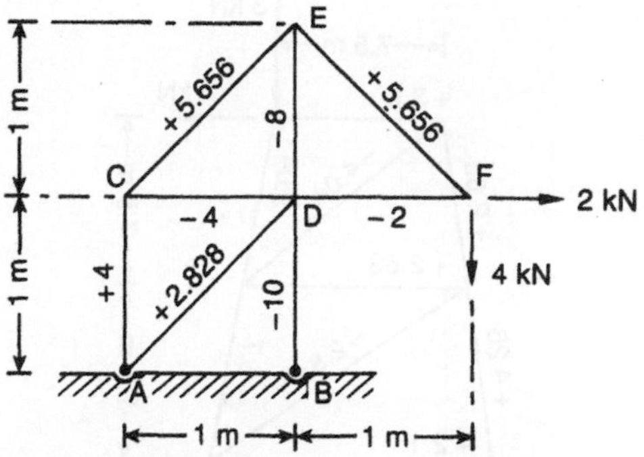

Figure 1.76

6. An overhanging pin-jointed roof truss, which may be assumed to be pinned rigidly to the wall at the joints A and B, is subjected to the loading shown in Figure 1.77.

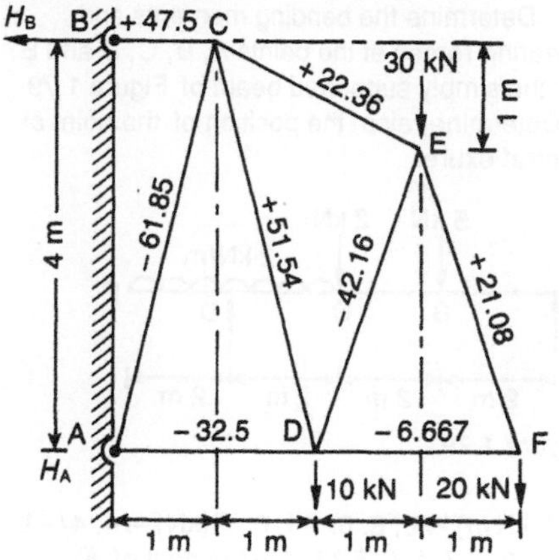

Figure 1.77

Determine the forces in the members of the truss, stating whether they are tensile or compressive. Determine, also, the reactions.

{$H_A = 47.5$, $V_A = 60$, $H_B = -47.5$, $V_B = 0$}

7. Determine the forces in the symmetrical pin-jointed truss of Figure 1.78 stating whether they are tensile or compressive.

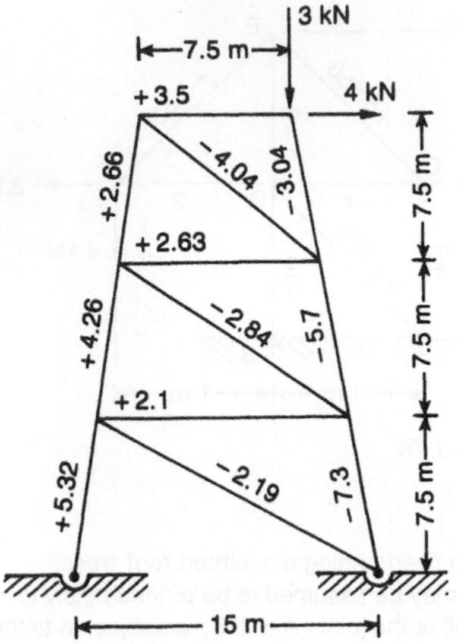

Figure 1.78

8. Determine the bending moments and shearing forces at the points A, B, C, D and E for the simply supported beam of Figure 1.79.

Determine, also, the position of the point of contraflexure.

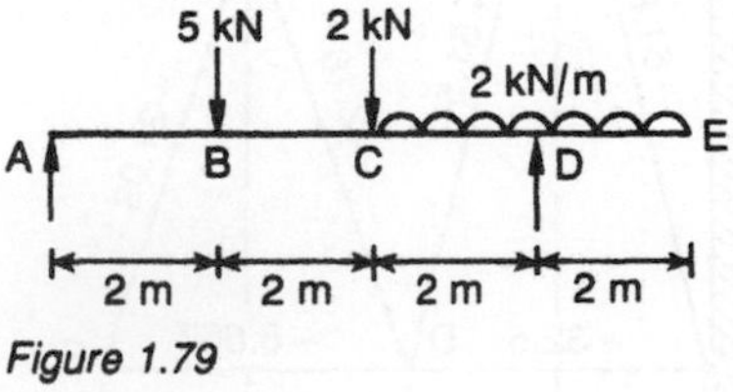

Figure 1.79

{M (kN m)→0, 8, 6, −4, 0; F (kN)→4, 4/−1, −1/−3, −7/4, 0; 5.73 m to the right of A}

9. Determine the bending moments and shearing forces at the points A, B, C and D on the cantilever of Figure 1.80.

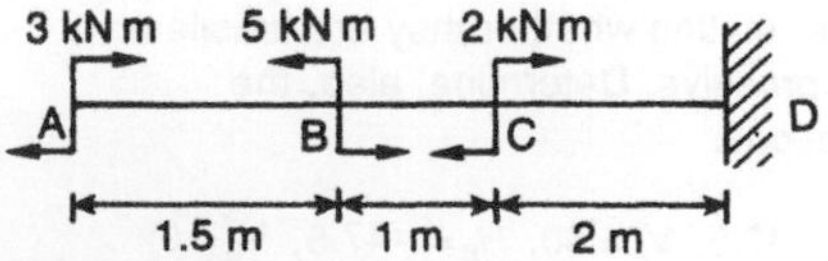

Figure 1.80

{M (kN m)→3, 3/−2, −2/0, 0; F (kN)→0, 0, 0, 0}

10. Determine the bending moments and shearing forces at the points A, B, C and D on the beam of Figure 1.81.

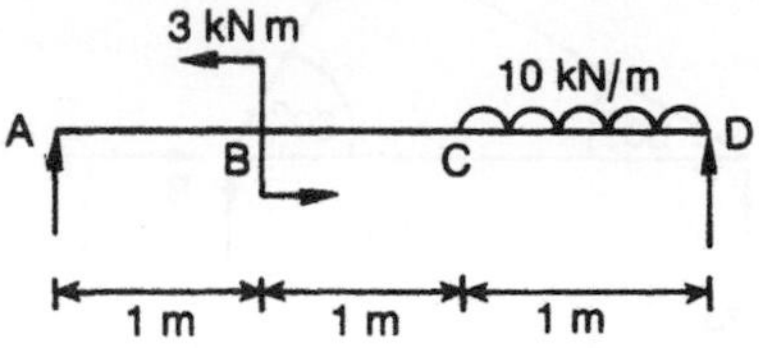

Figure 1.81

{M (kN m)→0, 2.667/−0.333, 2.333, 0; F (kN)→2.667, 2.667, 2.667, −7.333}

11. A uniform-section beam is simply supported at A and B, as shown in Figure 1.82. Determine the bending moments and shearing forces at the points C, A, D, E and B.

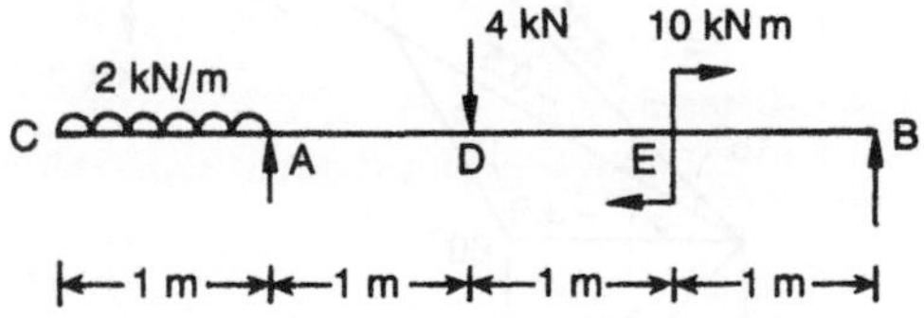

Figure 1.82

{M (kN m)→0, −1, −1.33, −5.67/4.33, 0; F (kN)→0, −2/0.333, 0.333/−4.33, −4.33, −4.33}

12. A simply supported beam supports a distributed load, as shown in Figure 1.83. Obtain an expression for the value of a, so that the bending moment at the support will be of the same magnitude as that at mid-span.

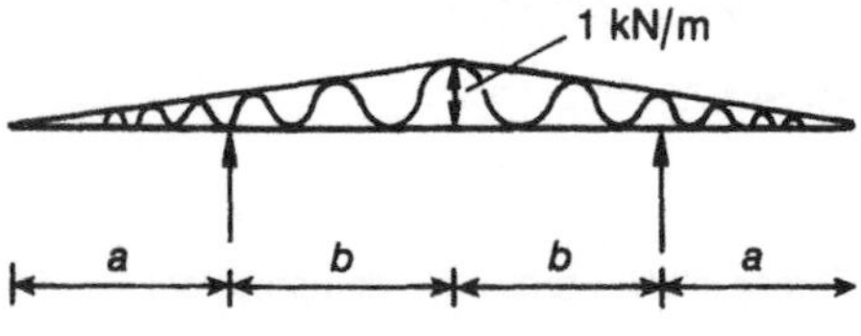

Figure 1.83

{$a^3/18 + 1.5a = 3$, $a = 1.788$ m}

13. Determine the maximum tensile force in the cable of Figure 1.84, and the vertical reactions at its ends, given the following:

$w = 200$ N/m and $H = 30$ kN

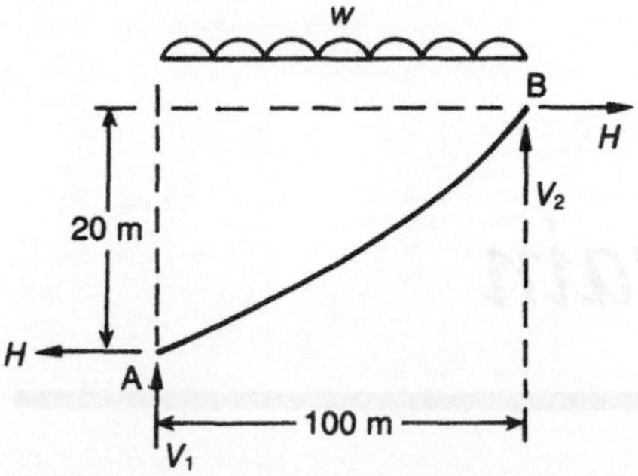

Figure 1.84 Cable with uniformly distributed load

{$\hat{T}$ (@B) $= 34.0$ kN, $V_1 = 4$ kN, $V_2 = 16$ kN}

14. Determine the tensile forces in the cable of Figure 1.85, together with the end reactions.

{$T_1 = 28.76$ kN, $T_2 = 29.40$ kN, $T_3 = 38.6$ kN;
$H = 27.28$ kN, $V_1 = 9.09$ kN, $V_2 = 27.29$ kN}

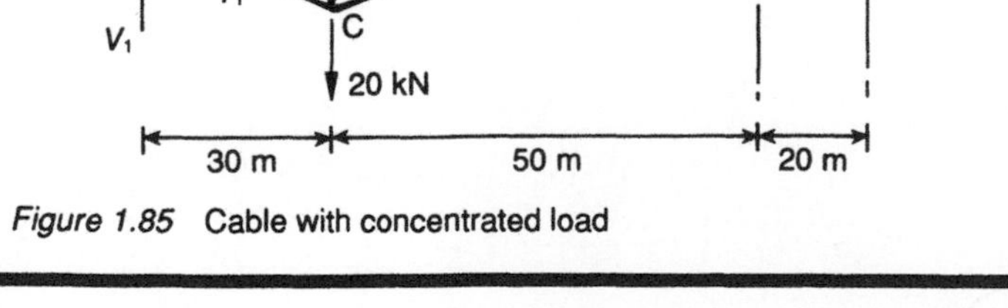

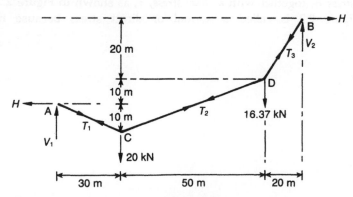

Figure 1.85 Cable with concentrated load

2

Stress and Strain

2.1 Introduction

The most elementary definition of stress is that it is the *load per unit area* acting on a surface, rather similar to pressure, except that it can be either tensile or compressive and it does not necessarily act normal to the surface, i.e.

$$\text{Stress} = \text{load/area} \tag{2.1}$$

In its simplest form, stress acts at an angle to the surface, as shown in Figure 2.1.

However, in the form shown in Figure 2.1, it is difficult to apply stress analysis to practical problems, and because of this, the resultant stress is represented by a *normal or direct stress* σ, together with a *shear stress*, τ, as shown in Figure 2.2. The stress, σ, in Figure 2.2 is called a normal or direct stress because it acts

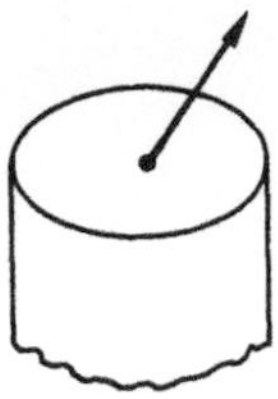

Figure 2.1

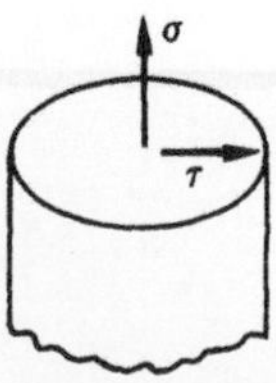

Figure 2.2 Normal and shear stress

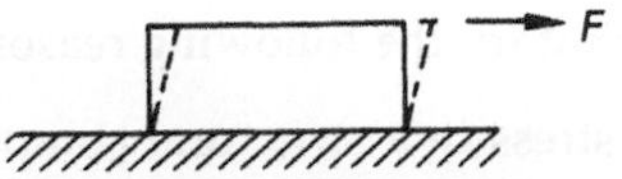

Figure 2.3 Shearing action of F

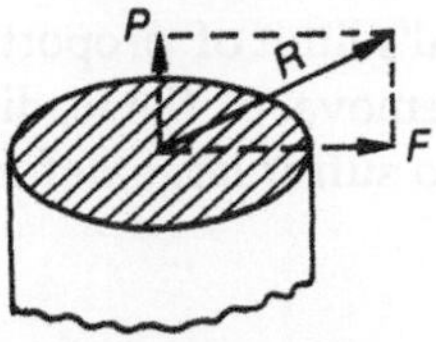

Figure 2.4 Components of R

perpendicularly to the surface under consideration, and the stress, τ, is called a shearing stress because it acts tangentially to the surface, causing shearing action, as shown in Figure 2.3. Thus, if a flat surface is subjected to a force, R, acting at an angle to the surface, it is convenient to represent this resultant force by its two perpendicular components, namely P and F, where P acts normal to the surface and F acts tangentially to the surface, as shown in Figure 2.4.

The effect of P will be to increase the length of the structural component and to cause a normal or direct stress, σ, where

$$\sigma = P/A \tag{2.2}$$

and A is the cross-sectional area. Similarly, the effect of F will be to cause the component to suffer shear deformation, as shown in Figure 2.3, and to cause a shear stress, τ, where

$$\tau = F/A \tag{2.3}$$

The *sign convention for direct stress* is as follows:

Tensile stresses are positive

Compressive stresses are negative

2.2 Hooke's Law

If a length of wire, made from steel or aluminium alloy, is tested in tension, the wire will be found to increase its length linearly with increase in load, for 'smaller' values of load, so that

$$\text{load} \propto \text{extension} \tag{2.4}$$

Expression (2.4) was discovered by Robert Hooke (1635–1703), and it applies to many materials up to the *limit of proportionality* of the material.

In structural design, Hooke's law is very important for the following reasons:

(a) In general, it is not satisfactory to allow the stress in a structural component to exceed the limit of proportionality. This is because, if the stress exceeds this value, it is likely that certain parts of the structural component will suffer permanent deformation.
(b) If the stress in a structure does not exceed the material's limit of proportionality, the structure will return to its undeformed shape on removal of the loading.
(c) For many structures it is undesirable to allow them to suffer large deformations under normal loading.

2.3 Load–extension relationships

A typical load–extension curve for mild steel, in tension, is shown in Figure 2.5.

2.3.1 *Some important points on Figure 2.5*

Limit of proportionality (A) Up to this point, the load–extension curve is linear and elastic.

Elastic limit (B) Up to this point, the material will recover its original shape on removal of the load. The section of the load–extension curve between the limit of proportionality and the elastic limit can be described as non-linear elastic.

Yield point (C/D) This is the point of the load–extension curve where the material suffers permanent deformation, i.e. the material behaves plastically beyond this point, and Poisson's ratio is approximately equal to 0.5. The extension of the specimen from C to D is approximately 40 times greater than the extension of the specimen up to B.

Strain hardening (D/E) After the point D, the material strain hardens, where

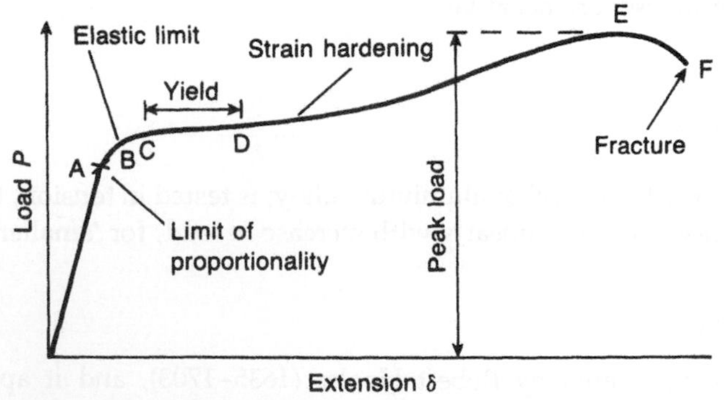

Figure 2.5 Load–extension curve for mild steel

the slope of the load–extension curve, just above D, is about 1/50th of the slope between the origin and A.

Peak load (E) This load is used for calculating the *ultimate tensile stress* or *tensile strength* of the material. After this point, the specimen 'necks' and eventually fractures at F. ('Necking' means that a certain section of the specimen suffers a local decrease in its cross-sectional area.)

2.3.2 *Stress–strain curve*

Normally, Figure 2.6 is preferred to Figure 2.5 where

Nominal stress, $\sigma = P/A$

Nominal strain, $\varepsilon = \delta/\text{original length}$

where

A = original cross-sectional area

δ = deflection due to P

In Figure 2.6,

σ_{yp} = Yield stress, and for most structural designs the stress–strain relationship is assumed to be linear up to this point

σ_{UTS} = ultimate tensile stress or nominal peak stress = peak load/A

In the design of structures, the stress is normally not allowed to exceed the limit of proportionality, where the relationship between σ and ε can be put in the form

$$\frac{\text{Stress }(\sigma)}{\text{Strain }(\varepsilon)} = E \qquad (2.5)$$

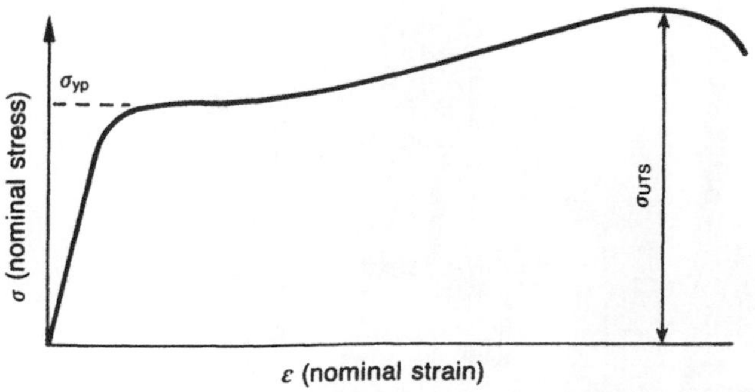

Figure 2.6 Nominal stress–nominal strain relationship

where

E = Young's (or the elastic) modulus
ε = nominal strain
= extension per unit length (unitless)

To determine E for construction materials, such as mild steel, aluminium alloy, etc., it is normal to make a suitable specimen from the appropriate material and to load it in tension in a universal testing machine, as shown in Figure 2.7. Prior to loading the specimen, a small extensometer is connected to the specimen, and measurements are made of the extension of the specimen over the gauge length l of the extensometer against increasing load, up to the limit of proportionality. The cross-section of the

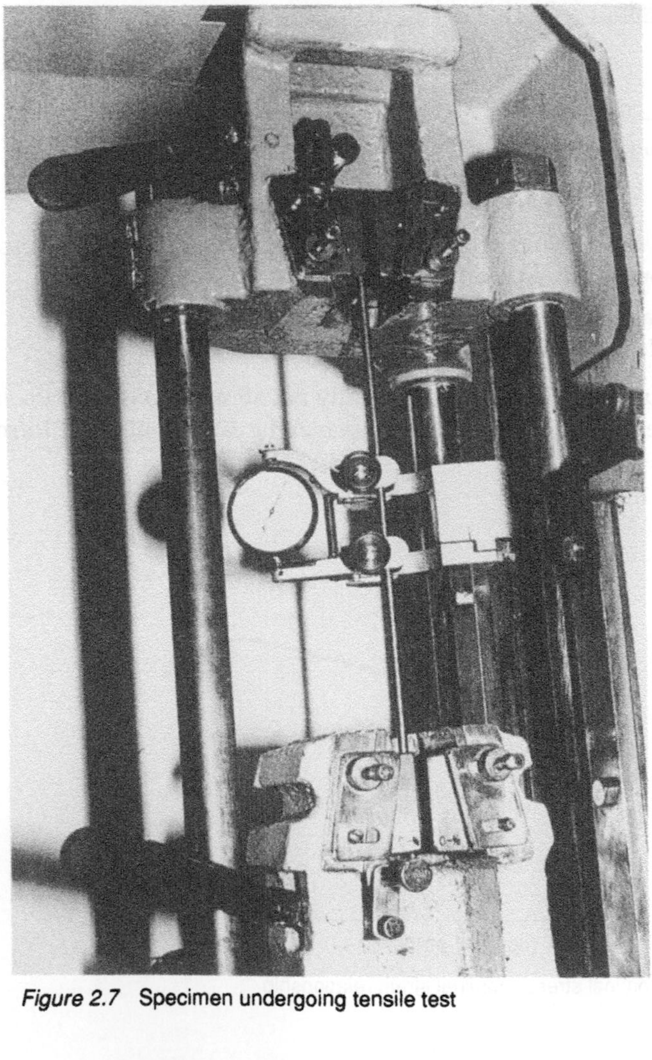

Figure 2.7 Specimen undergoing tensile test

Table 2.1 **Young's modulus E (N/m²)**

Steel	Aluminium alloy	Copper	Concrete 'new'	Concrete 'old'	Oak (with grain)
2.1×10^{11}	7×10^{10}	1.2×10^{11}	1.9×10^{10}	3.6×10^{10}	1.2×10^{10}

specimen is usually circular and is sensibly constant over the gauge length. In general, E is approximately the same in tension as it is in compression for most structural materials, and some typical values are given in Table 2.1.

2.4 Proof stress

It should be noted that certain materials, such as aluminium alloy and high tensile steel, do not exhibit a definite yield point, such as that shown in Figure 2.8. For such cases, a 0.1% or 0.2% proof stress is used instead of a yield stress, as shown in Figure 2.8.

To determine the 0.1% proof stress, a strain of 0.1% is set off along the horizontal axis of Figure 2.8, and a straight line is drawn from this point, parallel to the bottom section of the straight line part of the stress–strain relationship. The 0.1% proof stress is measured where the straight line intersects the stress–strain curve, as shown in Figure 2.8. A similar process is used to determine the 0.2% proof stress.

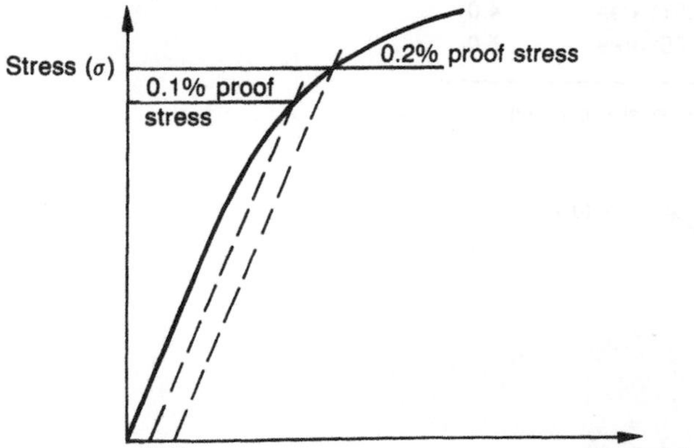

Figure 2.8 Stress–strain curve for aluminium alloy or a typical high tensile steel

2.5 Ductility

Another important material property is ductility. This can be described as the ability of a material to suffer plastic deformation while still resisting increasing load. The more the material can suffer plastic deformation, the more ductile it is said to be.

Ductility can be measured either by the percentage reduction in area or by the percentage elongation of a measured length of the specimen namely its *gauge length*:

$$\text{Percentage reduction in area} = \frac{A_I - A_F}{A_I} \times 100\%$$

$$\text{Percentage elongation} = \frac{L_I - L_F}{L_I} \times 100\%$$

where

A_I = initial cross-sectional area of the tensile specimen
A_F = final cross-sectional area of the tensile specimen
L_I = initial gauge length of the tensile specimen
L_F = final gauge length of the tensile specimen

It should be emphasized that comparisons between various values of percentage elongation will be dependent on whether or not the tensile specimens are geometrically similar.

Table 2.2 **Circular cylindrical tensile specimen**

Place	L_I	L_I/D_I
UK	$4\sqrt{\text{area}}$	3.54
USA	$4.51\sqrt{\text{area}}$	4.0
Europe	$5.65\sqrt{\text{area}}$	5.0

D_I = initial diameter of tensile specimen.

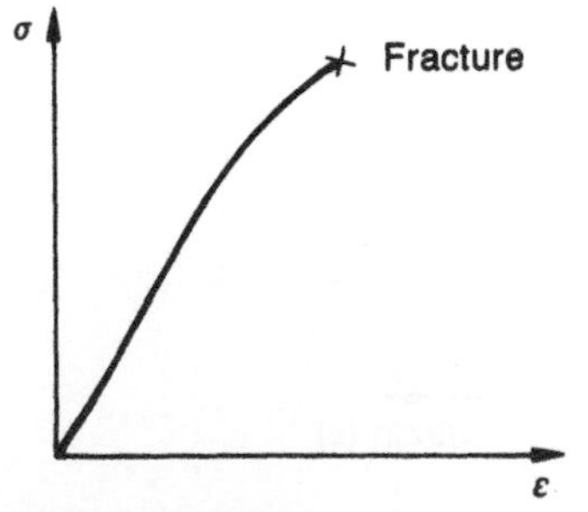

Figure 2.9 Stress–strain curve for a brittle material

Typical values used in various parts of the world for circular cylindrical specimens are shown in Table 2.2, where L_1 is usually taken as 50 mm.

Brittle materials, such as cast iron, have low ductility. A stress–strain curve for a typical brittle material is shown in Figure 2.9.

EXAMPLE 2.1

Two structural components are joined together by a steel bolt with a screw thread, which has a diameter of 12.5 mm and a pitch of 1 mm, as shown in Figure 2.10. Assuming that the bolt is initially stress free and that the structural components are inextensible, determine the stress in the bolt; if it is tightened by rotating it clockwise by one-eighth of a turn, the bolt will increase its length by an amount δ over the length 50 mm, where

$$\delta = \tfrac{1}{8} \times 1 \text{ mm} = 1.25 \times 10^{-4} \text{ m}$$

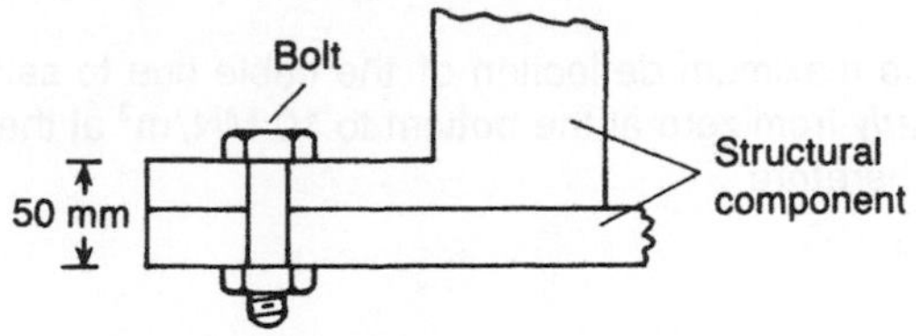

Figure 2.10 Bolt under tensile stress

SOLUTION

Now,

$$\varepsilon = \text{strain} = \frac{\delta}{l} = \frac{1.25 \times 10^{-4}}{50 \times 10^{-3}} = 2.5 \times 10^{-3}$$

$$\sigma = \text{stress} = E\varepsilon = 2 \times 10^{11} \times 2.5 \times 10^{-3}$$

$$\underline{\sigma = 500 \text{ MN/m}^2 \text{ (tensile)}}$$

EXAMPLE 2.2

A copper cable hangs down a vertical mineshaft. Determine the maximum permissible length of the cable, if its maximum permissible stress, due to self-weight, must not exceed 10 MN/m^2. Hence, or otherwise, determine the maximum vertical deflection of this cable due to self-weight.

$$E = 1 \times 10^{11} \text{N/m}^2 \quad \rho = 8900 \text{ kg/m}^3 \quad g = 9.81 \text{ m/s}^2$$

SOLUTION

The maximum stress in the cable will be at the top. Let $\hat{\sigma}$ = maximum stress = 10 MN/m^2.

$$\text{Weight of cable} = \rho A g l,$$

where

A = cross-sectional area

l = length of cable

Therefore

$$\hat{\sigma} = \frac{\rho A g l}{A} = \rho g l$$

or

$$10 \times 10^6 \,\frac{\text{N}}{\text{m}^2} = 8900 \times 9.81 \times l$$

i.e.

$$l = 114.5 \text{ m}$$

Now to determine δ, where δ is the maximum deflection of the cable due to self-weight. The stress in the cable varies *linearly* from zero at the bottom to 10 MN/m² at the top; the strain varies in a similar manner. Therefore

$$\text{Average stress} = 5 \text{ MN/m}^2$$

and

$$\text{Average strain} = \frac{5 \times 10^6}{1 \times 10^{11}} = 5 \times 10^{-5}$$

Hence

$$\underline{\delta = 5 \times 10^{-5} \times 114.5\ m = 5.73 \text{ mm}}$$

EXAMPLE 2.3

Determine the profile of a vertical pillar which is to have a constant normal stress due to self-weight

SOLUTION

Consider an element of the bar at any distance x from the bottom, as shown in Figure 2.11, where the cross-sectional area is A and the stress is σ.

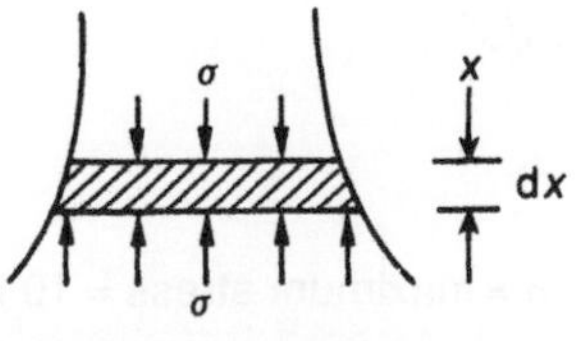

Figure 2.11 Constant strength pillar in compression

Resolving vertically,

$$\sigma(A+\mathrm{d}A) = \sigma A + \rho g A\,\mathrm{d}x \tag{2.6}$$

where

ρ = density of material

g = acceleration due to gravity

and

$\rho g A\,\mathrm{d}x$ = weight of element (shaded area)

Simplifying equation (2.6), the following is obtained:

$$\frac{\mathrm{d}A}{A} = \frac{\rho g}{\sigma}\,\mathrm{d}x$$

or

$$\ln A = \frac{\rho g x}{\sigma} + \ln C \tag{2.7}$$

where $\ln C$ is an arbitrary constant.

Taking the antilogarithm of equation (2.7),

$$\underline{A = Ce^{(\rho g x/\sigma)}} \tag{2.8}$$

i.e. if A is known for any x, C can be determined.

It can be seen from equation (2.8) that the cross-sectional area A is of exponential form, as shown by Figure 2.11; this is necessary for uniform stress to occur in a pillar under self-weight.

2.6 Shear stress and shear strain

From Figure 2.2, it can be seen that shear stress acts tangentially to the surface and that this shear stress (τ) causes the shape of a body to deform, as shown in Figure 2.12. Although shear stress causes a change of shape (or shear strain γ), it does not cause a change in volume.

By experiment, it has been found that for many materials, the relationship

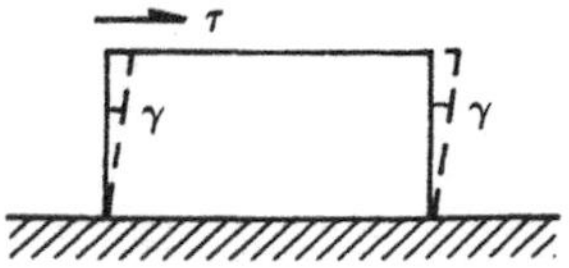

Figure 2.12 Shear stress τ and shear strain γ

between shear stress and shear strain is given by the equation

$$\frac{\tau}{\gamma} = G \tag{2.9}$$

where

G = modulus of rigidity or shear modulus (N/m^2 N/mm^2, MN/m^2, etc.)

γ = shear strain (unitless)

2.6.1 Complementary shear stress

The effect of shearing action on an element of material, as shown in Figure 2.13, will be to cause the system of shearing stresses (τ) in Figure 2.14.

Let t be the thickness of elemental lamina. By considerations of horizontal equilibrium, it is evident that τ will act on the top and bottom surfaces in the manner shown. The effect of these shearing stresses will be to cause a clockwise couple of $\tau \times t \times dx \times dy$, and from equilibrium considerations, the system of shearing stresses τ' must act in the direction shown. Hence, by taking moments about the bottom left-hand corner of the elemental lamina,

$$\tau \times t \times dx \times dy = \tau' \times t \times dy \times dx$$

or

$$\underline{\tau = \tau'}$$

That is, the systems of shearing stresses are complementary and equal. Positive and negative shearing stresses are shown in Figure 2.15. Negative shearing stresses are

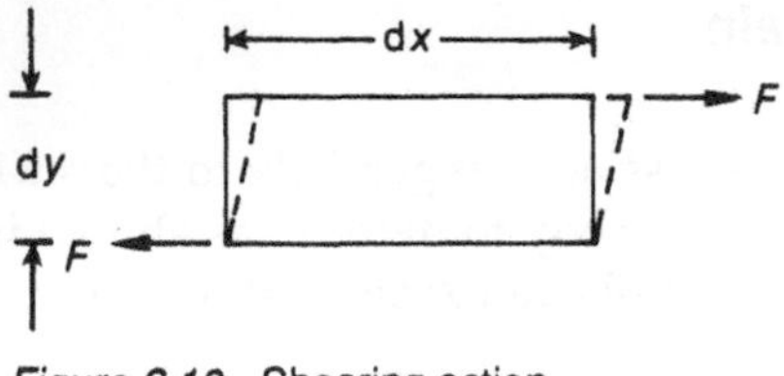

Figure 2.13 Shearing action

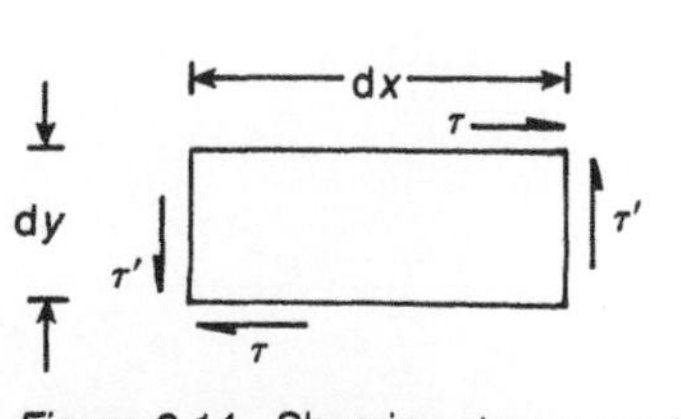

Figure 2.14 Shearing stresses acting on an element

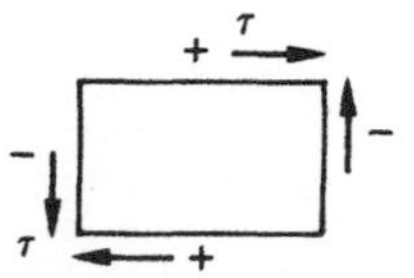

Figure 2.15 Positive and negative complementary shearing stresses

said to cause a counter-clockwise couple, as shown on the vertical surfaces of Figure 2.15, and positive shearing stresses are said to cause a clockwise couple, as shown on the horizontal surfaces of Figure 2.15.

2.6.2 Failure due to exceeding the shear stress

Failure of a component can also take place if the shear stress of the material exceeds the shear stress in 'yield'.

For *ductile materials*

$$\text{Yield shear stress} \doteqdot 0.577 \times \text{tensile yield stress } (\sigma_{yp})$$

(see Section 10.6 in Chapter 10.)

For *brittle materials*

$$\text{Failure shear stress} \doteqdot 0.5 \times 0.1\% \text{ proof stress}$$

EXAMPLE 2.4

Two lengths of a material are connected together by a single rivet, as shown in Figure 2.16. If failure of this joint is due to shearing of the rivet, determine the force T to cause failure. The material for the rivet may be assumed to be ductile.

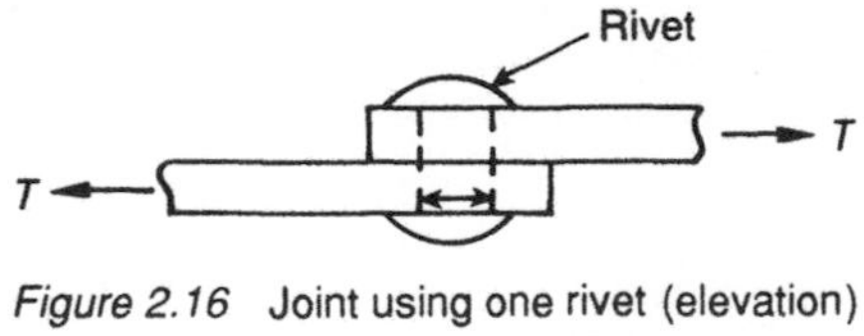

Figure 2.16 Joint using one rivet (elevation)

SOLUTION

If the rivet fails in shear, it will fail as shown in Figure 2.17.

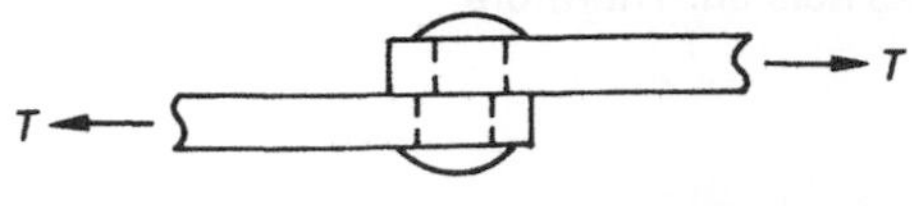

Figure 2.17 Failure mode of rivet (elevation)

Shear stress

$$\tau_{yield} = 0.577\sigma_{yp}$$

$$= \frac{T}{\text{cross-sectional area of rivet}} = \frac{T \times 4}{\pi d^2}$$

Therefore

$$T = \frac{0.577\sigma_{yp} \times \pi d^2}{4} \quad (2.10)$$

where d is the diameter of the rivet.

EXAMPLE 2.5

Determine the downward force P required to punch a hole in the plate of Figure 2.18.

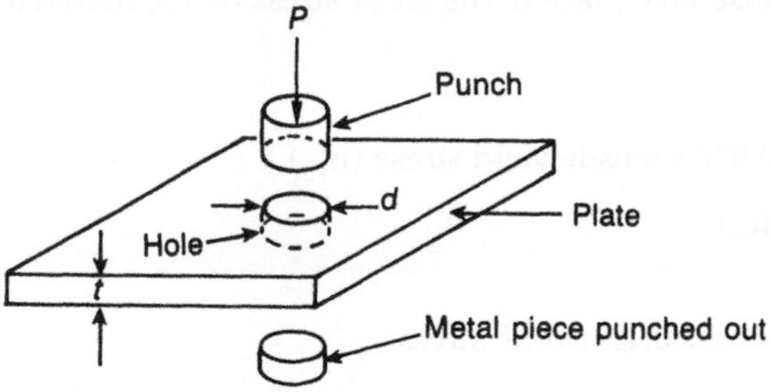

Figure 2.18 Hole punched in metal

SOLUTION

The shear stress τ_{yp} that acts on the circular cylindrical metal piece, as the piece is being punched out, is shown in Figure 2.19.

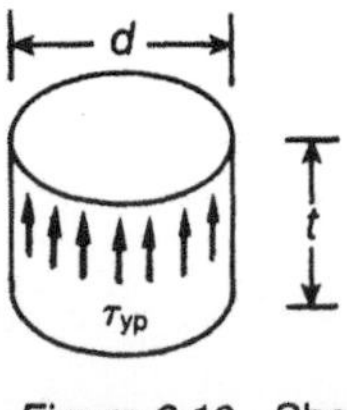

Figure 2.19 Shear stress acting on the metal piece

$P = \tau_{yp}$ x the area that the shear stress acts on. Therefore

$$\underline{P = \tau_{yp} \times \pi d \times t} \quad (2.11)$$

where τ_{yp} is the yield shear stress.

2.7 Poisson's ratio (ν)

If a length of wire or rubber or similar material is subjected to axial tension, as shown in Figure 2.20, then in addition to its length increasing, its lateral dimension will decrease owing to this axial stress.

The relationship between the lateral strain and the axial strain, due to the axial stress, is known as Poisson's ratio (ν), where

$$\nu = \frac{-\text{lateral strain}}{\text{longitudinal strain}}$$

$$\text{Lateral strain} = \frac{-\delta b}{b}$$

$$\text{Longitudinal strain} = \frac{\delta l}{l}$$

where

b = breadth

δb = increment of b

l = length

δl = increment of l

From Figure 2.20, it can be seen that although there is lateral strain, there is no lateral stress; thus, the relationship $\sigma/\varepsilon = E$ only applies for uniaxial stress in the direction of the uniaxial stress. Now,

$$\varepsilon_x = \text{longitudinal strain} = \sigma/E$$

Therefore

$$\varepsilon_y = \text{lateral strain} = -\nu\sigma/E \tag{2.12}$$

Thus, for two- and three-dimensional systems of stress, equation (2.5) does not apply, and for such cases, the *Poisson effect* of equation (2.12) must also be included (see Chapter 7).

Typical values of Possion's ratio are 0.3 for steel, 0.33 for aluminium alloy, 0.1 for concrete. It will be shown in the next section that ν cannot exceed 0.5.

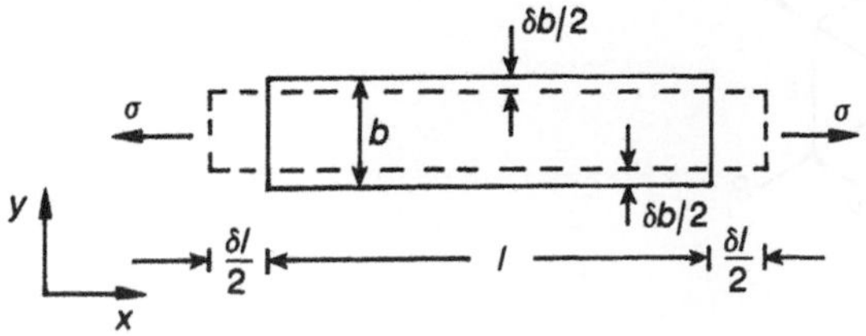

Figure 2.20 Axially loaded element

2.8 Hydrostatic stress

If a solid piece of material were dropped into the ocean, it would be subjected to a uniform external pressure P, caused by the weight of the water above it. If an internal elemental cube from this piece of material were examined, it would be found to be subjected to a three-dimensional system of stresses, where there is no shear stress, as shown in Figure 2.21. Such a state of stress is known as *hydrostatic stress*, and the stress everywhere is normal and equal to $-P$. If the dimensions of the cube are $x \times y \times z$, and the displacements due to P, corresponding to these dimensions, are δx, δy and δz, respectively, then from equation (2.12)

$$\varepsilon_x = \frac{\delta x}{x} = -\frac{P}{E}(1 - \nu - \nu)$$

$$\varepsilon_y = \frac{\delta y}{y} = -\frac{P}{E}(1 - \nu - \nu) \qquad (2.13)$$

$$\varepsilon_z = \frac{\delta z}{z} = -\frac{P}{E}(1 - \nu - \nu)$$

From equations (2.13), it can be seen that if $x = y = z$, and $\nu > 0.5$, then δx, δy and δz will be positive, which is impossible, i.e. ν *cannot be greater than 0.5*. In fact, the stress system of Figure 2.1 will cause the volume of the element to decrease by an amount δV, so that

$$\text{Volumetric strain} = \frac{\text{change in volume } (\delta V)}{\text{original volume } (V)} = \frac{\delta V}{V}$$

If the material obeys Hooke's law, then

$$\frac{\text{Volumetric stress}}{\text{Volumetric strain}} = K \qquad (2.14)$$

where K is the bulk modulus. In this case, volumetric stress equals $-P$.

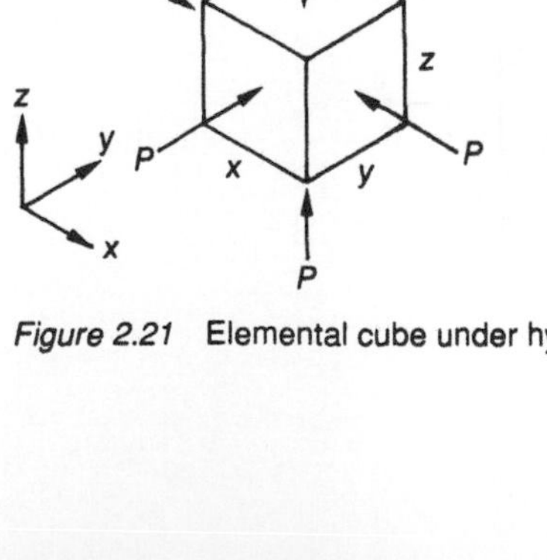

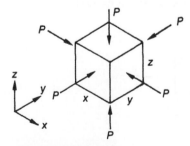

Figure 2.21 Elemental cube under hydrostatic stress

2.8.1 To determine volumetric strain

Now the original volume is

$$V = x \times y \times z$$

and the new volume is

$$V + \delta V = x(1+\varepsilon_x) \times y(1+\varepsilon_y) \times z(1+\varepsilon_z)$$

Assuming the deflections are small and neglecting higher-order terms,

$$V + \delta V = xyz(1 + \varepsilon_x + \varepsilon_y + \varepsilon_z)$$

and

$$\text{Volumetric strain} = \frac{\delta V}{V} = (\varepsilon_x + \varepsilon_y + \varepsilon_z)$$

i.e. the volumetric strain is the sum of three co-ordinate strains ε_x, ε_y and ε_z.

2.9 Relationship between the material constants *E, G, K* and ν

It will be shown in Chapter 7 that the relationships between the elastic constants are given by

$$G = E/[2(1+\nu)] \tag{2.15}$$

$$K = E/[3(1-2\nu)] \tag{2.16}$$

Some typical values of G and K (N/m^2) are shown in Table 2.3.

2.10 Three-dimensional stress

Stress is a tensor, which can be represented diagrammatically as in Figure 2.22.

If the front bottom-left corner of the cubic element of Figure 2.22 is sliced off, to yield the tetrahedral sub-element of Figure 2.23, it can be seen that equilibrium is achieved by the stress tensor σ_{ij} acting at an angle to the generic plane *abc*. It is

Table 2.3

Material	Steel	Aluminium alloy	Copper
G	8×10^{10}	2.6×10^{10}	4.4×10^{10}
K	1.67×10^{11}	6.68×10^{10}	1.33×10^{11}

K for water $= 2 \times 10^9$.

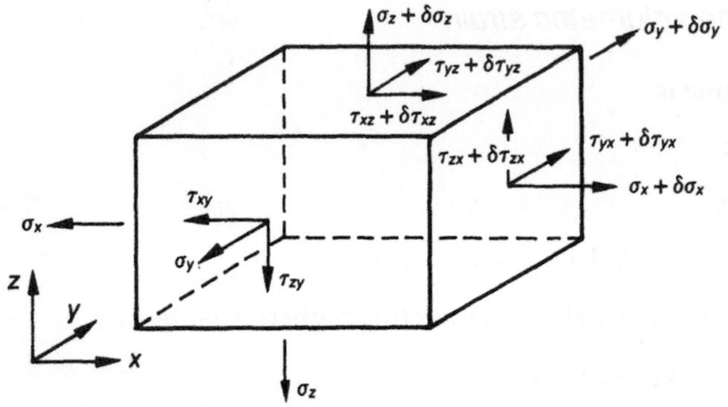

Figure 2.22 Three-dimensional stress system

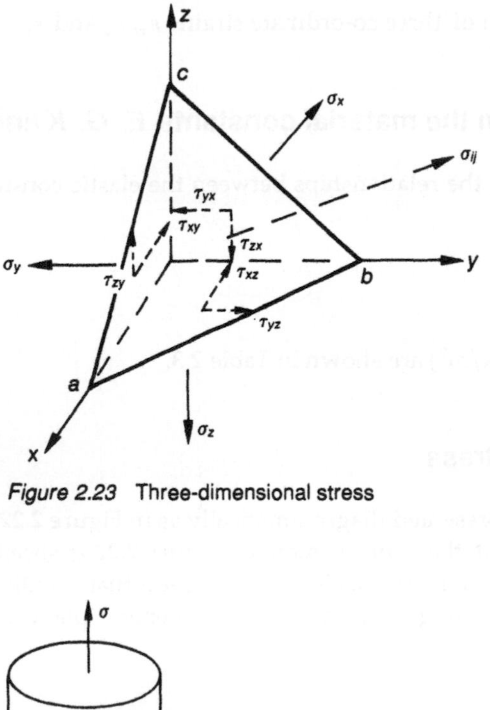

Figure 2.23 Three-dimensional stress

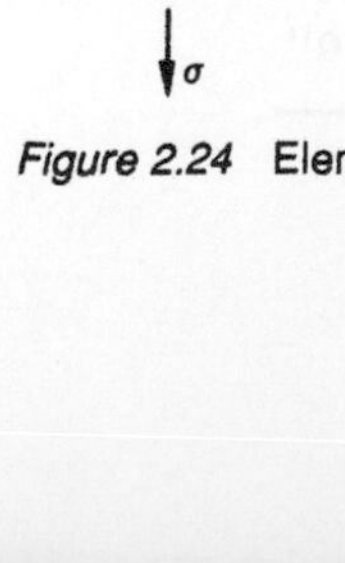

Figure 2.24 Elemental cube

evident, therefore, that even for a tensile specimen undergoing uniaxial stress, a three-dimensional system of stress can be obtained by examining an elemental cube, tilted at an angle to the axis, as shown in Figure 2.24.

2.10.1 *Stress–strain relationship in three dimensions*

In three dimensions, the stress–strain relationships are given by

$$\varepsilon_x = \frac{\sigma_x}{E} - \frac{\nu\sigma_y}{E} - \frac{\nu\sigma_z}{E}$$

$$\varepsilon_y = \frac{\sigma_y}{E} - \frac{\nu\sigma_x}{E} - \frac{\nu\sigma_z}{E} \qquad (2.17)$$

$$\varepsilon_z = \frac{\sigma_z}{E} - \frac{\nu\sigma_x}{E} - \frac{\nu\sigma_y}{E}$$

where ε_x, ε_y and ε_z are the co-ordinate strains in the x, y and z directions.

NB The stress–strain relationships quoted in equations (2.5), (2.9), (2.13), etc., are known as the *constitutive laws.*

2.11 Composite materials

So far we have only discussed traditional structural materials, such as metals, etc. In recent years, man-made materials have proved more popular than traditional materials for a number of applications. It is likely that in the future, these man-made materials will probably prove even more popular, especially when they become relatively cheaper to produce.

One reason for the popularity of man-made materials is that they prove more resistant to certain forms of corrosion. Another reason for their increasing popularity is that they have a better strength to weight ratio, as can be seen from the figures in Table 2.4.

Table 2.4 **Material properties of man-made fibers**

Material	Density (kg/m^3)	Tensile strength (GPa)	Tensile modulus (GPa)
E-glass fiber	2550	3.4	72
Boron fiber	2600	3.8	380
Carbon fiber (UHS)	1750	5.2	270
Piano wire	7860	3.0	210

UHS = Ultra High Strength.

This improved strength to weight ratio is of much importance when the weight of a structure is at a premium, such as in the cases of aircraft and submarine structures. It should also be noted from Table 2.4 that for composites, *tensile modulus* is used in place of Young's modulus.

For two-dimensional structures, most composite materials are assumed to have orthogonal properties. That is, the material properties of the two-dimensional composite are different in directions perpendicular to each other. Some typical material properties of composites, made from a combination of fibers and resins, are shown in Table 2.5. For these composites, the fibers were laid in the x direction in a resin.

$$\text{NB} \quad \nu_x E_y = \nu_y E_x \tag{2.18}$$

To achieve more even values for the material properties of composites in all directions, it is normal to lay the fibers in several layers, as shown in Figure 2.25, the whole being set in a resin.

For two-dimensional orthotropic materials, the stress–strain relationships are as follows:

$$\begin{aligned} \varepsilon_x &= \frac{\sigma_x}{E_x} - \frac{\nu_y \sigma_y}{E_y} \\ \varepsilon_y &= \frac{\sigma_y}{E_y} - \frac{\nu_x \sigma_x}{E_x} \end{aligned} \tag{2.19}$$

$$\gamma_{xy} = \frac{\tau_{xy}}{G_{xy}} \tag{2.20}$$

where

ε_x = direct strain in the x direction

ε_y = direct strain in the y direction

γ_{xy} = shear strain in the x–y plane

Table 2.5 **Material properties of fiber composites**

Material	Density (kg/m³)	E_x (GPa)	E_y (GPa)	ν_x	G_{xy} (GPa)
Glass fiber/resin	1800	40	8.5	0.25	4.1
Carbon fiber/resin	1600	180	10	0.28	7.2
Kevlar/resin	1430	75	5.5	0.34	2.3
Mild steel	7860	200	200	0.3	7.7

E_x = tensile modulus in the x direction; E_y = tensile modulus in the y direction; G_{xy} = shear modulus in the x–y plane; ν_x = Poisson's ration due to σ_x; ν_y = Poisson's ratio due to σ_y.

Figure 2.25 Five layers of fiber reinforcement

Equations (2.19) can be rearranged in the following form, which is more convenient for calculating stresses from measured strains:

$$\sigma_x = \frac{E}{(1 - \nu_x \nu_y)} (\varepsilon_x + \nu_y \varepsilon_y)$$
$$\sigma_y = \frac{E_y}{(1 - \nu_x \nu_y)} (\varepsilon_y + \nu_x \varepsilon_x) \qquad (2.21)$$

2.12 Thermal strain

If a bar of length l and coefficient of linear expansion α is subjected to a temperature rise T, its length will increase by an amount αlT, as shown in Figure 2.26. Thus, at this temperature, the natural length of the bar is $l(1 + \alpha T)$. In this condition, although the bar has a thermal strain of αT, it has no thermal stress, but if the free expansion αlT is prevented from taking place, compressive thermal stresses will occur.

Such problems are of great importance in a number of practical situations, including railway lines and pipe systems, and the following two examples will be used to demonstrate the problem.

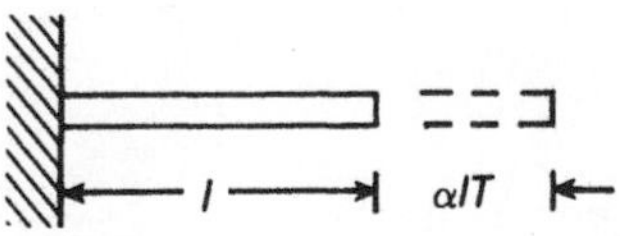

Figure 2.26 Free thermal expansion of a bar

EXAMPLE 2.6

A steel prop is used to stabilize a building, as shown in Figure 2.27. If the compressive stress in the prop at a temperature of 20 °C is 50 MN/m², what will the stress be in the prop if the temperature is raised to 30 °C? At what temperature will the prop cease to be

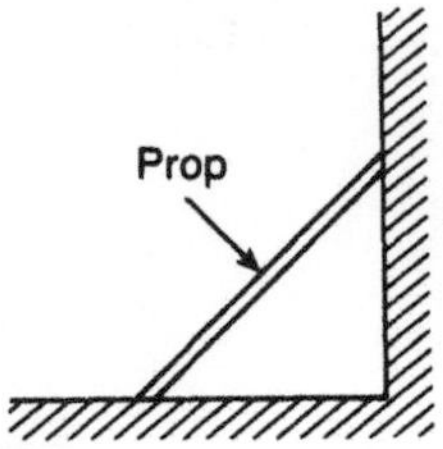

Figure 2.27 Prop

effective? It may be assumed that the floor, the building or the ends of the prop do not move.

$E = 2 \times 10^{11} \mathrm{N/m^2} \quad \alpha = 15 \times 10^{-6}\ °\mathrm{C}$

SOLUTION

$$\text{Additional thermal strain} = \frac{15 \times 10^{-6} \times l \times 10}{l}$$

$$= -150 \times 10^{-6}$$

$$\text{Thermal stress} = -2 \times 10^{11} \times 150 \times 10^{-6}$$

$$= -30\ \mathrm{MN/m^2}$$

$$\text{Stress at } 30\ °\mathrm{C} = -30 - 50 = -80\ \mathrm{MN/m^2}$$

For the prop to be ineffective, it will be necessary for the temperature to drop, so that the initial compressive stress of 50 MN/m² is nullified, i.e.

$$\text{Thermal stress} = 50\ \mathrm{MN/m^2}$$

$$\text{Thermal strain} = \frac{50 \times 10^6}{2 \times 10^{11}} = 2.5 \times 10^{-4}$$

or

$$2.5 \times 10^{-4} = -\alpha T$$

where T is the temperature fall. Therefore

$$\underline{T = -16.67\ °\mathrm{C}}$$

i.e.

$$\underline{\text{Temperature for prop to be ineffective} = 20 - 16.67 = 3.33\ °\mathrm{C}}$$

EXAMPLE 2.7

A steel rail may be assumed to be in a stress-free condition at 10 °C. If the stress required to cause buckling is $-75\ \mathrm{MN/m^2}$, what temperature rise will cause the rail to buckle, assuming that the rail is rigidly restrained at its ends and that its material properties are as in Example 2.6?

SOLUTION

$$\varepsilon = \frac{\sigma}{E} = \alpha T \tag{2.22}$$

or

$$\frac{75 \times 10^6}{2 \times 10^{11}} = 15 \times 10^{-6} T$$

Therefore

Temperature rise = $T = 25$ °C

and

Temperature to cause buckling = 35 °C

2.13 Compound bars

Compound bars are of much importance in a number of different branches of engineering, including reinforced concrete pillars, bimetallic bars, etc., and in this section the solution of such problems usually involves two important considerations, namely

(a) compatibility (or consideration of displacements)
(b) equilibrium.

NB It is necessary to introduce compatibility in this section as *compound bars are,* in general, *statically indeterminate.* To demonstrate the method of solution, the next two examples will be considered.

EXAMPLE 2.8

A solid bar of cross-sectional area A_1, elastic modulus E_1 and coefficient of linear expansion α_1 is surrounded co-axially by a hollow tube of cross-sectional area A_2, elastic modulus E_2 and coefficient of linear expansion α_2, as shown in Figure 2.28. If the two

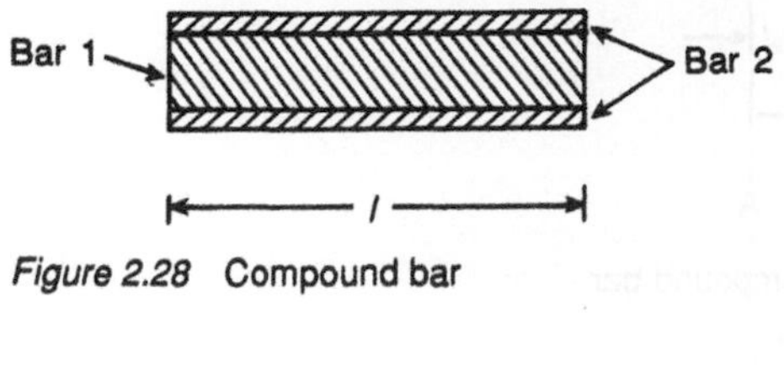

Figure 2.28 Compound bar

bars are secured firmly to each other, so that no slipping takes place with temperature change, determine the thermal stresses due to a temperature rise T. Both bars have an initial length l.

SOLUTION

Compatibility considerations

There are two unknowns; therefore two simultaneous equations will be required. The first equation can be obtained by considering the *compatibility* (i.e. 'deflections') of the bars, with the aid of Figure 2.29:

$$\text{Free expansion of bar 1} = \alpha_1 lT$$

$$\text{Free expansion of bar 2} = \alpha_2 lT$$

In practice, however, the final resting position of the compound bar will be *somewhere* between these two positions (i.e. at the position A–A). To achieve this, it will be necessary for bar 2 to be pulled out by a distance $\varepsilon_2 l$ and for bar 1 to be pushed in by a distance $\varepsilon_1 l$, where

ε_1 = compressive strain in 1

ε_2 = tensile strain in 2

From consideration of compatibility ('deflections') in Figure 2.29;

$$\alpha_1 lT - \varepsilon_1 l = \alpha_2 lT + \varepsilon_2 l$$

or

$$\varepsilon_1 = (\alpha_1 - \alpha_2)T - \varepsilon_2$$

or

$$\sigma_1 = (\alpha_1 - \alpha_2)E_1 T - \sigma_2 E_1/E_2 \qquad (2.23)$$

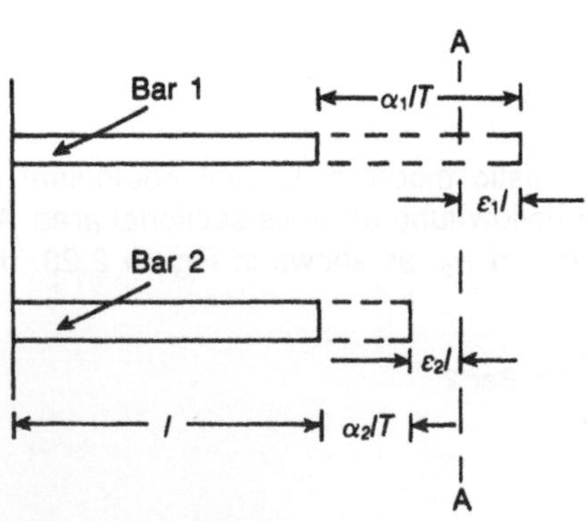

Figure 2.29 'Deflections' of compound bar

Equilibrium considerations

To obtain the second simultaneous equation, it will be necessary to consider *equilibrium.* Let

F_1 = compressive force in bar 1

F_2 = tensile force in bar 2

Now

$$F_1 = F_2$$

or

$$\sigma_1 A_1 = \sigma_2 A_2$$

Therefore

$$\sigma_1 = \sigma_2 A_2 / A_1 \tag{2.24}$$

Equating (2.23) and (2.24),

$$\sigma_2 A_2 / A_1 = (\alpha_1 - \alpha_2) E_1 T - \sigma_2 E_1 / E_2$$

Therefore

$$\sigma_2 = \frac{(\alpha_1 - \alpha_2) E_1 T}{(E_1/E_2 + A_2/A_1)}$$

or

$$\sigma_2 = \frac{(\alpha_1 - \alpha_2) E_1 E_2 A_1 T}{(A_1 E_1 + A_2 E_2)} \text{ (tensile)} \tag{2.25}$$

and

$$\sigma_1 = \frac{(\alpha_1 - \alpha_2) E_1 E_2 A_2 T}{(A_1 E_1 + A_2 E_2)} \text{ (compressive)} \tag{2.26}$$

EXAMPLE 2.9

If the solid bar of Example 2.8 did not undergo a temperature change, but instead was subjected to a tensile axial force *P*, as shown in Figure 2.30, determine σ_1 and σ_2

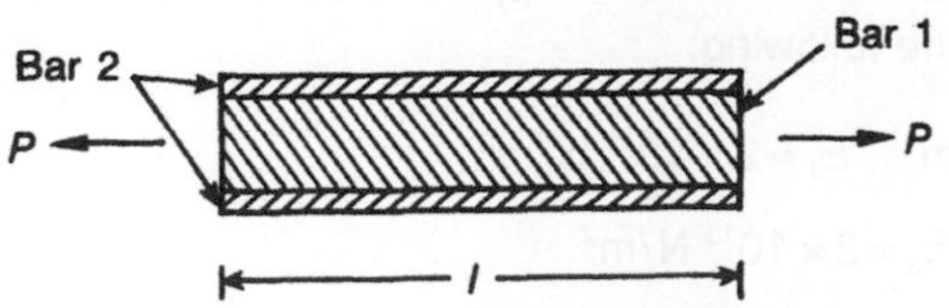

Figure 2.30 Compound bar under axial tension

SOLUTION

There are two unknowns; therefore two simultaneous equations will be required. The first of these simultaneous equations can be obtained by considering *compatibility*, i.e.

Deflection of bar 1 = deflection of bar 2

$$\varepsilon_1 l = \varepsilon_2 l$$

$$\frac{\sigma_1}{E_1} = \frac{\sigma_2}{E_2}$$

Therefore

$$\sigma_1 = \sigma_2 E_1 / E_2 \tag{2.27}$$

The second simultaneous equation can be obtained by considering *equilibrium*. Let

F_1 = tensile force in bar 1

F_2 = tensile force in bar 2

Now

$$P = F_1 + F_2$$

$$= \sigma_1 A_1 + \sigma_2 A_2 \tag{2.28}$$

Substituting (2.27) into (2.28),

$$\sigma_2 = \frac{PE_2}{(A_1 E_1 + A_2 E_2)} \tag{2.29}$$

and

$$\sigma_1 = \frac{PE_1}{(A_1 E_1 + A_2 E_2)} \tag{2.30}$$

NB If P is a compressive force, then both σ_1 and σ_2 will be compressive stresses.

EXAMPLE 2.10

A concrete pillar, which is reinforced with steel rods, supports a compressive axial load of 1 MN. Determine σ_1 and σ_2, given the following:

Steel: $A_1 = 3 \times 10^{-3}$ m^2 $E_1 = 2 \times 10^{11}$ N/m^2

Concrete: $A_2 = 0.1$ m^2 $E_2 = 2 \times 10^{10}$ N/m^2

What percentage of the total load does the steel reinforcement take?

SOLUTION

From equation (2.30),

$$\sigma_1 = \frac{-1 \times 10^6 \times 2 \times 10^{11}}{(6 \times 10^8 + 2 \times 10^9)} = \underline{-76.92\ \text{MN/m}^2} \tag{2.31}$$

$$\sigma_2 = -\frac{1 \times 10^6 \times 2 \times 10^{10}}{2.6 \times 10^9} = \underline{-7.69\ \text{MN/m}^2} \tag{2.32}$$

and

$$F_1 = -76.92 \times 10^6 \times 3 \times 10^{-3} = \underline{2.308 \times 10^5\ \text{N}}$$

Therefore the percentage total load taken by the steel reinforcement equals $\underline{23.08}$.

EXAMPLE 2.11

If the pillar of Example 2.10 were subjected to a temperature rise of 30 °C, what would be the values of σ_1 and σ_2?

$$\alpha_1 = 15 \times 10^{-6}/°\text{C (steel)} \quad \alpha_2 = 12 \times 10^{-6}/°\text{C (concrete)}$$

SOLUTION

As α_1 is larger than α_2, the effect of a temperature rise will cause the 'thermal stresses' in 1 to be compressive and those in 2 to be tensile.

From equation (2.26),

$$\sigma_1(\text{thermal}) = -\frac{(15 \times 10^{-6} - 12 \times 10^{-6}) \times 2 \times 10^{11} \times 2 \times 10^{10} \times 0.1 \times 30}{(2.6 \times 10^9)}$$

$$= \underline{-13.85\ \text{MN/m}^2} \tag{2.33}$$

From equation (2.25),

$$\sigma_2(\text{thermal}) = \underline{0.42\ \text{MN/m}^2} \tag{2.34}$$

From equations (2.31) to (2.34),

$$\sigma_1 = -76.92 - 13.85 = \underline{-90.77\ \text{MN/m}^2}$$

$$\sigma_2 = -7.69 + 0.42 = \underline{-7.27\ \text{MN/m}^2}$$

EXAMPLE 2.12

A rigid horizontal bar is supported by three rods, where the outer rods are made from aluminium alloy and the middle rod from steel, as shown in Figure 2.31. If the temperature of all three rods is raised by 50 °C, what will be the thermal stresses in the rods?

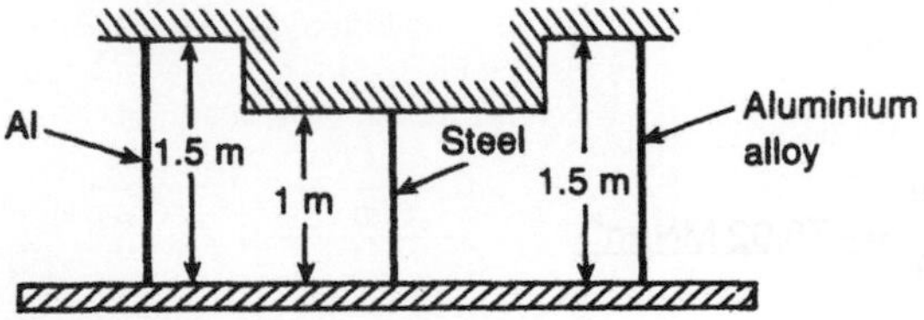

Figure 2.31 Compound bar

The following may be assumed:

$A_a = 3 \times 10^{-3}\ m^2$ = sectional area of one aluminium rod

$E_a = 7 \times 10^{10}\ N/m^2$ = elastic modulus of aluminium

$\alpha_a = 25 \times 10^{-6}/°C$ = coefficient of linear expansion of aluminium

$A_s = 2 \times 10^{-3}\ m^2$ = sectional area of steel rod

$E_s = 2 \times 10^{11}\ N/m^2$ = elastic modulus of steel

$\alpha_s = 15 \times 10^{-6}/°C$ = coefficient of linear expansion of steel

SOLUTION

Free expansion of aluminium $= 25 \times 10^{-6} \times 1.5 \times 50$

$= \underline{1.875 \times 10^{-3}}$

Free expansion of steel $= 15 \times 10^{-6} \times 1 \times 50$

$= \underline{7.5 \times 10^{-4}}$

That is, as the free expansion of the aluminium is greater than that of steel, the aluminium will be in compression and the steel in tension, owing to a temperature rise.

Let

ε_a = compressive strain in aluminium

ε_s = tensile strain in steel

Now there are two unknowns; therefore two simultaneous equations will be required. The first of these can be obtained by considering *compatibility*, with the aid of Figure 2.32. From this figure, the final resting place of the compound bar will be at the position A-A, so that

$$\alpha_a l_a T - \varepsilon_a l_a = \alpha_s l_s T + \varepsilon_s l_s$$

or

$$\varepsilon_a = (\alpha_a l_a - \alpha_s l_s)T/l_a - \varepsilon_s l_s/l_a$$

$$\sigma_a = E_a\{(\alpha_a l_a - \alpha_s l_s)T - \sigma_s l_s/E_s\}/l_a$$

$$= 7 \times 10^{10}(1.125 \times 10^{-3} - 5 \times 10^{-12}\sigma_s)/1.5$$

$$\sigma_a = (7.875 \times 10^7 - 0.35\sigma_s)/1.5$$

$$\sigma_a = 5.25 \times 10^7 - 0.233\sigma_s \qquad (2.35)$$

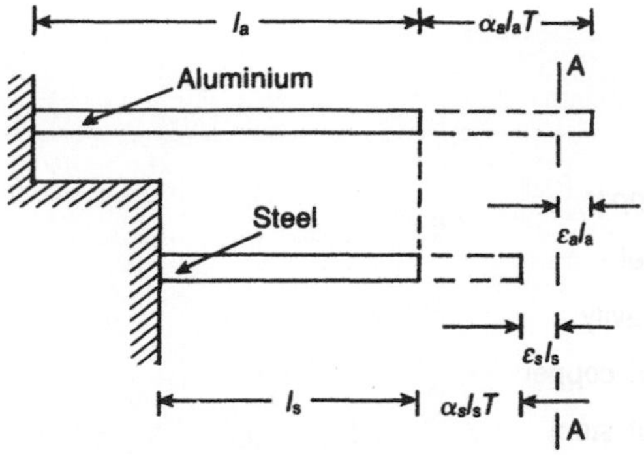

Figure 2.32

The second equation can be obtained from *equilibrium* considerations, where

Tensile force in steel (F_s) = compressive force in aluminium (F_a)

or

$$F_s = F_a$$

$$\sigma_s A_s = 2 x \sigma_a A_a$$

Therefore

$$\sigma_a = 0.333\sigma_s \tag{2.36}$$

Equating (2.35) and (2.36),

$$5.25 \times 10^7 - 0.233\sigma_s = 0.333\sigma_s$$

Therefore

$$\underline{\sigma_s = 92.76 \text{ MN/m}^2 \text{ (tensile)}}$$

and

$$\underline{\sigma_s = 30.9 \text{ MN/m}^2 \text{ (compressive)}}$$

EXAMPLE 2.13

An electrical cable consists of a copper core surrounded co-axially by a steel sheath, so that the whole acts as a compound bar. If this cable hangs vertically down a mineshaft, prove that the maximum stresses in the copper and the steel are given by

$$\hat{\sigma}_c = \text{maximum stress in copper} = \frac{E_c(\rho_c A_c + \rho_s A_s)gl}{(A_c E_c + A_s E_s)}$$

$$\hat{\sigma}_s = \text{maximum stress in steel} = \frac{E_s(\rho_c A_c + \rho_s A_s)gl}{(A_c E_c + A_s E_s)}$$

where

ρ_c = density of copper

ρ_s = density of steel

E_c = elastic modulus of copper

E_s = elastic modulus of steel

g = acceleration due to gravity

A_c = cross-sectional area of copper

A_s = cross-sectional area of steel

l = length of cable

SOLUTION

Consider compatibility

Let

δ_c = maximum deflection of the copper core

δ_s = maximum deflection of the steel sheath (δ_s)

Now,

$$\delta_c = \delta_s$$

or

$$l\varepsilon_c = l\varepsilon_s$$

where

ε_c = maximum strain in copper
ε_s = maximum strain in steel

or

$$\frac{\hat{\sigma}_c}{E_c} = \frac{\hat{\sigma}_s}{E_s}$$

$$\hat{\sigma}_c = \hat{\sigma}_s E_c / E_s \tag{2.37}$$

Consider equilibrium

Weight of cable = $(\rho_c A_c + \rho_s A_s) g l$

Resolve vertically

$$(\rho_c A_c + \rho_s A_s) g l = \hat{\sigma}_c \times A_c + \hat{\sigma}_s \times A_s \tag{2.38}$$

Substituting equation (2.37) into (2.38),

$$(\rho_c A_c + \rho_s A_s) gl = \hat{\sigma}_s A_c E_c / E_s + \hat{\sigma}_s A_s$$

$$\hat{\sigma}_s \left(\frac{A_c E_c}{E_s} + A_s \right) = (\rho_c A_c + \rho_s A_s) gl$$

$$\hat{\sigma}_s = \frac{(\rho_c A_c + \rho_s A_s) gl}{(A_c E_c / E_s + A_s)}$$

$$\hat{\sigma}_s = \frac{E_s(\rho_c A_c + \rho_s A_s) gl}{(A_c E_c + A_s E_s)} \tag{2.39}$$

Substituting equation (2.39) into (2.37),

$$\hat{\sigma}_c = \frac{E_c(\rho_c A_c + \rho_s A_s) gl}{(A_c E_c + A_s E_s)} \tag{2.40}$$

EXAMPLE 2.14

A compound bar consists of a steel bolt surrounded co-axially by an aluminium-alloy tube, as shown in Figure 2.33. Assuming that the nut on the end of the steel bolt is initially 'just' hand-tight, determine the strains in the bolt and the tube if the nut is rotated clockwise by an angle θ, relative to the other end.

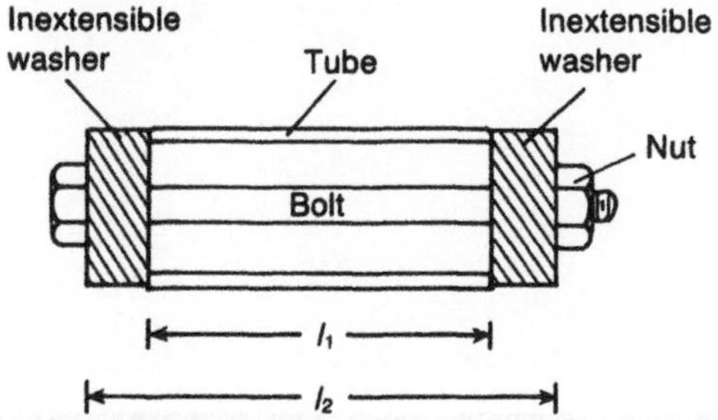

Figure 2.33 Bolt/tube compound bar

SOLUTION

Let

ε_1 = strain in the aluminium-alloy tube = δ_1 / l_1

ε_2 = strain in the steel bolt = δ_2 / l_2

a_1 = sectional area of aluminium alloy

a_2 = sectional area of steel

l_1 = length of aluminium-alloy tube

l_2 = length of steel bolt

$\delta = \theta \times$ pitch of thread/360°

θ = rotation in degrees

E_1 = elastic modulus for aluminium alloy

E_2 = elastic modulus for steel

Now, when the nut is turned clockwise by an angle θ, the aluminium alloy tube will decrease its length by δ_1, and the steel bolt will increase its length by δ_2, where

$$\delta = \delta_1 + \delta_2$$

$$\delta = \varepsilon_1 l_1 + \varepsilon_2 l_2 \tag{2.41}$$

From equilibrium considerations,

$$\sigma_1 a_1 = \sigma_2 a_2$$

or

$$E_1 \varepsilon_1 a_1 = E_2 \varepsilon_2 a_2$$

i.e.

$$\varepsilon_1 = \left(\frac{a_2 E_2}{a_1 E_1}\right)\varepsilon_2 \tag{2.42}$$

Substituting equation (2.42) into (2.41),

$$\varepsilon_2\left(\frac{l_1 a_2 E_2}{a_1 E_1} + l_2\right) = \delta$$

$$\varepsilon_2 = \frac{\delta}{(l_1 a_2 E_2 / a_1 E_1 + l_2)} \tag{2.43}$$

Hence

$$\varepsilon_1 = \frac{-\delta}{(l_1 + l_2 a_1 E_1 / a_2 E_2)} \tag{2.44}$$

2.14 Failure by fatigue

Under repeated cyclic loading, many structures are known to fail, despite the fact that the maximum calculated stress in the structure is well below the elastic limit. The reason for this mode of failure is that most structures have microscopic cracks in them, together with other stress concentrations. Under repeated cyclic loading, the maximum stresses in these cracks exceed the elastic limit and this causes the cracks to grow. The larger the crack becomes, the more rapidly it grows, until catastrophic failure takes place.

Materials such as mild steel and titanium will not fail through fatigue if the maximum stress in these materials does not exceed certain values. These values are

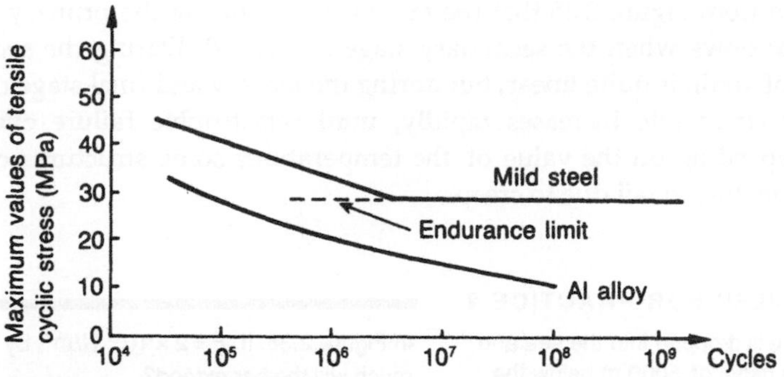

Figure 2.34 Fatigue failure

called the **endurance limits**. Materials such as aluminium alloy and most other non-ferrous materials do not exhibit an endurance limit, as shown in Figure 2.34.

2.15 Failure due to creep

Under high temperatures, most materials lose some of their resistance to deformation and will yield at stresses lower than the so-called yield stress. Additionally, if the same load is continuously applied to the material, when it is subjected to this high temperature, the material will continue to deform with time, until failure takes place. This mode of failure is known as creep. Many plastic materials **creep** at room temperature, including polymethylmethacrylate (PMMA or 'perspex').

The rate of growth of the structure, when failing due to creep, is usually in three stages, namely the primary stage, the secondary stage and the tertiary stage, as shown in Figure 2.35.

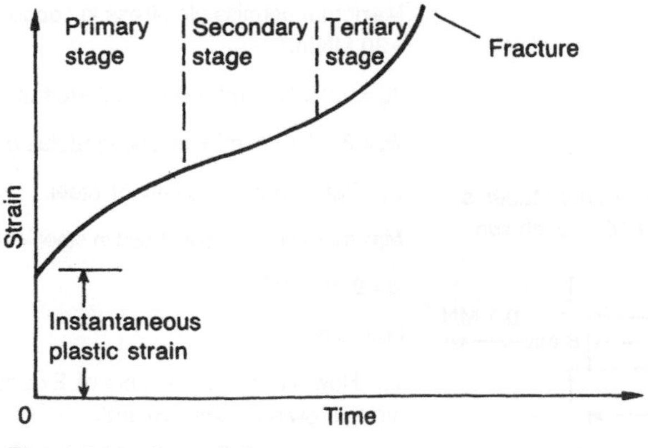

Figure 2.35 Creep failure

It can be seen from Figure 2.35 that the rate of strain during the primary stage is quite rapid, but slows when the secondary stage is reached. During the secondary stage, the rate of strain is quite linear, but during the tertiary and final stage of creep behaviour, the strain rate increases rapidly, until catastrophic failure eventually takes place. Depending on the value of the temperatures some structures can take days, or even months, to fail due to creep.

EXAMPLES FOR PRACTICE 2

1. If a solid stone is dropped into the sea and comes to rest at a depth of 5000 m below the surface of the sea, what will be the stress in the stone?

Density of sea water = 1020 kg/m^3 g = 9.81 m/s^2
{−50 MN/m^2}

2. A solid bar of length 1 m consists of three shorter sections firmly joined together. Assuming the following apply, determine the change in length of the bar when it is subjected to an axial pull of 50 kN.

$E = 2 \times 10^{11}$ N/m^2

Section	Length (m)	Diameter (mm)
1	0.2	15
2	0.3	20
3	0.5	30

{0.699 mm}

3. If the bar of Example 2 were made from three different materials with the following elastic moduli, determine the change in length of the bar:

Section	E (N/m^2)
1	2×10^{11}
2	7×10^{10}
3	1×10^{11}

{1.319 mm}

4. A circular-section solid bar of linear taper is subjected to an axial pull of 0.1 MN, as shown in Figure 2.36. If $E = 2 \times 10^{11}$ N/m^2, by how much will the bar extend?

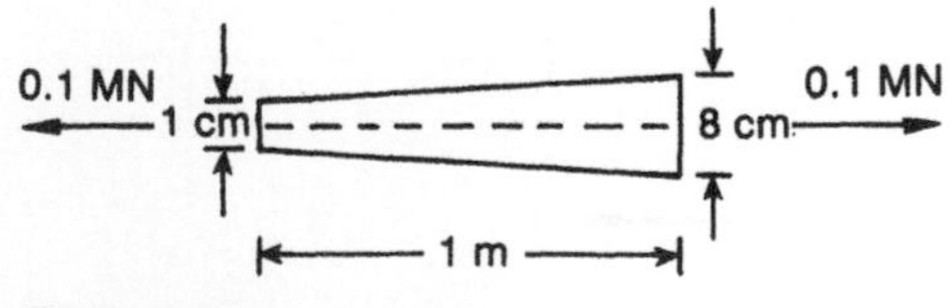

Figure 2.36

{0.796 mm}

5. If the bar of Example 4 were prevented from moving axially by two rigid walls and subjected to a temperature rise of 10 °C, what would be the maximum stress in the bar? Assume the 0.1 MN load is not acting.

$\alpha = 15 \times 10^{-6}$/°C

{−240 MN/m^2 at the smaller end}

6. An electrical cable consists of a copper core surrounded co-axially by a steel sheath, so that the two can be assumed to act as a compound bar. If the cable hangs down a vertical mineshaft, determine the maximum permissible length of the cable, assuming the following apply:

$A_c = 1 \times 10^{-4}$ m^2 = sectional area of copper

$E_c = 1 \times 10^{11}$ N/m^2 = elastic modulus of copper

ρ_c = 8960 kg/m^3 = density of copper

Maximum permissible stress in copper = 30 MN/m^2

$A_s = 0.2 \times 10^{-4}$ m^2 = sectional area of steel

$E_s = 2 \times 10^{11}$ N/m^2 = elastic modulus of steel

ρ_s = 7860 kg/m^3 = density of steel

Maximum permissible stress in steel = 100 MN/m^2

g = 9.81 m/s^2

{406.5 m}

7. How much will the cable of Example 6 stretch, owing to self-weight?

{61 mm}

8. If a weight of 100 kN were lowered into the sea, via a steel cable of cross-sectional area 8×10^{-4} m², what would be the maximum permissible depth that the weight could be lowered if the following apply?

Density of steel = 7860 kg/m³

Density of sea water = 1020 kg/m³

Maximum permissible stress in steel = 200 MN/m²

g = 9.81 m/s²

Any buoyancy acting on the weight itself may be neglected.

{1118 m}

9. A weightless rigid horizontal beam is supported by two vertical wires, as shown in Figure 2.37. If the following apply, determine the position from the left that a weight W can be suspended, so that the bar will remain horizontal when the wires stretch.

Left wire

cross-sectional area = $2A$
elastic modulus = E
length = $2l$

Right wire

cross-sectional area = A
elastic modulus = $3E$
length = l

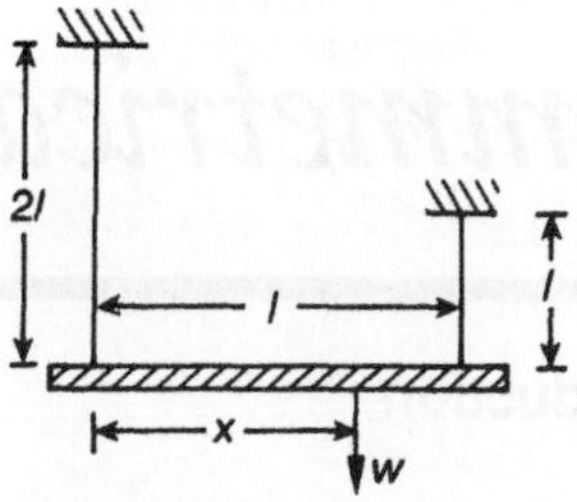

Figure 2.37

{$0.75l$}

3

Geometrical properties of symmetrical sections

3.1 Introduction

The geometrical properties of sections are of great importance in a number of different branches of engineering, including stress analysis. For example, if a beam is subjected to bending action, its bending stiffness will depend not only on the material properties of the beam, but also on the geometrical properties of its cross-section. Some typical cross-sections for symmetrical sections are shown in Figure 3.1, where it is evident that providing the material properties of the section are the same, the bending resistances of the sections are dependent on their geometrical properties.

After many years of experience, structural engineers have found that if beams are made from steel or aluminium alloy, the cross-sections of Figure 3.1(d) and (e) usually provide a better strength to weight ratio than do the sections of Figures 3.1(a) to (c).

The section of Figure 3.1(d) is known as a *rolled steel joist* (RSJ) and that of Figure 3.1(e) is known as a *tee section*.

3.2 Centroid

The centroid is the centre of the *moment of area* of a plane figure; or if the plane figure is of uniform thickness, this is the same position as the centre of gravity. This position is very important in elastic stress analysis.

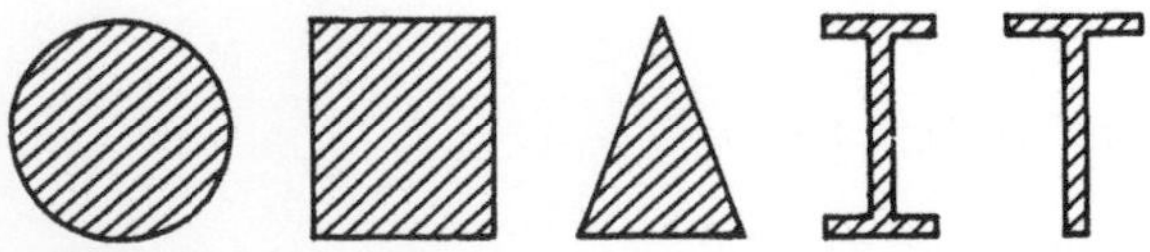

Figure 3.1 Some symmetrical cross-sections of beams: (a) circular section; (b) rectangular section; (c) triangular section; (d) 'I' section (e) 'T' section

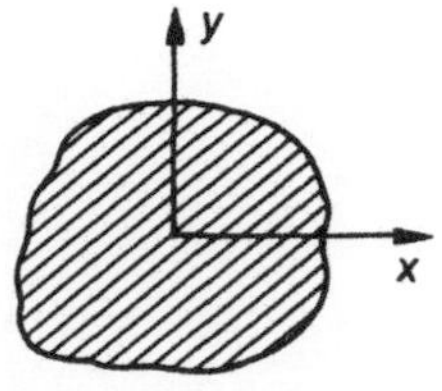

Figure 3.2 Plane figure

At the centroid, the following equations apply:

$$\int y\,dA = 0 \quad \text{and} \quad \int x\,dA = 0 \tag{3.1}$$

where x is the horizontal axis, and y is the vertical axis, mutually perpendicular to x, as shown in Figure 3.2.

3.2.1 Centroidal axes

These are lines that pass through the centroid.

3.2.2 Centre of area

For a plane figure, the centre of area is obtained from the following considerations:

Area above the horizontal central axis = area below the horizontal central axis (3.2)

Area to the left of the vertical central axis = area to the right of the vertical central axis (3.3)

3.2.3 Central axes

These are lines that pass through the centre of area.

3.3 Second moment of area

The second moment of area of the section of Figure 3.3 about XX is

$$I_{XX} = \int y^2\,dA \tag{3.4}$$

The second moment of area of the section of Figure 3.3 about YY is

$$I_{YY} = \int x^2\,dA \tag{3.5}$$

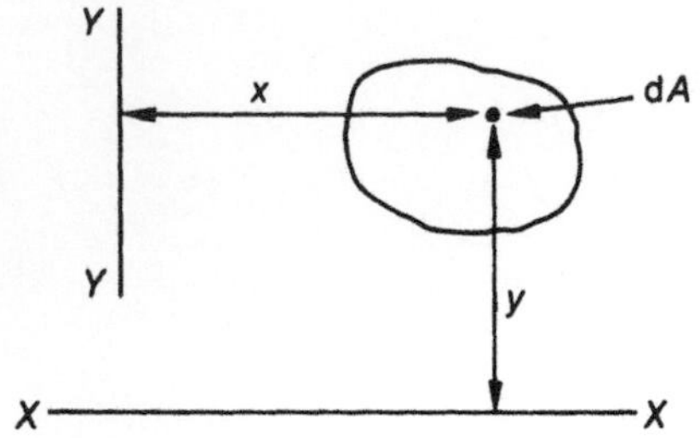

Figure 3.3

where

$$\int dA = \text{area of section}$$

3.4 Polar second moment of area

The polar second moment of area of the circular section of Figure 3.4, about its centre O is given by

$$J = \int_0^R r^2 \, dA \tag{3.6}$$

where

$$dA = 2\pi r \, dr$$

J is of importance in the torsion of circular sections.

3.5 Parallel axes theorem

Consider the plane section of Figure 3.5, which has a second moment of area about its centroid equal to I_{XX}, and suppose that it is required to determine the second moment of area about XX, where XX is parallel to xx, and that the perpendicular

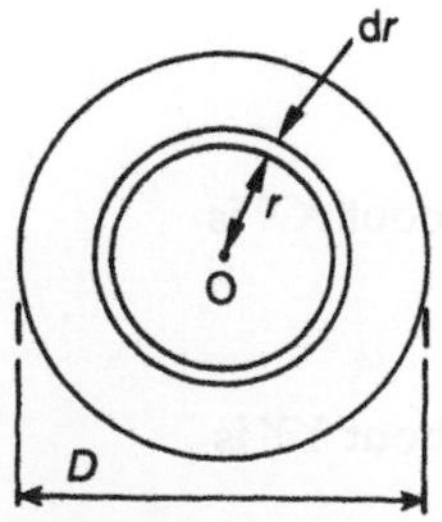

Figure 3.4 Circular section

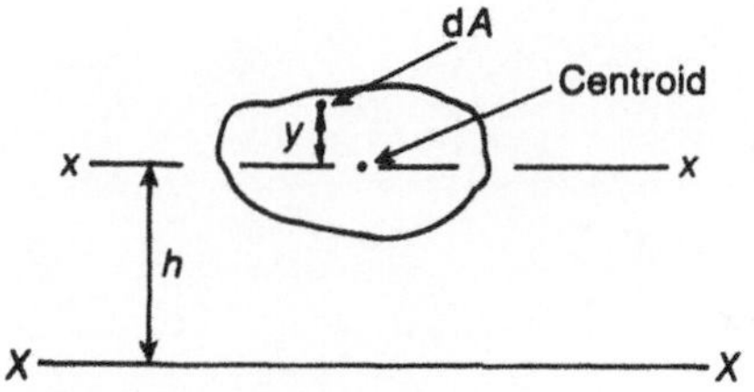

Figure 3.5 Parallel axes

distance between the two axes is h. Now,

$$I_{xx} = \int y^2 \, dA$$

and

$$I_{XX} = \int (y + h)^2 \, dA$$
$$= \int (y^2 + 2hy + h^2) \, dA$$

or

$$I_{XX} = \int y^2 \, dA + \int 2hy \, dA + \int h^2 \, dA \tag{3.7}$$

but as xx is at the centroid,

$$\int 2hy \, dA = 0$$

Therefore

$$\underline{I_{XX} = I_{xx} + h^2 \int dA} \tag{3.8}$$

Equation (3.8) is known as the *parallel axes theorem* and it is important in determining second moments of area for 'built-up' sections, such as RSJs, tees, etc.

3.6 Perpendicular axes theorem

From Figure 3.6, it can be seen that

$$I_{xx} = \int y^2 \, dA$$
$$I_{yy} = \int x^2 \, dA$$

Now from equation (3.6),

$$J = \int r^2 \, dA$$

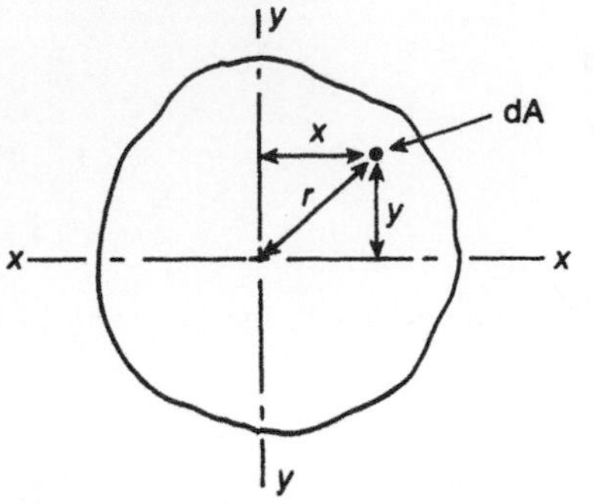

Figure 3.6

but as $x^2 + y^2 = r^2$,

$$\underline{J = I_{xx} + I_{yy}} \tag{3.9}$$

Equation (3.9) is known as the *perpendicular axes theorem*, which states that the sum of the second moments of area of two mutually perpendicular axes is equal to the polar second moment of area about the point where these two axes cross.

To demonstrate the theories described in this section, the following examples will be considered.

EXAMPLE 3.1

Determine the positions of the centroidal and central axes for the isosceles triangle of Figure 3.7. Hence, or otherwise, determine the second moments of area about the centroid and the base *XX*. Verify the parallel axes theorem by induction.

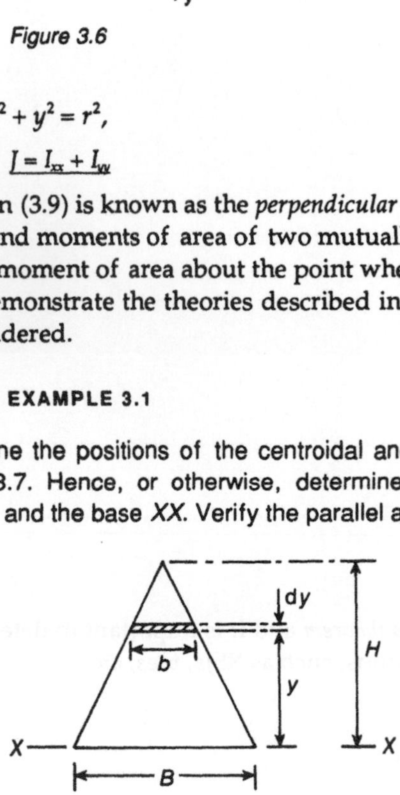

Figure 3.7 Isosceles triangle

SOLUTION

To find the area

$$\int dA = \int b\,dy$$

but

$$b = B(1 - y/H)$$

Therefore

$$\int b\,dy = B\int_0^H (1 - y/H)\,dy$$
$$= B[y - y^2/2H]_0^H$$

or the area is

$$\underline{A = BH/2} \qquad (3.10)$$

To find the centroidal axis

Let $\bar{y}$ be the distance of the centroid from XX. Therefore the first moment of area about XX is

$$A\bar{y} = \int_0^H yb\,dy$$
$$= B\int_0^H (y - y^2/H)\,dy$$
$$= B[y^2/2 - y^3/3H]_0^H$$

or

$$A\bar{y} = BH^2/6$$

Therefore

$$\bar{y} = H/3 \qquad (3.11)$$

i.e. the distance of the centroid above XX is $H/3$.

To find the central axis

Let $\bar{Y}$ be the distance of the central axis above XX. Consider the isosceles triangle of Figure 3.8 and equate areas above and below the central axis.

Area above the central axis = area below the central axis

$$\frac{b(H - \bar{Y})}{2} = \frac{(B + b)\bar{Y}}{2}$$

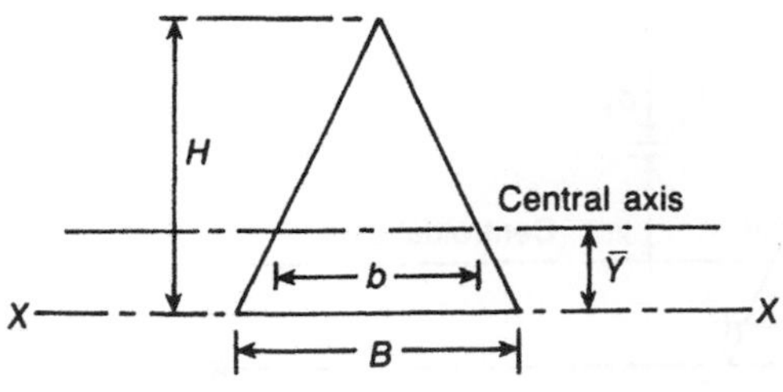

Figure 3.8 Central axis

but $b = B(1 - \bar{Y}/H)$. Therefore,

$$B(H - \bar{Y} - \bar{Y} + \bar{Y}^2/H) = B\bar{Y} + B(\bar{Y} - \bar{Y}^2/H)$$

or

$$2\bar{Y}^2 - 4\bar{Y}H + H^2 = 0$$

Therefore,

$$\bar{Y} = \frac{4H \pm \sqrt{(16-8)}H}{4}$$

$$\bar{Y} = 0.293H \tag{3.12}$$

i.e. for this case, the central axis is 12.1% below the centroidal axis.

In stress analysis, the position of the central axis is only of importance in plasticity; so for the remainder of the present chapter, we will restrict our interest to the centroidal axis, which is of much interest in elastic theory.

To find I_{xx}

Let I_{xx} be the second moment of area about the centroidal axis, as in Figure 3.9. Now,

$$b = 2B/3 - By/H$$

and

$$I_{xx} = \int_{-H/3}^{2H/3} y^2 b\, dy$$

$$= B\int_{-H/3}^{2H/3} \left(\frac{2y^2}{3} - \frac{y^3}{H}\right) dy$$

$$= B\left[\frac{2y^3}{9} - \frac{y^4}{4H}\right]_{-H/3}^{2H/3}$$

$$= BH^3\left\{\left[\frac{16}{27\times 9} - \frac{16}{4\times 81}\right] - \left[\frac{-2}{9\times 27} - \frac{1}{4\times 81}\right]\right\}$$

$$\underline{I_{xx} = BH^3/36} \tag{3.13}$$

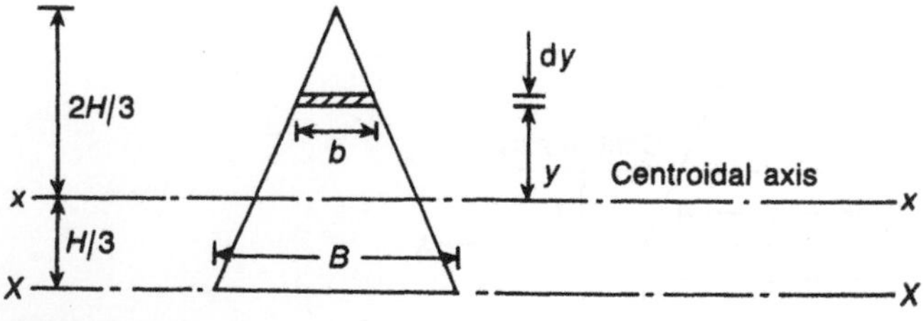

Figure 3.9

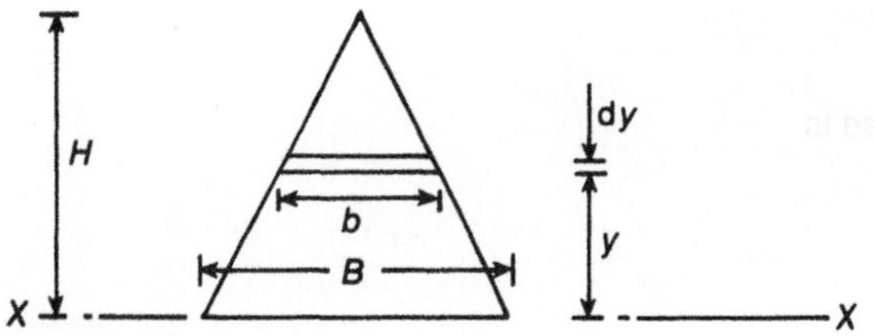

Figure 3.10

Similarly, from Figure 3.10,

$$I_{XX} = \int_0^H y^2 b \, dy$$

but

$$b = B(1 - y/H)$$

$$I_{XX} = B\int_0^H (y^2 - y^3/H) \, dy$$

$$= B\left[\frac{y^3}{3} - \frac{y^4}{4H}\right]_0^H$$

$$\underline{I_{XX} = BH^3/12} \qquad (3.14)$$

Check on parallel axes theorem

From equation (3.8),

$$I_{XX} = I_{xx} + \frac{BH}{2} \times \left(\frac{H}{3}\right)^2$$

$$= \frac{BH^3}{12} \text{ (as required)}$$

EXAMPLE 3.2

Determine the area and second moments of area about the major and minor axes for the elliptical section of Figure 3.11.

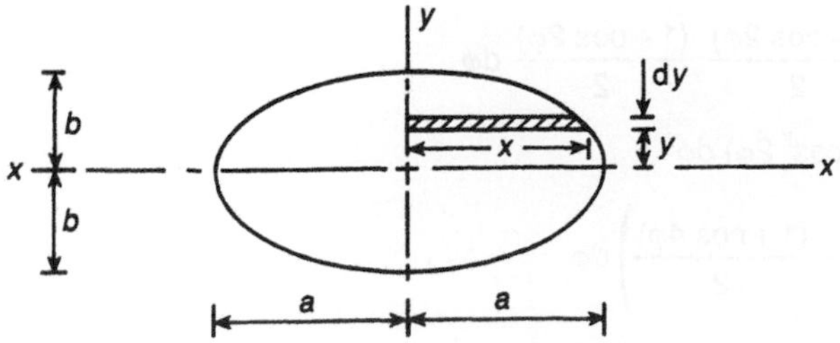

Figure 3.11 Elliptical section

SOLUTION

Now the equation of an ellipse is

$$\frac{x^2}{a^2} + \frac{y^2}{b^2} = 1$$

Therefore

$$y^2 = (1 - x^2/a^2)b^2$$

Let

$$x = a\cos\phi \tag{3.15}$$

Therefore

$$y = (1 - a^2\cos^2\phi/a^2)^{1/2}b$$

$$\therefore\ y = b\sin\phi \tag{3.16}$$

$$dy = b\cos\phi\, d\phi \tag{3.17}$$

Now,

$$\begin{aligned} A &= \text{area of elliptical figure} \\ &= 4\int x\,dy = 4\int_0^{\pi/2} a\cos\phi\, b\cos\phi\, d\phi \\ &= 4ab\int_0^{\pi/2} \cos^2\phi\, d\phi \\ &= 4ab\int_0^{\pi/2} \frac{(1+\cos 2\phi)}{2}\, d\phi \end{aligned}$$

Therefore

$$\underline{A = \pi ab} \tag{3.18}$$

From Figure 3.11, it can be seen that

$$\begin{aligned} I_{xx} &= 4\int y^2 x\, dy \\ &= 4\int b^2\sin^2\phi\, a\cos\phi\, b\cos\phi\, d\phi \\ &= 4ab^3\int \frac{(1-\cos 2\phi)}{2}\,\frac{(1+\cos 2\phi)}{2}\, d\phi \\ &= ab^3\int (1-\cos^2 2\phi)\, d\phi \\ &= ab^3\int_0^{\pi/2}\left(1 - \frac{(1+\cos 4\phi)}{2}\right) d\phi \end{aligned}$$

$$\underline{I_{xx} = \pi ab^3/4} \tag{3.19}$$

Similarly, it can be proven that

$$\underline{I_{yy} = \pi a^3 b/4} \tag{3.20}$$

For a circle of radius R (or diameter D),

$$R = a = b$$

Therefore

$$I_{xx} = I_{yy} = \pi R^4/4 = \pi D^4/64 \tag{3.21}$$

and

$$J = \pi D^4/32 \tag{3.22}$$

EXAMPLE 3.3

Determine the second moments of area about xx and XX for the rectangle in Figure 3.12, and verify the parallel axes theorem by induction.

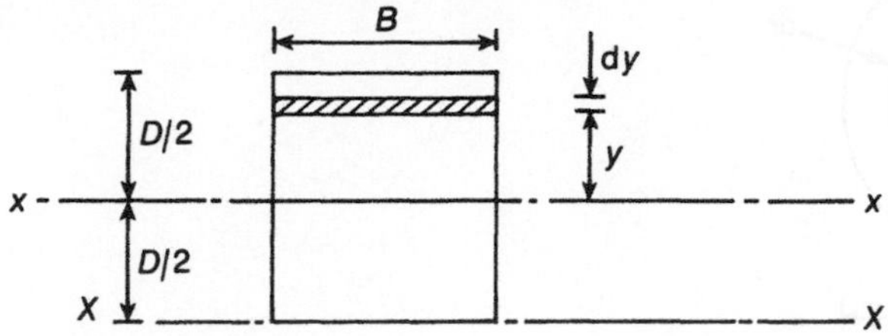

Figure 3.12 Rectangle

SOLUTION

$$I_{xx} = \int_{-D/2}^{D/2} y^2 B \, dy$$

$$= B[y^3/3]_{-D/2}^{D/2}$$

$$\underline{I_{xx} = BD^3/12} \tag{3.23}$$

$$I_{XX} = B\int_{-D/2}^{D/2} (y + D/2)^2 \, dy$$

$$= B\int_{-D/2}^{D/2} (y^2 + Dy + D^2/4) \, dy$$

$$= B[y^3/3 + Dy^2/2 + D^2y/4]_{-D/2}^{D/2}$$

$$\underline{I_{XX} = BD^3/3} \tag{3.24}$$

Check on parallel axes theorem

$$I_{XX} = I_{xx} + A(D/2)^2$$

$$= \underline{BD^3/3 \text{ (as required)}}$$

For the parallelogram of Figure 3.13, it can be proven that

$\underline{I_{XX} = BD^3/3}$

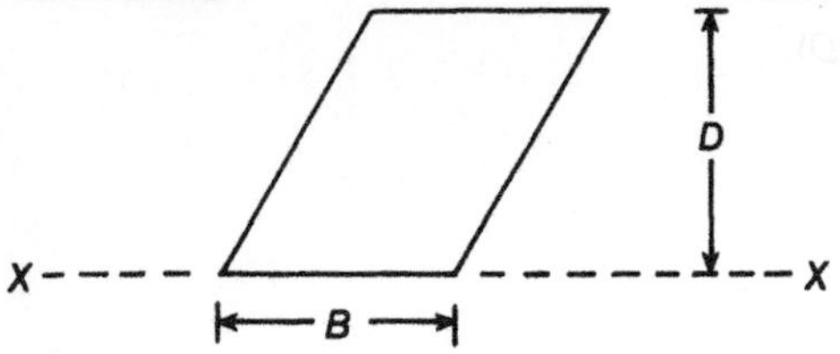

Figure 3.13 Parallelogram

EXAMPLE 3.4

Determine the polar second moment of area for the circular section in Figure 3.14.

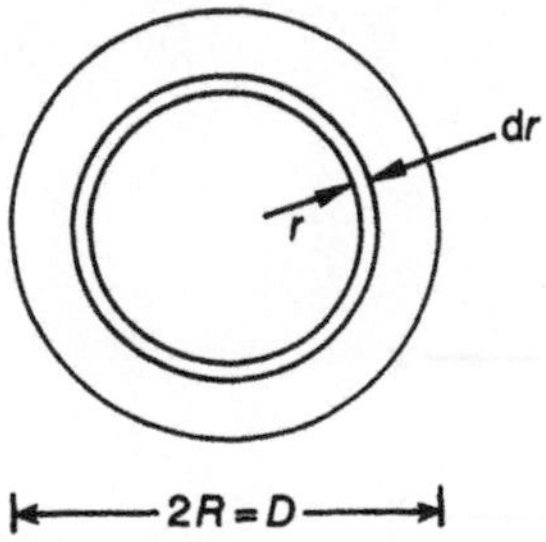

Figure 3.14 Circle

SOLUTION

$$J = \int_0^R r^2\, 2\pi r\, dr$$

$$= 2\pi[r^4/4]_0^R$$

$$\underline{J = \pi R^4/2 = \pi D^4/32} \quad (3.25)$$

EXAMPLE 3.5

Determine the position of the centroidal axis *xx* and the second moment of area about this axis for the tee-bar in Figure 3.15.

SOLUTION

Table 3.1 will be used to determine the geometrical properties of the tee-bar, where the rows of the table refer to the two rectangular elements ① and ②.

a = area of an individual rectangular element

y = distance of local centroid of an individual rectangular element from XX.

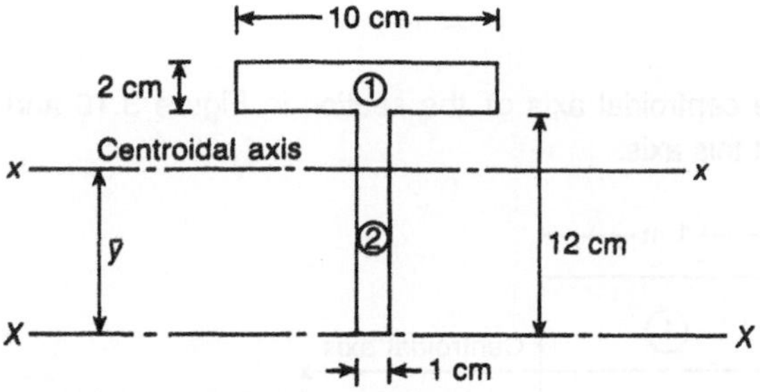

Figure 3.15 Tee-bar

Table 3.1 **Geometrical calculations for tee-bar**

Section	a (m²)	y (m)	ay (m³)	ay^2 (m⁴)	i (m⁴)
①	2×10^{-3}	0.13	2.6×10^{-4}	3.38×10^{-5}	$\frac{0.1 \times 0.02^3}{12} = 6.67 \times 10^{-8}$
②	1.2×10^{-3}	0.06	7.2×10^{-5}	4.32×10^{-6}	$\frac{0.01 \times 0.12^3}{12} = 1.44 \times 10^{-6}$
	3.2×10^{-3}	–	3.32×10^{-4}	3.812×10^{-5}	1.51×10^{-6}

i = second moment of area of an individual rectangular element about its local centroid and parallel to XX.

ay = the product $a \times y$

ay^2 = the product $a \times y \times y$

$\sum$ = summation of the appropriate column

From Table 3.1,

$$\bar{y} = \sum ay \Big/ \sum a = 3.32 \times 10^{-4}/3.2 \times 10^{-3} \tag{3.26}$$

$$\underline{\bar{y} = 0.104 \text{ m}}$$

$$I_{XX} = \sum ay^2 + \sum i \tag{3.27}$$

$$= \underline{3.812 \times 10^{-5} + 1.51 \times 10^{-6}}$$

$$\underline{I_{XX} = 3.963 \times 10^{-5} \text{ m}^4}$$

$$I_{xx} = I_{XX} - (\bar{y})^2 \sum a \tag{3.28}$$

$$= 3.963 \times 10^{-5} - (0.104)^2 \times 3.2 \times 10^{-3}$$

$$\underline{I_{xx} = 5.02 \times 10^{-6} \text{ m}^4}$$

EXAMPLE 3.6

Determine the position of the centroidal axis of the section in Figure 3.16 and also the second moment of area about this axis.

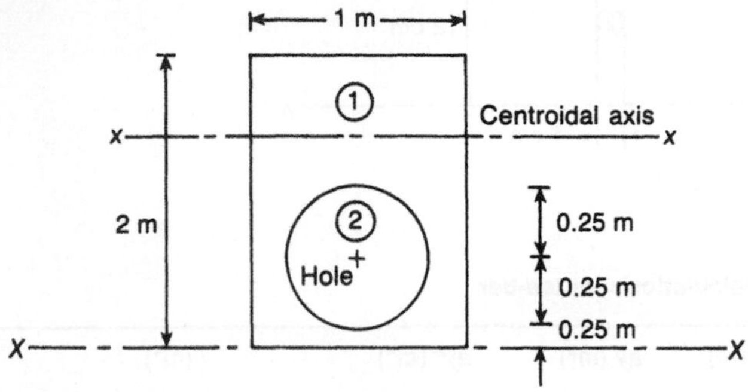

Figure 3.16

SOLUTION

The determination of the geometrical properties of the section in Figure 3.16 will be aided with the calculations of Table 3.2, where the symbols are defined as in Example 3.5. From Table 3.2,

$$\bar{y} = \sum ay \Big/ \sum a = \underline{1.054\ \text{m}}$$

$$I_{XX} = \sum ay^2 + \sum i = \underline{2.615\ \text{m}^4}$$

$$I_{xx} = I_{XX} - (\bar{y})^2 \sum a = \underline{0.611\ \text{m}^4}$$

Table 3.2 **Geometrical calculations**

Section	a (m²)	y (m)	ay (m³)	ay^2 (m⁴)	i (m⁴)
①	2	1	2	2	$1 \times 2^3/12 = 0.66667$
②	−0.1963	0.5	−0.098	−0.0491	$-\pi \times (5 \times 10^{-1})^4/64 = -3.07 \times 10^{-3}$
	1.8037	—	1.902	1.9509	0.664

3.7 Calculation of I through numerical integration

If I is required for an arbitrarily shaped section, such as that shown in Figure 3.17, the calculation for I can be carried out through numerical integration.

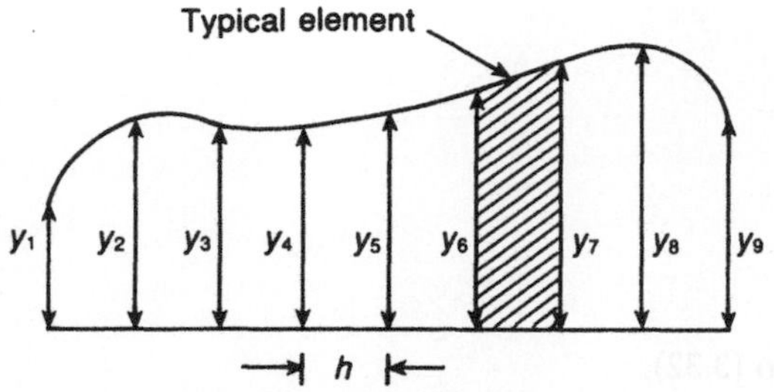

Figure 3.17 Arbitrarily shaped section

If the numerical integration is based on a Simpson's rule approach, then the section must be divided into an equal number of elements, where the number of elements must be even.

3.7.1 Proof of Simpson's rule

Simpson's rule is based on employing a parabola to describe the function over any three stations, as shown in Figure 3.18. The equation is

$$y = a + bx + cx^2 \tag{3.29}$$

where a, b and c are arbitrary constants.

To obtain the three unknown constants, it will be necessary to obtain three simultaneous equations by putting in the *boundary values* for y in equation (3.29), as follows:

$\underline{@\, x = 0} \quad y = y_2 = a$

Therefore

$$\underline{a = y_2} \tag{3.30}$$

or

$\underline{@\, x = -h} \quad y = y_1$

$$y_1 - y_2 = -bh + ch^2 \tag{3.31}$$

$\underline{@\, x = h} \quad y = y_3$

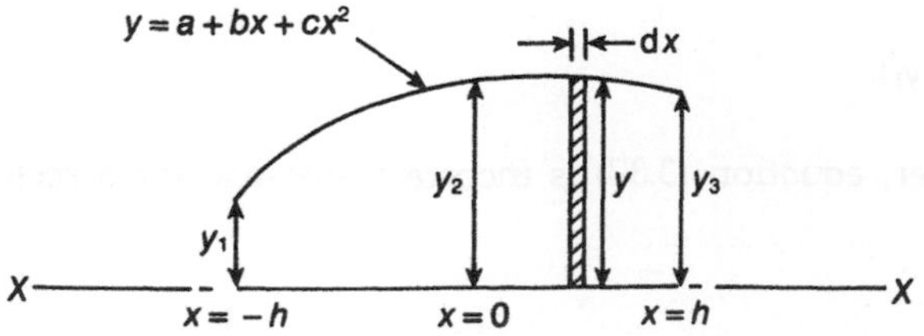

Figure 3.18 Parabolic variation

or

$$y_3 - y_2 = bh + ch^2 \tag{3.32}$$

Adding (3.31) and (3.32),

$$c = \frac{(y_1 - 2y_2 + y_3)}{2h^2} \tag{3.33}$$

Substituting equation (3.33) into (3.32),

$$b = \frac{(-y_1 + y_3)}{2h} \tag{3.34}$$

Now,

$$A = \text{area of section}$$
$$= \int_{-h}^{h} y \, dx$$
$$= \int_{-h}^{h} (a + bx + cx^2) \, dx$$
$$= \left(ah + \frac{ch^3}{3}\right) \times 2 \tag{3.35}$$

Substituting equations (3.30) and (3.33) into (3.35),

$$A = 2h[y_2 + \tfrac{1}{6}(y_1 - 2y_2 + y_3)]$$

$$\underline{A = \frac{h}{3}(y_1 + 4y_2 + y_3)} \tag{3.36}$$

Equation (3.36) is known as *Simpson's rule* for calculating areas.

3.7.2 *Naval architect's method of numerically calculating I_{xx} for a ship's water plane*

The naval architect's method of calculating I_{XX}, which is based on Simpson's rule, is given by equation (3.37). This expression is reasonable for gentle curves with relatively small values of h:

$$I_{XX} = \frac{h}{9}(y_1^3 + 4y_2^3 + y_3^3) \tag{3.37}$$

Strictly speaking, however, equation (3.37) is incorrect, because for a rectangle of height y and width dx,

$$I_{XX} = \int_{-h}^{h} \frac{y^3}{3} \, dx$$

where y is a function of x. Hence, by substitution,

$$I_{XX} = \frac{1}{3}\int_{-h}^{h} (a + bx + cx^2)^3\, dx$$

which is very different to equation (3.37), i.e.

$$I_{XX} = \frac{2}{3}\left(a^3h + \frac{c^3h^7}{7} + ab^2h^3 + a^2ch^3 + \frac{3}{5}\, b^2ch^5 + \frac{3ac^2}{5}\, h^5\right)$$

$$= \frac{2h}{3}\Bigg(y_2^3 + \frac{1}{7 \times 8}\,(y_1^3 - 6y_1^2y_2 + 3y_1^2y_3$$

$$+ 12y_1y_2^2 - 12y_1y_2y_3 + 3y_1y_3^2 - 8y_2^3$$

$$+ 12y_2^2y_3 - 6y_2y_3^2 + y_3^3)$$

$$+ \frac{y_2}{4}\,(y_1^2 - 2y_1y_3 + y_3^2) + \frac{y_2^2}{2}\,(y_1 - 2y_2 + y_3)$$

$$+ \frac{3}{40}\,(y_1^2 - 2y_1y_3 + y_3^2)(y_1 - 2y_2 + y_3)$$

$$+ \frac{3}{20}\, y_2(y_1^2 - 4y_1y_2 + 2y_1y_3 + 4y_2^2 - 4y_2y_3 + y_3^2)\Bigg)$$

which, when rearranged, becomes

$$I_{XX} = \frac{2h}{3}\,[\tfrac{13}{140}\,(y_1^3 + y_3^3) + \tfrac{16}{35}\,y_2^3 + \tfrac{1}{7}\,(y_1^2y_2 + y_2y_3^2)$$

$$- \tfrac{3}{140}\,(y_1^2y_3 + y_1y_3^2) + \tfrac{4}{35}\,(y_1y_2^2 + y_2^2y_3) - \tfrac{4}{35}\,(y_1y_2y_3)] \qquad (3.38)$$

EXAMPLE 3.7

Calculate I_{XX} for the section in Figure 3.19 by equations (3.37) and (3.38), and then compare the two results.

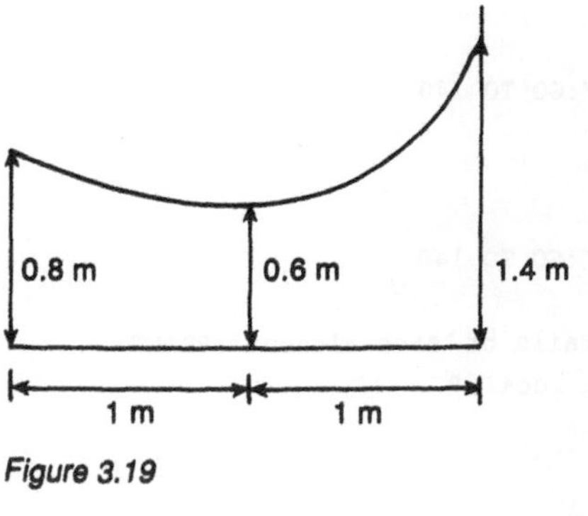

Figure 3.19

SOLUTION

From equation (3.37),

$$I_{XX} = \tfrac{1}{9}(0.8^3 + 4 \times 0.6^3 + 1.4^3) = 0.4578\ \text{m}^4$$

From equation (3.38),

$$I_{XX} = \tfrac{2}{3}\left(\tfrac{13}{140} \times 3.256 + \tfrac{16}{35} \times 0.216 + \tfrac{1}{7} \times 1.56 - \tfrac{3}{140} \times 2.464 + \tfrac{4}{35} \times 0.792 - \tfrac{4}{35} \times 0.672\right)$$

$$\underline{I_{XX} = 0.39\ m^4}$$

i.e.

$$\underline{\text{percentage difference} = 17.41}$$

3.7.3 *Computer program for calculating $\bar{y}$ and I_{xx}*

Listing 3.1 gives a computer program, in BASIC, for calculating $\bar{y}$ and I_{xx} for symmetrical sections, Listings 3.2 and 3.3 give the outputs for this program for problems 2(b) and 2(d), respectively, from the examples for practice at the end of this chapter.

It should be noted from Listings 3.2 and 3.3 that the units being used must not be mixed (i.e. do not use a mixture of centimetres with metres or millimetres, etc.).

Listing 3.1 **Computer program for calculating $\bar{y}$ and I_{xx}**

```
100  CLS
110  REMark program for second moments of area for symmetrical sections
120  PRINT:PRINT"program for second moments of area for symmetrical
     sections"
130  PRINT:PRINT"copyright of Dr.C.T.F.ROSS"
140  PRINT:PRINT"type in the number of sections"
150  INPUT n
160  IF n>0 THEN GO TO 180
170  PRINT:PRINT"incorrect data":GO TO 140
180  absn=ABS (n)
190  in=INT (absn)
200  IF in=n THEN GO TO 220
210  PRINT:PRINT"incorrect data":GO TO 140
220  DIM a(n) ,y(n), i0 (n)
230  PRINT:PRINT"type in the details of each element":PRINT
240  area=0:moment=0:second=0:i local=0
250  FOR i=1 TO n
```

Listing 3.1 ***Continued***

```
260 PRINT"elemental area (";i;")=";:INPUT a (i)
270 PRINT"element centroid from XX _ i.e. y (";i;")=";:INPUT y (i)
280 PRINT"elemental local 2nd moment of area i0 (";i;")=";:INPUT i0 (i)
290 area=area+a(i)
300 moment=moment+a(i) *y(i)
310 second=second+a(i) *y(i) *y(i)
320 ilocal=ilocal+i0(i)
330 NEXT i
340 ybar=moment/area
350 ixx=ilocal+second
360 ina=ixx-ybar^2*area
380 PRINT"number of elements=";n
390 FOR I=1 to n
400 PRINT"element area a(";i;")=";a(i)
410 PRINT"element centroid y (";i;")=";y(i)
420 PRINT"elemental local 2nd moment of area=";i0(i)
430 NEXT
440 PRINT
450 PRINT
460 PRINT"sectional area=";area
470 PRINT"sectional centroid (ybar) from XX=";ybar
480 PRINT"2nd moment of area of section about its centroid=";ina
500 STOP
```

Listing 3.2 **Computer output for problem 2(b)**

```
number of elements=3
element area a(1) =8
element centroid y(1) =14.5
elemental local 2nd moment of area=.666667
element area a(2) =12
element centroid y(2) =8
elemental local 2nd moment of area=144
element area a(3) =20
element centroid y(3) =1
elemental local 2nd moment of area=6.6667

sectional area=40
section centroid (ybar) from XX=5.8
2nd moment of area of section about its centroid=1275.733
```

Listing 3.3 **Computer output for problem 2(d)**

```
number of elements=2
element area a(1) =154
element centroid y(1) =7
elemental local 2nd moment of area=2515.3
element area a(2) =-78.54
element centroid y(2) =6
elemental local 2nd moment of area=-490.9

sectional area=75.46
section centroid (ybar) from XX=8.040816
2nd moment of area of section about its centroid=1864.1144
```

3.7.4 Program input

INPUT n – number of elements
FOR i = 1 TO *n*
INPUT *a*(*i*) – area of element *i*
INPUT *y*(*i*) – distance of centroid of element *i* from XX
INPUT i0(*i*) – local second moment of area of element *i*
NEXT *i*

3.7.5 Program output

area = ΣA
ybar = $\bar{y}$
ina = I_{xx}

EXAMPLES FOR PRACTICE 3

1. Determine the second moments of area about the centroid of the squares shown in Figures 3.20(a) and (b).

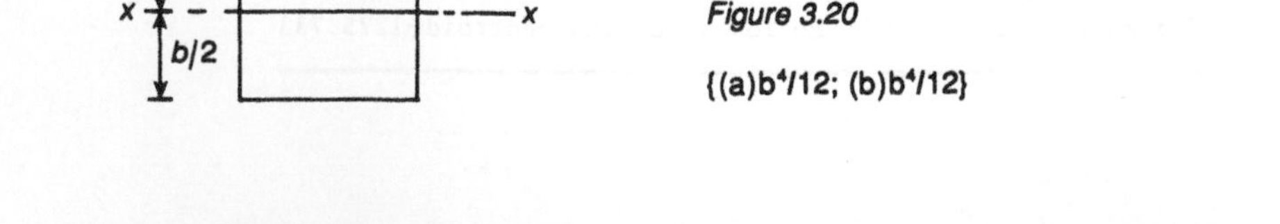

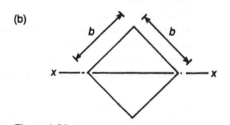

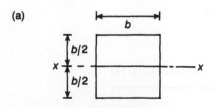

Figure 3.20

{(a)$b^4/12$; (b)$b^4/12$}

2. Determine the positions of the centroidal axes *xx* and the second moments of area about these axes, for the sections of Figures 3.21 (a)–(d).

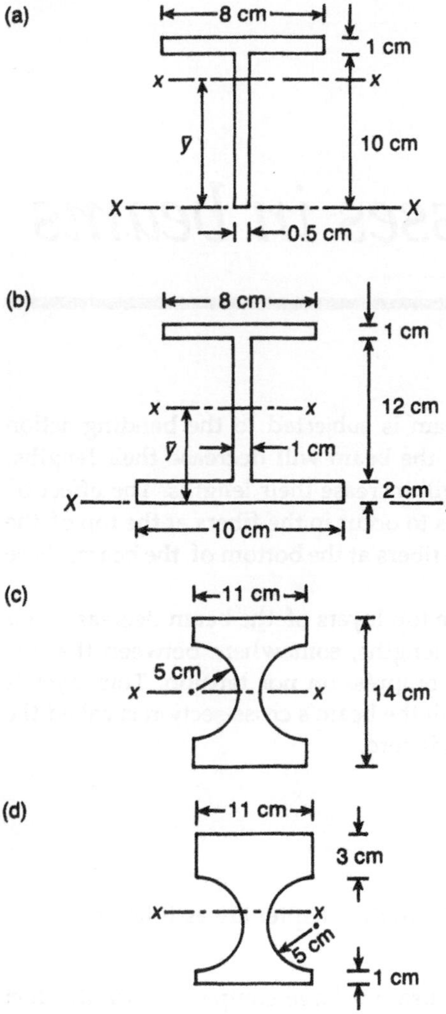

Figure 3.21

{(a)8.38 cm, 1.354×10^{-6} m^4; (b) 5.8 cm, 1.275×10^{-5} m^4; (c) 2.024×10^{-5} m^4; (d) 8.04 cm, 1.864×10^{-5} m^4}

3. Determine the second moment of area of the section shown in Figure 3.22 about an axis passing through the centroid and parallel to the XX axis.

What would be the percentage reduction in second moment of area if the bottom flange were identical to the top flange? (Portsmouth, June 1982).

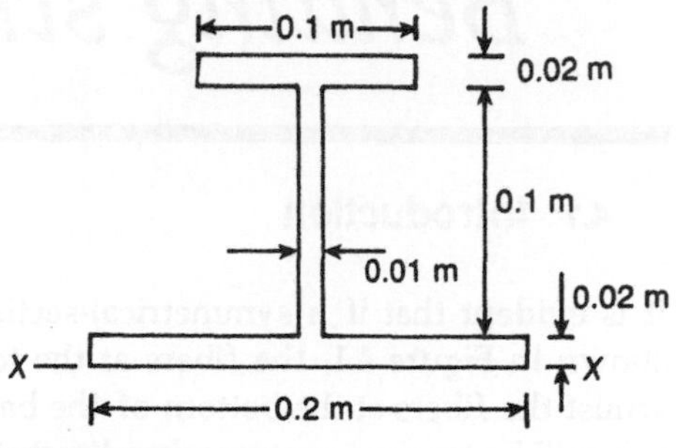

Figure 3.22

{2.058×10^{-5} m^4, 24.9%}

4

Bending stresses in beams

4.1 Introduction

It is evident that if a symmetrical-section beam is subjected to the bending action shown in Figure 4.1, the fibers at the top of the beam will decrease their lengths, whilst the fibers at the bottom of the beam will increase their lengths. The effect of this will be to cause compressive direct stresses to occur in the fibers at the top of the beam and tensile direct stresses to occur in the fibers at the bottom of the beam, these stresses being parallel to the axis of the beam.

It is evident also from Figure 4.1 that, as the top layers of the beam decrease their lengths and the bottom fibers increase their lengths, somewhere between the two there will be a layer that will be in neither compression nor tension. This layer is called the *neutral layer*, and its intersection with the beam's cross-section is called the *neutral axis* (NA). The stress at the neutral axis is zero.

4.1.1 Assumptions made

The assumptions made in the theory of bending in this chapter are as follows:

(a) Deflections are small.
(b) The radius of curvature of the deformed beam is large compared with its other dimensions.
(c) The beam is initially straight.
(d) The cross-section of the beam is symmetrical.
(e) The effects of shear are negligible.
(f) Transverse sections of the beam, which are plane and normal before bending, remain plane and normal during bending.
(g) Elastic theory is obeyed, and the elastic modulus of the beam is the same in tension as it is in compression.
(h) The beam material is homogeneous and isotropic.

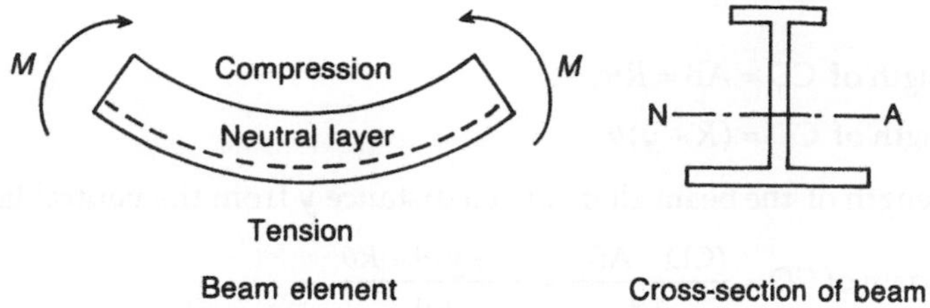

Figure 4.1 Beam element in bending

4.2 Proof of $\sigma/y = M/I = E/R$

Consider the equation

$$\frac{\sigma}{y} = \frac{M}{I} = \frac{E}{R} \tag{4.1}$$

where

σ = stress (due to the bending moment M) occurring at a distance y from the neutral axis
M = bending moment
I = second moment of area of the cross-section about its neutral axis (centroidal axis)
E = elastic modulus
R = radius of curvature of the neutral layer of the beam, when M is applied

Equation (4.1) is the fundamental expression that is used in the bending theory of beams, and it will be proven with the aid of Figure 4.2, which shows the deformed shape of an initially straight beam under the action of a sagging bending moment M. Initially, all layers of the beam element will be the same length as the neutral layer,

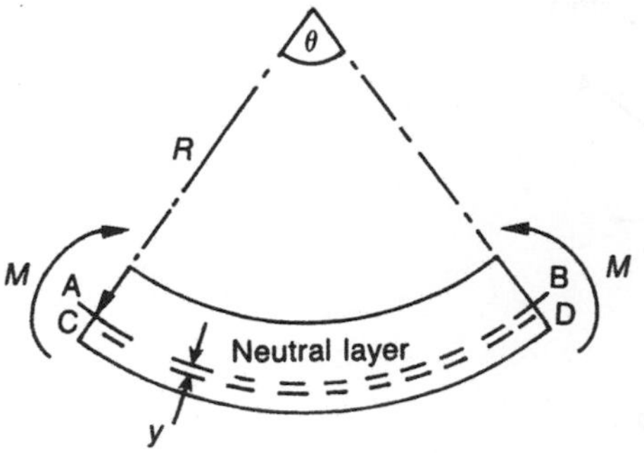

Figure 4.2 Beam element in bending

AB, so that

$$\text{Initial length of CD} = \text{AB} = R\theta$$

$$\text{Final length of CD} = (R + y)\theta$$

where CD is the length of the beam element at a distance y from the neutral layer.

$$\text{Tensile strain of CD} = \frac{(\text{CD} - \text{AB})}{\text{AB}} = \frac{(R+y)\theta - R\theta}{R\theta}$$

$$= \frac{y}{R}$$

and

$$\text{Stress in the layer CD} = \sigma = Ey/R$$

Therefore

$$\frac{\sigma}{y} = \frac{E}{R} \tag{4.2}$$

Equation (4.2) shows that the bending stress σ varies linearly with y and it will act on the section, as shown in Figure 4.3, where NA is the position of the neutral axis. Equation (4.2) also shows that the *largest stress* in magnitude occurs in the fiber which is the *furthest distance from the neutral axis*.

It is evident also, from equilibrium considerations, that the longitudinal tensile force caused by the tensile stresses due to bending must be equal and opposite to the longitudinal compressive force caused by the compressive stresses due to bending, so that

$$\int \sigma \, dA = 0$$

where dA is the area of an element of the cross-section at a distance y from the neutral axis, but

$$\sigma = Ey/R$$

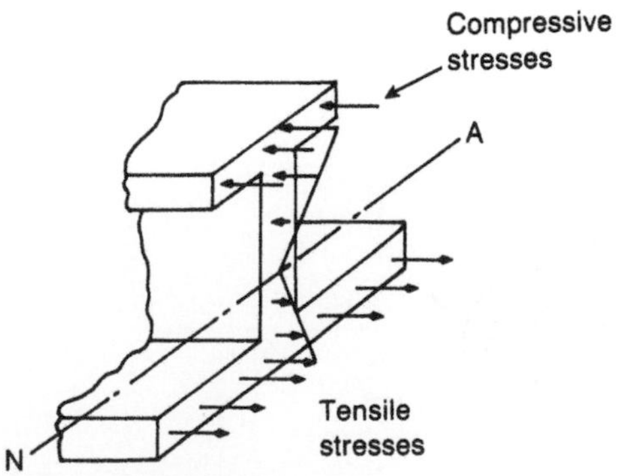

Figure 4.3 Bending stress distribution

Therefore

$$\frac{E}{R}\int y\,dA = 0$$

i.e.

$$\int y\,dA = 0$$

or *the neutral axis* is at the *centroid of the beam's cross-section.*

Now it can be seen from Figure 4.3 that the bending stresses cause a couple which, from equilibrium considerations, must be equal and opposite to the externally applied moment M at the appropriate section, i.e.

$$M = \int \sigma y\,dA$$

$$= \frac{E}{R}\int y^2\,dA$$

but

$$\int y^2\,dA = I$$

Therefore

$$\frac{M}{I} = \frac{E}{R} \tag{4.3}$$

From equations (4.2) and (4.3),

$$\frac{\sigma}{y} = \frac{M}{I} = \frac{E}{R} \tag{4.4}$$

4.3 Sectional modulus (Z)

From equation (4.2), it can be seen that the maximum stress due to bending occurs in the fiber which is the greatest distance from the neutral axis.

Let

$\bar{y}$ = distance of the fiber in the cross-section of the beam which is the furthest distance from NA

$$Z = I/\bar{y} = \text{sectional modulus} \tag{4.5}$$

Therefore the maximum bending stress is

$$\hat{\sigma} = \frac{M}{Z} \tag{4.6}$$

EXAMPLE 4.1

A solid circular-section steel bar of diameter 2 cm and length 1 m is subjected to a pure bending moment of magnitude M. If the maximum permissible stress in the bar is 100 MN/m^2, determine the maximum permissible value of M. If the lateral deflection at the mid-point of this beam, relative to its two ends, is 6.25 mm, what will be the elastic modulus of the beam?

SOLUTION

$$I = \frac{\pi d^4}{64} = \frac{\pi \times (2 \times 10^{-2})^4}{64} = 7.854 \times 10^{-9}\,\text{m}^4$$

The maximum stress will occur at the fiber in the cross-section which is the furthest distance from the neutral axis, i.e.

$$\bar{y} = 1\text{ cm} = 1 \times 10^{-2}\text{ m}$$

$$Z = I/\bar{y} = 7.854 \times 10^{-7}\text{ m}^3$$

From equation (4.6),

$$M = \hat{\sigma} Z = 100 \times 10^6 \times 7.854 \times 10^{-7} = \underline{78.54\text{ N m}}$$

Now under pure bending, the beam will bend into a perfect arc of a circle, as shown in Figure 4.4. In the figure

l = length of beam

δ = central deflection

Now from the properties of a circle,

$$\delta(2R - \delta) = \frac{l}{2} \times \frac{l}{2}$$

or

$$2R\delta - \delta^2 = l^2/4$$

but as deflections are small, δ^2 is small compared with $2R\delta$. Therefore

$$\delta = l^2/8R$$

or

$$R = l^2/8\delta$$

$$= 1/(8 \times 6.25 \times 10^{-3})$$

$$\underline{R = 20\text{ m}}$$

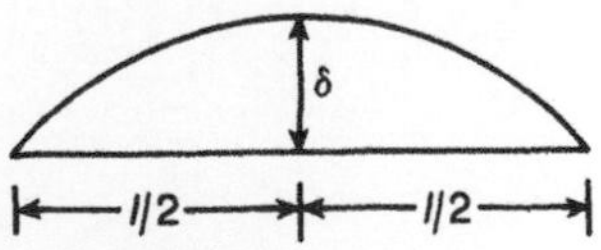

Figure 4.4

From equation (4.4),

$$E = MR/I = 78.54 \times 20/(7.854 \times 10^{-9})$$

$$\underline{E = 2 \times 10^{11}\ \text{N/m}^2}$$

EXAMPLE 4.2

A beam of length 2 m and with the cross-section shown in Figure 3.15 is simply supported at its ends and carries a uniformly distributed load *w*, spread over its entire length, as shown in Figure 4.5.

Determine a suitable value for *w*, given that the maximum permissible tensile stress is 100 MN/m² and the maximum permissible compressive stress is 30 MN/m².

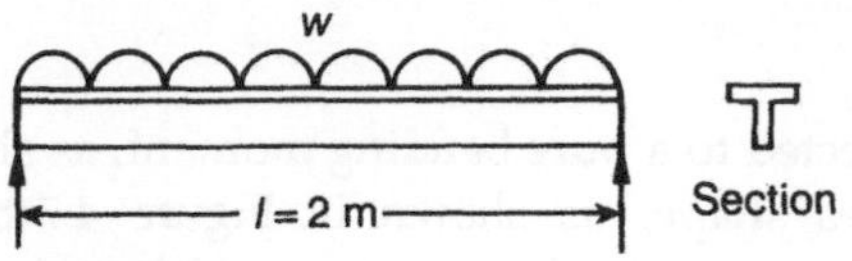

Figure 4.5

SOLUTION

The maximum bending moment is

$$\hat{M} = wl^2/8 \quad \text{at mid-span} \tag{4.7}$$

The bottom of the beam will be in tension, and the top will be in compression.

Now from equation (3.26), the distance of the neutral axis is 0.104 m from the bottom, as shown in Figure 4.6. In the figure

y_1 is used to determine the maximum compressive stress

y_2 is used to determine the maximum tensile stress

To determine the design criterion

If the tensile stress of 100 MN/m² is used in conjunction with y_2, then the maximum permissible compressive stress which is at the top is

$$\frac{0.036}{0.104} \times 100 = 34.6\ \text{MN/m}^2$$

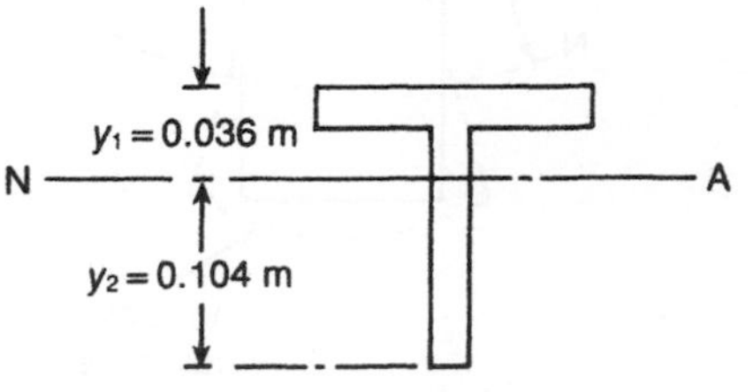

Figure 4.6 Beam cross-section

That is, the compressive stress in the top of the flange will be exceeded if the tensile stress of 100 MN/m^2 is adopted as the design criterion; hence, the design criterion is the *30 MN/m^2 in the top flange*. If the top flange is under a compressive stress of 30 MN/m^2 then the tensile stress at the bottom of the web is $30 \times 0.104/0.036 = 86.67$ MN/m^2 in the bottom flange. Therefore

$$\hat{M} = \sigma \times I/y = 30 \times 10^6 \times 5.02 \times 10^{-6}/0.036$$

$$\underline{\hat{M} = 4183.3 \text{ N m}} \qquad (4.8)$$

Equating (4.7) and (4.8),

$$\underline{w = 8.37 \text{ kN/m}}$$

4.4 Anticlastic curvature

If a beam of rectangular section is subjected to a pure bending moment, as shown in Figure 4.7(a), its cross-section changes shape, as shown in Figure 4.7(b). This curvature of the cross-section is known as anticlastic curvature and it is due to the Poisson effect.

If the radius of curvature of the beam is R, then the radius of curvature of the neutral axis of the beam due to anticlastic curvature is R/ν.

4.5 Composite beams

Composite beams occur in a number of different branches of engineering, and appear in the form of reinforced concrete beams, flitched beams, ship structures, glass-reinforced plastics, etc.

In the case of *reinforced concrete beams*, it is normal practice to reinforce the concrete with steel rods on the section of the beam where tensile stresses occur, leaving the unreinforced section of the beam to withstand compressive stresses, as shown in Figure 4.8. The reason for this practice is that, whereas concrete is strong in

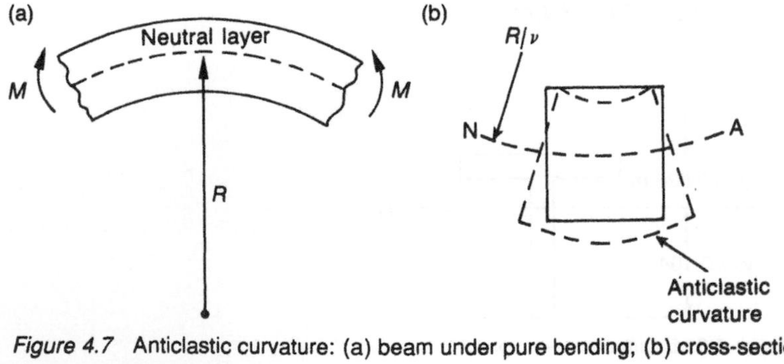

Figure 4.7 Anticlastic curvature: (a) beam under pure bending; (b) cross-section of beam

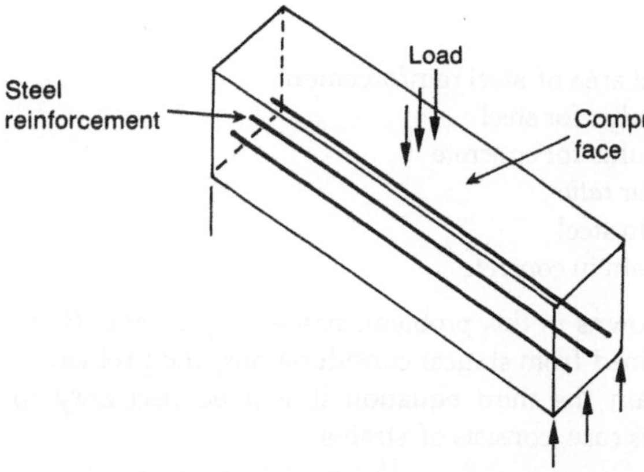

Figure 4.8 Reinforced concrete beam

compression, it is weak in tension, but because the elastic modulus of steel is about 15 times greater than concrete, the steel will absorb the vast majority of the load on the tensile side of the beam. Furthermore, the alkaline content of the concrete reacts with the rust on the steel, causing the rust to form a tight protective coating around the steel reinforcement, thereby preventing further rusting. Thus, steel and concrete form a mutually compatible pair of materials which, when used together, actually improve each other's performance.

The method of analyzing reinforced concrete beams of the type shown in Figure 4.8 is to assume that all the tensile load is taken by the steel reinforcement and that all the compressive load is taken by that part of the concrete above the neutral axis, so that the stress distribution will be as shown in Figure 4.9. In this figure

σ_s = tensile stress in steel
$\hat{\sigma}_c$ = maximum compressive stress in concrete
H = distance of the neutral axis of the beam from its top face
B = breadth of beam
D = distance between the steel reinforcement and the top of the beam

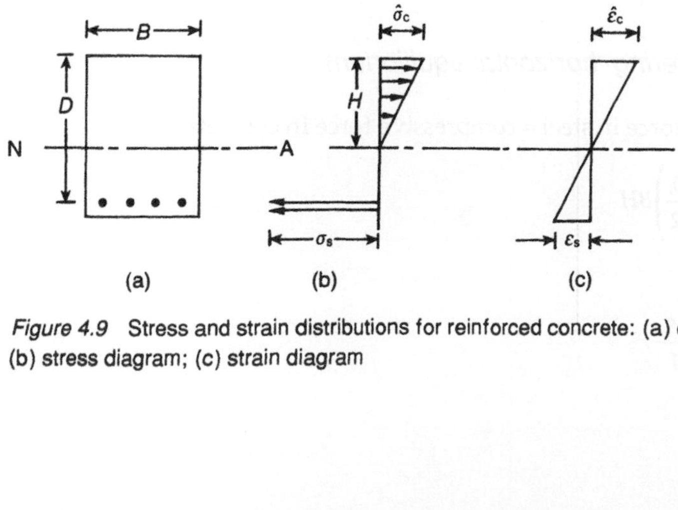

Figure 4.9 Stress and strain distributions for reinforced concrete: (a) cross-section; (b) stress diagram; (c) strain diagram

Let

A_s = cross-sectional area of steel reinforcement
E_s = Young's modulus for steel
E_c = Young's modulus for concrete
$m = E_s/E_c$ = *modular ratio*
ε_s = tensile strain in steel
$\hat{\varepsilon}_c$ = maximum strain in concrete

Now there are three unknowns in this problem, namely σ_s, $\hat{\sigma}_c$ and H, and as only two equations can be obtained from statical considerations, the problem is statically indeterminate, i.e. to obtain the third equation it will be necessary to consider compatibility, which, in this case, consists of strains.

4.5.1 *Compatibility considerations*

Consider similar triangles in the strain diagrams of Figure 4.9(c):

$$\frac{\hat{\varepsilon}_c}{H} = \frac{\varepsilon_s}{(D-H)}$$

or

$$\frac{\hat{\sigma}_c}{E_c H} = \frac{\sigma_s}{E_s(D-H)}$$

or

$$\hat{\sigma}_c = \frac{E_c}{E_s}\,\frac{H\sigma_s}{(D-H)}$$

Therefore

$$\hat{\sigma}_c = \frac{H\sigma_s}{m(D-H)} \tag{4.9}$$

4.5.2 *Considering 'horizontal equilibrium'*

Tensile force in steel = compressive force in concrete

$$\sigma_s A_s = \left(\frac{\hat{\sigma}_c}{2}\right)BH$$

or

$$\hat{\sigma}_c = \frac{2\sigma_s A_s}{BH} \tag{4.10}$$

4.5.3 Considering 'rotational equilibrium'

Externally applied moment M = moment of resistance of the section

or

$$M = \sigma_s A_s \times (D - H) + \left(\frac{\hat{\sigma}_c}{2}\right) \times BH \times \frac{2H}{3}$$

$$\underline{M = \sigma_s A_s (D - H) + \hat{\sigma}_c BH^2/3} \qquad (4.11)$$

Equating (4.9) and (4.10),

$$\frac{H\sigma_s}{m(D-H)} = \frac{2\sigma_s A_s}{BH}$$

or

$$BH^2 = 2m(D - H)A_s$$

Therefore

$$H^2 + 2mA_sH/B - 2mDA_s/B = 0$$

i.e.

$$H = \frac{-2mA_s/B + \sqrt{(2mA_s/B)^2 + 8mDA_s/B}}{2}$$

$$H = \sqrt{(mA_s/B)^2 + 2mDA_s/B} - mA_s/B \qquad (4.12)$$

Substituting equation (4.10) into (4.11),

$$M = \sigma_s A_s (D - H) + \frac{2\sigma_s A_s}{BH} \frac{BH^2}{3}$$

Therefore

$$\sigma_s = \frac{M}{A_s[(D-H) + 2H/3]} = \frac{M}{A_s(D - H/3)} \qquad (4.13)$$

and from equation (4.10)

$$\hat{\sigma}_c = \frac{2M}{BH(D - H/3)} \qquad (4.14)$$

EXAMPLE 4.3

A reinforced concrete beam, of rectangular section, is subjected to a bending moment, such that the steel reinforcement is in tension.

Given the following, determine the maximum permissible value of this bending moment.

$$D = 0.4\text{ m} \quad B = 0.3\text{ m} \quad m = 15$$

Maximum permissible compressive stress in concrete = 10 MN/m^2
Maximum permissible tensile stress in steel = 150 MN/m^2
Diameter of each steel reinforcing rod = 2 cm
n = number of rods = 8

SOLUTION

$$A_s = 2.513 \times 10^{-3}\text{ m}^2$$

From equation (4.12),

$$H = \sqrt{0.0158 + 0.1005} - 0.125\,65$$

$$\underline{H = 0.215\text{ m}}$$

From equation (4.13),

$$M = \sigma_s A_s (D - H/3)$$

$$\underline{M = 0.124\text{ MN m}}$$

From equation (4.14),

$$M = \hat{\sigma}_c BH(D - H/3)/2$$

$$\underline{M = 0.106\text{ MN m}}$$

i.e. the maximum stress in the concrete is the design criterion. Therefore

$$\underline{\text{Maximum permissible bending moment} = 0.106\text{ MN m}}$$

NB The overall dimensions of the beam's cross-section should allow for the steel reinforcement to be covered by at least 5 cm of concrete.

4.6 Flitched beams

A flitched beam is a common type of composite beam, where the reinforcements are relatively thin compared with the depth of the beam, and are usually attached to its outer surfaces, as shown in Figure 4.10. Typical materials used for flitched beams

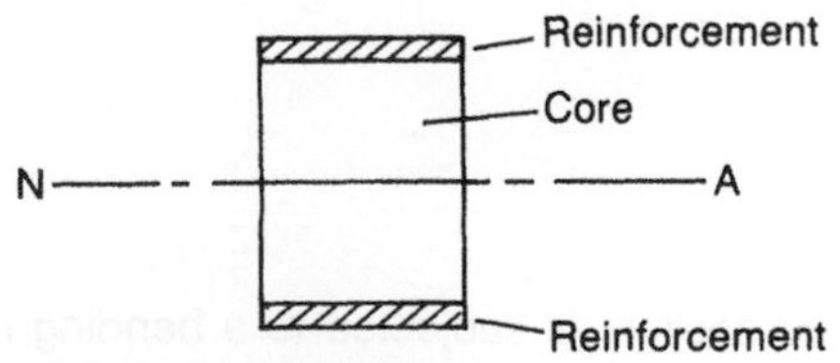

Figure 4.10 Cross-section of beam with horizontal reinforcement

include a wooden core combined with external steel reinforcement and various types of plastic reinforcement combined with a synthetic porous core of low density.

Let

M = applied moment at the section
M_r = moment of resistance of external reinforcement
M_c = moment of resistance of core

so that

$$M = M_r + M_c \tag{4.15}$$

The main assumption made is that the *radius of curvature, R, is the same for the core as it is for the reinforcement*, i.e.

$$R = \frac{E_c I_c}{M_c} = \frac{E_r I_r}{M_r}$$

or

$$M_r = \frac{E_r I_r}{E_c I_c} M_c \tag{4.16}$$

where

I_r = second moment of area of the external reinforcement about the neutral axis of the composite beam
I_c = second moment of area of the core about the neutral axis of the composite beam
E_r = Young's modulus for the external reinforcement
E_c = Young's modulus for the core

Substituting equation (4.16) into (4.15),

$$M = \frac{M_c}{E_c I_c}(E_r I_r + E_c I_c)$$

Therefore

$$M_c = \frac{E_c I_c M}{(E_r I_r + E_c I_c)} \tag{4.17}$$

and,

$$M_r = \frac{E_r I_r M}{(E_r I_r + E_c I_c)} \tag{4.18}$$

Now,

$$\sigma_r = M_r / Z_r$$

and

$$\sigma_c = M_c / Z_c$$

where

σ_r = maximum stress in the external reinforcement
σ_c = maximum stress in the core
Z_r = sectional modulus of the external reinforcement about NA
Z_c = sectional modulus of the core

Hence,

$$\sigma_r = E_r y_r M/(E_r I_r + E_c I_c) \tag{4.19}$$

and

$$\sigma_c = E_c y_c M/(E_r I_r + E_c I_c) \tag{4.20}$$

where

y_r = distance of the outermost fiber of the external reinforcement from the neutral axis of the composite beam
y_c = distance of the outermost fiber of the core from the neutral axis of the composite beam

EXAMPLE 4.4

(a) A wooden beam of rectangular section is of depth 10 cm and width 5 cm. Determine the moment of resistance of this section given the following:

Young's modulus for wood = 1.4×10^{10} N/m^2
Maximum permissible stress in wood = 20 MN/m^2

(b) What percentage increase will there be in the moment of resistance of the beam section if it is reinforced by two 5 mm thick galvanized steel plates attached to the top and bottom surfaces of the beam?

Young's modulus for steel = 2×10^{11} N/m^2
Maximum permissible stress in steel = 150 MN/m^2

SOLUTION

(a) Now,

$$M = \frac{\sigma I}{y}$$

I_w = second moment of area of wood about its neutral axis

$= 5 \times 10^{-2} \times (10 \times 10^{-2})^3/12$

$I_w = 4.167 \times 10^{-6}\text{m}^4$

$\bar{y} = 5 \times 10^{-2}$ m

Therefore

M_w = moment of resistance of wood

$$= \frac{20 \times 10^6 \times 4.167 \times 10^{-6}}{5 \times 10^{-2}}$$

$\underline{M_w = 1666.7 \text{ N m}}$

(b)

R_w = radius of curvature of wood

R_s = radius of curvature of steel

but

$$R_s = R_w$$

Therefore

$$\frac{\sigma_s}{E_s y_s} = \frac{\sigma_w}{E_w y_w}$$

or

$$\sigma_s = \sigma_w \frac{E_s y_s}{E_w y_w}$$

$$= \frac{\sigma_w \times 2 \times 10^{11} \times 5.25 \times 10^{-2}}{1.4 \times 10^{10} \times 5 \times 10^{-2}}$$

$\underline{\sigma_s = 15\sigma_w}$

i.e.

$\underline{\sigma_s = 150 \text{ MN/m}^2 \text{ is the design criterion}}$

where

σ_s = maximum stress in steel
E_s = elastic modulus of steel
E_w = elastic modulus of wood
y_s = distance of steel from NA
y_w = distance of outermost fiber of wooden core from NA

Now, as

$$\sigma_s = 15\sigma_w$$

$$\sigma_w = 150/15 = 10 \text{ MN/m}^2$$

then

$$M_w = 1666.7 \times 10/20 = 833.4 \text{ N m}$$

$$I_s = 5 \times 10^{-2} \times 5 \times 10^{-3} \times (5.25 \times 10^{-2})^2 \times 2$$

$$= 1.378 \times 10^{-6}$$

and

$$M_s = \frac{150 \times 10^6 \times 1.378 \times 10^{-6}}{5.25 \times 10^{-2}} = \underline{3937.5 \text{ N m}}$$

$$M = M_s + M_w = \underline{4770.9 \text{ N m}}$$

i.e. the percentage increase in moment of resistance of the flitched beam over the wooden beam is 186.2.

NB In this case, the chosen thickness for the steel plate was too small, as the stress in the wood was well below its permissible value.

Another type of horizontal flitched beam is shown in Figure 4.11, where the reinforcement is vertical.

In this case, the composite beam can be regarded as an equivalent wooden beam, as shown in Figure 4.12(b), or as an equivalent steel beam, as shown in Figure 4.12(c).

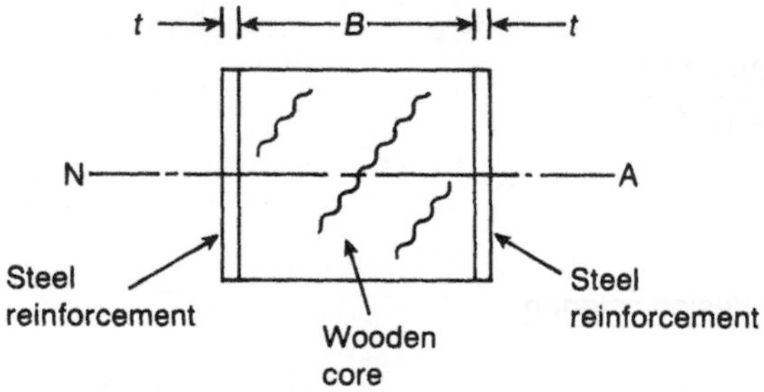

Figure 4.11 Cross-section of a horizontal beam with vertical reinforcement

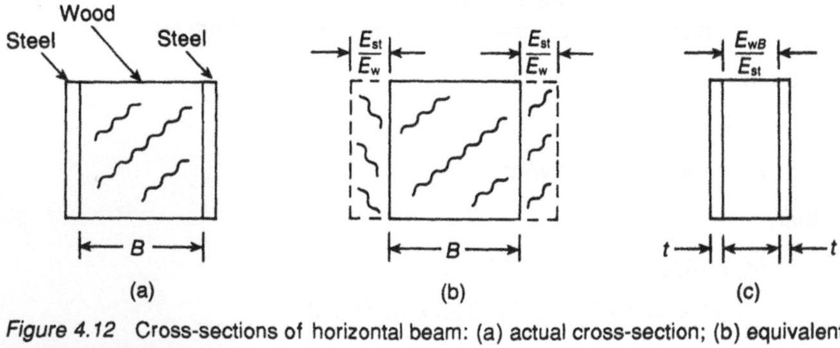

Figure 4.12 Cross-sections of horizontal beam: (a) actual cross-section; (b) equivalent wooden cross-section; (c) equivalent steel cross-section

If

I_s = total second moment of area of the two steel beam reinforcements about NA
I_w = second moment of area of the wooden core about NA
M_s = total bending moment of resistance of the two steel reinforcements
M_w = bending moment of resistance of the wooden core

then from equation (4.1),

$$M_s = \frac{E_s I_s}{R} \tag{4.21}$$

and

$$M_w = \frac{E_w I_w}{R} \tag{4.22}$$

The total bending moment of resistance of the composite beam is

$$M = M_s + M_w$$

Therefore

$$M = \frac{1}{R}(E_s I_s + E_w I_w) \tag{4.23}$$

where

E_s = Young's modulus for steel reinforcement
E_w = Young's modulus for wooden core

Let E_I be the bending stiffness of the composite beam. Then

$$E_I = E_s I_s + E_w I_w \tag{4.24}$$

From equation (4.23),

$$\frac{1}{R} = \frac{M}{(E_s I_s + E_w I_w)} \tag{4.25}$$

and from equation (4.21),

$$\frac{1}{R} = \frac{M_s}{E_s I_s} \tag{4.26}$$

Equating (4.25) and (4.26)

$$M_s = \frac{E_s I_s M}{(E_s I_s + E_w I_w)}$$

$$= \frac{M}{(1 + E_w I_w / E_s I_s)} \tag{4.27}$$

From equation (4.1)

$$\sigma_s = \frac{M_s y}{I_s} \tag{4.28}$$

Substituting equation (4.27) into (4.28),

$$\sigma_s = \frac{My}{I_s(1 + E_w I_w / E_s I_s)} = \frac{My}{(I_s + E_w I_w / E_s)} \tag{4.29}$$

Now the bending strain on any fiber that is a distance y from NA is

$$\varepsilon = \frac{\sigma_s}{E_s} \tag{4.30}$$

or

$$\varepsilon = \frac{My}{(E_s I_s + E_w I_w)} \tag{4.31}$$

but the strain will be the same in both the steel reinforcement and the wooden core at distance y from NA, so that $\sigma_w = E_w \varepsilon$. Therefore

$$\sigma_w = \frac{My}{(E_s I_s / E_w + I_w)} \tag{4.32}$$

From equations (4.29) and (4.32), it can be seen that the denominators are equivalent to an equivalent steel second moment of area for the first equation, and to an equivalent wooden second moment of area for the second equation, as shown in Figures 4.12(c) and (b), respectively.

4.7 Composite ship structures

Composite ship structures appear in the form of a steel hull, together with an aluminium-alloy superstructure. The reason for this combination is that steel is a suitable material for the main hull of a ship because of its ductility and good welding properties, but in order to keep the centre of gravity of a ship as low as possible, for the purposes of ship stability, it is convenient to use a material with a lower density than steel for the superstructure. In general, it is not suitable to use aluminium for the main hull, because aluminium has poor corrosion resistance to salt water.

Under longitudinal bending moments, caused by the self-weight of the ship and buoyant forces due to waves, as shown in Figure 4.13, the longitudinal strength of a

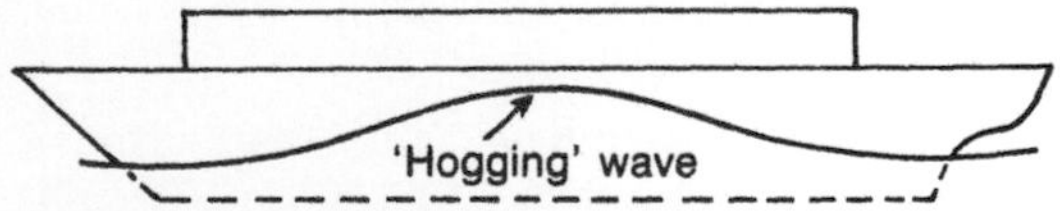

Figure 4.13 Longitudinal bending of a ship

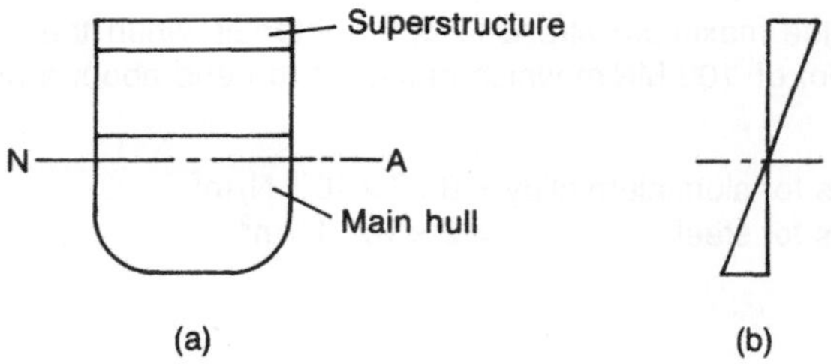

Figure 4.14 Cross-section of a ship: (a) transverse section; (b) strain distribution

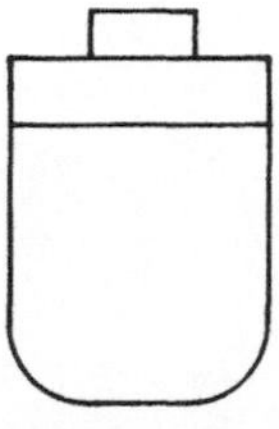

Figure 4.15 Equivalent section

ship can be based on beam theory, where the cross-section of the 'equivalent beam' is in fact the cross-section of the ship, as shown in Figure 4.15. The strain distribution across the transverse section of the ship is as shown in Figure 4.14(b), but as stress = $E \times$ strain, the equivalent moment of resistance of the aluminium alloy will be equivalent to E_a/E_s of a steel section of the same size. Thus, to calculate the position of the neutral axis (NA) and the second moment of area, the aluminium-alloy superstructure can be assumed to be equivalent to the form shown in Figure 4.15, where

E_a = Young's modulus for aluminium alloy
E_s = Young's modulus for steel

EXAMPLE 4.5

A box-like cross-section consists of two parts, namely a steel bottom and an aluminium-alloy top, as shown in Figure 4.16. If the plate thickness of both the aluminium alloy and

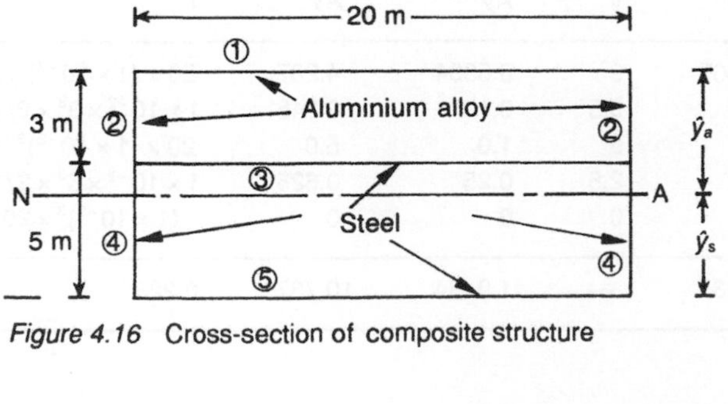

Figure 4.16 Cross-section of composite structure

the steel is 1 cm, determine the maximum stress in both materials when the section is subjected to a bending moment of 100 MN m which causes it to bend about a horizontal plane (NA).

E_a = Young's modulus for aluminium alloy = 6.67×10^{10} N/m²
E_s = Young's modulus for steel = 2×10^{11} N/m²

SOLUTION

To determine the position of the neutral axis and the second moment of area of the equivalent steel section, use will be made of Table 4.1:

$$\hat{y}_s = \sum Ay / \sum A = 1.9134/0.5867$$

$$\underline{\hat{y}_s = 3.261 \text{ m}}$$

Therefore

$$\underline{\hat{y}_a = 4.739 \text{ m}}$$

$$I_{XX} = \sum Ay^2 + \sum i = 10.737 + 0.223$$

$$\underline{I_{XX} = 10.96 \text{ m}^4}$$

$$I_{NA} = I_{XX} - (\hat{y}_s)^2 \sum A = 10.96 - 3.261^2 \times 0.5867$$

$$\underline{I_{NA} = 4.721 \text{ m}^4}$$

$\hat{\sigma}_a$ = maximum stress in the aluminium alloy

$$= \frac{M\hat{y}_a}{I_{NA}} \times \left(\frac{E_a}{E_s}\right) = \frac{100 \times 4.739}{4.721 \times 3}$$

$$\underline{\hat{\sigma}_a = 33.46 \text{ MN/m}^2}$$

$\hat{\sigma}_s$ = maximum stress in the steel

$$= \frac{M\hat{y}_s}{I_{NA}} = \frac{100 \times 3.261}{4.721}$$

$$\underline{\hat{\sigma}_s = 69.08 \text{ MN/m}^2}$$

Table 4.1

Section	A	y	Ay	Ay^2	i
①	0.0667	8	0.5334	4.267	$20 \times (1 \times 10^{-2})^3/36 = 0$
②	0.02	6.5	0.13	0.845	$1 \times 10^{-2} \times 3^3 \times 2/36 = 0.015$
③	0.2	5	1.0	5.0	$20 \times (1 \times 10^{-2})^3/12 = 0$
④	0.1	2.5	0.25	0.625	$1 \times 10^{-2} \times 5^3 \times 2/12 = 0.208$
⑤	0.2	0	0	0	$(1 \times 10^{-2})^3 \times 20/3 = 0$
Σ	0.5867	—	1.9134	10.737	0.223

It can be seen from the above calculations that, despite the fact that the aluminium-alloy deck is further away from NA than is the steel bottom, the stress in the aluminium alloy is less than in the steel, because its elasticity is three times greater than that of the steel.

4.8 Composite structures

The use of composites is of much interest in structures varying from car bodies to boat hulls and from chairs to ship superstructures. Composites appear in many and various forms from glass-reinforced plastics (GRP) and carbon-fiber-reinforced plastics (CFRP) to metal matrix composites (MMC), etc. Analysis of such structures is beyond the scope of the present book and for further study the reader should consult references 2 and 3.

This chapter has shown that the sensible use of composites can improve the structural efficiency of many types of structure in various engineering applications.

4.9 Combined bending and direct stress

The case of combined bending and direct stress occurs in a number of engineering situations, including the eccentric loading of short columns, as shown in Figure 4.17. By placing the equal and opposite forces on the centre line of the strut, as shown in Figure 4.17(b), the loading condition of Figure 4.17(a) is unaltered. However, it can be seen that the column of Figure 4.17(b) is in fact subjected to a centrally applied force P and a couple $P\Delta$, as shown in Figure 4.17(c). Furthermore, from Figure 4.17(c), it can be seen that owing to the centrally applied direct load P, the whole of the strut will be subjected to a direct compressive stress σ_d, but owing to the couple $P\Delta$, the side AB will be in tension and the side CD will be in compression.

Thus, the effect of M will be to cause the stress to be increased in magnitude on

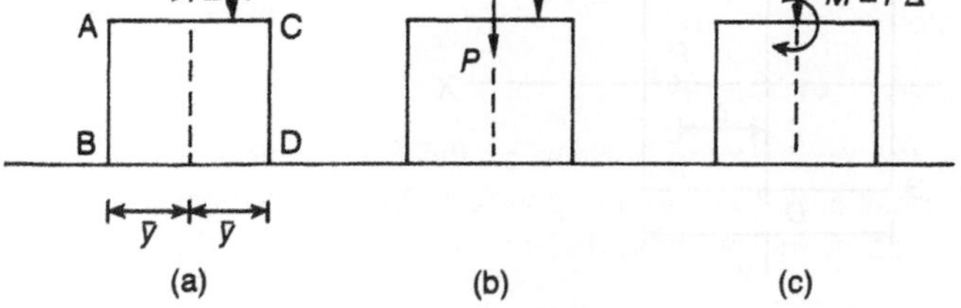

Figure 4.17 Eccentrically loaded short column

face CD, and to be decreased in magnitude on face AB, i.e.

$$\text{Stress on face AB} = \sigma_{AB} = -\frac{P}{A} + \frac{M\bar{y}}{I} \tag{4.33}$$

$$\text{Stress on face CD} = \sigma_{CD} = -\frac{P}{A} - \frac{M\bar{y}}{I} \tag{4.34}$$

It is evident, therefore, that in general the stress due to the combined effects of a bending moment M and a tensile load P will be given by

$$\sigma = \sigma_d \pm \sigma_b$$

where

σ_d = direct stress (tensile is positive)
σ_b = bending stress

4.9.1 Eccentrically loaded concrete columns

As concrete is weak in tension, it is desirable to determine how eccentric a load can be so that no part of a short column is in tension.

The next two examples will be used to demonstrate the calculations usually associated with eccentrically loaded short columns.

EXAMPLE 4.6

Determine the position in which an eccentrically applied vertical compressive load can be placed, so that no tension occurs in a short vertical column of square cross-section.

SOLUTION

By applying a compressive force P at the point shown in Figure 4.18 it can be seen that the face AB is likely to develop tensile stresses due to the bending action about the *YY* axis.

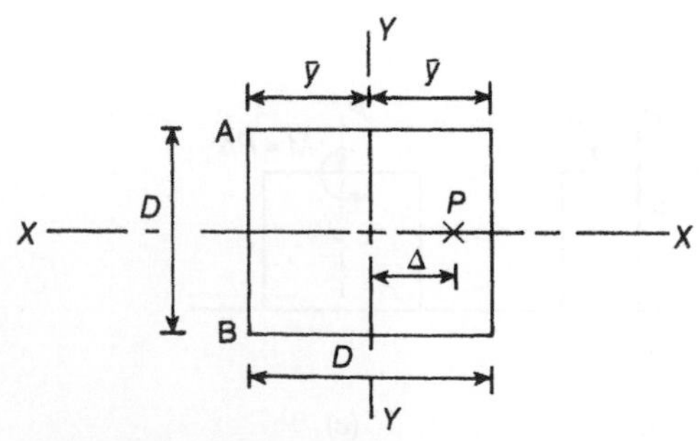

Figure 4.18 Cross-section of concrete column

To satisfy the requirements of the example, let the stress on the face AB equal zero, so that

$$0 = -\frac{P}{A} + \frac{P\Delta\bar{y}}{I}$$

$$= -\frac{1}{D^2} + \frac{6\Delta}{D^3}$$

Therefore

$$\underline{\Delta = D/6} \tag{4.35}$$

From equation (4.35), it can be seen that for no tension to occur in the short column, owing to an eccentrically applied compressive load, the eccentrically applied load must be applied within the mid-third area of the centre of the square section. For this reason, this rule is known as the *mid-third rule.*

EXAMPLE 4.7

Determine the position in which an eccentrically applied vertical compressive load can be placed, so that no tension occurs in a short vertical column of circular cross-section.

SOLUTION

By applying a compressive force P to the point shown in Figure 4.19, it can be seen that tension is most likely to occur at the point C, so that to satisfy the requirements of the example, the stress at C equals zero.

$$0 = -\frac{4P}{\pi d^2} + 32\,\frac{P\Delta}{\pi d^3}$$

or

$$\underline{\Delta = d/8} \tag{4.36}$$

Equation (4.36) shows that for no tensile stress to occur in a short concrete column of circular cross-section, the eccentricity of the load must not exceed $d/8$. For this reason, this is known as the *mid-quarter rule.*

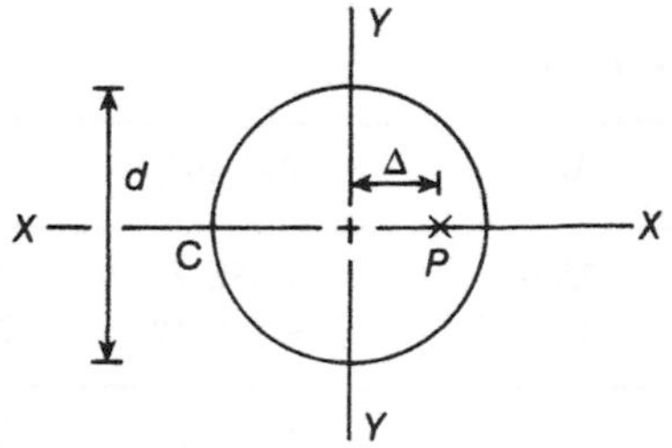

Figure 4.19 Cross-section of concrete column

EXAMPLE 4.8

Determine the maximum tensile and compressive stresses in the clamp of Figure 4.20.

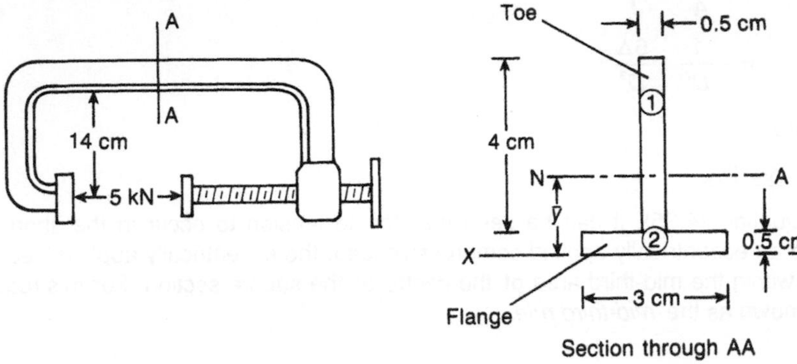

Figure 4.20 Clamp under loading

SOLUTION

The relevant geometrical properties of the tee-bar are calculated with the aid of Table 4.2:

$$\bar{y} = \frac{\sum Ay}{\sum A} = \frac{5.375\text{E-}6}{3.5\text{E-}4} = \underline{0.0154 \text{ m}}$$

$$I_{XX} = \sum Ay^2 + \sum i_0 = 1.259\text{E-}7 + 2.7\text{E-}8 = \underline{1.529\text{E-}7 \text{ m}^4}$$

$$I_{NA} = I_{XX} - \bar{y}^2 \sum A = 1.529\text{E-}7 - 0.0154^2 \times 3.5\text{E-}4$$

$$\underline{I_{NA} = 6.989\text{E-}8 \text{ m}^4}$$

Let

$\hat{M}$ = the maximum bending moment on the 'top' member of the clamp (or at AA)

$$= 5 \text{ kN} \times (14 \times 10^{-2} \text{ m} + \bar{y})$$

$$= 5 \text{ kN} \times (14 \times 10^{-2} + 0.0154)$$

$$\hat{M} = 0.777 \text{ kN m}$$

Table 4.2 **Geometrical properties of the tee-bar**

Section	A (m²)	y (m)	Ay (m³)	Ay^2 (m⁴)	i_0(m⁴)
①	2E-4	2.5E-2	5E-6	1.25E-7	2.667E-8
②	1.5E-4	0.25E-2	3.75E-7	9.38E-10	3.125E-10
Σ	3.5E-4	—	5.375E-6	1.259E-7	2.7E-8

σ_T = the stress in the 'top' (toe) of the tee

$$= -\frac{0.777 \text{ kN m} \times 0.0296 \text{ m}}{6.989\text{E-8 m}^4} + \frac{5 \text{ kN}}{A}$$

$$= -329\,077 \frac{\text{kN}}{\text{m}^2} + \frac{5 \text{ kN}}{3.5\text{E-4 m}^2}$$

$$= -329\,077 + 14\,286 = -314\,791 \text{ kN/m}^2$$

$\underline{\sigma_T = -314.8 \text{ MN/m}^2}$

σ_B = the maximum stress in the 'bottom' (flange) of the tee

$$= \frac{0.777 \times 0.0154}{6.989\text{E-8}} + 14\,286 = 171\,209 + 14\,286$$

$$= 185\,495 \text{ kN/m}^2$$

σ_B = 185.5 MN/m²

NB By placing the flange of the tee-bar on the inner part of the clamp, the maximum stresses have been reduced. This has been achieved by 'lowering' the neutral axis towards the tensile face, where the bending and direct stresses are additive. Movement of the neutral axis towards the tensile face also had the effect of lowering the maximum bending moment, as the lever arm that the load was acting on was decreased.

EXAMPLES FOR PRACTICE 4

1. A concrete beam of uniform square cross-section, as shown in Figure 4.21 is to be lifted by its ends, so that it may be regarded as being equivalent to a horizontal beam, simply supported at its ends and subjected to a uniformly distributed load due to its self-weight.

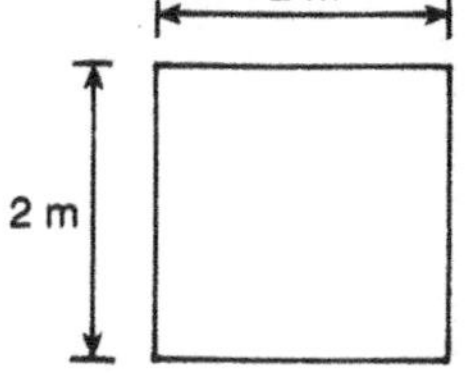

Figure 4.21 Cross-section of concrete beam

Determine the maximum permissible length of this beam, given the following:

Density of concrete = 2400 kg/m³

Maximum permissible tensile stress in the concrete = 1 MN/m²

g = 9.81 m/s²

{10.64 m}

2. If the concrete beam of Example 1 had a hole in the bottom of its cross-section, as shown in Figure 4.22, what would be the maximum permissible length of the beam?

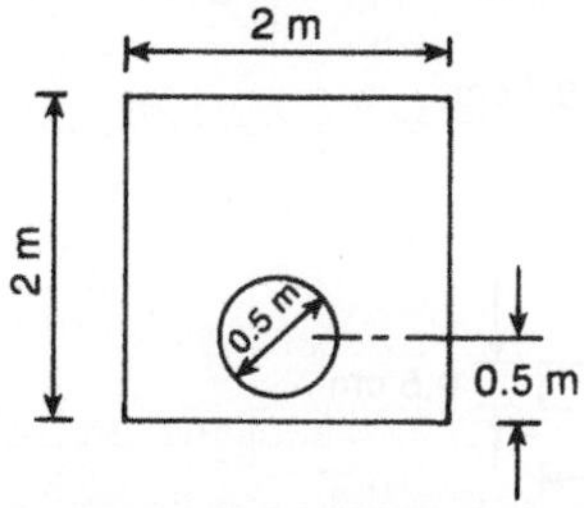

Figure 4.22 Cross-section with hole

{10.55 m}

3. What would be the maximum permissible length of the beam, if the hole were at the top?

{10.83 m}

4. A horizontal beam, of length 4 m, is simply supported at its ends and subjected to a vertically applied concentrated load of 10 kN at mid-span. Assuming that the width of the beam is constant and equal to 0.03 m, and neglecting the self-weight of the beam, determine an equation for the depth of the beam, so that the beam will be of uniform strength. The maximum permissible stress in the beam is 100 MN/m².

{$d = 0.1x^{1/2}$ from 0 to 2 m}

5. A short steel column, of circular cross-section, has an external diameter of 0.4 m and a wall thickness of 0.1 m and carries a compressive but axially applied eccentric load. Two linear strain gauges, which are mounted longitudinally at opposite sides on the external surface of the column but in the plane of the load, record strains of 50×10^{-6} and -200×10^{-6}. Determine the magnitude and eccentricity of the axial load.
$E = 2 \times 10^{11}$ N/m²

{−1.413 MN, 0.104 m}

6. Determine the maximum tensile and compressive stresses in the clamp of Figure 4.23.

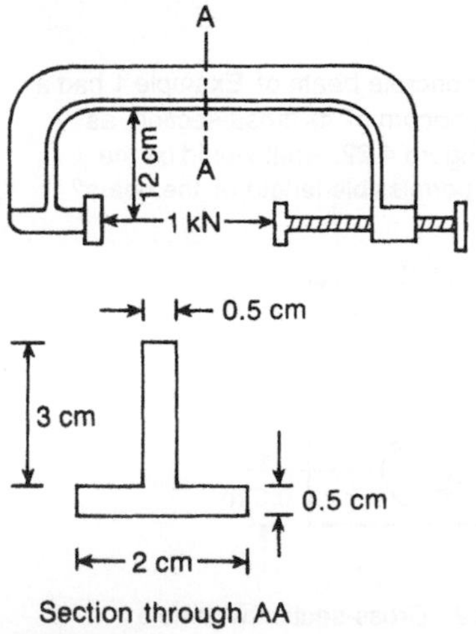

Figure 4.23

{61.9 MN/m², −94 MN/m²}

7. The cross-section of a reinforced concrete beam is as shown in Figure 4.24.

Determine the maximum bending moment that this beam can sustain, assuming that the steel reinforcement is on the tensile side and that the following apply:

Maximum permissible compressive stress in concrete = 10 MN/m²

Maximum permissible tensile stress in steel = 200 MN/m²

Modular ratio = 15

Diameter of a steel rod = 2 cm

n = number of steel rods = 6

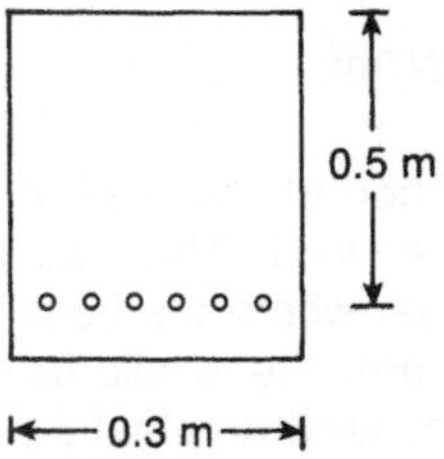

Figure 4.24

{0.144 MN m}

5 Beam deflections due to bending

5.1 Introduction

Beam deflections are usually due to bending and shear, but only those due to the former will be considered in this chapter.

The radius of curvature R of a beam in terms of its deflection y, at a distance x along the length of the beam, is given by

$$\frac{1}{R} = \frac{d^2y/dx^2}{[1 + (dy/dx)^2]^{3/2}} \tag{5.1}$$

However, if the deflections are small, as is the usual requirement in structural design, then $(dy/dx)^2$ is negligible compared with unity, so that equation (5.1) can be approximated by

$$\frac{1}{R} = \frac{d^2y}{dx^2} \tag{5.2}$$

but

$$\frac{M}{I} = \frac{E}{R}$$

Therefore

$$EI\frac{d^2y}{dx^2} = M \tag{5.3}$$

Equation (5.3) is a very important expression for the bending of beams, and there are a number of different ways of solving it, but only three methods will be considered here, namely repeated integration, moment–area methods and use of the slope–deflection equations.

5.2 Repeated integration method

This is a boundary value method which depends on integrating equation (5.3) twice and then substituting boundary conditions to determine the arbitrary constants, together with other unknowns. It will be demonstrated through detailed solutions of the following section of examples.

EXAMPLE 5.1

Determine an expression for the maximum deflection of a cantilever of uniform section, under a concentrated load at its free end, as shown in Figure 5.1.

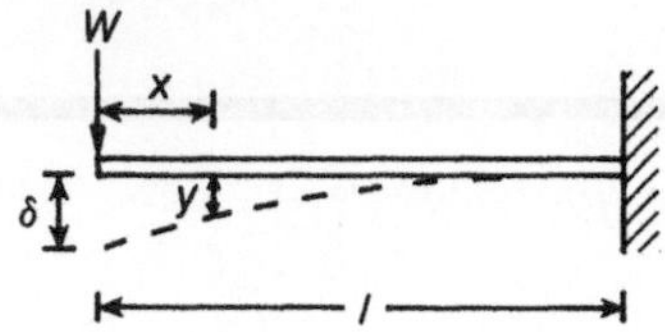

Figure 5.1 Cantilever with end load

SOLUTION

$$EI\frac{d^2y}{dx^2} = M$$

$$= -Wx$$

$$EI\frac{dy}{dx} = -\frac{Wx^2}{2} + A \tag{5.4}$$

$$EIy = -\frac{Wx^3}{6} + Ax + B \tag{5.5}$$

There are two unknowns, namely the arbitrary constants A and B; therefore, two boundary conditions will be required, as follows:

$$\text{At } x = l \quad \frac{dy}{dx} = 0$$

(i.e. the slope at the built-in end is zero). Hence from equation (5.4),

$$0 = -\frac{Wl^2}{2} + A$$

Therefore

$$A = Wl^2/2 \tag{5.6}$$

The other boundary condition is that

$$\text{At } x = l \quad y = 0$$

(i.e. the deflection y is zero at the built-in end). Hence from equation (5.5),

$$0 = -\frac{Wl^3}{6} + Al + B$$

or

$$\underline{B = -Wl^3/3} \qquad (5.7)$$

Substituting equations (5.6) and (5.7) into equation (5.5), the deflection y at any distance x along the length of the beam is given by

$$\underline{y = -\frac{W}{EI}(x^3/6 - l^2x/2 + l^3/3)} \qquad (5.8)$$

By inspection, the maximum deflection δ occurs at $x = 0$, i.e.

$$\delta = -\frac{Wl^3}{3EI} \qquad (5.9)$$

The negative sign denotes that the deflection is downward.

EXAMPLE 5.2

Determine an expression for the maximum deflection of a cantilever under a uniformly distributed load w, as shown in Figure 5.2.

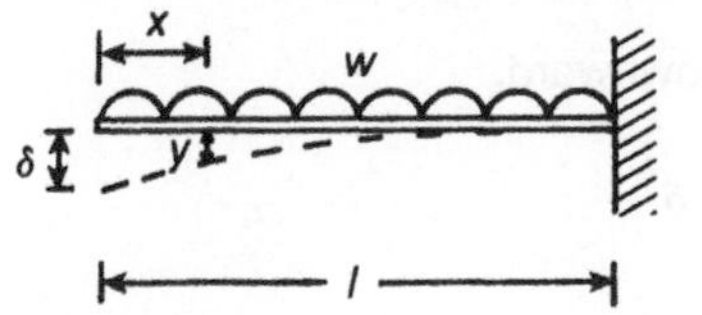

Figure 5.2 Cantilever with a UDL

SOLUTION

At any distance x along the length of the beam, the bending moment M is given by

$$M = -wx^2/2$$

so that

$$EI\frac{d^2y}{dx^2} = -\frac{wx^2}{2} \qquad (5.10)$$

$$EI\frac{dy}{dx} = -\frac{wx^3}{6} + A$$

$$EIy = -\frac{wx^4}{24} + Ax + B \qquad (5.11)$$

There are two unknowns, namely A and B, and therefore two boundary conditions will be required, as follows:

$$\text{At } x = l \quad \frac{dy}{dx} = 0$$

Therefore

$$A = wl^3/6 \tag{5.12}$$

$$\text{At } x = l \quad y = 0$$

or

$$0 = -\frac{wl^4}{24} + \frac{wl^4}{6} + B$$

Therefore

$$B = -wl^4/8 \tag{5.13}$$

Substituting equations (5.12) and (5.13) into equation (5.11), the following expression is obtained for the deflection y of the beam at any distance x from its free end:

$$\underline{y = -\frac{w}{EI}(x^4/24 - l^3x/6 + l^4/8)} \tag{5.14}$$

By inspection, the maximum deflections δ occurs at $x = 0$, where

$$\underline{\delta = -wl^4/(8EI)} \tag{5.15}$$

The negative sign denotes that the deflection is downward.

Alternative method for determining δ

From Section 1.11

$$\frac{d^2M}{dx^2} = w$$

but

$$EI\frac{d^2y}{dx^2} = M$$

Therefore

$$EI\frac{d^4y}{dx^4} = w \tag{5.16}$$

In this case, w, *is downward*; therefore equation (5.16) becomes

$$EI\frac{d^4y}{dx^4} = -w$$

$$EI\frac{d^3y}{dx^3} = F \text{ (the shearing force)} = -wx + A$$

$$EI\frac{d^2y}{dx^2} = M = -\frac{wx^2}{2} + Ax + B$$

At $x = 0$ $F = 0$; therefore $A = 0$

At $x = 0$ $M = 0$; therefore $B = 0$

i.e.

$$EI\frac{d^2y}{dx^2} = -\frac{wx^2}{2} \tag{5.17}$$

which is identical to equation (5.10).

Equation (5.16) is particularly useful for beams with distributed loads of complex form, such as those met in determining the longitudinal strengths of ships, owing to the combined effects of self-weight and buoyant forces caused by waves.

EXAMPLE 5.3

Determine an expression for the maximum deflection of a uniform-section beam, simply supported at its ends and subjected to a centrally placed concentrated load W, as shown in Figure 5.3.

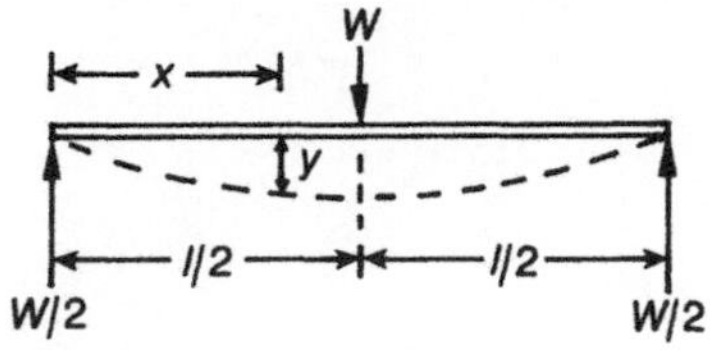

Figure 5.3

SOLUTION

In this case, there is a discontinuity for bending moment distribution at mid-span, hence, equation (5.3), together with its boundary conditions, can only be applied between $x = 0$ and $x = l/2$, i.e.

$$EI\frac{d^2y}{dx^2} = M = \frac{W}{2}x$$

$$EI\frac{dy}{dx} = \frac{Wx^2}{4} + A$$

$$EIy = \frac{Wx^3}{12} + Ax + B \tag{5.18}$$

At $x = 0$ $y = 0$; therefore $0 = B$

At $x = l/2$ $\frac{dy}{dx} = 0$; therefore $\underline{A = -Wl^2/16}$

i.e.

$$y = \frac{W}{EI}(x^3/12 - l^2x/16)$$

By inspection, the maximum deflection δ occurs at $x = l/2$, where

$$\delta = \frac{-Wl^3}{48EI} \tag{5.19}$$

EXAMPLE 5.4

Determine an expression for the maximum deflection of a uniform-section beam, simply supported at its ends and subjected to a uniformly distributed load w, as shown in Figure 5.4.

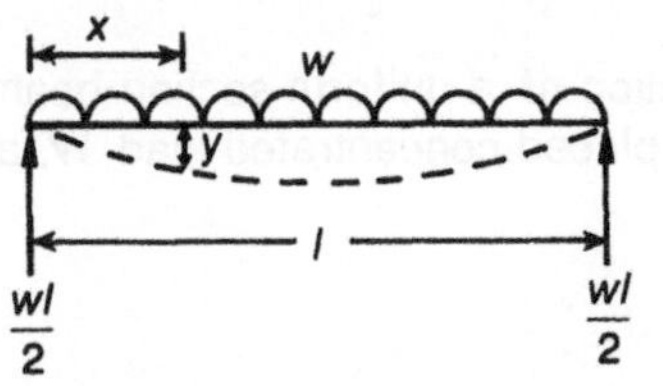

Figure 5.4

SOLUTION

$$\text{At } x \quad M = \frac{wlx}{2} - \frac{wx^2}{2}$$

so that

$$EI\frac{d^2y}{dx^2} = \frac{wl}{2}x - \frac{wx^2}{2} \tag{5.20}$$

$$EI\frac{dy}{dx} = \frac{wlx^2}{4} - \frac{wx^3}{6} + A$$

$$EIy = \frac{wlx^3}{12} - \frac{wx^4}{24} + Ax + B$$

At $x = 0$ $y = 0$; therefore $\underline{B = 0}$

At $x = l$ $y = 0$; therefore $\underline{A = -wl^3/24}$

i.e.

$$y = \frac{w}{EI}(lx^3/12 - x^4/24 - l^3x/24) \tag{5.21}$$

By inspection, δ occurs at $x = l/2$, where

$$\delta = -\frac{5wl^4}{384EI}$$

5.2.1 Alternative method of solving Example 5.4

From equation (5.16),

$$EI\frac{d^4y}{dx^4} = -w$$

On integrating once,

$$EI\frac{d^3y}{dx^3} = F = -wx + A$$

At $x = 0$ $F = wl/2$; therefore $\underline{A = wl/2}$

or

$$EI\frac{d^3y}{dx^3} = -wx + \frac{wl}{2}$$

$$EI\frac{d^2y}{dx^2} = M = -\frac{wx^2}{2} + \frac{wlx}{2} + B$$

At $x = 0$ $M = 0$; therefore $\underline{B = 0}$

Therefore

$$EI\frac{d^2y}{dx^2} = -\frac{wx^2}{2} + \frac{wlx}{2} \qquad (5.22)$$

which is identical to equation (5.20).

5.3 Macaulay's method

This method will be given without proof, as a number of proofs already exist in numerous texts [3,4]. In the case of Example 5.3, it can be seen that the equation for M only applied between $x = 0$ and $x = l/2$, and that both boundary conditions had to be applied within these limits. Furthermore, because of symmetry, the boundary condition $dy/dx = 0$ at $x = l/2$ also applied.

However, if the beam were not symmetrically loaded, it would not be possible to obtain the second boundary condition, namely $dy/dx = 0$ at $x = l/2$.

To demonstrate how to overcome this problem, the following examples will be considered, which are based on Macaulay's method.

EXAMPLE 5.5

Determine an expression for the deflection under the load W for the uniform-section beam of Figure 5.5.

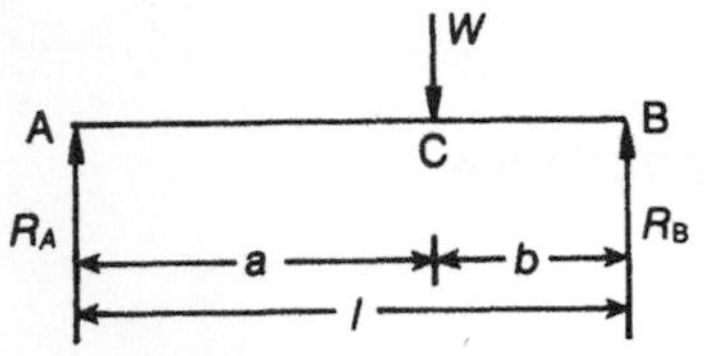

Figure 5.5

SOLUTION

First, it will be necessary to determine the value of R_A, which can be obtained by taking moments about the point B:

$$R_A l = Wb$$

Therefore

$$\underline{R_A = Wb/l}$$

Macaulay's method is to use separate bending moment equations for each section of the beam, but to integrate the equations via the Macaulay brackets, so that the constants of integration apply to all sections of the beam. It must, however, be emphasized that if the term within the *Macaulay bracket is negative*, then *that part of the expression does not apply* for boundary conditions, etc.

For the present problem, the bending moment between the points A and C will be different to that between the points C and B; hence, it will be necessary to separate the two expressions by the dashed line, as shown below.

$$EI\frac{d^2y}{dx^2} = R_A x \quad \Big| \quad -W[x-a]$$

$$EI\frac{d^2y}{dx^2} = \frac{Wbx}{l} \quad \Big| \quad -W[x-a] \tag{5.23}$$

$$EI\frac{dy}{dx} = \frac{Wbx^2}{2l} + A \quad \Big| \quad -\frac{W}{2}[x-a]^2 \tag{5.24}$$

$$EIy = \frac{Wbx^3}{6l} + Ax + B \quad \Big| \quad -\frac{W}{6}[x-a]^3 \tag{5.25}$$

The brackets, [], which appear in equations (5.23) to (5.25) are known as Macaulay brackets, and their integration must be carried out in the manner shown in equations (5.24) and (5.25), so that the arbitrary constants A and B apply to both sides of the beam.

Now in setting the boundary conditions and in obtaining values for dy/dx and y, *if the terms within the Macaulay brackets become negative, then they do not apply.*

Boundary conditions

The first boundary condition is as follows: at $x=0$, $y=0$ which, when applied to equation (5.25), reveals that

$$\underline{B=0}$$

NB The expression $[0-a]^3$ does not apply when the above boundary condition is substituted into equation (5.25), because the term within the Macaulay brackets, [ ], is negative.

The second boundary condition is as follows: at $x=l$, $y=0$; therefore

$$0=\frac{Wbl^2}{6}+Al-\frac{Wb^3}{6}$$

or

$$\underline{A=-\frac{Wab(l+b)}{l}}$$

i.e. the deflection y at a distance x along the length of the beam is given by equation (5.26), providing the term within the Macaulay brackets, [], does not become negative:

$$EIy=\frac{Wbx^3}{6l}-\frac{Wab(l+b)x}{l}\ \Bigg|\ -\frac{W}{6}[x-a]^3 \qquad (5.26)$$

The deflection under the load δ_c is given by

$$\delta_c=\frac{W}{EI}\left(\frac{ba^3}{6l}-\frac{ab(l+b)a}{l}\right)$$

$$\underline{\delta_c=-\frac{Wa^2b^2}{3EIl}} \qquad (5.27)$$

When $a=b=l/2$ in equation (5.27),

$$\underline{\delta_c=-\frac{Wl^3}{48EI}\ \text{(as required)}}$$

EXAMPLE 5.6

Determine an expression for the deflection distribution for the simply supported beam of Figure 5.6. Hence, or otherwise, obtain the position and value of the maximum deflection,

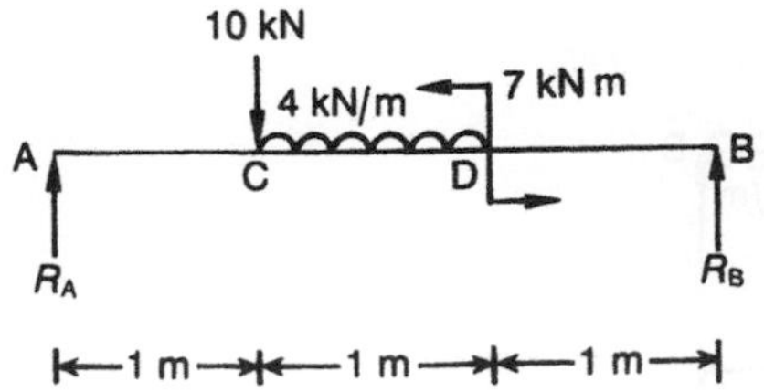

Figure 5.6

given the following:

$$E = 2 \times 10^{11}\ \text{N}/m^2 \quad I = 2 \times 10^{-8}\ \text{m}^4$$

SOLUTION

First, it is necessary to determine R_A, which can be obtained by taking moments about the point B:

$$R_A \times 3 = 10 \times 2 + 4 \times 1 \times 1.5 + 7$$

$$\underline{R_A = 11\ \text{kN}}$$

In applying Macaulay's method to this beam, and remembering that the negative terms inside the Macaulay brackets must be ignored, it is necessary to make the distributed load of Figure 5.6 equivalent to that of Figure 5.7, which is essentially the same as that of Figure 5.6.

As the bending moment expression is different for sections AC, CD and DB, it will be necessary to apply Macaulay's method to each of these sections, as follows:

$$EI\frac{d^2y}{dx^2} = 11x \quad \Big|\ -10[x-1] -\tfrac{4}{2}[x-1]^2 \quad \Big|\ +\tfrac{4}{2}[x-2]^2 -7[x-2]^0 \qquad (5.28)$$

$$EI\frac{dy}{dx} = \frac{11x^2}{2} + A \quad \Big|\ -\tfrac{10}{2}[x-1]^2 -\tfrac{2}{3}[x-1]^3 \quad \Big|\ +\tfrac{2}{3}[x-2]^3 -7[x-2] \qquad (5.29)$$

$$EIy = \frac{11x^3}{6} + Ax + B \quad \Big|\ -\tfrac{5}{3}[x-1]^3 -\tfrac{1}{6}[x-1]^4 \quad \Big|\ +\tfrac{1}{6}[x-2]^4 -\tfrac{7}{2}[x-2]^2 \qquad (5.30)$$

NB The Macaulay bracket for the couple must be written as in equation (5.28), so that integration can be carried out as in equations (5.29) and (5.30).

Boundary conditions

A suitable boundary condition is as follows:

At $x = 0$ $y = 0$; therefore $\underline{B = 0}$

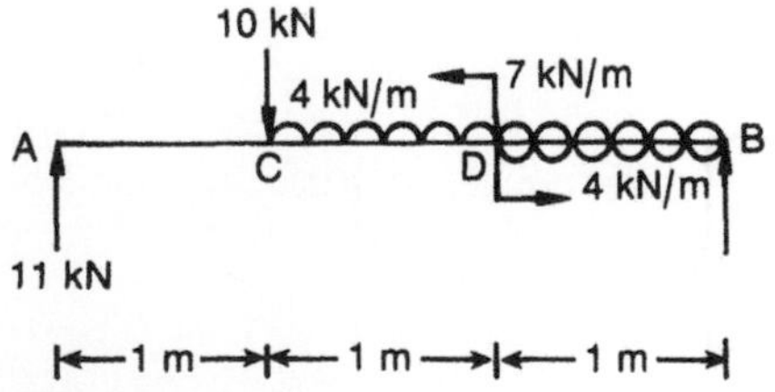

Figure 5.7

NB As the terms in the Macaulay brackets in the second and third columns are negative, they must be ignored when applying the above boundary condition to equation (5.30).

Another suitable boundary condition is as follows:

At $x=3$ $y=0$; therefore $\underline{A=-10.056}$

Substituting the above boundary conditions into equation (5.30), the following is obtained for the deflection y at any point x along the length of the beam:

$$EIy=\frac{11x^3}{6}$$

$$-10.056x \;\Big|\; -\tfrac{5}{3}[x-1]^3-\tfrac{1}{6}[x-1]^4 \;\Big|\; +\tfrac{1}{6}[x-2]^4$$

$$\Big|\; -\tfrac{7}{2}[x-2]^2$$

The maximum deflection may occur in the span CD, where the condition $dy/dx=0$ must be satisfied, i.e.

$$EI\frac{dy}{dx}=\frac{11x^2}{2}-10.056-5[x-1]^2-\tfrac{2}{3}[x-1]^3$$

or

$$0=5.5x^2-10.056-5(x^2-2x+1)-\tfrac{2}{3}(x^3-3x^2+3x-1)$$

or

$$-0.667x^3+2.5x^2+8x-14.389=0$$

which has three real roots, as follows:

$$x_1=-2.913\text{ m}$$

$$x_2=5.252\text{ m}$$

$$x_3=1.411\text{ m}$$

It is evident that the root of interest is $x_3=1.411$ m, as this is the only one that applies within the span CD, i.e.

$$\delta=\text{maximum deflection}$$

$$=\frac{1}{2\times10^{11}\times2\times10^{-8}}\times\left(\frac{11\times1.411^3}{6}-10.05\times1.411-\tfrac{5}{3}(0.411)^3-\frac{0.411^4}{6}\right)$$

$\underline{\delta=-2.288\text{ mm}}$

5.4 Statically indeterminate beams

So far, the beams that have been analyzed were statically determinate; that is their reactions and bending moments were determined solely from statical considerations. For statically indeterminate beams, their analysis is more difficult, as their reactions and bending moments cannot be obtained from statical considerations alone. To demonstrate the method of analyzing statically indeterminate beams, the following two simple cases will be considered.

EXAMPLE 5.7

Determine the end fixing moments, M_F, and the maximum deflection for the encastré beam of Figure 5.8.

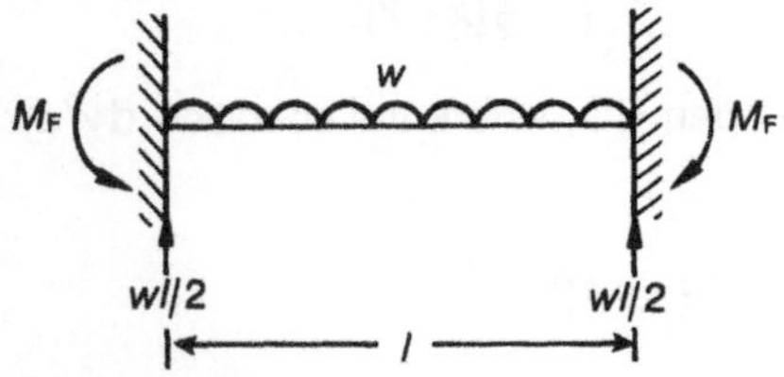

Figure 5.8

SOLUTION

$$EI\frac{d^2y}{dx^2} = -M_F + \frac{wl}{2}x - \frac{wx^2}{2}$$

$$EI\frac{dy}{dx} = -M_F x + \frac{wlx^2}{4} - \frac{wx^3}{6} + A \tag{5.31}$$

At $x = 0$ $\quad \frac{dy}{dx} = 0$; therefore $\underline{A = 0}$

At $x = l$ $\quad \frac{dy}{dx} = 0$; therefore $0 = -M_F l + \frac{wl^3}{4} - \frac{wl^3}{6}$

or

$$\underline{M_F = \frac{wl^2}{12}} \tag{5.32}$$

i.e. the end fixing moment $M_F = wl^2/2$.

On integrating equation (5.31),

$$EIy = -\frac{M_F x^2}{2} + \frac{wlx^3}{12} - \frac{wx^4}{24} + B$$

At $x = 0$ $\quad y = 0$; therefore $\underline{B = 0}$

i.e.

$$y = \frac{w}{EI}\left(-\frac{l^2x^2}{24} + \frac{lx^3}{12} - \frac{x^4}{24}\right)$$

By inspection, the maximum deflection δ occurs at $x = l/2$, where

$$\underline{\delta = -\frac{wl^4}{384EI}} \qquad (5.33)$$

Equation (5.33) shows that the central deflection of an encastré beam is only one-fifth of that of the simply supported case.

EXAMPLE 5.8

Determine the end fixing moments and reactions and the deflection under the load for the encastré beam of Figure 5.9.

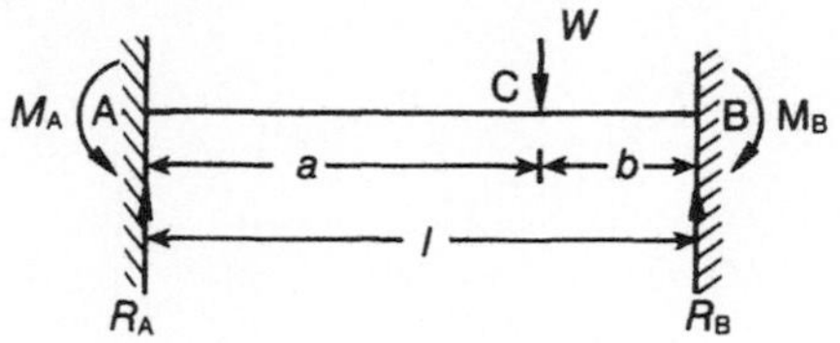

Figure 5.9

SOLUTION

$$EI\frac{d^2y}{dx^2} = -M_A + R_Ax \quad \Big| \quad -W[x-a] \qquad (5.34)$$

$$EI\frac{dy}{dx} = -M_Ax + \frac{R_Ax^2}{2} + A \quad \Big| \quad -\frac{W}{2}[x-a]^2 \qquad (5.35)$$

$$EIy = -\frac{M_Ax^2}{2} + \frac{R_Ax^3}{6} + \quad x + B \quad \Big| \quad -\frac{W}{6}[x-a]^3 \qquad (5.36)$$

To determine A, B, R_A and M_A, it will be necessary to apply four boundary conditions to equations (5.35) and (5.36), as follows:

At $x = 0$ $y = 0$; therefore $\underline{B = 0}$

At $x = 0$ $\frac{dy}{dx} = 0$; therefore $\underline{A = 0}$

At $x = l$ $\frac{dy}{dx} = 0$ and $y = 0$; therefore

$$0 = -M_Al + \frac{R_Al^2}{2} - \frac{Wb^2}{2} \qquad (5.37)$$

and

$$0 = -\frac{M_A l^2}{2} + \frac{R_A l^3}{6} - \frac{Wb^3}{6} \tag{5.38}$$

From equation (5.37),

$$-M_A = -\frac{R_A l}{2} + \frac{Wb^2}{2l} \tag{5.39}$$

and from equation (5.38),

$$-M_A = -\frac{R_A l}{3} + \frac{Wb^3}{3l^2} \tag{5.40}$$

Equating (5.39) and (5.40),

$$-\frac{R_A l}{2} + \frac{Wb^2}{2l} = -\frac{R_A l}{3} + \frac{Wb^3}{3l^2}$$

or

$$\underline{R_A = \frac{Wb^2}{l^3}(l + 2a)} \tag{5.41}$$

and from equation (5.39),

$$M_A = \frac{Wb^2}{2l^2}(l + 2a) - \frac{Wb^2}{2l}$$

$$\underline{M_A = \frac{Wab^2}{l^2}} \tag{5.42}$$

Expressions for R_B and M_B can now be obtained from statical considerations, as follows.

Resolving vertically

$$R_A + R_B = W$$

Therefore

$$\underline{R_B = \frac{Wa^2(l + 2b)}{l^3}} \tag{5.43}$$

and

$$\underline{M_B = \frac{Wa^2 b}{l^2}} \tag{5.44}$$

Substituting equations (5.41) and (5.42) into equation (5.36), with a value of $x = a$,

$$\delta_c = -\frac{Wa^3 b^3}{3EIl^3} \tag{5.45}$$

If the beam of Figure 5.9 were loaded symmetrically, so that $a = b = l/2$, then

$$M_A = M_B = Wl/8 \tag{5.46}$$

and

$$\delta_c = -\frac{Wl^3}{192EI} \tag{5.47}$$

From equation (5.47), it can be seen that the central deflection for a centrally loaded beam, with encastré ends, is one-quarter of the value for a similar beam with simply supported ends.

5.5 Moment–area method

This method is particularly useful if there is step variation with the sectional properties of the beam

Now,

$$EI\frac{d^2y}{dx^2} = M$$

or

$$\frac{d^2y}{dx^2} = \frac{M}{EI} \tag{5.48}$$

Consider the deformed beam of Figure 5.10, and apply equation (5.48) between the points A and B:

$$\int_{x_A}^{x_B} \frac{d^2y}{dx^2}\,dx = \int_{x_A}^{x_B}\left(\frac{M}{EI}\right)dx$$

$$\left[\frac{dy}{dx}\right]_{x_A}^{x_B} = \int_{x_A}^{x_B}\left(\frac{M}{EI}\right)dx$$

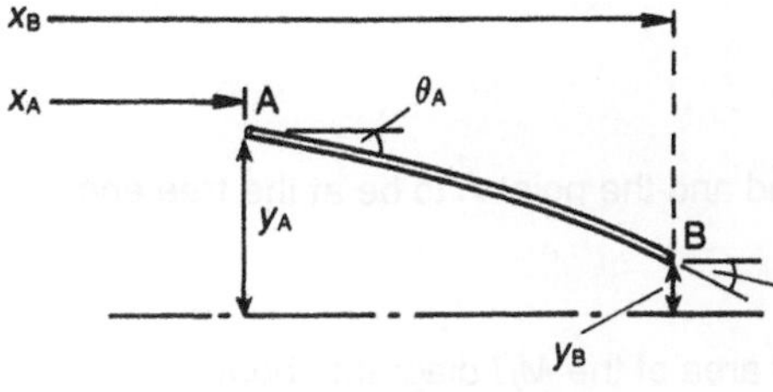

Figure 5.10 Deformed beam element

i.e.

$$\theta_B - \theta_A = \int_{x_A}^{x_B} \left(\frac{M}{EI}\right) dx \tag{5.49}$$

$$= \text{area of } (M/EI) \text{ between } x_A \text{ and } x_B$$

and if EI is constant

$$\theta_B - \theta_A = (1/EI) \times \text{area of bending moment diagram between } x_A \text{ and } x_B$$

Furthermore, it can be seen that if both sides of equation (5.48) are multiplied by x, then

$$x\frac{d^2y}{dx^2} = \frac{M}{EI}x$$

or

$$\int_{x_A}^{x_B} x\frac{d^2y}{dx^2}\,dx = \int_{x_A}^{x_B} \frac{M}{EI}x\,dx$$

or

$$\left[x\frac{dy}{dx} - y\right]_{x_A}^{x_B} = \int_{x_A}^{x_B} \left(\frac{M}{EI}\right) x\,dx \tag{5.50}$$

Suitable use of equations (5.49) and (5.50) can be used to solve many problems, and if EI is constant, then

$$\left[x\frac{dy}{dx} - y\right]_{x_A}^{x_B} = \frac{1}{EI} \times \text{moment of area of the bending moment diagram about the point A}$$

EXAMPLE 5.9

Determine the end deflection for the cantilever loaded with a point load at its free end, as shown in Figure 5.11.

SOLUTION

Take the point B to be at the built-in end and the point A to be at the free end.

From equation (5.50),

$$\left[x\frac{dy}{dx} - y\right]_0^l = \frac{1}{E} \times \text{moment of area of the } M/I \text{ diagram about A}$$

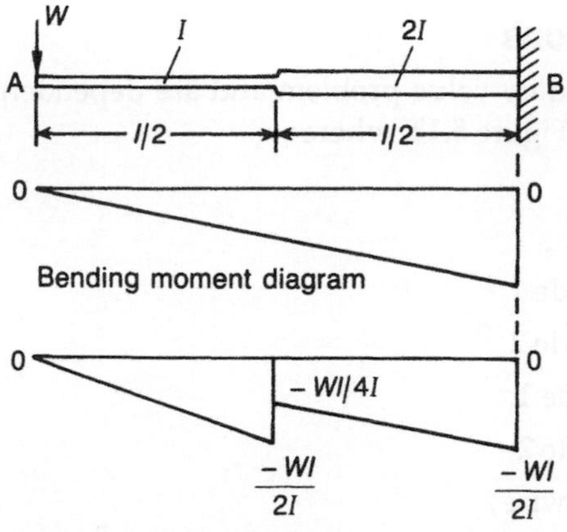

Figure 5.11 M/I diagram

Therefore

$$[(l\theta_B - y_B) - (0 \times \theta_A - y_A)] = -\frac{1}{E}\left[\frac{Wl}{2I} \times \frac{3}{3} \times \frac{l}{2} \times \frac{l}{4} + \frac{Wl}{4I} \times \frac{3l}{4} \times \frac{l}{2} + \frac{Wl}{4I} \times \frac{l}{4} \times \left(\frac{l}{4} + \frac{2}{3} \times \frac{l}{2}\right)\right]$$

but

$$\theta_B = y_B = 0$$

Therefore

$$y_A = -\frac{Wl^3}{EI}\left(\frac{1}{24} + \frac{3}{32} + \frac{5}{96}\right)$$

i.e.

$$\underline{\delta = y_A = -\frac{9Wl^3}{48EI} = -\frac{3Wl^3}{16EI}}$$

If I were constant throughout the length of the cantilever,

$$\delta = -\frac{Wl^3}{96EI}(4 + 18 + 10)$$

$$\underline{\delta = -\frac{Wl^3}{3EI}}$$

To solve beam problems, other than statically determinate ones or the simpler cases of statically indeterminate ones, is difficult and cumbersome, and for such cases it is better to use computer methods [1,5,6]. Some of these computer methods are based on the *slope–deflection equations*, which are now derived.

5.6 Slope–deflection equations

Slope deflection equations are a boundary value problem and are dependent on the displacement boundary conditions of Figure 5.12, where

Y_1 = vertical force at node 1

Y_2 = vertical force at node 2

M_1 = clockwise moment at node 1

M_2 = clockwise moment at node 2

y_1 = vertical deflection at node 1

y_2 = vertical deflection at node 2

θ_1 = rotation at node 1 (clockwise)

θ_2 = rotation at node 2 (clockwise)

NB A node is defined as a point.

Resolving vertically,

$$Y_1 = -Y_2 \tag{5.51}$$

Taking moments about the right end,

$$Y_1 = -(M_1 + M_2)/l \tag{5.52}$$

and

$$Y_2 = (M_1 + M_2)/l$$

From equation (5.3),

$$EI\frac{d^2y}{dx^2} = Y_1x + M_1$$

$$= -(M_1 + M_2)x/l + M_1$$

$$EI\frac{dy}{dx} = -\frac{(M_1 + M_2)x^2}{2l} + M_1x + A$$

$$\text{At } x = 0 \quad \frac{dy}{dx} = -\theta_1; \quad \text{therefore } A = -EI\theta_1 \tag{5.53}$$

Hence

$$EI\frac{dy}{dx} = -\frac{(M_1 + M_2)x^2}{2l} + M_1x - EI\theta_1$$

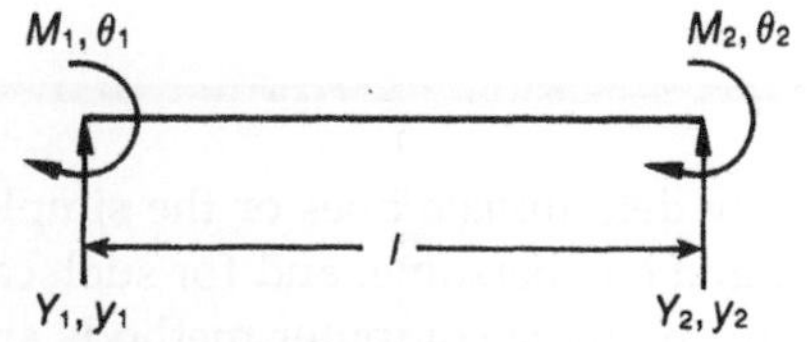

Figure 5.12

and

$$EIy = -\frac{(M_1 + M_2)x^3}{6l} + \frac{M_1x^2}{2} - EI\theta_1 x + B$$

At $x = 0$ $y = y_1$; therefore $B = EIy_1$ (5.54)

At $x = l$ $\frac{dy}{dx} = -\theta_2$; therefore

$$-EI\theta_2 = -\frac{(M_1 + M_2)l}{2} + M_1 l - EI\theta_1 \quad (5.55)$$

At $x = l$ $y = y_2$; therefore

$$EIy_2 = -\frac{(M_1 + M_2)l^2}{6} + \frac{M_1 l}{2} - EI\theta_1 l + EIy_1 \quad (5.56)$$

From equations (5.51)–(5.56), the *slope–deflection equations* are obtained as follows:

$$M_1 = \frac{4EI\theta_1}{l} + \frac{2EI\theta_2}{l} - \frac{6EI}{l^2}(y_1 - y_2)$$

$$Y_1 = -\frac{6EI\theta_1}{l^2} - \frac{6EI\theta_2}{l^2} + \frac{12EI}{l^3}(y_1 - y_2) \quad (5.57)$$

$$M_2 = \frac{2EI\theta_1}{l} + \frac{4EI\theta_2}{l} - \frac{6EI}{l^2}(y_1 - y_2)$$

$$Y_2 = \frac{6EI\theta_1}{l^2} + \frac{6EI\theta_2}{l^2} - \frac{12EI}{l^3}(y_1 - y_2)$$

Equations (5.57) lend themselves to satisfactory computer analysis and form the basis of the finite element method, which is beyond the scope of the present book but is described in detail in a number of other texts [5–7].

EXAMPLES FOR PRACTICE 5

1. Obtain an expression for the deflection y at any distance x from the left end of the uniform-section beam of Figure 5.13.

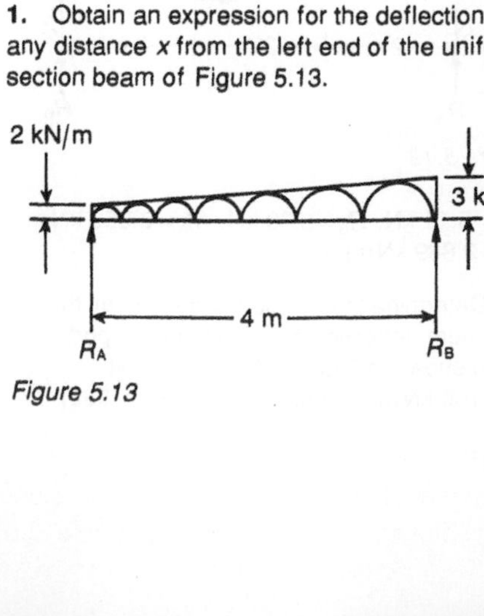

Figure 5.13

$$\left\{y = \frac{1}{EI}(0.778x^2 - 0.0833x^4 - 2.083 \times 10^{-3}x^5 - 6.579x)\right\}$$

2. Determine the value of the reactions and end fixing moments for the uniform-section beam of Figure 5.14. Hence, or otherwise, obtain an expression for the deflection y at any distance x from the left end of the beam.

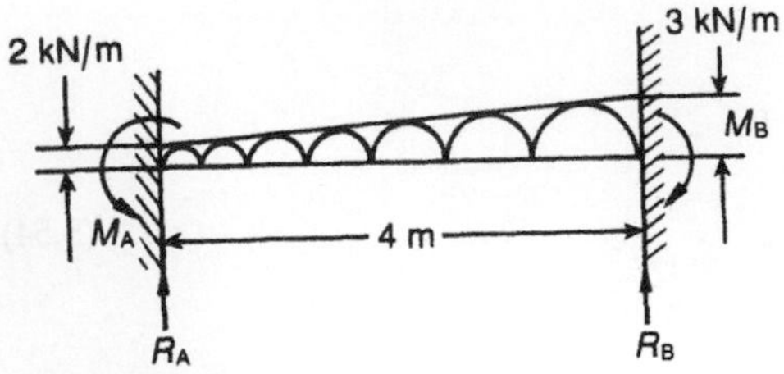

Figure 5.14

$$\left\{R_A = 4.6 \text{ kN},\ M_A = 3.2 \text{ kN m},\ R_B = 5.4 \text{ kN},\ M_B = 3.467 \text{ kN m};\ y = \frac{1}{EI}(0.777x^3 - 0.0833x^4 - 2.083 \times 10^{-3}x^5 - 1.6x^2)\right\}$$

3. Determine the position (from the left end) and the value of the maximum deflection for the uniform-section beam of Figure 5.15.

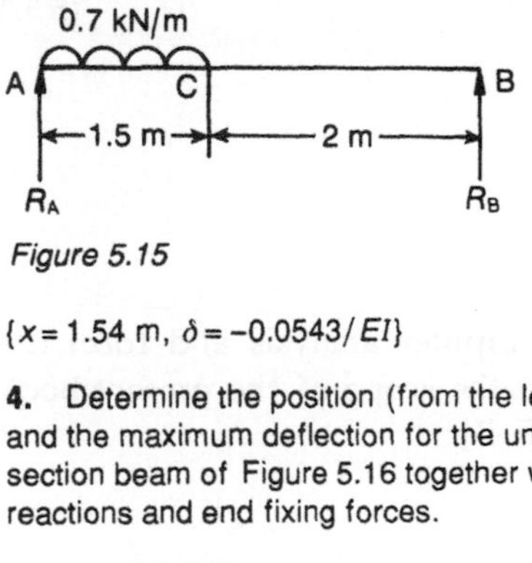

Figure 5.15

$\{x = 1.54 \text{ m},\ \delta = -0.0543/EI\}$

4. Determine the position (from the left end) and the maximum deflection for the uniform-section beam of Figure 5.16 together with the reactions and end fixing forces.

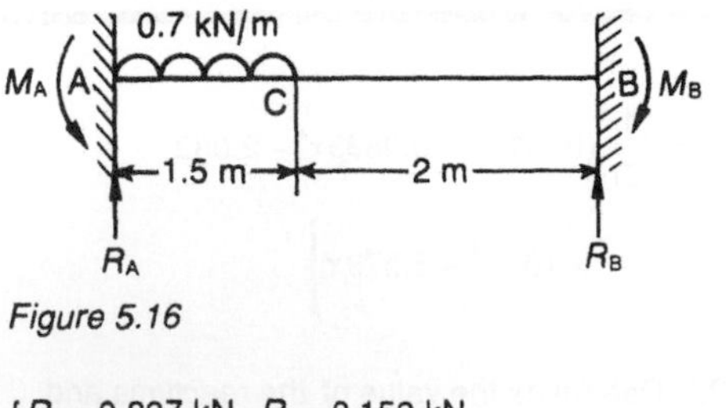

Figure 5.16

$\{R_A = 0.897 \text{ kN},\ R_B = 0.153 \text{ kN},$
$M_A = 0.407 \text{ kN m},\ M_B = 0.155 \text{ kN m};\ x = 1.47 \text{ m},$
$\delta = -0.103/EI\}$

5. Determine the deflections at the points C and D for the uniform-section beam of Figure 5.17, given that $EI = 4300 \text{ kN m}^2$. (Portsmouth, June 1982).

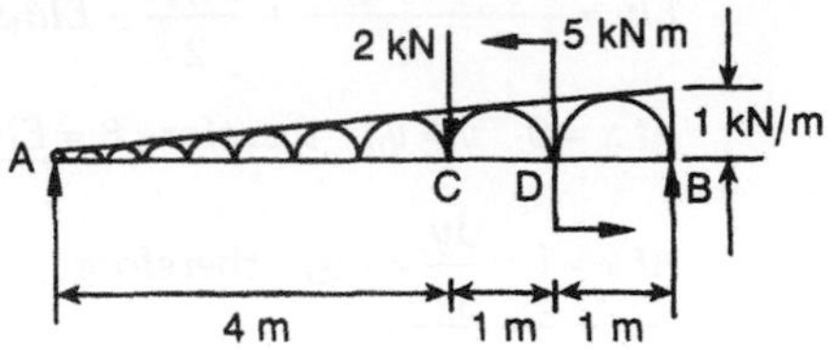

Figure 5.17

$\{\delta_C = -5.58 \times 10^{-3} \text{ m},\ \delta_D = -3.32 \times 10^{-3} \text{ m}\}$

6. Determine the end fixing moments and reactions for the encastré beam shown in Figure 5.18. The beam may be assumed to be of uniform section.

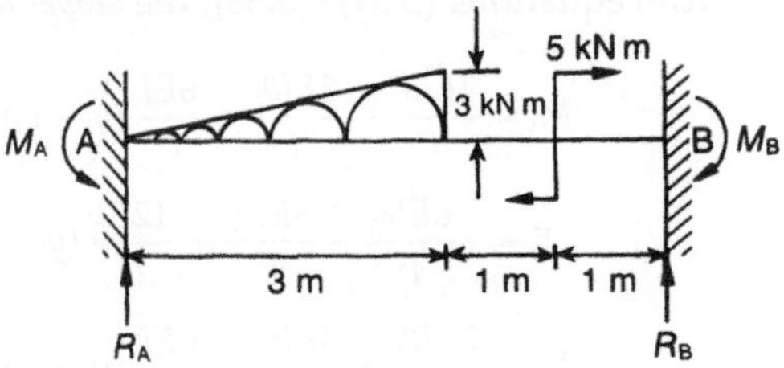

Figure 5.18

$\{R_A = 1.888 \text{ kN},\ R_B = 2.612 \text{ kN},$
$M_A = 1.444 \text{ kN m},\ M_B = 0.506 \text{ kN m}\}$

7. Determine the end fixing moments and reactions for the uniform-section encastré beam of Figure 5.19.

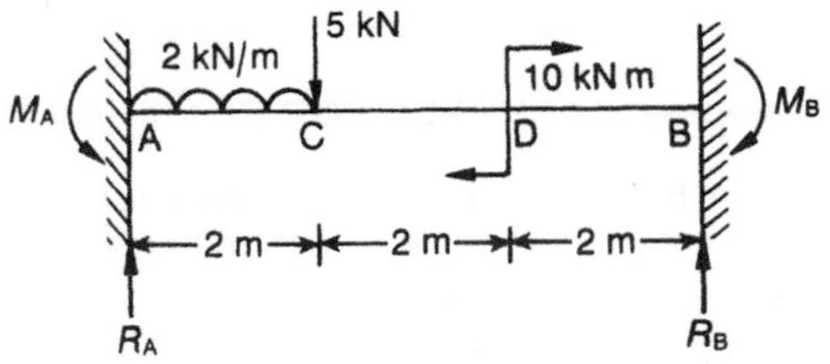

Figure 5.19

$\{R_A = 5.11 \text{ kN},\ R_B = 3.89 \text{ kN},\ M_A = 3.556 \text{ kN m},$
$M_B = 2.889 \text{ kN m}\}$

8. Determine the position and value of the maximum deflection of the simply supported beam shown in Figure 5.20, given that $EI = 100 \text{ kN m}^2$. (Portsmouth, March 1982)

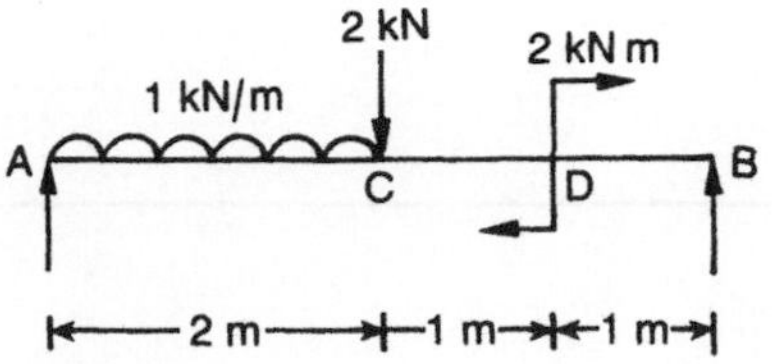

Figure 5.20

{$x = 1.87$ m, $\delta = -0.0285$ m}

9. The beam CAB is simply supported at the points A and B and is subjected to a concentrated load W at the point C, together with a uniformly distributed load w between the points A and B, as shown in Figure 5.21.

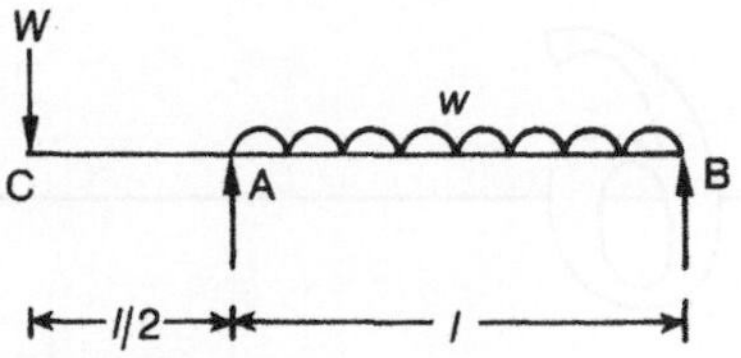

Figure 5.21

Determine the relationship between W and wl, so that no deflection will occur at the point C.

{$w = 6\,W/l$}

6 Torsion

6.1 Introduction

In engineering, it is often required to transmit power via a circular-section shaft, and some typical examples of this are given below:

(a) Propulsion of a ship or a boat by a screw propeller, via a shaft.
(b) Transmission of power to the rear wheels of an automobile, via a shaft.
(c) Transmission of power from an electric motor to various types of machinery, via a shaft.

6.2 Torque (T)

A torque is defined as a twisting moment that acts on the shaft in an axial direction, as shown in Figure 6.1, where T is according to the right-hand screw rule. This torque causes the end B to rotate by an angle θ, relative to A, where

θ = angle of twist

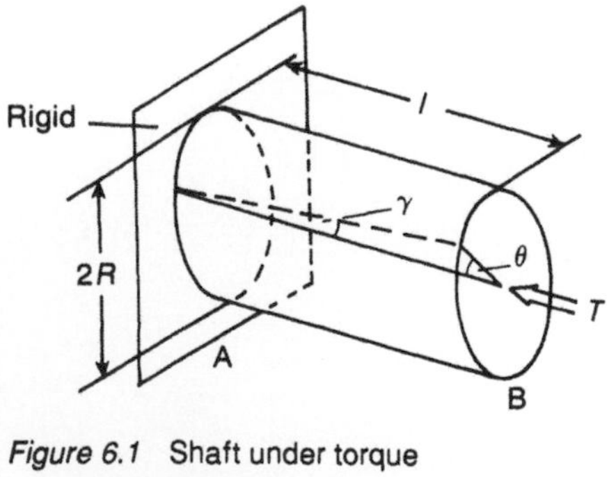

Figure 6.1 Shaft under torque

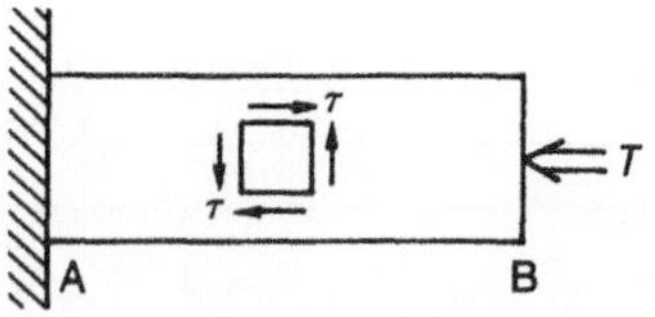

Figure 6.2 Shaft in pure shear

Assuming that no other forces act on the shaft, the effect of T will be to cause the shaft to be subjected to a system of shearing stresses, as shown in Figure 6.2. The system of shearing stresses acting on an element of the shaft of Figure 6.2 is known as *pure shear*, as these shearing stresses are unaccompanied by direct stresses.

Later, it will be proven that, providing the angle of twist is small and the limit of proportionality is not exceeded, the shearing stresses τ will have a maximum value on the external surface of the shaft and their magnitude will decrease linearly to zero at the centre of the shaft.

6.3 Assumptions made

The following assumptions are made in the theory of circular section shafts:

(a) The shaft is of circular cross-section.
(b) The cross-section of the shaft is uniform throughout its length.
(c) The shaft is straight.
(d) The material is homogeneous, isotropic and obeys Hooke's law.
(e) Rotations are small and the limit of proportionality is not exceeded.
(f) Plane cross-sections remain plane during twisting.
(g) Radial lines across the shaft's cross-section remain radial during twisting.

6.4 Proof of $\tau/r = T/J = G\theta/l$

The following relationships, which are used in the torsional theory of *circular-section shafts*, will now be proven:

$$\frac{\tau}{r} = \frac{T}{J} = \frac{G\theta}{l}$$

where

τ = shearing stress at any radius r
T = applied torque
J = polar second moment of area
G = rigidity or shear modulus
θ = angle of twist over a length l

From Figure 6.1, it can be seen that

$$\gamma = \text{shear strain}$$

and that

$$\gamma l = R\theta$$

provided θ is small, or

$$\left(\frac{\tau}{G}\right) l = R\theta$$

Therefore

$$\frac{\tau}{R} = \frac{G\theta}{l}$$

If the radial lines across the section remain radial on twisting, then it follows that the shearing stress is proportional to any radius r, so that

$$\frac{\tau}{r} = \frac{G\theta}{l} \tag{6.1}$$

Similarly,

$$\frac{\tau}{r} = G\frac{d\theta}{dx} \tag{6.2}$$

where $d\theta/dx$ is the change of the angle of twist over a length dx.

Consider a cylindrical shell element of radius r and thickness dr, as shown by the shaded area of Figure 6.3. The shearing stresses acting on the cross-section of this cylindrical shell are shown in Figure 6.4, where they can be seen to act tangentially to the cross-section of the cylindrical shell.

From Figure 6.4, it can be seen that these shearing stresses cause an elemental torque, δT, where

$$\delta T = \tau \times 2\pi r \, dr \times r$$

but the total torque T is the sum of all the elemental torques acting on the section,

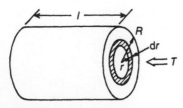

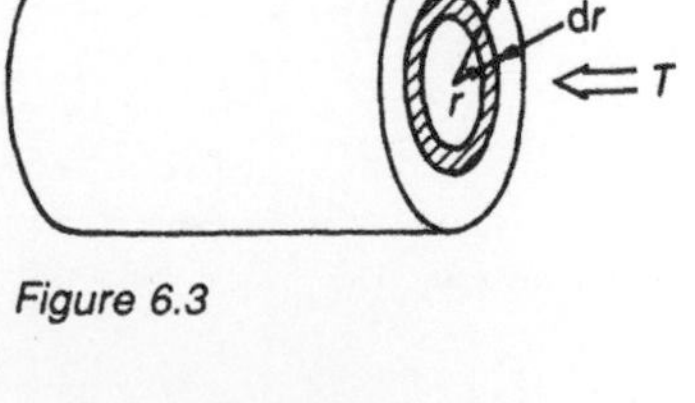

Figure 6.3

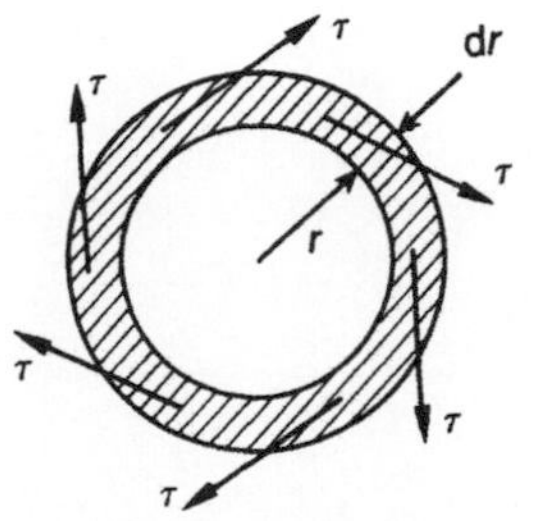

Figure 6.4 Shearing stresses acting on annular element

i.e.

$$T = \sum \delta T$$
$$= 2\pi \int r^2 \tau \, dr \qquad (6.3)$$

Substituting τ from equation (6.1) into equation (6.3),

$$T = 2\pi \int_0^R \frac{G\theta}{l} r^3 \, dr$$
$$= \frac{G\theta}{l}\left(\frac{\pi R^4}{2}\right)$$

but

$$\frac{\pi R^4}{2} = J$$

is the polar second moment of area of a circular section. Therefore

$$\frac{T}{J} = \frac{G\theta}{l} \qquad (6.4)$$

From equations (6.1) and (6.4),

$$\frac{\tau}{r} = \frac{T}{J} = \frac{G\theta}{l} \qquad (6.5)$$

From equation (6.5), it can be seen that τ is linearly proportional to the radius r, so that the shear stress at the centre is zero and it reaches a maximum value on the outermost surface of the shaft, as shown in Figure 6.5.

6.4.1 *Hollow circular sections*

Equation (6.5) is also applicable to circular-section tubes of uniform thickness, except

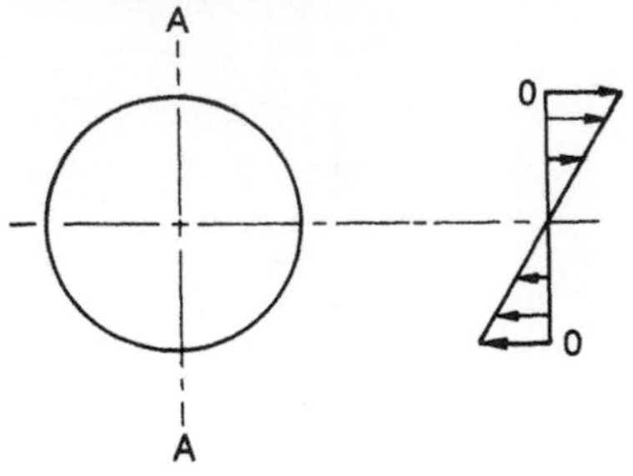

Figure 6.5 Distribution of τ: (a) section; (b) shear stress distribution at AA

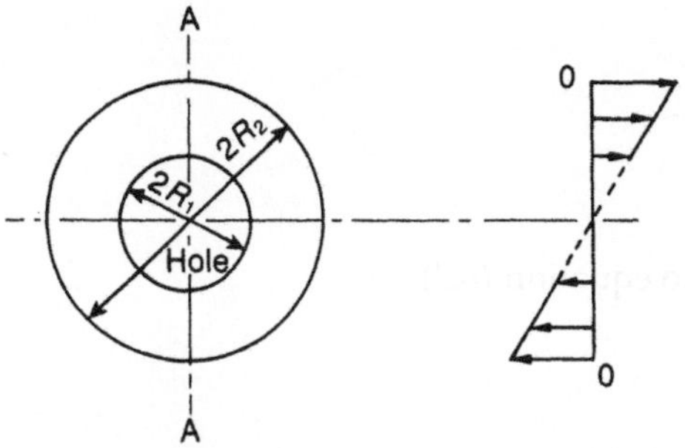

Figure 6.6 Distribution of τ across a hollow circular section: (a) section; (b) distribution of τ across AA

that

$$J = \pi(R_2^4 - R_1^4)/2$$

is the polar second moment of area of an annulus, where

R_2 = external radius of tube
R_1 = internal radius of tube

Using the above value of J, equation (6.5) can be applied to a circular-section tube, between the radii R_1 and R_2, and the shear stress distribution will have the form shown in Figure 6.6.

The *torsional stiffness of a shaft* is given by

$$\underline{k = GJ/l} \qquad (6.6)$$

To demonstrate the use of equation (6.5), several worked examples will be considered.

EXAMPLE 6.1

An internal combustion engine transmits 40 horse-power (hp) at 200 rev/min (rpm) to the rear wheels of an automobile. Neglecting transmission losses, determine the minimum diameter of a solid circular-section shaft, if the maximum permissible shear stress in the shaft is 60 MN/m^2. Hence, or otherwise, determine the angle of twist over a length of 2 m, given that $G = 7.7 \times 10^{10}$ N/m^2.

SOLUTION

$$40 \text{ hp} = 40 \text{ hp} \times 745.7 \frac{\text{W}}{\text{hp}}$$
$$= 29.83 \text{ kW}$$

Now,

$$\text{Power} = T\omega \tag{6.7}$$

where power is in units of watts:

T = torque (N m)

ω = speed of rotation (rad/s)

In this case,

$$\omega = 2\pi \frac{\text{rad}}{\text{rev}} \times 200 \frac{\text{rev}}{\text{min}} \times \frac{1 \text{ min}}{60 \text{ s}}$$
$$\underline{\omega = 20.94 \text{ rad/s}}$$

i.e.

$$29.83 \times 10^3 = T \times 20.94$$

Therefore

$$\underline{T = 1.424 \text{ kN m}} \tag{6.8}$$

Now the maximum permissible shearing stress is 60 MN/m^2, which occurs at the external radius R. Hence, from $\tau/r = T/J$,

$$\frac{60 \times 10^6}{R} = \frac{1.424 \times 10^3}{(\pi R^4/2)}$$
$$R^3 = \frac{1.424 \times 10^3 \times 2}{\pi \times 60 \times 10^6}$$
$$\underline{R = 0.0247 \text{ m}}$$

i.e.

$$\underline{\text{Shaft diameter} = 4.94 \text{ cm}}$$

From

$$\frac{\tau}{r} = \frac{G\theta}{l}$$
$$\theta = \frac{\tau l}{GR} = \frac{60 \times 10^6 \times 2}{7.7 \times 10^{10} \times 0.0247}$$
$$\underline{\theta = 0.0631 \text{ rad} = 3.62^\circ}$$

EXAMPLE 6.2

If the shaft of Example 6.1 were in the form of a circular-section tube, where the external diameter had twice the magnitude of the internal diameter, what would be the weight saving if this hollow shaft were adopted in place of the solid one, assuming that both shafts had the same maximum shearing stress?

SOLUTION

Let R be the external radius of the hollow shaft. Then

$$J = \pi[R^4 - (R/2)^4]/2$$

$$\underline{J = 1.473R^4}$$

From

$$\frac{\tau}{r} = \frac{T}{J}$$

$$\frac{60 \times 10^6}{R} = \frac{1.424 \times 10^3}{1.473R^4}$$

$$R^3 = \frac{1.424 \times 10^3}{1.473 \times 60 \times 10^6}$$

$$\underline{R = 0.0253 \text{ m}}$$

i.e.

$$\text{External diameter of hollow shaft} = \underline{5.05 \text{ cm}}$$

$$\text{Internal diameter of hollow shaft} = \underline{2.53 \text{ cm}}$$

$$\text{Weight of solid shaft} = \rho g \times \pi \times 0.0247^2 \times l$$

$$\text{Weight of hollow shaft} = \rho g \times \pi[R^2 - (R/2)^2] \times l$$

Therefore

$$\text{Percentage weight saving} = \frac{0.0247^2 - (0.0253^2 - 0.012\,65^2)}{0.0247^2} \times 100$$

$$= \underline{21.3}$$

From

$$\frac{\tau}{r} = \frac{G\theta}{l}$$

$$\theta = \frac{60 \times 10^6 \times 2}{0.0253 \times 7.7 \times 10^{10}}$$

$$\underline{\theta = 0.0616 \text{ rad} = 3.53°}$$

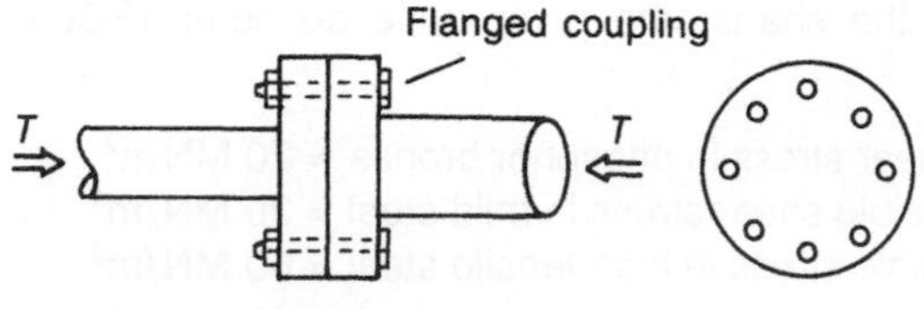

Figure 6.7 Flanged coupling

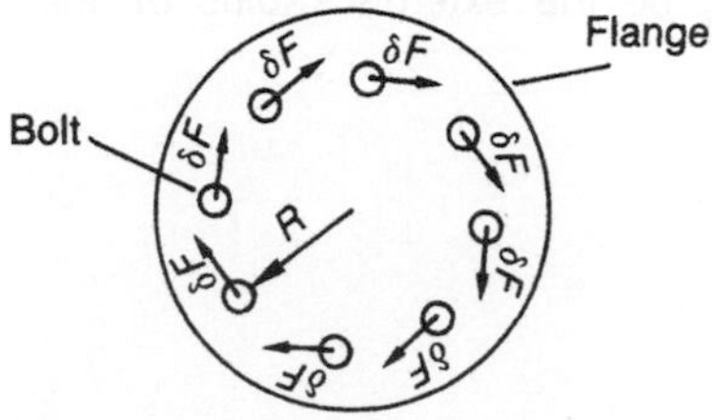

Figure 6.8 Shearing forces on bolts

6.5 Flanged couplings

In some engineering situations, it is necessary to join together two shafts made from dissimilar materials. In such cases, it is usually undesirable to weld together the two shafts, and one method of overcoming this problem is by the use of a flanged coupling, as shown in Figure 6.7. In this case the torque is transmitted from one shaft to the other by the action of shearing forces δF acting on the bolts, as shown in Figure 6.8. Thus, if there are n bolts,

$$T = n \times \delta F \times R \tag{6.9}$$

where

$$\delta F = \tau_b \times \pi d^2/4 \tag{6.10}$$

R = pitch circle radius of bolt

τ_b = shearing stress in bolt

d = diameter of bolt

In Figure 6.8, it is assumed that all the bolts carry an equal shearing force δF; this may not be the case in practice, owing to manufacturing imperfections.

EXAMPLE 6.3

A torque of 10 kN m is to be transmitted from a hollow phosphor bronze shaft, whose external diameter is twice its internal diameter, to a solid mild-steel shaft, through a flanged coupling with 12 bolts made from a high tensile steel.

Determine the dimensions of the shafts, the pitch circle diameter (PCD), and the diameters of the bolts, given that

Maximum permissible shear stress in phosphor bronze = 20 MN m^2
Maximum permissible shear stress in mild steel = 30 MN/m^2
Maximum permissible stress in high tensile steel = 60 MN/m^2

SOLUTION

Consider the phosphor bronze shaft. Let R_p be the external radius of the phosphor bronze shaft. Then

$$J = \frac{\pi[R_p^4 - (R_p/2)^4]}{2}$$

$$\underline{J = 1.473R_p^4}$$

From

$$\frac{\tau}{r} = \frac{T}{J}$$

$$R_p^3 = \frac{10 \times 10^3}{1.473 \times 20 \times 10^6}$$

$$\underline{R_p = 0.0698 \text{ m}}$$

i.e.

External diameter of phosphor bronze shaft = 0.1396 *m* = 13.96 *cm, say* 14 *cm*
Internal diameter of phosphor bronze shaft = 0.0698 *m* = 6.98 *cm, say* 7 *cm*

Consider the steel shaft. Let R_s be the external radius of the steel shaft. Then

$$J = \frac{\pi \times R_s^4}{2} = 1.571R_s^4$$

From

$$\frac{\tau}{r} = \frac{T}{J}$$

$$R_s^3 = \frac{10 \times 10^3}{1.571 \times 30 \times 10^6}$$

$$\underline{R_s = 0.0596 \text{ m}}$$

i.e.

Diameter of steel shaft = 0.1193 *m* = 11.9 *cm, say* 12 *cm*

Bolts on flanged coupling

As the external diameter of the hollow shaft is 14 cm, it will be necessary to assume that the PCD of the bolts on the flanged coupling is larger than this, so that the bolts can be accommodated.

Assume that

$D = \text{PCD}$

$= 20$ cm (to allow for fitting)

n = number of bolts = 12

From equation (6.9),

$$\delta F = \frac{T}{nR} = \frac{10 \text{ kN m}}{12 \times 0.1}$$

$\underline{\delta F = 8.33 \text{ kN}}$

From equation (6.10),

$$d^2 = \frac{4 \times \delta F}{\pi \times \tau_b} = \frac{4 \times 8.33 \times 10^3}{\pi \times 60 \times 10^6}$$

$d = 0.0133$ m

$\underline{d = 1.33 \text{ cm, say } 1.5 \text{ cm}}$

6.6 Keyed couplings

Another method of transmitting power through shafts of dissimilar materials is through the use of keyed couplings, as shown in Figure 6.9. In such cases, the key is the male portion of the coupling, and the keyway, which is the female portion of the coupling, is on the shaft. It is evident that the manufacture of the 'key' and the 'keyway' has to be precise, so that the former fits snugly into the latter. Precise analysis of keyed couplings is difficult and beyond the scope of this book.

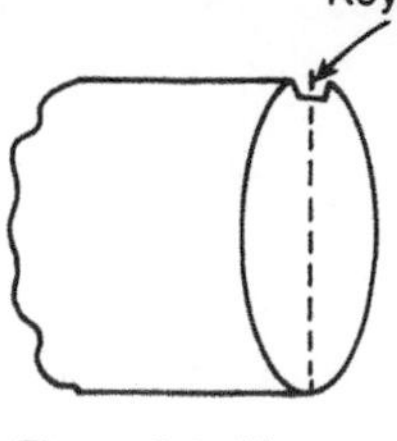

Figure 6.9 Keyway on a shaft

EXAMPLE 6.4

Determine the torque diagram for the shaft AB, which is subjected to an intermediate torque of magnitude T at the point C, as shown in Figure 6.10. The shaft may be assumed to be firmly fixed at its ends. From Newton's third law of motion, it is evident that the intermediate torque T will be resisted by the end torques T_1 and T_2, acting in an opposite direction to T.

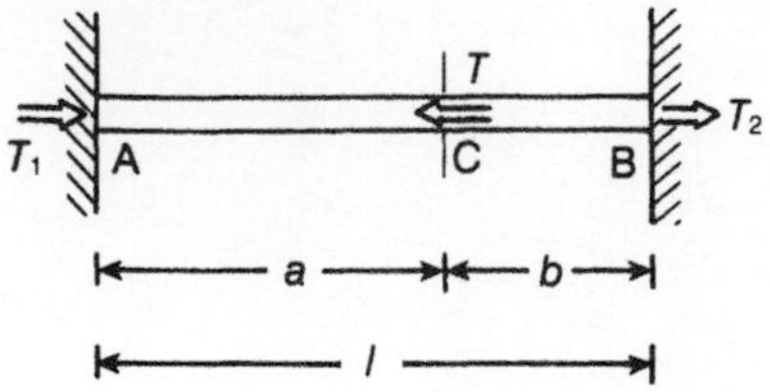

Figure 6.10 Shaft with intermediate torque

SOLUTION

$$T_1 + T_2 = T \quad (6.11)$$

Let θ_C be the rotation at the point C. Then, from

$$\frac{T}{J} = \frac{G\theta}{l}$$

$$T_1 = \frac{GJ\theta_C}{a} \quad (6.12)$$

and

$$T_2 = \frac{GJ\theta_C}{b} \quad (6.13)$$

Dividing (6.12) by (6.13),

$$\frac{T_1}{T_2} = \frac{b}{a}$$

or

$$T_1 = bT_2/a \quad (6.14)$$

Substituting equation (6.14) into equation (6.11),

$$T_2 = \frac{T}{(1 + b/a)} = \frac{Ta}{l} \quad (6.15)$$

and

$$T_1 = \frac{Tb}{l} \quad (6.16)$$

From equations (6.15) and (6.16), it can be seen that the torque diagram is as shown in Figure 6.11.

If the shaft of Figure 6.10 were made from two different sections, joined together at C, so that

J_1 = polar second moment of area of shaft AC
J_2 = polar second moment CB of area
R_1 = external radius of shaft AC

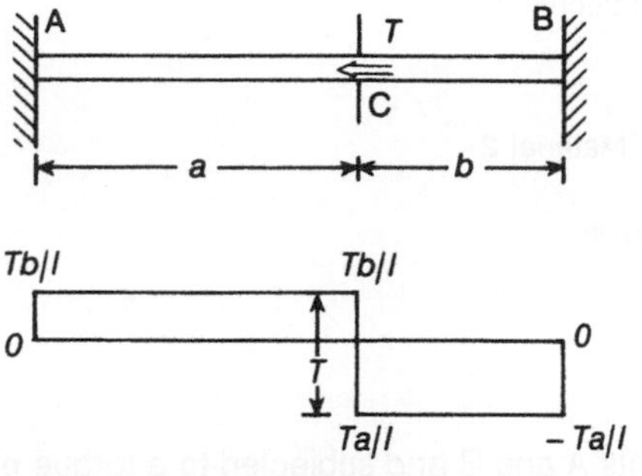

Figure 6.11 Torque diagram

R_2 = external radius CB of shaft
τ_1 = maximum shear stress in shaft AC
τ_2 = maximum shear stress in shaft CB

then

$$\frac{\tau_1}{R_1} = \frac{T_1}{J_1} = \frac{G\theta_C}{a} \tag{6.17}$$

and

$$\frac{\tau_2}{R_2} = \frac{T_2}{J_2} = \frac{G\theta_C}{b} \tag{6.18}$$

From equations (6.17) and (6.18), the two shafts can be designed.

6.7 Compound shafts

Composite shafts are made up from several smaller shafts with different material properties. The use of different materials for the manufacture of shafts is required when a shaft has to pass through different fluids, some of which are hostile to certain materials.

Composite shafts are usually either in series, as shown in Figure 6.12, or in parallel as shown in Figure 6.13.

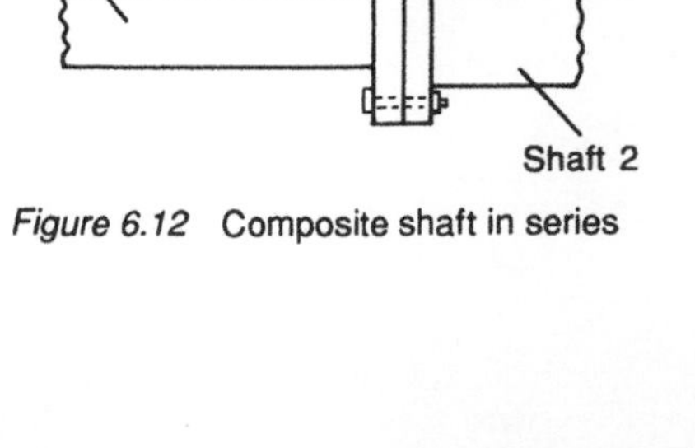

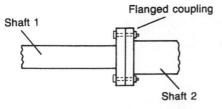

Figure 6.12 Composite shaft in series

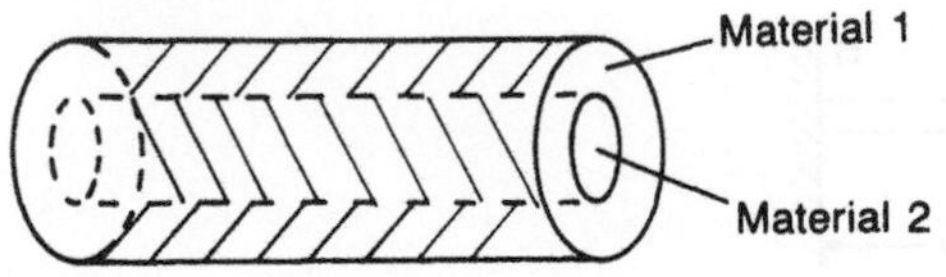

Figure 6.13 Composite shafts in parallel

EXAMPLE 6.5

A compound shaft ACB is fixed at the points A and B and subjected to a torque of 5 kN m at the point C, as shown in Figure 6.14. If both parts of the shaft are of solid circular sections, determine the angle of twist at the point C and the maximum shearing stresses in each section, assuming the following apply:

Shaft 1

$$G_1 = 2.6 \times 10^{10}\ \text{N/m}^2$$

$$R_1 = 6\ \text{cm}$$

$$a = 1\ \text{m}$$

Shaft 2

$$G_2 = 7.8 \times 10^{10}\ \text{N/m}^2$$

$$R_2 = 4\ \text{cm}$$

$$b = 2\ \text{m}$$

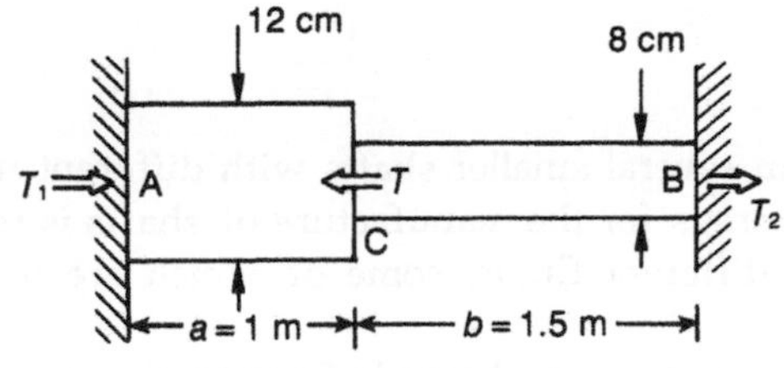

Figure 6.14 Compound shaft

SOLUTION

$$J_1 = \frac{\pi \times 0.06^4}{2} = 2.036 \times 10^{-5}\ \text{m}^4$$

$$J_2 = \frac{\pi \times 0.04^4}{2} = 4.021 \times 10^{-6}\ \text{m}^4$$

Let θ_C be the angle of twist at the point C. Then, from

$$\frac{T}{J} = \frac{G\theta}{l}$$

$$\theta_C = \frac{T_1 a}{G_1 J_1} = \frac{T_1 \times 1}{2.6 \times 10^{10} \times 2.036 \times 10^{-5}}$$

$$\underline{\theta_C = 1.889 \times 10^{-6} T_1} \tag{6.19}$$

Similarly,

$$\theta_C = \frac{T_2 b}{G_2 J_2} = \frac{T_2 \times 1.5}{7.8 \times 10^{10} \times 4.021 \times 10^{-6}}$$

$$\underline{\theta_C = 4.783 \times 10^{-6} T_2} \tag{6.20}$$

Equating (6.19) and (6.20),

$$T_1 = \frac{T_2 \times 4.783 \times 10^{-6}}{1.889 \times 10^{-6}}$$

$$\underline{T_1 = 2.532 T_2} \tag{6.21}$$

Now,

$$T = T_1 + T_2$$

$$5 \text{ kN m} = T_2(1 + 2.532)$$

$$\underline{T_2 = 1.416 \text{ kN m}}$$

$$\underline{T_1 = 3.584 \text{ kN m}}$$

From equation (6.20),

$$\theta_C = 4.783 \times 10^{-6} \times 1.416 \times 10^3$$

$$\underline{\theta_C = 6.773 \times 10^{-3} \text{ rad} = 0.388^\circ}$$

From

$$\frac{\tau}{r} = \frac{T}{J}$$

$$\tau_1 = \frac{T_1 R_1}{J_1} = \frac{3.584 \times 10^3 \times 0.06}{2.036 \times 10^{-5}}$$

$$\underline{\tau_1 = 10.56 \text{ MN/m}^2}$$

$$\tau_2 = \frac{T_2 R_2}{J_2} = \frac{1.416 \times 10^3 \times 0.04}{4.021 \times 10^{-6}}$$

$$\underline{\tau_2 = 14.09 \text{ MN/m}^2}$$

EXAMPLE 6.6

The compound shaft of Figure 6.15 is subjected to a torque of 5 kN m. Determine the angle of twist and the maximum shearing stresses in materials 1 and 2, given the following values for the moduli of rigidity:

$$G_1 = 2.5 \times 10^{10}\ \text{N/m}^2 \text{ – material 1}$$

$$G_2 = 7.8 \times 10^{10}\ \text{N/m}^2 \text{ – material 2}$$

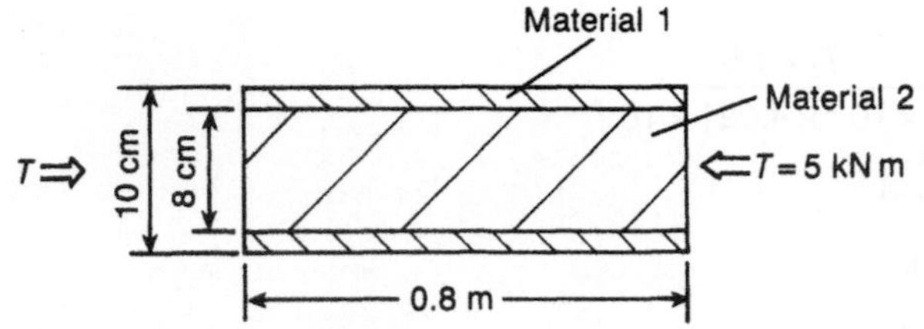

Figure 6.15 Compound shaft

SOLUTION

Assumptions made

(a) No slipping take place at the common interface.
(b) Radial lines remain straight on twisting.

From (b),

$$\underline{\theta_1 = \theta_2}$$

$$J_1 = \frac{\pi(0.05^4 - 0.04^4)}{2} = \underline{5.796 \times 10^{-6}\ \text{m}^4}$$

$$J_2 = \frac{\pi \times 0.04^4}{2} = \underline{4.021 \times 10^{-6}\ \text{m}^4}$$

NB Although the thickness of material 1 is only 1 cm, $J_1 > J_2$, which shows that a hollow shaft has a better strength to weight ratio than a solid one.

From

$$\frac{T}{J} = \frac{G\theta}{l}$$

$$\theta = \frac{T_1 l}{G_1 J_1} = \frac{T_1 \times 0.8}{2.5 \times 10^{10} \times 5.796 \times 10^{-6}}$$

$$\underline{\theta = 5.521 \times 10^{-6} T_1} \tag{6.22}$$

but

$$\theta = \frac{T_2 l}{G_2 J_2} = \frac{T_2 \times 0.8}{7.8 \times 10^{10} \times 4.021 \times 10^{-6}}$$

$$\underline{\theta = 2.551 \times 10^{-6} T_2} \tag{6.23}$$

Equating (6.22) and (6.23),

$$\underline{T_1 = 0.462 T_2}$$

but

$$T = T_1 + T_2$$

Therefore

$$T_2 = \frac{5}{1.462} = 3.42 \text{ kN m}$$

and

$$\underline{T_1 = 1.58 \text{ kN m}}$$

Hence, from equation (6.23),

$$\underline{\theta = 8.724 \times 10^{-3} \text{ rad} = 0.5°}$$

Let τ_1 be the maximum shearing stress in material 1. Then

$$\tau_1 = \frac{T_1 \times R_1}{J_1} = \frac{1.58 \times 10^3 \times 0.05}{5.796 \times 10^{-6}}$$

$$\underline{\tau_1 = 13.63 \text{ MN/m}^2}$$

Let τ_2 be the maximum shearing stress in material 2. Then

$$\tau_2 = \frac{T_2 \times R_2}{J_2} = \frac{3.42 \times 10^3 \times 0.04}{4.021 \times 10^{-6}}$$

$$\underline{\tau_2 = 34.02 \text{ MN/m}^2}$$

NB From the practical point of view, if the shaft of Example 6.6 were used out of doors, in a *normal UK atmosphere*, convenient materials might have been aluminium alloy for material 1 and steel for material 2. The reasons for such a choice would be as follows:

(a) Both materials are relatively inexpensive.
(b) Mild steel is, in general, stronger and stiffer than aluminium alloy.
(c) Aluminium alloy does not rust; hence its use as a sheath over a steel core.

To *manufacture* the compound shaft of Figure 6.15, it is necessary to make the external diameter of the steel shaft slightly larger than the internal diameter of the aluminium-alloy shaft and to 'join' the two together by either heating the aluminium-alloy shaft and/or cooling the steel shaft.

6.8 Tapered shafts

There are a number of texts that treat the analysis of tapered shafts by using a simple extension of equation (6.5), but this method of analysis is incorrect, as the longitudinal shearing stresses produced on torsion are not parallel to the axis of the shaft.

For an analysis of tapered shafts, see reference 8.

6.9 Close-coiled helical springs

Equation (6.5) can be used for the stress analysis of close-coiled helical springs, providing the angle of helix and the deflections are small.

For the close-coiled helical spring of Figure 6.16, most of the coils will be under torsion, where the torque T equals $WD/2$, so that the maximum deflection at B, caused by W, would be due to the combined effect of all the rotations of all the circular sections of the spring. Thus, the close-coiled helical spring can be assumed to be equivalent to a long shaft of length πDn, as shown in Figure 6.17, where

δ = deflection of spring at B due to W
D = mean coil diameter = $2R$
d = wire diameter
n = number of coils

From Figure 6.17,

$$\theta = \frac{T}{J} \times \frac{\pi Dn}{G}$$

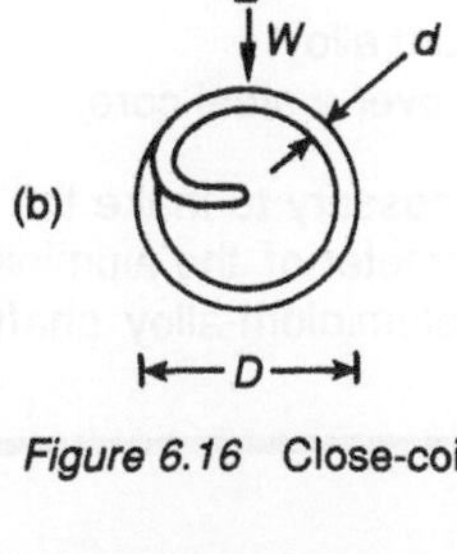

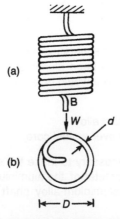

Figure 6.16 Close-coiled helical spring: (a) front elevation; (b) plan (looking upwards)

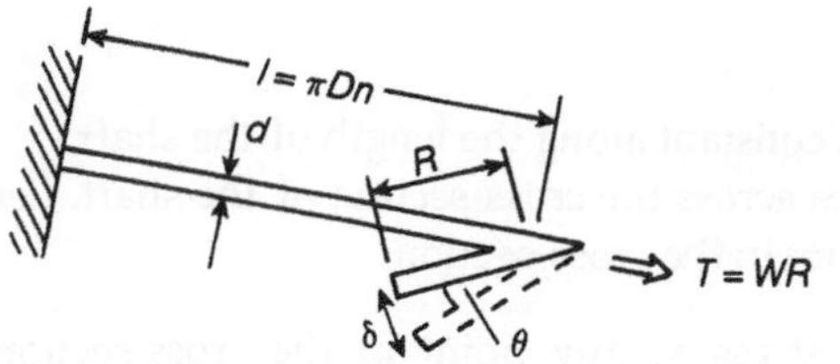

Figure 6.17

or

$$\theta = \frac{W \times R \times \pi \times 2Rn}{\pi \times (d^4/32) \times G}$$

$$\theta = \frac{64WR^2n}{GD^4}$$

$$\delta = R\theta = \frac{64WR^3n}{GD^4}$$

If τ is the maximum shearing stress in the spring, then

$$\tau = \frac{T \times (d/2)}{\pi \times (d^4/32)} = \frac{16WR}{\pi d^3}$$

which occurs throughout the coils on the external surface of the spring. For the torsion of non-circular sections, see references 8–11.

6.10 Torsion of thin-walled non-circular sections

The torsion of non-circular sections involves the solution of Poisson's equation [8–11], which is beyond the scope of the present book. However, for the torsion of thin-walled non-circular sections, a much simpler theory is available, which is now presented.

Consider a thin-walled non-circular-section tube to be subjected to the torques shown in Figure 6.18.

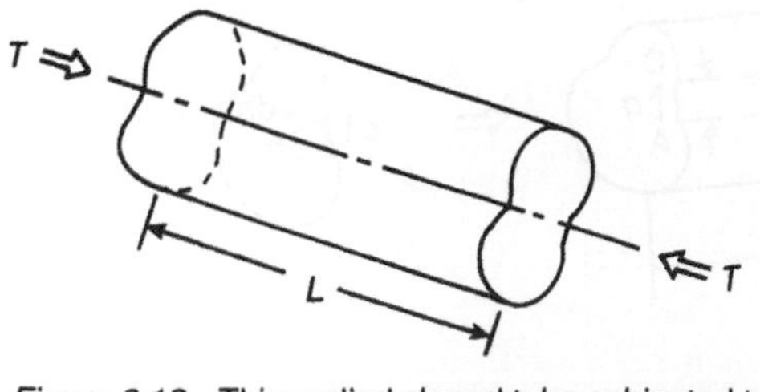

Figure 6.18 Thin-walled closed tube subjected to torques

6.10.1 Assumptions

(a) The cross-section of the tube is constant along the length of the shaft.
(b) The thickness of the tube varies across the cross-section of the shaft. Let t be the thickness of the tube at any point in the cross-section.

Under a torque T, the shearing stress at any point in the cross-section can be assumed to be τ. Now the shearing force/unit circumferential length at any point in the cross-section equals

$$q = \tau t = \text{shear flow}$$

Consider the equilibrium of q on the shell ABCD, as shown in Figure 6.19(a).

From this figure it can be seen from equilibrium considerations that q is the same on faces AC and BD, and that q^1 is the same on faces AB and CD.

Taking moments about A

Clockwise moments = counter-clockwise moments

i.e.

$$q^1 \times L \times b = q \times b \times L$$

Therefore

$$q^1 = q$$

That is, the shear flow q is constant at any point on the section of the tube.

6.10.2 To determine the relationship between T and q

From Figure 6.19(b), it can be seen that the torsional resistance of the tube is

$$T = \oint q \, ds \, r$$

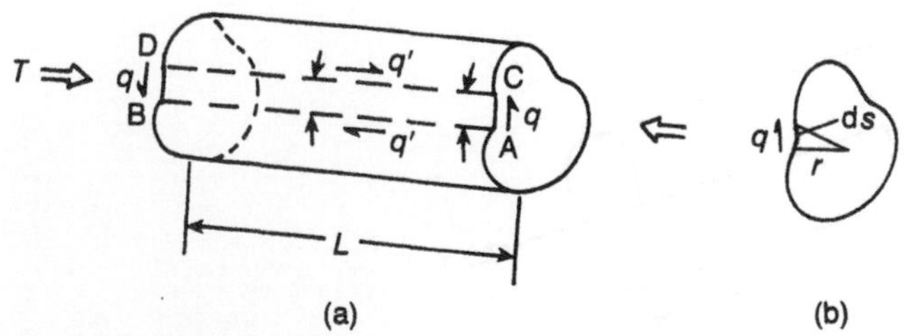

Figure 6.19 Shear flow in a closed tube: (a) shear flow in tube; (b) cross-section of tube

but

$$q = \tau t = \text{a constant}$$

Therefore

$$T = \tau t \oint r\,\mathrm{ds}$$

However,

$$\oint r\,\mathrm{ds} = 2A$$

where A is the enclosed area of the tube. Therefore

$$T = 2A\tau t$$

or

$$\tau = \frac{T}{2At} \tag{6.24}$$

From equation (6.24), it can be seen that the shearing stress in a closed non-circular section is maximum where the thickness is the smallest.

6.10.3 *To obtain the relationship between* T *and* θ

From equation (9.23), the strain energy per unit volume due to τ is

$$U = \frac{\tau^2}{2G} \times \text{volume}$$

$$= \oint \frac{\tau^2}{2G} \times L \times t \times \mathrm{ds}$$

$$= \oint \frac{q^2}{2G} \times L\,\frac{\mathrm{ds}}{t}$$

but

$$\frac{q^2 L}{2G} = \text{constant}$$

Therefore

$$U = \frac{q^2 L}{2G} \oint \frac{\mathrm{ds}}{t} \tag{6.25}$$

From equation (6.24)

$$q = \frac{T}{2A} \tag{6.26}$$

Substituting equation (6.26) into (6.25)

$$U = \frac{T^2L}{8A^2G}\oint\frac{ds}{t} \tag{6.27}$$

If θ is the rotation of the cross-section over a length L then

$$U = \tfrac{1}{2}T\theta \tag{6.28}$$

Equating (6.27) to (6.28),

$$\theta = \frac{TL}{4A^2G}\oint\frac{ds}{t} \tag{6.29}$$

If t is constant over the whole circumference and S is the length of the perimeter of the cross-section, then

$$\theta = \frac{TLS}{4A^2Gt} \tag{6.30}$$

Therefore

$$J = \frac{4A^2t}{S} = \text{torsional constant} \tag{6.31}$$

It can be seen from equation (6.31) that the second polar moment of area **cannot** be used for the torsion or non-circular sections.

6.11 Torsion of thin-walled rectangular sections

Consider the thin-walled rectangular cross-section of Figure 6.20, where $b/t \geq 5$.

Assume that the shearing stress due to torsion varies as shown to the right of Figure 6.20; this will be proven to be correct a little later in this section. Analysis of the rectangular cross-section can be achieved if the theory for thin-walled closed tubes is applied to the thin tube, shown shaded in Figure 6.20.

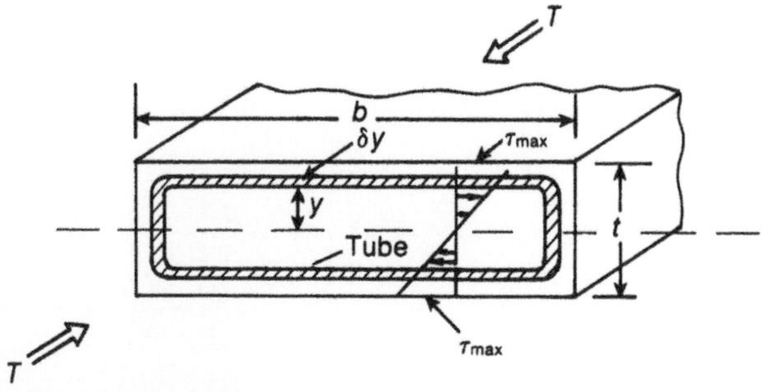

Figure 6.20 Rectangular cross-section

Now from equation (6.24)

$$\tau = \frac{T}{2At}$$

which, when applied to the thin tube of Figure 6.20, gives

$$\tau = \frac{\delta T}{2 \times b \times 2y \times \delta y}$$

or

$$\frac{dT}{dy} = 4by\tau \tag{6.32}$$

From equation (6.30)

$$\theta = \frac{TLS}{4A^2Gt}$$

which, when applied to our thin-walled tube, gives

$$\theta = \frac{\delta TL(2b + 4y)}{4 \times 4b^2y^2G\delta y}$$

but $2b \gg 4y$, so

$$\theta = \frac{\delta T}{\delta y}\,\frac{L}{8by^2G}$$

or

$$\frac{dT}{dy} = \frac{8by^2G\theta}{L} \tag{6.33}$$

Equating (6.32) and (6.33),

$$4b \times y\tau = \frac{8by^2G\theta}{L}$$

or

$$\tau = \frac{2Gy\theta}{L} \tag{6.34}$$

i.e. τ varies **linearly** with y and is a maximum when $y = t/2$. Therefore

$$\tau_{max} = Gt\left(\frac{\theta}{L}\right) \tag{6.35}$$

6.11.1 To obtain the relationship between T and θ

From equation (6.32)

$$dT = 4by\tau\, dy \tag{6.36}$$

Substituting equation (6.34) into (6.36),

$$dT = 4by \times 2Gy \frac{\theta}{L} dy$$

$$= 8bG \frac{\theta}{L} y^2 dy$$

$$T = 8bG\left(\frac{\theta}{L}\right)\int_{-t/2}^{+t/2} y^2 dy$$

$$= 8bG\left(\frac{\theta}{L}\right)\left[\frac{y^3}{3}\right]_{-t/2}^{+t/2}$$

or

$$T = G\left(\frac{bt^3}{3}\right)\left(\frac{\theta}{L}\right) \tag{6.37}$$

Now from equation (6.4),

$$T = GJ\left(\frac{\theta}{L}\right)$$

Therefore

$$J = \text{torsional constant} = bt^3/3 \tag{6.38}$$

which **does not equal** the polar second moment of area!

6.12 Torsion of thin-walled open sections

The theory of Section 6.11 can be used for built-up open sections, such as angle bars, channel and tee sections, as shown in Table 6.1.

For a built-up section,

$$\tau_{max} = Gt_{max}\left(\frac{\theta}{L}\right) \tag{6.39}$$

where (θ/L) is the same for each rectangular element of the cross-section.

Table 6.1 **J for built-up sections**

Section	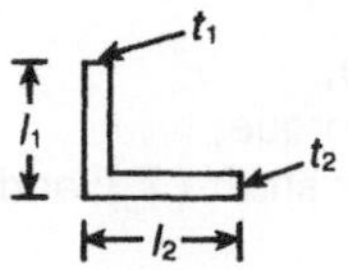	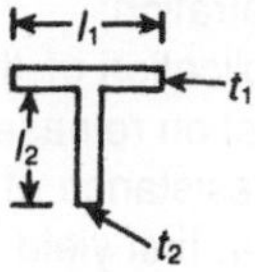	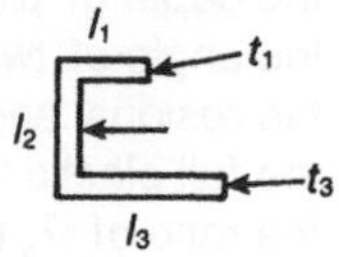
J	$l_1t_1^3/3 + l_2t_2^3/3$	$l_1t_1^3/3 + l_2t_2^3/3$	$l_1t_1^3/3 + l_2t_2^3/3 + l_3t_3^3/3$

Additionally

$$T = GJ\left(\frac{\theta}{l}\right) \tag{6.40}$$

where J is calculated as described in Table 6.1, so that

$$\tau_{max} = \frac{Tt_{max}}{J} \tag{6.41}$$

6.13 Elastic–plastic torsion of circular-section shafts

Equation (6.5) is based on elastic theory, but in practice a circular-section shaft can withstand a much larger torque than that predicted by this theory, because for most materials the shaft can become fully plastic before failure.

In this section the theory will be based on the material behaving in an ideally elastic–plastic material, as shown by Figure 6.21, where τ_{yp} is the shear yield stress.

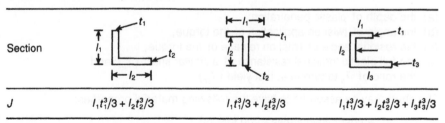

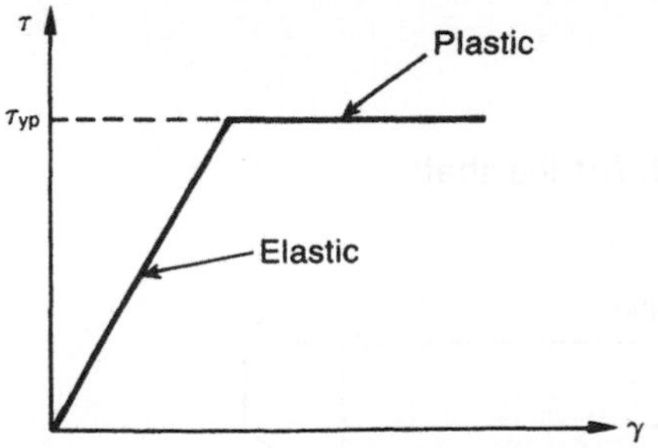

Figure 6.21 Shear stress–shear strain relationship

EXAMPLE 6.7

A solid steel shaft of diameter 0.1 m and length 0.7 m is subjected to a torque of 0.05 MN m, causing the shaft to suffer elastic–plastic deformation.

Determine:

(a) the depth of plastic penetration;
(b) the angle of twist on application of the torque;
(c) the residual angle of twist on release of the torque;
(d) the full plastic torsional resistance of a similar shaft (T_p); and
(e) the ratio of T_p to torque at first yield (T_{yp}).

The shaft may be assumed to have the following material properties:

τ_{yp} = shear yield stress = 200 MN/m²

G = rigidity modulus = 7.7E10 N/m²

SOLUTION

As the shaft section is partially elastic and partially plastic, the shear stress distribution will be as shown in Figure 6.22, where it can be seen that there is an elastic core.

Let

R = radius of shaft
R_p = outer radius of elastic core
T = applied torque = $T_e + T_{p'}$
T_e = torsional resistance of elastic core
$T_{p'}$ = torsional resistance of plastic portion of section (shaded in Figure 6.22(a))
J_e = polar second moment of area of elastic core
τ_{yp} = shear stress at yield

To calculate angle of twist θ

First, it will be necessary to calculate T_e and $T_{p'}$.

From reference 1, the torsional equation for circular-section shafts can be applied to the elastic core, i.e.

$$\frac{\tau_{yp}}{R_p} = \frac{T_e}{J_e} = \frac{G\theta}{l} \quad (6.42)$$

where

G = rigidity or shear modulus
θ = angle of twist over the length l of the shaft
$J_e = \pi R_p^4/2$

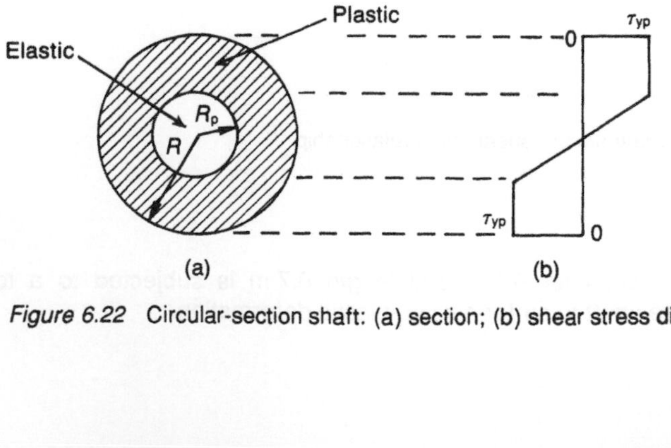

Figure 6.22 Circular-section shaft: (a) section; (b) shear stress distribution

Therefore

$$T_e = \frac{\tau_{yp}}{R_p} \times \frac{\pi R_p^4}{2} = \frac{200\pi R_p^3}{2}$$

or

$$\underline{T_e = 314.16 R_p^3} \qquad (6.43)$$

To calculate $T_{p'}$, consider an annular element of radius r and the thickness dr in the plastic zone, where the shear stress is of constant value τ_{yp}, as shown in Figure 6.23.

From Figure 6.22,

$$T_{p'} = \int_{R_p}^{R} \tau_{yp} \times (2\pi r \times dr) \times r$$

$$= 2\pi\tau_{yp}\left[\frac{r^3}{3}\right]_{R_p}^{R}$$

$$= \frac{2\pi}{3}\tau_{yp}(R^3 - R_p^3) \qquad (6.44)$$

$$\underline{T_{p'} = 418.9(1.25\text{E-}4 - R_p^3)}$$

Now

$$T = T_e + T_{p'}$$

or

$$0.05 \text{ MN m} = 314.16 R_p^3 + 418.9(1.25\text{E-}4 - R_p^3)$$

$$= 0.0524 - 104.7 R_p^3$$

$$\underline{R_p = 0.028 \text{ m}} \qquad (6.45)$$

Therefore *depth of plastic penetration* is 0.05 – 0.028 = $\underline{0.022 \text{ m}}$.

To calculate θ₁, the angle of twist due to T

From (6.42)

$$\theta_1 = \frac{\tau_{yp} \times l}{R_p \times G} = \frac{200\text{E}6 \times 0.7}{0.028 \times 7.7\text{E}10} = 0.064 \text{ rad}$$

$$\underline{\theta_1 = 3.67°}$$

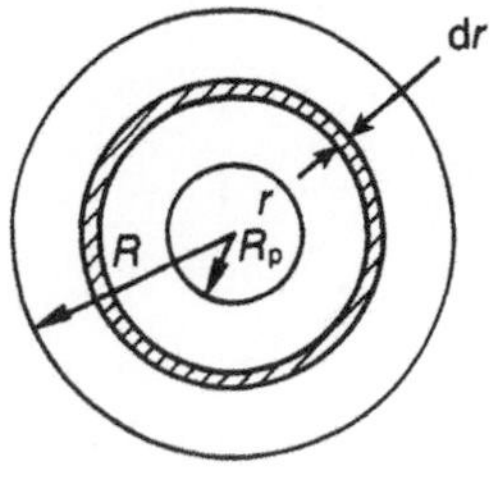

Figure 6.23 Plastic section of shaft

To calculate the residual angle of twist (θ_R)

The $T-\theta$ relationship on loading and on unloading is shown in Figure 6.24 where on removal of the applied torque, the shaft behaves elastically.

From Figure 6.24, it can be seen that θ_2 can be calculated by applying the full torque T to the whole section, and assuming elastic behaviour on unloading, i.e.

$$\theta_2 = \frac{Tl}{GJ} = \frac{0.5E5 \times 0.7}{7.7E10 \times J}$$

but

$$J = \pi \times R^4/2 = 9.817E\text{-}6 \text{ m}^4$$

Therefore

$$\underline{\theta_2 = 0.046 \text{ rad} = 2.65°}$$

Therefore the residual angle of twist is

$$\underline{\theta_R = \theta_1 - \theta_2 = 1.02°}$$

To determine fully plastic torsional resistance (T_p)

From equation (6.44)

$$T_p = \frac{2\pi}{3} \times \tau_{yp} \times R^3$$

$$= \frac{2\pi \times 200 \times 0.05^3}{3}$$

$$\underline{T_p = 0.0524 \text{ MN m}}$$

To determine the ratio T_p/T_{yp}

T_{yp} = maximum torsional resistance up to first yield

$$= \frac{\tau_{yp} \times J}{R} = \frac{200 \times \pi \times 0.1^4}{0.05 \times 32}$$

$$\underline{T = 0.0393 \text{ MN m}}$$

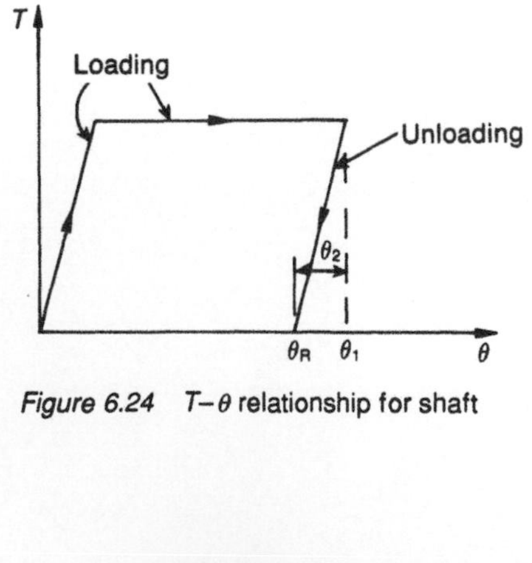

Figure 6.24 $T-\theta$ relationship for shaft

Therefore

$$\underline{\frac{T_p}{T_{yp}} = \frac{0.0524}{0.0393} = 1.334}$$

EXAMPLE 6.8

A compound shaft of length 0.7 m consists of an aluminium-alloy core, of diameter 0.07 m, surrounded co-axially by a steel tube of external diameter 0.1 m. Determine:

(a) the torque that can be applied without causing yield;
(b) the resulting angle of twist on applying a torque of 36 000 N m;
(c) the residual angle of twist remaining on removal of this torque;
(d) the fully plastic torsional resistance of this shaft; and
(e) the ratio of the torque obtained from (d) to that obtained from (c).

The following may be assumed to apply:

Steel

τ_{yps} = shear yield stress in steel = 200 MN/m^2

G_s = rigidity modulus = 7.7E10 N/m^2

Aluminium alloy

τ_{ypa} = shear yield stress in aluminium alloy = 100 MN/m^2

G_a = rigidity modulus = 2.6E10 N/m^2

SOLUTION

The assumption can be made that radial lines remain straight on application or removal of the torque, i.e. θ, the angle of twist for the steel and the aluminium alloy, is constant.

To determine T_{yp}

Prior to determining T_{yp}, it will be necessary to determine whether the aluminium alloy or the steel will yield first.

Considering the angle of twist in the steel

$$\theta = \frac{\tau_{yps} \times l}{R \times G} = \frac{200\text{E}6 \times 0.7}{0.05 \times 7.7\text{E}10}$$

$$\underline{\theta = 0.0364 \text{ rad}} \qquad (6.46)$$

Considering the angle of twist in the aluminium alloy

$$\theta = \frac{\tau_{ypa} \times l}{G \times r} = \frac{100\text{E}6 \times 0.7}{2.6\text{E}10 \times 0.035}$$

$$\underline{\theta = 0.0769 \text{ rad}} \qquad (6.47)$$

That is, if the aluminium alloy were allowed to reach yield, then the steel would become plastic; therefore, the yield stress in the steel is the design criterion.

$$\underline{\text{'Design' } \theta = 0.0364 \text{ rad}} \tag{6.48}$$

Let

J_s = polar second moment of area of the steel tube
J_a = polar second moment of the aluminium-alloy shaft
T_s = elastic torque in steel due to the application of T_{yp}
T_a = elastic torque in the aluminium alloy due to the application of T_{yp}

$$J_s = \frac{\pi \times (0.1^4 - 0.07^4)}{32} = \underline{7.46\text{E-}6 \text{ m}^4}$$

$$J_a = \frac{\pi \times 0.07^4}{32} = \underline{2.357\text{E-}6 \text{ m}^4}$$

From (6.42)

$$T_a = \frac{G_a \theta \times J_a}{l} = \frac{2.6\text{E}10 \times 0.0364 \times 2.357\text{E-}6}{0.7}$$

$$\underline{T_a = 3187 \text{ N m}}$$

Similarly,

$$T_s = \frac{G_s \theta \times J_s}{l} = \frac{7.7\text{E}10 \times 0.0364 \times 7.46\text{E-}6}{0.7}$$

$$\underline{T_s = 29\,871 \text{ N m}}$$

Now,

$$T_{yp} = T_a + T_s = 3187 + 29\,871$$

$$\underline{T_{yp} = 33\,058 \text{ N m}}$$

To determine the angle of twist θ due to a torque of 36 000 N m

On application of the torque, both the steel and the aluminium alloy may be assumed to go plastic, as shown in Figure 6.25.

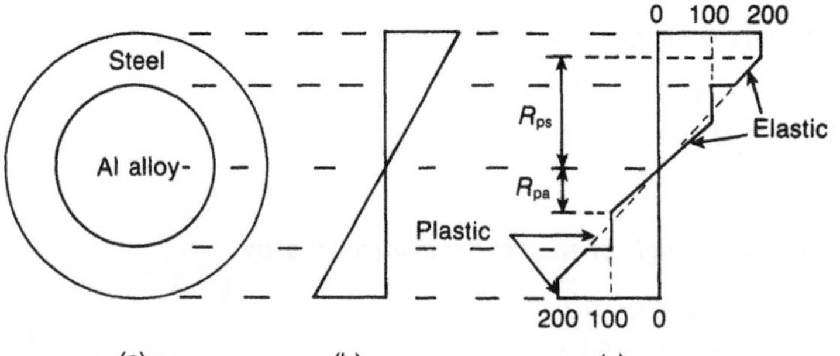

Figure 6.25 Shear stress and shear strain distributions in compound shaft: (a) section; (b) shear strain distribution; (c) shear stress distribution

From equation (6.44),

$T_{p's}$ = torque contribution from the part of the steel that becomes plastic

$$= \frac{2\pi}{3} \times \tau_{yps} \times (0.05^3 - R_{ps}^3)$$

$$\underline{T_{p's} = 418.9 \times (1.25\text{E-}4 - R_{ps}^3)} \qquad (6.49)$$

Also from equation (6.44)

$T_{p'a}$ = torque contribution from that part of the aluminium alloy that becomes plastic

$$= \frac{2\pi}{3} \times \tau_{ypa} \times (0.035^3 - R_{pa}^3)$$

$$\underline{T_{p'a} = 209.4 \times (4.288\text{E-}5 - R_{pa}^3)} \qquad (6.50)$$

The elastic components of torque in the steel (T_{es}) and in the aluminium alloy (T_{ea}) can be calculated from elementary elastic theory, as follows:

$$T_{es} = \frac{\tau_{yps}}{R_{ps}} \times \frac{\pi \times (R_{ps}^4 - 0.035^4)}{2} \qquad (6.51)$$

and

$$T_{ea} = \frac{\tau_{ypa}}{R_{pa}} \times \frac{\pi \times R_{pa}^4}{2}$$

giving

$$\underline{T_{es} = 314.2 \times (R_{ps}^4 - 1.5\text{E-}6)/R_{ps}} \qquad (6.52)$$

and

$$\underline{T_{ea} = 157.1 R_{pa}^3} \qquad (6.53)$$

Now, the total torque is

$$T = T_{p's} + T_{p'a} + T_{es} + T_{ea}$$

Therefore

$$\frac{36\,000}{1\text{E}6} = 418.9 \times (1.25\text{E-}4 - R_{ps}^3) + 209.4 \times (4.288\text{E-}5 - R_{pa}^3)$$
$$+ 314.2 \times (R_{ps}^4 - 1.5\text{E-}6)/R_{ps} + 157.1 R_{pa}^3$$

Therefore

$$0.036 = 0.0524 - 418.9 R_{ps}^3 + 8.979\text{E-}3$$
$$-209.4 R_{pa}^3 + 314.2 R_{ps}^3 - 4.713\text{E-}4/R_{ps} + 157.1 R_{pa}^3$$

$$0.036 = 0.0614 - 104.7 R_{ps}^3 - 52.3 R_{ps}^3 - 4.713\text{E-}4/R_{ps}$$

or

$$0.0254 = 104.7 R_{ps}^3 + 52.3 R_{pa}^3 + 4.713\text{E-}4/R_{ps} \qquad (6.54)$$

This problem is statically indeterminate; hence, it will be necessary to consider compatibility, i.e.

$$\theta = \text{constant} = \frac{\tau}{R} \times \frac{l}{G}$$

Therefore

$$\theta = \frac{\tau_{yps}}{R_{ps}} \times \frac{l}{G_s} = \frac{\tau_{ypa}}{R_{pa}} \times \frac{l}{G_a}$$

or

$$\frac{200\text{E}6}{R_{ps} \times 7.7\text{E}10} = \frac{100\text{E}6}{R_{pa} \times 2.6\text{E}10}$$

Therefore

$$R_{ps} = R_{pa} \times 2 \times 2.6/7.7$$

$$\underline{R_{ps} = 0.675 R_{pa}} \qquad (6.55)$$

However, R_{ps} cannot be less than R_{pa}; therefore the aluminium alloy must be completely elastic, so that

$$\underline{R_{pa} = 0.035 \text{ m} \quad \text{and} \quad T_{p'a} = 0} \qquad (6.56)$$

Furthermore, it can no longer be assumed that the maximum shear stress in the aluminium alloy will reach τ_{ypa}, so that it will be necessary to determine a new expression for T_{ea}.

T_{ea} can be obtained in terms of R_{ps} by considering the compatibility condition that

$$\theta = \text{constant} = \frac{\tau \times l}{G \times r}$$

Therefore

$$\frac{\tau_{yps} \times l}{G_s \times R_{ps}} = \frac{\tau_a \times l}{G_a \times 0.035}$$

where τ_a is the maximum shear stress in the aluminium alloy. Therefore

$$\tau_a = \frac{0.035 \times \tau_{yps} \times G_a}{G_s \times R_{ps}}$$

$$= \frac{0.035 \times 200 \times 2.6\text{E}10}{7.7\text{E}10 \times R_{ps}}$$

$$\underline{\tau_a = 2.364/R_{ps}} \qquad (6.57)$$

and

$$T_{ea} = \frac{\tau_a}{0.035} \times \frac{\pi \times 0.035^4}{2}$$

$$\underline{T_{ea} = 1.592\text{E-}4/R_{ps}} \qquad (6.58)$$

Hence, from (6.49), (6.51) and (6.58),

$$0.036 = 418.9(1.25\text{E-}4 - R_{ps}^3) + \frac{\tau_{yps}}{R_{ps}} \times \frac{\pi(R_{ps}^4 - 0.035^4)}{2} + 1.592\text{E-}4/R_{ps}$$

$$0.036 = 0.0524 - 418.9R_{ps}^3 + 314.2R_{ps}^3 - 4.714\text{E-}4/R_{ps} + 1.592\text{E-}4/R_{ps}$$

Therefore

$$0 = 0.0164 - 104.7R_{ps}^3 - 3.122\text{E-}4/R_{ps}$$

or

$$104.7R_{ps}^4 - 0.0164R_{ps} + 3.122\text{E-}4 = 0 \qquad (6.59)$$

Solving equation (6.59), the four roots of the quartic were found to be as follows:

$$R_{ps} = 0.02\text{ m} \quad 0.0448\text{ m} \quad (-0.0325 \pm 0.0475j)$$

The root of interest is

$$\underline{R_{ps} = 0.0448\text{ m}}$$

Hence, from (6.58),

$$\underline{T_{ea} = 3.554\text{E-}3\text{ MN m}}$$

Therefore

$$\theta = \frac{3.554\text{E-}3 \times 0.7}{(\pi \times 0.035^4/2) \times 2.6\text{E}4} = 0.0406\text{ rad}$$

$$\underline{\theta = 2.326°}$$

To determine the residual angle of twist θ_R

On release of the torque of 36 000 N m, the shaft is assumed to behave elastically. Thus (see Figure 6.26)

$$\theta_1 = \frac{Tl}{GJ}$$

$$= \frac{T_s l}{G_s J_s} = \frac{T_a l}{G_a J_a}$$

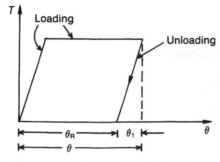

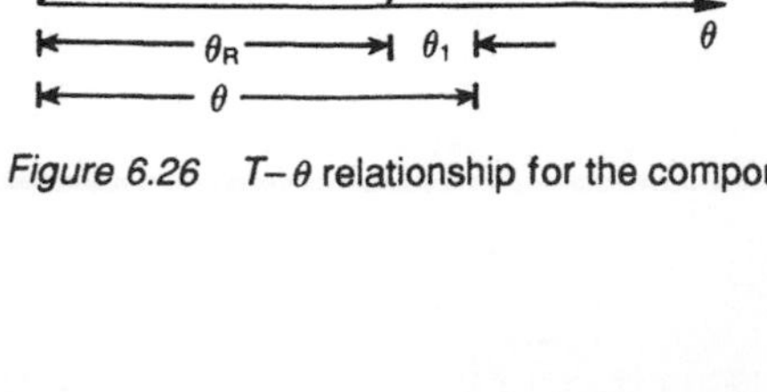

Figure 6.26 $T-\theta$ relationship for the compound shaft

Therefore

$$T_s = \frac{T_a G_s J_s}{G_a J_a} = \frac{T_a \times 7.7\text{E}10 \times 7.46\text{E-}6}{2.6\text{E}10 \times 2.357\text{E-}6}$$

$$\underline{T_s = 9.373 T_a}$$

and

$$T_a = \frac{36\,000}{10.373} = \underline{3470 \text{ N m}}$$

Therefore

$$\theta_1 = \frac{3470 \times 0.7}{2.6\text{E}10 \times 2.357\text{E-}6} = 0.0396 \text{ rad}$$

$$\underline{\theta_1 = 2.27°}$$

and θ_R, the residual angle of twist, is given by

$$\underline{\theta_R = \theta - \theta_1 = 0.056°}$$

To determine the fully plastic torsional resistance of the compound shaft (T_p)

The steel tube will first become fully plastic and then it will rotate, until the aluminium alloy becomes fully plastic.

From equation (6.44)

$$T_{pa} = \frac{2\pi}{3} \times \tau_{ypa} \times 0.035^3$$

$$= \frac{2\pi}{3} \times 100 \times 0.035^3$$

$$\underline{T_{pa} = 8980 \text{ N m}}$$

Similarly,

$$T_{ps} = \frac{2\pi}{3} \times \tau_{yps} \times (0.05^3 - 0.035^3)$$

$$\underline{T_{ps} = 34\,400 \text{ N m}}$$

Therefore the total fully plastic moment of resistance is

$$\underline{T_p = 8980 + 34\,400 = 43\,380 \text{ N m}}$$

and

$$\frac{T_p}{T_{yp}} = \frac{43\,380}{33\,058} = 1.312$$

EXAMPLES FOR PRACTICE 6

1. A circular-section steel shaft consists of three elements, two solid and one hollow, as shown in Figure 6.27, and it is subjected to a torque of 3 kN m.

Determine:

(a) the angle of twist of one end relative to the other;

(b) the maximum shearing stresses in each element.

It may be assumed that $G = 7.7 \times 10^{10}$ N/m^2.

{(a) 1.91°; (b) 70.75 MN/m^2, 29.84 MN/m^2, 31.83 MN/m^2}

2. If the shaft of Example 1 were constructed from three separate materials, namely steel in element ①, aluminium alloy in element ② and manganese bronze in element ③, determine

(a) the total angle of twist;

(b) the maximum shearing stresses in each element.

$G_1 = 7.7 \times 10^{10}$ N/m^2 $G_2 = 2.5 \times 10^{10}$ N/m^2
$G_3 = 3.9 \times 10^{10}$ N/m^2

{(a) 3.24°; (b) 70.75 MN/m^2, 29.84 MN/m^2, 31.83 MN/m^2}

3. Determine the output torque of an electric motor which supplies 5 kW at 25 rev/s.

(a) If this torque is transmitted through a tube of external diameter 20 mm, determine the internal diameter if the maximum permissible shearing stress in the shaft is 35 MN/m^2.

(b) If this shaft is to be connected to another one, via a flanged coupling, determine a suitable bolt diameter if four bolts are used on a pitch circle diameter of 30 mm, and the maximum permissible shearing stresses in the bolts equal 45 MN/m^2.

{31.83 N m; (a) 16.11 mm, say 16 mm; (b) 3.88 mm, say 4 mm}

4. A compound shaft consists of two equal-length hollow shafts joined together in series and subjected to a torque T, as shown in Figure 6.28.

If the shaft on the left is made from steel, and the shaft on the right of the figure is made from aluminium alloy, determine the maximum permissible value of T, given the

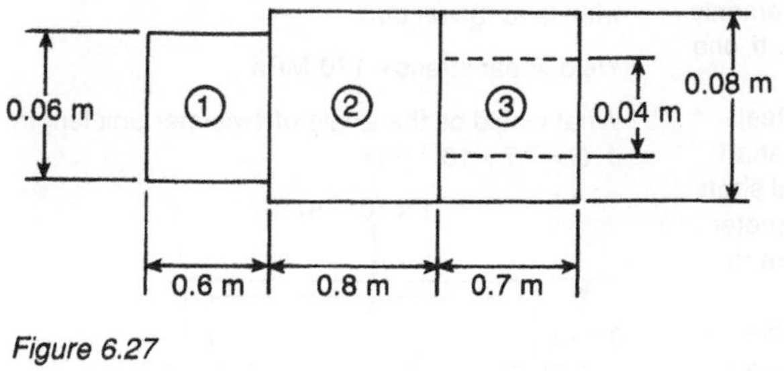

Figure 6.27

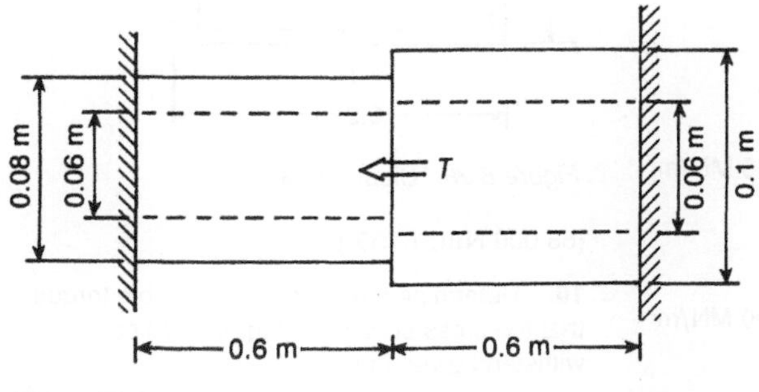

Figure 6.28

following:

For steel

$G = 7.7 \times 10^{10}$ N/m^2
Maximum permissible shear stress = 140 MN/m^2

For aluminium alloy

$G = 2.6 \times 10^{10}$ N/m^2
Maximum permissible shear stress = 90 MN/m^2

(Portsmouth, June 1980)

{19.72 kN m}

5. A compound shaft consists of a solid aluminium-alloy cylinder of length 1 m and diameter 0.1 m, connected in series to a steel tube of the same length and external diameter, and of thickness 0.02 m. The shaft is fixed at its ends and is subjected to an intermediate torque of 9 kN m at the joint.

Determine the angle of twist and the maximum shear stress in the two halves.

$G_{(steel)} = 7.7 \times 10^{10}$ N/m^2

$G_{(Al\ alloy)} = 2.6 \times 10^{10}$ N/m^2

(Portsmouth, June 1982)

{0.569°, $\tau_{(Al\ alloy)} = 12.9$ MN/m^2, $\tau_{(steel)} = 37.82$ MN/m$_2$}

6. A compound shaft consists of two elements of equal length, joined together in series. If one element of the shaft is constructed from gunmetal tube and the other from solid steel, where the external diameter of the steel shaft and the internal diameter of the gunmetal shaft equal 50 mm, determine the external diameter of the gunmetal shaft if the two shafts are to have the some torsional stiffness.

Determine also the maximum permissible torque that can be applied to the shaft, given the following:

For gunmetal

$G = 3 \times 10^{10}$ N/m^2
Maximum permissible shearing stress = 45 MN/m^2

For steel

$G = 7.5 \times 10^{10}$ N/m^2
Maximum permissible shearing stress = 90 MN/m^2

(Portsmouth, June 1977)

{68.4 mm, 2018 N m}

7. A compound shaft consists of a solid steel core, which is surrounded co-axially by an aluminium bronze sheath.

Determine suitable values for the diameters of the shafts if the steel core is to carry two-thirds of a total torque of 500 N m and if the limiting stresses are not exceeded.

For aluminium alloy

$G = 3.8 \times 10^{10}$ N/m^2
Maximum permissible shear stress = 18 MN/m^2

For steel

$G = 7.7 \times 10^{10}$ N/m^2
Maximum permissible shear stress = 36 MN/m^2

{45.4 mm, 38.1 mm, where the stress in the bronze shaft is the design criterion}

8. A circular-section shaft of diameter 0.2 m and length 1 m is subjected to a torque that causes an angle of twist of 3.5°. Determine this torque and the residual angle of twist on removal of this torque.

$G = 7.7$E10 N/m^2 $\tau_{yp} = 180$ MN/m^2

{0.37 MN m, 1.75°}

9. Determine the maximum possible torque that the cross-section of Figure 6.29 can withstand, given that

Yield shear stress = 170 MPa

What would be the angle of twist per unit length if $G = 7.7 \times 10^{10}$ Pa?

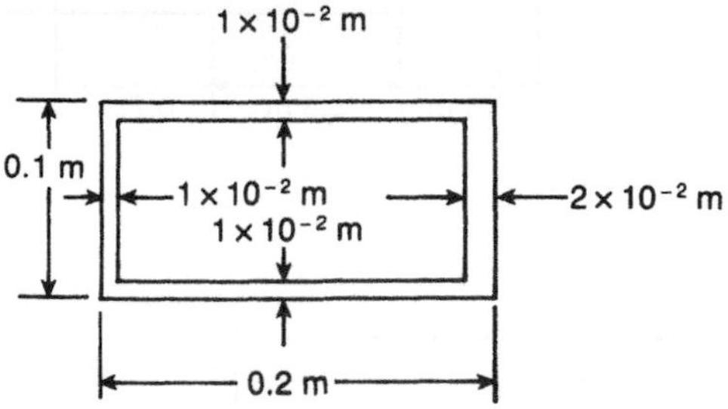

Figure 6.29 Closed tube

{68 000 N m, 1.739°}

10. Determine the maximum possible torque that the cross-section of Figure 6.30 can withstand given that

Yield shear stress = 170 MPa

What would be the angle of twist per unit length if $G = 7.7 \times 10^{10}$ Pa?

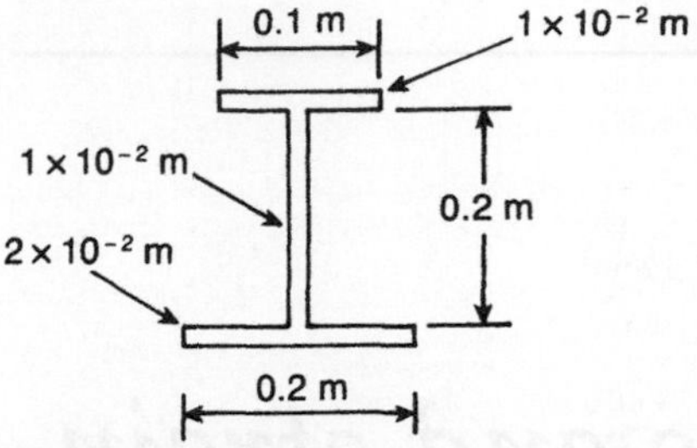

Figure 6.30 Open section

{5383 N m, 6.325°}

7

Complex stress and strain

7.1 Introduction

A typical system of two-dimensional complex stresses, acting on an infinitesimally small rectangular lamina, is shown in Figure 7.1, where the direct stresses σ_x and σ_y are accompanied by a set of shearing stresses τ_{xy}, acting in the x–y plane. These stresses are known as *co-ordinate stresses*, and Figure 7.1 shows the positive signs for the direct stresses σ_x and σ_y.

Positive shearing stresses are said to act clockwise, as shown by the horizontal faces of Figure 7.1, and negative shearing stresses are said to act counter-clockwise, as shown by the vertical faces of Figure 7.1.

The reason for choosing a lamina of rectangular shape is to simplify mathematical computation. That is, σ_x has no component in the y direction and σ_y has no component in the x direction, and also because τ_{xy} is complementary and equal.

In Figure 7.1, the existence of the direct stresses is self-evident, but the reader may not as readily accept the system of shearing stresses. These, however, can be explained with the aid of Section 2.3.2, where it can be seen that shearing stresses are

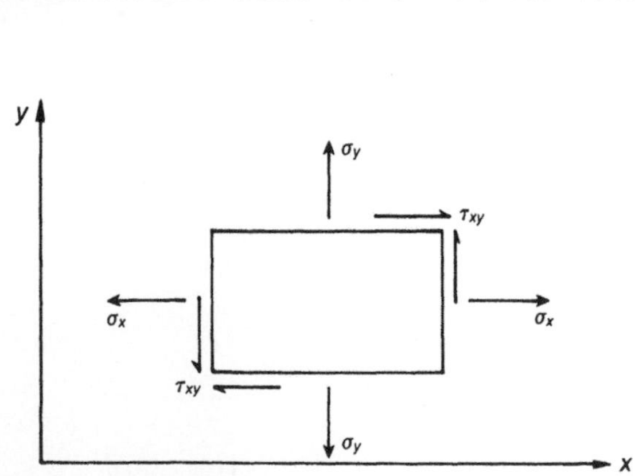

Figure 7.1 Complex stress system

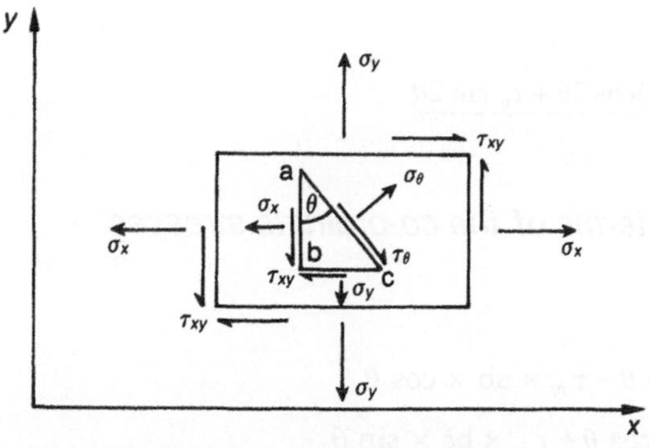

Figure 7.2 Stress system on sub-element abc

complementary and equal; thus, for completeness, it is necessary to assume the shear stress system of Figure 7.1.

In practice, however, it will be useful to have relationships for the direct and shearing stresses at any angle θ, in terms of the co-ordinate stresses.

Consider the stress system acting on the sub-element abc of Figure 7.2, where ac is at an angle θ to the y axis and σ_θ is at an angle θ (anticlockwise) to the x axis.

Let

t = thickness of lamina

σ_θ = direct stress acting on the plane ac

τ_θ = shear stress acting on the plane ac

7.2 To obtain σ_θ in terms of the co-ordinate stresses

Resolving perpendicular to ac,

$$\sigma_\theta \times \text{ac} \times t = \sigma_x \times \text{ab} \times t \times \cos\theta + \sigma_y \times \text{bc} \times t \times \sin\theta$$

$$+ \tau_{xy} \times \text{ab} \times t \times \sin\theta + \tau_{xy} \times \text{bc} \times t \times \cos\theta$$

Therefore

$$\sigma_\theta = \sigma_x \times \frac{\text{ab}}{\text{ac}} \cos\theta + \sigma_y \times \frac{\text{bc}}{\text{ac}} \sin\theta + \tau_{xy} \times \frac{\text{ab}}{\text{ac}} \sin\theta$$

$$+\tau_{xy} \times \frac{\text{bc}}{\text{ac}} \cos\theta$$

$$= \sigma_x \cos^2\theta + \sigma_y \sin^2\theta + 2\tau_{xy} \sin\theta\cos\theta$$

$$= \frac{\sigma_x}{2}(1 + \cos 2\theta) + \frac{\sigma_y}{2}(1 - \cos 2\theta) + 2\tau_{xy}\sin\theta\cos\theta$$

Thus

$$\sigma_\theta = \tfrac{1}{2}(\sigma_x + \sigma_y) + \tfrac{1}{2}(\sigma_x - \sigma_y)\cos 2\theta + \tau_{xy}\sin 2\theta \qquad (7.1)$$

7.2.1 *To determine τ_θ in terms of the co-ordinate stresses*

Resolving parallel to ac,

$$\tau_\theta \times \text{ac} = \sigma_x \times \text{ab} \times \sin\theta - \tau_{xy} \times \text{ab} \times \cos\theta$$

$$- \sigma_y \times \text{bc} \times \cos\theta + \tau_{xy} \times \text{bc} \times \sin\theta$$

NB The effects of t can be ignored, as t appears on both sides of the equation and therefore cancels out.

Therefore

$$\tau_\theta = \sigma_x \times \frac{\text{ab}}{\text{ac}}\sin\theta - \tau_{xy} \times \frac{\text{ab}}{\text{ac}}\cos\theta$$

$$-\sigma_y \times \frac{\text{bc}}{\text{ac}}\cos\theta + \tau_{xy} \times \frac{\text{bc}}{\text{ac}}\sin\theta$$

$$= \sigma_x \cos\theta \sin\theta - \sigma_y \sin\theta\cos\theta - \tau_{xy}\cos^2\theta + \tau_{xy}\sin^2\theta$$

$$= \frac{\sigma_x}{2}\sin 2\theta - \frac{\sigma_y}{2}\sin 2\theta - \frac{\tau_{xy}}{2}(1 + \cos 2\theta)$$

$$+ \frac{\tau_{xy}}{2}(1 - \cos 2\theta)$$

Thus

$$\tau_\theta = \tfrac{1}{2}(\sigma_x - \sigma_y)\sin 2\theta - \tau_{xy}\cos 2\theta \qquad (7.2)$$

One of the main reasons for obtaining equations (7.1) and (7.2) is to determine the magnitudes and directions of the maximum values of σ_θ and τ_θ.

7.3 Principal stresses (σ_1 and σ_2)

The maximum and minimum values of σ_θ occur when

$$\frac{d\sigma_\theta}{d\theta} = 0$$

Hence, from equation (7.1),

$$-(\sigma_x - \sigma_y)\sin 2\theta + 2\tau_{xy}\cos 2\theta = 0$$

or

$$\tan 2\theta = \frac{2\tau_{xy}}{(\sigma_x - \sigma_y)} \tag{7.3}$$

7.3.1 *To determine the maximum and minimum values of σ_θ (i.e. $\hat{\sigma}_\theta$)*

If equation (7.3) is represented by the mathematical triangle of Figure 7.3 then

$$\sin 2\theta = \pm \frac{2\tau_{xy}}{\sqrt{(\sigma_x - \sigma_y)^2 + 4\tau_{xy}^2}} \tag{7.4}$$

and

$$\cos 2\theta = \pm \frac{(\sigma_x - \sigma_y)}{\sqrt{(\sigma_x - \sigma_y)^2 + 4\tau_{xy}^2}} \tag{7.5}$$

Substituting equations (7.4) and (7.5) into equation (7.1),

$$\hat{\sigma}_\theta = \tfrac{1}{2}(\sigma_x + \sigma_y) \pm \frac{1}{2}\frac{(\sigma_x - \sigma_y)(\sigma_x - \sigma_y)}{\sqrt{(\sigma_x - \sigma_y)^2 + 4\tau_{xy}^2}}$$

$$\pm \tau_{xy}\frac{2\tau_{xy}}{\sqrt{(\sigma_x - \sigma_y)^2 + 4\tau_{xy}^2}}$$

or

$$\hat{\sigma}_\theta = \tfrac{1}{2}(\sigma_x + \sigma_y) \pm \tfrac{1}{2}\sqrt{(\sigma_x - \sigma_y)^2 + 4\tau_{xy}^2}$$

i.e. $\hat{\sigma}_\theta$ has two values: a maximum value σ_1 and a minimum value σ_2, where

$$\underline{\sigma_1 = \text{maximum principal stress}}$$

$$\underline{= \tfrac{1}{2}(\sigma_x + \sigma_y) + \tfrac{1}{2}\sqrt{(\sigma_x - \sigma_y)^2 + 4\tau_{xy}^2}} \tag{7.6}$$

$$\underline{\sigma_2 = \text{minimum principal stress}}$$

$$\underline{= \tfrac{1}{2}(\sigma_x + \sigma_y) - \tfrac{1}{2}\sqrt{(\sigma_x - \sigma_y)^2 + 4\tau_{xy}^2}} \tag{7.7}$$

NB Even though σ_2 is the minimum principal stress, if it is negative it can be larger in magnitude than σ_1.

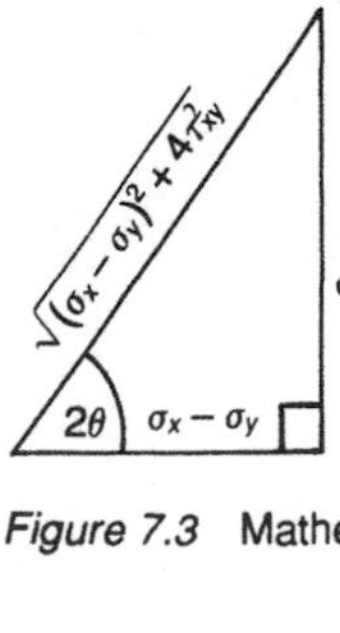

Figure 7.3 Mathematical triangle

7.3.2 Maximum shear stress ($\hat{\tau}$)

$\hat{\tau}$ occurs when

$$\frac{d\tau_\theta}{d_\theta} = 0$$

i.e. from equation (7.2),

$$0 = \tfrac{1}{2}(\sigma_x - \sigma_y)2\cos 2\theta + 2\tau_{xy}\sin 2\theta$$

$$\tan 2\theta = -\frac{(\sigma_x - \sigma_y)}{2\tau_{xy}} = \frac{(\sigma_y - \sigma_x)}{2\tau_{xy}} \quad (7.8)$$

Equation (7.8) can be represented by the mathematical triangle of Figure 7.4, where it can be seen that

$$\cos 2\theta = \frac{2\tau_{xy}}{\sqrt{(\sigma_x - \sigma_y)^2 + 4\tau_{xy}^2}} \quad (7.9)$$

$$\sin 2\theta = \frac{(\sigma_y - \sigma_x)}{\sqrt{(\sigma_x - \sigma_y)^2 + 4\tau_{xy}^2}} \quad (7.10)$$

Substituting equations (7.9) and (7.10) into equation (7.2),

$$\hat{\tau} = \pm\frac{1}{2}\frac{(\sigma_x - \sigma_y)(\sigma_y - \sigma_x)}{\sqrt{(\sigma_x - \sigma_y)^2 + 4\tau_{xy}^2}} \pm \frac{\tau_{xy}2\tau_{xy}}{\sqrt{(\sigma_x - \sigma_y)^2 + 4\tau_{xy}^2}}$$

$$\underline{\hat{\tau} = \pm\sqrt{\tfrac{1}{4}(\sigma_x - \sigma_y)^2 + \tau_{xy}^2}} \quad (7.11)$$

If Figure 7.4 is compared with Figure 7.3, it can be seen that $\hat{\tau}$ occurs on planes which are at 45° to the planes of the principal stresses. Furthermore, it can be seen that if equation (7.3) is substituted into equation (7.2), then $\underline{\tau_\theta = 0}$ on the principal planes. That is, *there are no shearing stresses on a principal plane*, and the maximum shearing stresses occur on planes at 45° to the principal planes.

It can also be seen from equations (7.6), (7.7) and (7.11) that

$$\underline{\hat{\tau} = (\sigma_1 - \sigma_2)/2} \quad (7.12)$$

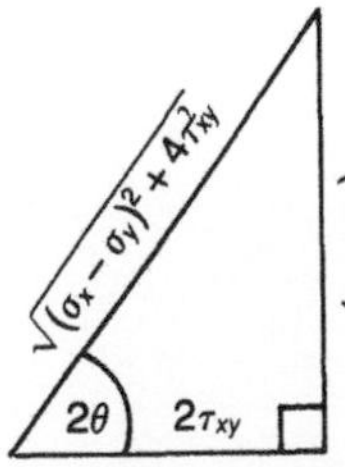

Figure 7.4 Mathematical triangle

7.4 Mohr's stress circle

Equations (7.1) and (7.2) can be represented in terms of principal stresses, by substituting

$$\sigma_x = \sigma_1$$

$$\sigma_y = \sigma_2$$

on a principal plane, and

$$\tau_{xy} = 0$$

so that

$$\sigma_\theta = \tfrac{1}{2}(\sigma_1 + \sigma_2) + \tfrac{1}{2}(\sigma_1 - \sigma_2)\cos 2\theta \tag{7.13}$$

and

$$\tau_\theta = \tfrac{1}{2}(\sigma_1 - \sigma_2)\sin 2\theta \tag{7.14}$$

A careful study of equations (7.13) and (7.14) reveals that they can be represented by a circle of radius $(\sigma_1 - \sigma_2)/2$, if σ_θ is the horizontal axis and τ_θ is the vertical axis, as shown in Figure 7.5. The centre of the circle is at a distance $(\sigma_1 + \sigma_2)/2$ from the origin as shown in Figure 7.5.

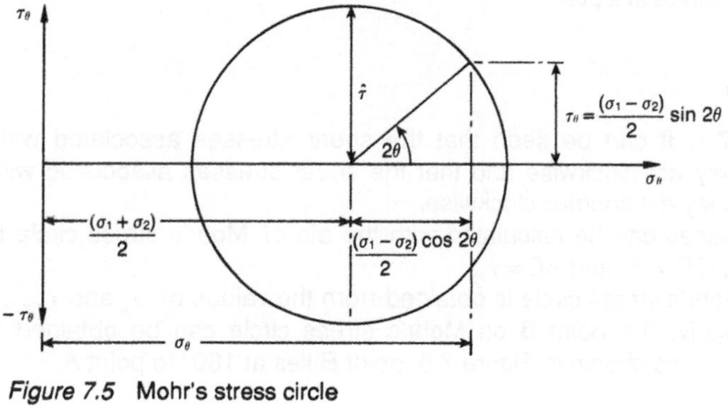

Figure 7.5 Mohr's stress circle

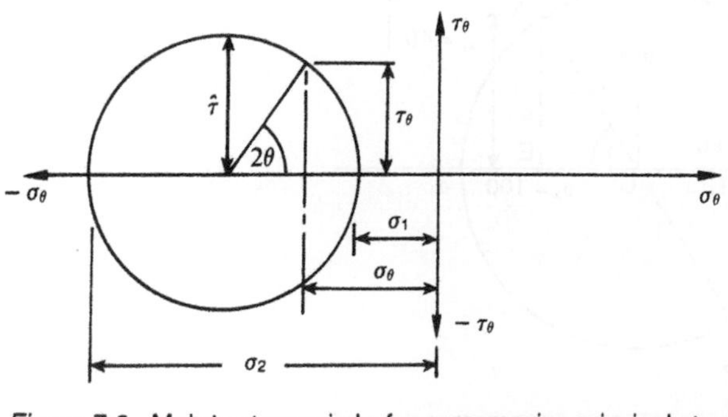

Figure 7.6 Mohr's stress circle for compressive principal stresses

From Figure 7.5, it can be seen that the maximum principal stress is on the right of the circle, and the minimum principal stress is on the left of the circle, but it must be pointed out that the magnitude of the minimum principal stress can be larger than the magnitude of the maximum principal stress, if the former is compressive, as shown in Figure 7.6.

EXAMPLE 7.1

The state of stress in a point of material is shown in Figure 7.7. Determine the direction and magnitudes of the principal stresses.

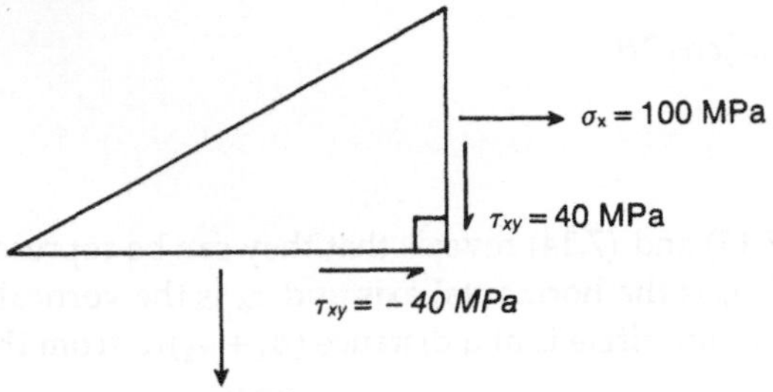

Figure 7.7 Stress at a point

SOLUTION

Now from Figure 7.7, it can be seen that the shear stresses associated with σ_x are positive because they act clockwise and that the shear stresses associated with σ_y are negative, because they act counter-clockwise.

The principal stresses can be calculated with the aid of Mohr's stress circle in Figure 7.8, where OE = σ_x, OD = σ_y and AE = τ_{xy}.

The point A on Mohr's stress circle is obtained from the values of σ_x and τ_{xy}, as shown in Figure 7.8. Similarly, the point B on Mohr's stress circle can be obtained from the values of σ_y and $-\tau_{xy}$, as shown in Figure 7.8; point B lies at 180° to point A.

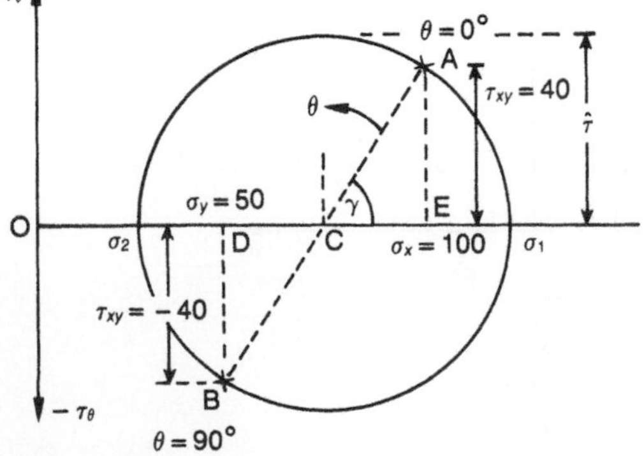

Figure 7.8

From Figure 7.8

$$OC = (OE + OD)/2 = 150/2 = 75$$

$$AC^2 = CE^2 + AE^2$$

where

$$CE = OE - OC = 25$$

Therefore

$$AC^2 = 25^2 + 40^2$$

$$AC = 47.17 \text{ MPa}$$

It is evident that

$$\sigma_1 = OC + AC = 75 + 47.17$$

$$\underline{\sigma_1 = 122.17 \text{ MPa}}$$

Similarly,

$$\sigma_2 = OC - AC = 75 - 47.17$$

$$\underline{\sigma_2 = 27.83 \text{ MPa}}$$

Also from Figure 7.8,

$$\gamma = \tan^{-1}(AE/CE)$$

$$\gamma = \tan^{-1}(1.6) = 58°$$

Therefore

$$\underline{\theta = -\gamma/2 = -29°}$$

from σ_x to σ_1 (Figure 7.9).

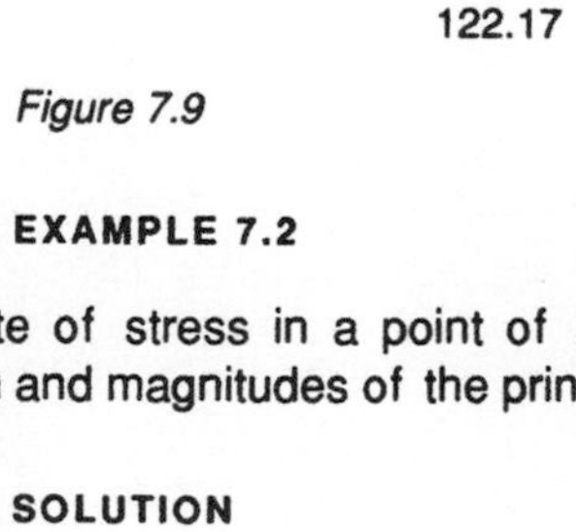

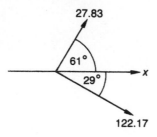

Figure 7.9

EXAMPLE 7.2

The state of stress in a point of material is as shown in Figure 7.10. Determine the direction and magnitudes of the principal stresses.

SOLUTION

From Figure 7.10, it can be seen that as σ_x is tensile it is positive, and also that as σ_y is

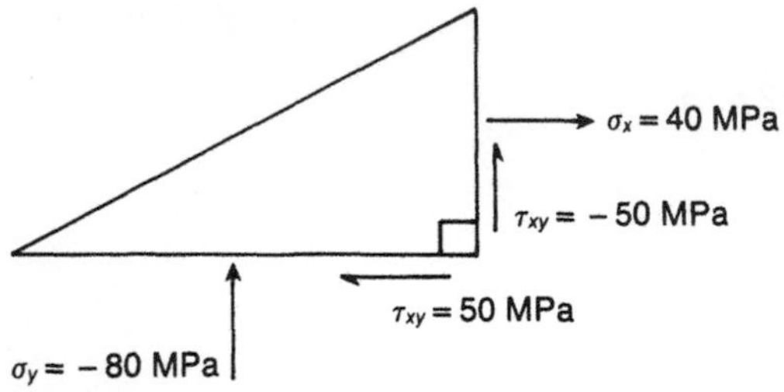

Figure 7.10 Stress acting at a point

compressive, it is negative. Additionally the shearing stress that is acting on the vertical face is causing a counter-clockwise couple, so that it is negative, and that the shearing stress acting on the horizontal face is causing a clockwise couple, so that it is positive.

Thus, point A on Mohr's stress circle is obtained by plotting a positive σ_x and a negative τ_{xy}, as shown in Figure 7.11. Similarly, point B on Mohr's stress circle, which lies at 180° to point A, is obtained by plotting a negative σ_y with a positive τ_{xy}, as shown in Figure 7.11, where OE = σ_x, OD = σ_y and AE = 50.

From Figure 7.11

$$OC = (OE + OD)/2$$

$$OC = (40 - 80)/2 = -20 \text{ MPa}$$

$$CE = -OC + OE = 60$$

$$AC^2 = CE^2 + AE^2$$

$$= 60^2 + 50^2$$

Therefore

$$AC = 78.1$$

Now,

$$\sigma_1 = CF + OC$$

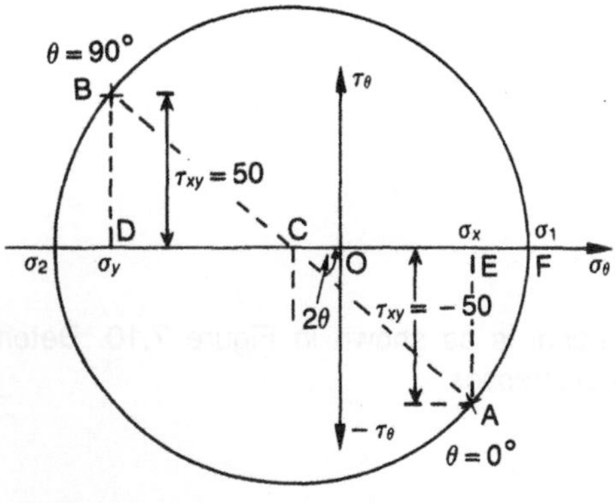

Figure 7.11

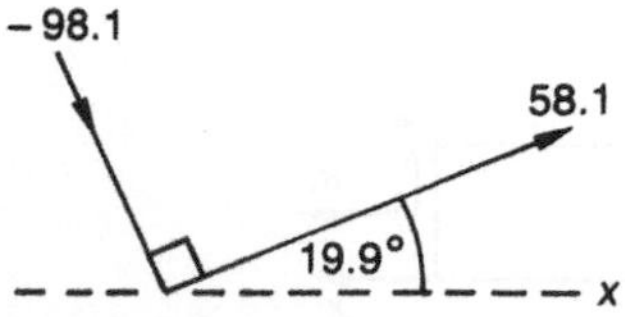

Figure 7.12

but CF = AC. Therefore

$$\underline{\sigma_1 = 78.1 - 20 = 58.1 \text{ MPa}}$$

From Figure 7.11, similarly,

$$\underline{\sigma_2 = -78.1 - 20 = -98.1 \text{ MPa}}$$

Also,

$$2\theta = \tan^{-1}(50/60)$$

Therefore

$$\theta = 19.9°$$

from σ_x to σ_1 (Figure 7.12). Thus

$$\tau_{max} = (\sigma_1 - \sigma_2)/2$$
$$= (58.1 + 98.1)/2$$
$$\underline{\tau_{max} = 78.1 \text{ MPa}}$$

7.5 Combined bending and torsion

Problems in this category frequently occur in circular-section shafts, particularly those of ships. In such cases, the combination of shearing stresses due to torsion and direct stresses due to bending causes a complex system of stress.

Consider a circular-section shaft subjected to a combined bending moment M and a torque T, as shown in Figure 7.13.

The largest bending stresses due to M will be at both the top and the bottom of the shaft, and the combined effects of bending stress and shear stress at those positions will be as shown in Figure 7.14.

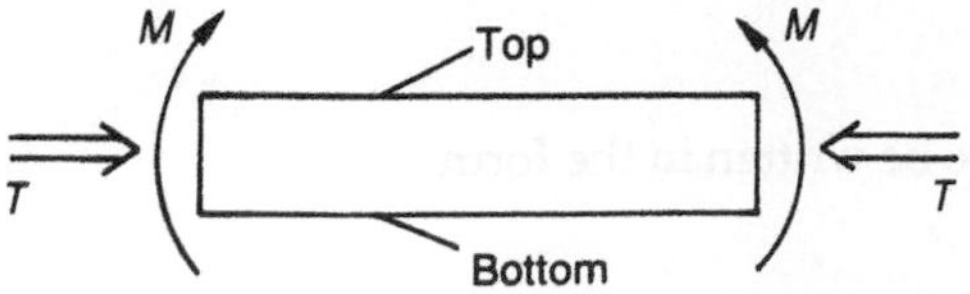

Figure 7.13 Shaft under combined bending and torsion

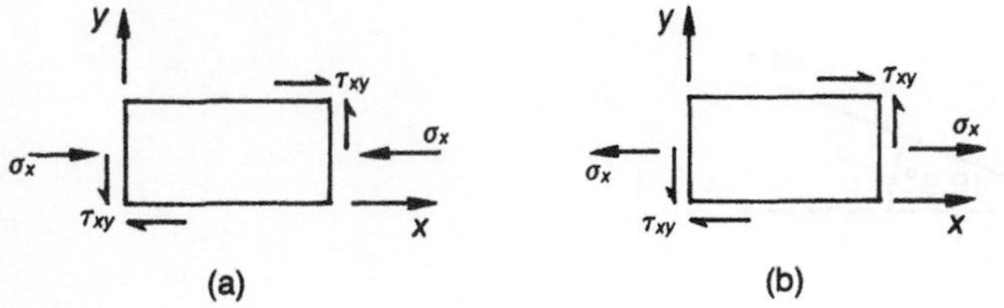

Figure 7.14 Complex stress system due to M and T: (a) top of shaft (looking down); (b) bottom of shaft (looking up)

Now σ_x is entirely due to M, and τ_{xy} is entirely due to T, so that

$$\sigma_x = M\left(\frac{64}{\pi d^4}\right)\frac{d}{2}$$

$$\sigma_x = \frac{32M}{\pi d^3} \tag{7.15}$$

and

$$\tau_{xy} = T\left(\frac{32}{\pi d^4}\right)\frac{d}{2}$$

$$\tau_{xy} = \frac{16T}{\pi d^3} \tag{7.16}$$

From equilibrium considerations,

$$\underline{\sigma_y = 0}$$

Substituting equations (7.15) and (7.16) into equations (7.6) and (7.7),

$$\sigma_1, \sigma_2 = \frac{16M}{\pi d^3} \pm \sqrt{\left(\frac{16M}{\pi d^3}\right)^2 + \left(\frac{16T}{\pi d^3}\right)^2}$$

$$\underline{\sigma_1, \sigma_2 = \frac{16}{\pi d^3}(M \pm \sqrt{M^2 + T^2})} \tag{7.17}$$

Similarly,

$$\underline{\hat{\tau} = \frac{16}{\pi d^3}(M^2 + T^2)} \tag{7.18}$$

Equations (7.17) and (7.18) can also be written in the form

$$\underline{\sigma_1, \sigma_2 = \frac{32M_e}{\pi d^3}} \tag{7.19}$$

and

$$\hat{\tau} = \frac{16T_e}{\pi d^3} \tag{7.20}$$

where

$$M_e = \text{equivalent bending moment}$$
$$= \tfrac{1}{2}\left(M \pm \sqrt{M^2 + T^2}\right) \tag{7.21}$$

and

$$T_e = \text{equivalent torque}$$
$$= \sqrt{M^2 + T^2} \tag{7.22}$$

EXAMPLE 7.3

A ship's propeller shaft is of a solid circular cross-section, of diameter 0.25 m. If the shaft is subjected to an axial thrust of 1 MN, together with a bending moment of 0.02 MN m and a torque of 0.05 MN m, determine the magnitude of the largest direct stress.

SOLUTION

$$I = \frac{\pi \times 0.25^4}{64} = 1.917\text{E-4 m}^4$$

$$J = 3.83\text{E-4 m}^4$$

$$A = \frac{\pi \times 0.25^2}{4} = 0.0491 \text{ m}^2$$

The value of σ_x due to bending, which is of interest, is the negative value, because the axial thrust causes a compressive stress, and if these two are added together, the value of the largest σ_x will be given by

$$\sigma_x = -\frac{1}{0.0491} - \frac{0.02}{1.917\text{E-4}} \times 0.125$$
$$= -20.37 - 13.04$$

$$\underline{\sigma_x = -33.41 \text{ MN/m}^2} \tag{7.23}$$

$$\underline{\sigma_y = 0} \tag{7.24}$$

$$\tau_{xy} = \frac{0.05}{3.83\text{E-4}} \times 0.125 = \underline{16.32 \text{ MN/m}^2} \tag{7.25}$$

Substituting equations (7.23)–(7.25) into equations (7.6) and (7.7),

$$\underline{\sigma_1 = 6.65\ \text{MN/m}^2}$$

$$\underline{\sigma_2 = -40.06\ \text{MN/m}^2}$$

i.e. the largest magnitude of direct stress is compressive and equal to $-40.06\ \text{MN/m}^2$.

EXAMPLE 7.4

The stresses σ_x, τ_{xy} and σ_1 are known on the triangular element abc of Figure 7.15. Determine the principal stresses, and also σ_y and α.

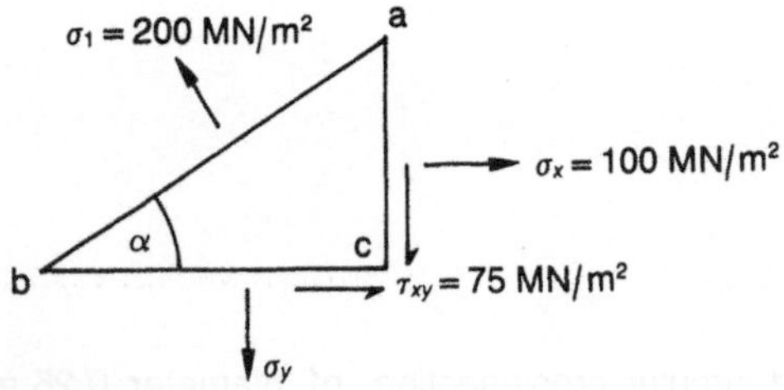

Figure 7.15

SOLUTION

As there is no shear stress on the plane ab, σ_1 is a principal stress, and as σ_1 is greater than σ_x, σ_1 must be the maximum principal stress:

$$\underline{\sigma_1 = 200\ \text{MN/m}^2}$$

Resolving horizontally

$$200 \times \text{ab} \sin\alpha = 75 \times \text{bc} + 100 \times \text{ac}$$

$$200 = 75 \times \frac{\text{bc}}{\text{ab} \sin\alpha} + 100 \times \frac{\text{ac}}{\text{ab} \sin\alpha}$$

$$= 75 \cot\alpha + 100$$

$$\cot\alpha = \tfrac{100}{75} \quad \text{or} \quad \tan\alpha = 0.75$$

Therefore

$$\underline{\alpha = 36.87^\circ}$$

Resolving vertically

$$200 \times \text{ab} \cos\alpha = 75 \times \text{ac} + \sigma_y \times \text{bc}$$

$$\sigma_y = 200 \times \frac{\text{ab}}{\text{bc}} \cos\alpha - 75 \times \frac{\text{ac}}{\text{bc}}$$

$$= 200 - 75 \tan\alpha$$

$$\underline{\sigma_y = 143.75\ \text{MN/m}^2}$$

From equation (7.7),

$$\sigma_2 = \frac{243.75}{2} - \tfrac{1}{2}\sqrt{43.75^2 + 4 \times 75^2}$$

$$= 121.88 - 78.13$$

$$\underline{\sigma_2 = 43.75 \text{ MN/m}^2}$$

7.6 Two-dimensional strain systems

In a number of practical situations, particularly with experimental strain analysis, it is more convenient to make calculations with equations involving strains. Hence, for such cases, it will be necessary to obtain the expressions for strains, rather similar to those obtained for stresses in Section 7.3.1.

Consider an infinitesimal rectangular elemental lamina of material OABC, in the x–y plane, which is subjected to an in-plane stress system that causes the lamina to strain, as shown by the deformed quadrilateral OA'B'C' in Figure 7.16. Let the co-ordinate strains ε_x, ε_y and γ_{xy} be defined as follows:

ε_x = direct strain in the x direction

ε_y = direct strain in the y direction

γ_{xy} = shear strain in the x–y plane

7.6.1 *To obtain ε_θ in terms of the co-ordinate strains*

Let

ε_θ = direct strain at angle θ

OB = r

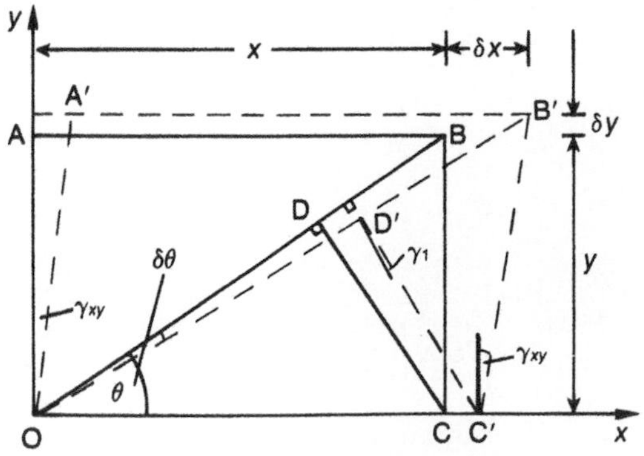

Figure 7.16 Strained quadrilateral

so that

$$BC = y = r \sin \theta$$

$$OC = x = r \cos \theta$$

Hence,

$$\delta_y = r \sin \theta \, \varepsilon_y$$

and

$$\delta x = r \cos \theta \, \varepsilon_x + r \sin \theta \, \gamma_{xy}$$

7.6.2 *Consider the movement of B parallel to OB*

Now,

$$(OB)^2 = x^2 + y^2 = r^2$$

and

$$(OB')^2 = (x + \delta x)^2 + (y + \delta y)^2 = (r + \delta r)^2$$

i.e.

$$x^2 + 2x\,\delta x + (\delta x)^2 + y^2 + 2y\,\delta y + (\delta y)^2 = r^2 + 2r\,\delta r + (dr)^2$$

or

$$x^2 + y^2 + 2x\,\delta x + 2y\,\delta y = r^2 + 2r\,\delta r$$

but as

$$x^2 + y^2 = r^2$$

then

$$\delta r = \frac{x}{r}\,\delta x + \frac{y}{r}\,\delta y$$

or

$$\delta r = \delta x \cos \theta + \delta y \sin \theta$$

Since the strains are small,

$$\varepsilon_\theta = \frac{\delta r}{r}$$

$$= \frac{\delta x}{r} \cos \theta + \frac{\delta y}{r} \sin \theta$$

Substituting for δx and δy,

$$\varepsilon_\theta = \varepsilon_x \cos^2\theta + \varepsilon_y \sin^2\theta + \gamma_{xy} \sin \theta \cos \theta$$

so that

$$\varepsilon_\theta = \tfrac{1}{2}(\varepsilon_x + \varepsilon_y) + \tfrac{1}{2}(\varepsilon_x - \varepsilon_y)\cos 2\theta + \tfrac{1}{2}\gamma_{xy}\sin 2\theta \qquad (7.26)$$

This is similar in form to the equation for stress at any angle θ.

7.6.3 *To determine the shearing strain γ_θ in terms of co-ordinate strains*

To evaluate shearing strain at an angle θ, we note that D is displaced to D′.

Now, as

$$BB' = \varepsilon_\theta r$$

and

$$OD = OC\cos\theta = r\cos^2\theta$$

then

$$DD' = \varepsilon_\theta OD = \varepsilon_\theta r\cos^2\theta$$

During straining, the line CD rotates anticlockwise through a small angle γ_1, where

$$\gamma_1 = \left(\frac{CC'\cos\theta - DD'}{CD}\right) = \left(\frac{\varepsilon_x\cos^2\theta - \varepsilon_\theta\cos^2\theta}{\cos\theta\sin\theta}\right)$$

Dividing the denominator into the numerator,

$$\gamma_1 = (\varepsilon_x - \varepsilon_\theta)\cot\theta$$

At the same time, OB rotates in a clockwise direction through a small angle, $\delta\theta$, as shown in Figure 7.17;

$$\theta = \tan^{-1}\left(\frac{y}{x}\right)$$

$$\delta\theta = -\left(\frac{\partial\theta}{\partial x}\delta x + \frac{\partial\theta}{\partial y}\delta y\right)$$

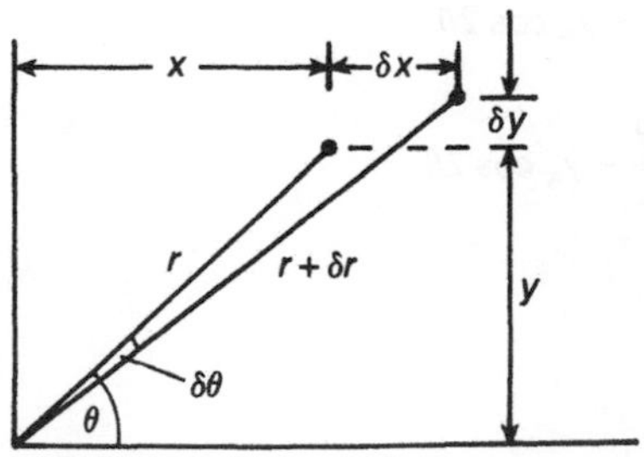

Figure 7.17

Now,

$$\frac{\partial \theta}{\partial x} = \frac{-y/x^2}{1+(y/x)^2} = \frac{-y}{x^2+y^2} = \frac{-y}{r^2} = \frac{-\sin\theta}{r}$$

$$\frac{\partial \theta}{\partial y} = \frac{1/x}{1+(y/x)^2} = \frac{x}{x^2+y^2} = \frac{x}{r^2} = \frac{\cos\theta}{r}$$

Therefore

$$\delta\theta = \frac{(\delta x \sin\theta - \delta y \cos\theta)}{r}$$

$$= \frac{[(r\cos\theta\,\varepsilon_x + \gamma_{xy}\, r\sin\theta)\sin\theta - (r\sin\theta\,\varepsilon_y)\cos\theta]}{r}$$

and the shear strain at any angle θ is

$$\gamma_\theta = \gamma_1 + \delta\theta$$

$$\gamma_\theta = (\varepsilon_x - \varepsilon_\theta)\cot\theta + (\varepsilon_x\cos\theta + \gamma_{xy}\sin\theta)\sin\theta - \varepsilon_y\sin\theta\cos\theta$$

Substituting for ε_θ from equation (7.26), we get

$$\gamma_\theta = \varepsilon_x\cot\theta - \tfrac{1}{2}(\varepsilon_x + \varepsilon_y)\cot\theta$$

$$-\tfrac{1}{2}(\varepsilon_x - \varepsilon_y)\cot\theta\cos 2\theta - \tfrac{1}{2}\gamma_{xy}\sin 2\theta\cot\theta$$

$$+\varepsilon_x\cos\theta\sin\theta + \gamma_{xy}\sin^2\theta - \varepsilon_y\sin\theta\cos\theta$$

$$= \tfrac{1}{2}(\varepsilon_x - \varepsilon_y)\cot\theta - \tfrac{1}{2}(\varepsilon_x - \varepsilon_y)\cot\theta\cos 2\theta$$

$$-\sin\theta\cos\theta(-\varepsilon_x + \varepsilon_y) - \gamma_{xy}\left(\frac{\sin 2\theta\cot\theta}{2} - \sin^2\theta\right)$$

$$= \tfrac{1}{2}(\varepsilon_x - \varepsilon_y)\cot\theta(1 - \cos 2\theta) + (\varepsilon_x - \varepsilon_y)\frac{\sin 2\theta}{2}$$

$$-\gamma_{xy}\left(\sin\theta\cos\theta\frac{\cos\theta}{\sin\theta} - \sin^2\theta\right)$$

$$= (\varepsilon_x - \varepsilon_y)\cot\theta\sin^2\theta + (\varepsilon_x - \varepsilon_y)\frac{\sin 2\theta}{2} - \gamma_{xy}\cos 2\theta$$

$$= (\varepsilon_x - \varepsilon_y)\frac{\cot\theta}{\sin\theta}\sin^2\theta + \frac{(\varepsilon_x - \varepsilon_y)\sin 2\theta}{2} - \gamma_{xy}\cos 2\theta$$

$$\gamma_\theta = (\varepsilon_x - \varepsilon_y)\sin 2\theta - \gamma_{xy}\cos 2\theta \qquad (7.27)$$

Equation (7.27) is often written in the form

$$\frac{\gamma_\theta}{2} = \tfrac{1}{2}(\varepsilon_x - \varepsilon_y)\sin 2\theta - \tfrac{1}{2}\gamma_{xy}\cos 2\theta \qquad (7.28)$$

7.7 Principal strains (ε_1 and ε_2)

Principal strains are the maximum and minimum values of direct strain, and they are obtained by satisfying the equation

$$\frac{d\varepsilon_\theta}{d\theta} = 0$$

Hence, from equation (7.26),

$$0 = -(\varepsilon_x - \varepsilon_y)\sin 2\theta + \gamma_{xy}\cos 2\theta$$

i.e.

$$\tan 2\theta = \frac{\gamma_{xy}}{(\varepsilon_x - \varepsilon_y)} \qquad (7.29)$$

Equation (7.29) can be represented by the mathematical triangle of Figure 7.18. From this figure

$$\sin 2\theta = \pm\frac{\gamma_{xy}}{\sqrt{(\varepsilon_x - \varepsilon_y)^2 + \gamma_{xy}^2}} \qquad (7.30)$$

and

$$\cos 2\theta = \pm\frac{\varepsilon_x - \varepsilon_y}{\sqrt{(\varepsilon_x - \varepsilon_y)^2 + \gamma_{xy}^2}} \qquad (7.31)$$

Substituting equations (7.30) and (7.31) into equation (7.28),

$$\underline{\gamma_\theta = 0}$$

i.e. the shear strain on a principal plane is zero, as is the case for shear stress. Furthermore, as $\tau = G\gamma$, it follows that the *planes for principal strains are the same for principal stresses.*

7.7.1 *To obtain the expressions for the principal strains in terms of the co-ordinate strains*

The values of the principal strains can be obtained by substituting equations (7.30)

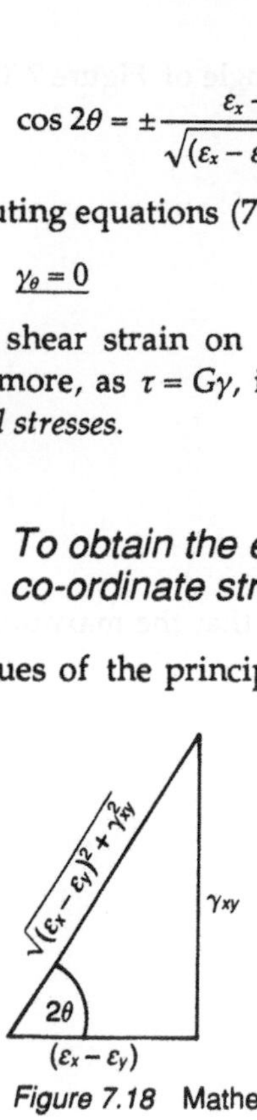

Figure 7.18 Mathematical triangle

and (7.31) into equation (7.26):

$$\underline{\varepsilon_1, \varepsilon_2 = \tfrac{1}{2}(\varepsilon_x + \varepsilon_y) \pm \tfrac{1}{2}\sqrt{(\varepsilon_x - \varepsilon_y)^2 + \gamma_{xy}^2}} \tag{7.32}$$

7.7.2 To determine the value and direction of the maximum shear strain ($\hat{\gamma}$)

The direction of $\hat{\gamma}$ can be obtained by satisfying the condition

$$\frac{d\gamma_\theta}{d\theta} = 0$$

Hence, from equation (7.28),

$$0 = (\varepsilon_x - \varepsilon_y)\cos 2\theta + \gamma_{xy} \sin 2\theta$$

Therefore

$$\tan 2\theta = -\frac{(\varepsilon_x - \varepsilon_y)}{\gamma_{xy}} \tag{7.33}$$

Equation (7.33) can be represented by the mathematical triangle of Figure 7.19. From this figure

$$\sin 2\theta = \pm(\varepsilon_y - \varepsilon_x)/\sqrt{(\varepsilon_x - \varepsilon_y)^2 + \gamma_{xy}^2} \tag{7.34}$$

and

$$\cos 2\theta = \pm\gamma_{xy}/\sqrt{(\varepsilon_x - \varepsilon_y)^2 + \gamma_{xy}^2} \tag{7.35}$$

Substituting equations (7.34) and (7.35) into equation (7.28),

$$\hat{\gamma} = \sqrt{(\varepsilon_x - \varepsilon_y)^2 + \gamma_{xy}^2} \tag{7.36}$$

which, when compared with equation (7.32), reveals that

$$\underline{\hat{\gamma} = \pm(\varepsilon_1 - \varepsilon_2)} \tag{7.37}$$

If Figure 7.19 is compared with Figure 7.18, it can be seen that the maximum shear strain occurs at 45° to the principal planes.

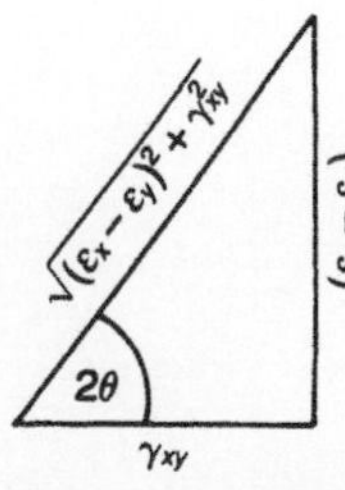

Figure 7.19 Mathematical triangle

7.8 Mohr's circle of strain

On a principal plane,

$$\varepsilon_x = \varepsilon_1$$

$$\varepsilon_y = \varepsilon_2$$

and

$$\gamma_{xy} = 0$$

which, when substituted into equations (7.26) and (7.28), yield the following expressions:

$$\varepsilon_\theta = \tfrac{1}{2}(\varepsilon_1 - \varepsilon_2) + \tfrac{1}{2}(\varepsilon_1 - \varepsilon_2)\cos 2\theta \qquad (7.38)$$

and

$$\frac{\gamma_\theta}{2} = \frac{(\varepsilon_1 - \varepsilon_2)}{2}\sin 2\theta \qquad (7.39)$$

Equations (7.38) and (7.39) can be represented by a circle if ε_θ is taken as the horizontal axis and $\gamma_\theta/2$ as the vertical axis, as shown in Figure 7.20.

Plane stress is a two-dimensional system of stress and a three-dimensional system of strain, where the stresses σ_x and σ_y act in the plane of the plate, as shown in Figure 7.21.

In addition to causing strains in the x and y directions, these stresses will cause an out-of-plane strain due to the Poisson effect.

7.9 Stress–strain relationships for plane stress

From Figure 7.21, it can be seen that the strain in the x direction due to σ_x is σ_x/E, and the strain in the x direction due to σ_y is $-\nu\sigma_y/E$, so that ε_x, the strain in the x

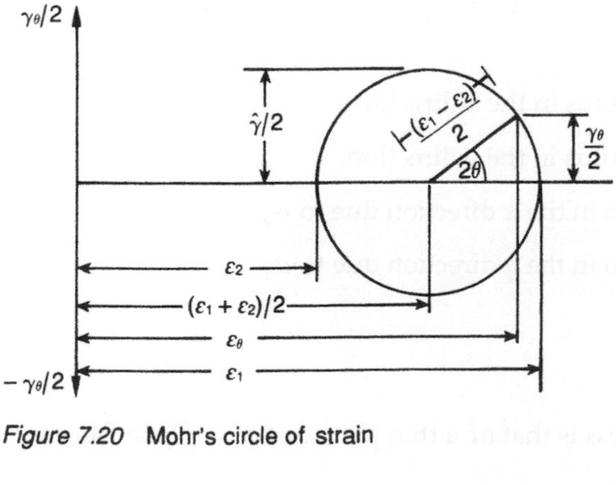

Figure 7.20 Mohr's circle of strain

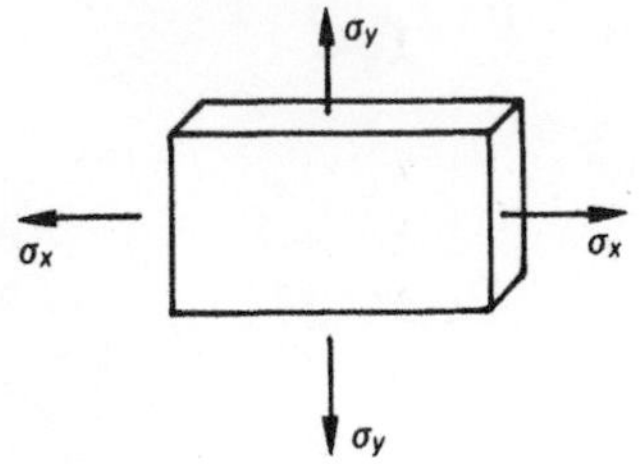

Figure 7.21 Plane stress

direction due to the combined stresses, is given by

$$\varepsilon_x = (\sigma_x - \nu\sigma_y)/E \tag{7.40}$$

Similarly, ε_y, the strain in the y direction due to the combined stress, is given by

$$\varepsilon_y = (\sigma_y - \nu\sigma_x)/E \tag{7.41}$$

The stress–strain relationships of equations (7.40) and (7.41) can be put in the alternative from of equations (7.42) and (7.43):

$$\sigma_x = \frac{E}{(1-\nu^2)}(\varepsilon_x + \nu\varepsilon_y) \tag{7.42}$$

$$\sigma_y = \frac{E}{(1-\nu^2)}(\varepsilon_y + \nu\varepsilon_x) \tag{7.43}$$

For an orthotropic material, equations (7.42) and (7.43) can be put in the form

$$\sigma_x = \frac{1}{(1-\nu_x\nu_y)}(E_x\varepsilon_x + \nu_x E_y\varepsilon_y) \tag{7.44}$$

$$\sigma_y = \frac{1}{(1-\nu_x\nu_y)}(E_y\varepsilon_y + \nu_y E_x\varepsilon_x) \tag{7.45}$$

where

E_x = Young's modulus in the x direction

E_y = Young's modulus in the y direction

ν_x = Poisson's ratio in the x direction due to σ_y

ν_y = Poisson's ratio in the y direction due to σ_x

and

$$\nu_x E_y = \nu_y E_x \tag{7.46}$$

A typical case for plane stress is that of a thin plate under in-plane forces.

7.10 Stress–strain relationships for plane strain

Plane strain is a three-dimensional system of stress and a two-dimensional system of strain, as shown in Figure 7.22, where the out-of-plane stress σ_z is related to the in-plane stresses σ_x and σ_y, and Poisson's ratio, so that the out-of-plane strain is zero. From Figure 7.22

$$\varepsilon_x = (\sigma_x - \nu\sigma_y - \nu\sigma_z)/E \tag{7.47}$$

$$\varepsilon_y = (\sigma_y - \nu\sigma_x - \nu\sigma_z)/E \tag{7.48}$$

$$\varepsilon_z = 0 = (\sigma_z - \nu\sigma_x - \nu\sigma_y)/E \tag{7.49}$$

From equation (7.49),

$$\sigma_z = \nu(\sigma_x + \sigma_y) \tag{7.50}$$

Substituting equation (7.50) into equations (7.47) and (7.48), these last two equations can be put in the form of equations (7.51) and (7.52), which is usually found to be more convenient:

$$\sigma_x = \frac{E}{(1+\nu)(1-2\nu)}[(1-\nu)\varepsilon_x + \nu\varepsilon_y] \tag{7.51}$$

$$\sigma_y = \frac{E}{(1+\nu)(1-2\nu)}[(1-\nu)\varepsilon_y + \nu\varepsilon_x] \tag{7.52}$$

For both plane stress and plane strain,

$$\tau_{xy} = G\gamma_{xy} \tag{7.53}$$

A typical case for plane strain occurs at the mid-length of the cross-section of a gravity dam.

7.11 Pure shear

A system of pure shear is shown by Figure 7.23, where the shear stresses τ are not accompanied by direct stresses on the same planes. As these shear stresses are

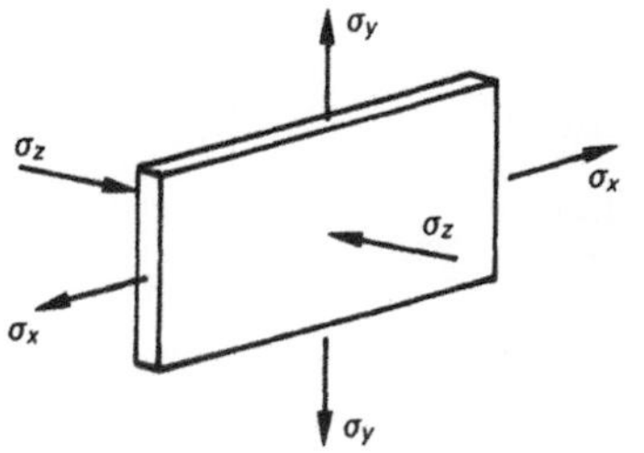

Figure 7.22 Plane strain

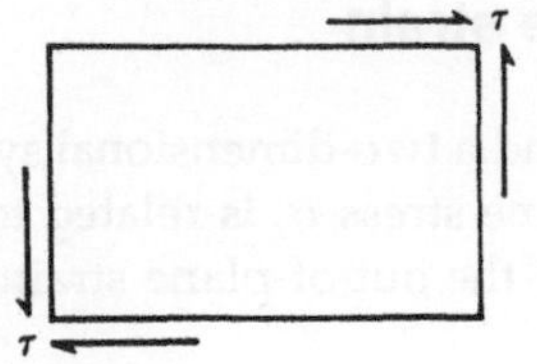

Figure 7.23 Pure shear

maximum shear stresses at the point, the principal stresses will lie at 45° to the direction of these shear stresses, as shown by Figure 7.24, where

σ_1 = maximum principal stress

σ_2 = minimum principal stress

By resolution it can be seen that

$$\sigma_1 = \tau \quad \text{and} \quad \sigma_2 = -\tau \tag{7.54}$$

i.e. the system of shear stresses of Figure 7.23 is equivalent to the system of Figure 7.25. Thus, if shear strain is required to be measured, the strain gauges have to be placed at 45° to the directions of shear stresses, as shown in Figure 7.26. These strains will measure the principal strains, ε_1 and ε_2, and such strain gauges are known as shear pairs (see Chapter 16).

From equation (7.37),

$$\gamma = \varepsilon_1 - \varepsilon_2 \tag{7.55}$$

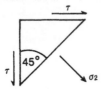

Figure 7.24

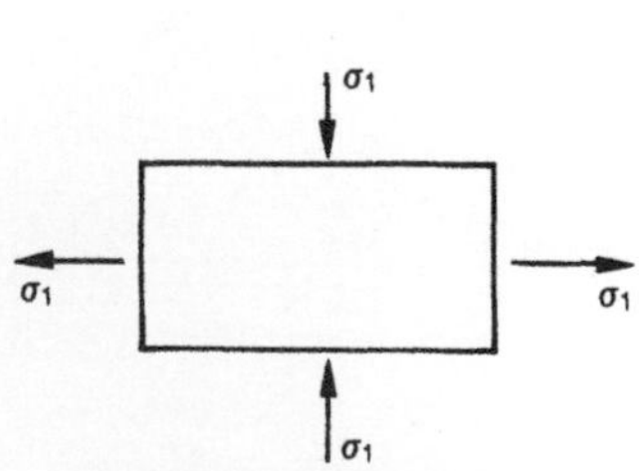

Figure 7.25 Pure shear

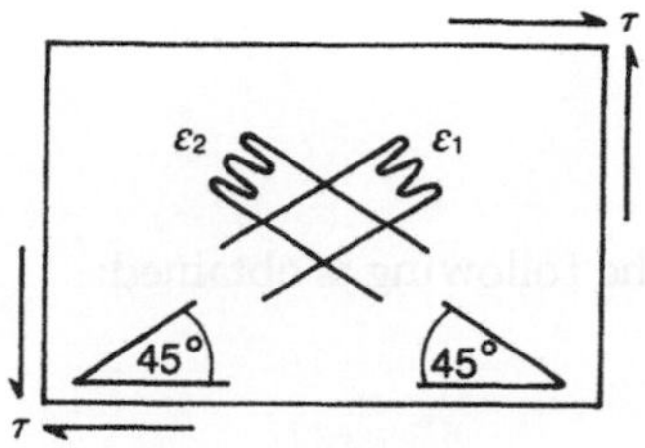

Figure 7.26 Pure shear

and

$$\tau = G\gamma \tag{7.56}$$

NB If strain gauges were placed in the direction of x or y, then the gauges would simply change their shapes, as shown in Figure 7.27 and would not measure strain.

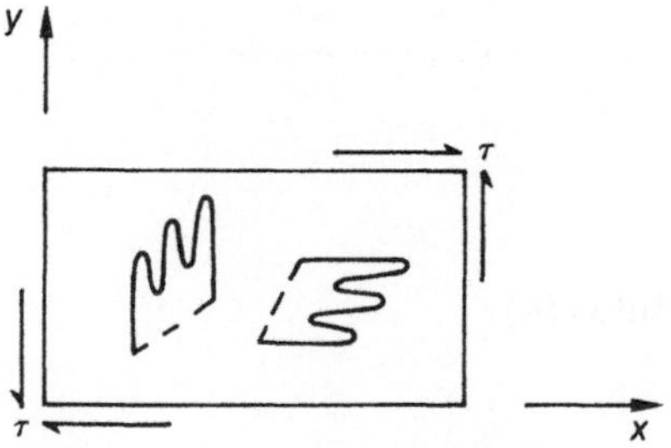

Figure 7.27 Incorrect method of shear strain measurement

7.11.1 *Relationships between elastic constants*

From equations (7.55) and (7.56),

$$\tau = G(\varepsilon_1 - \varepsilon_2) \tag{7.57}$$

but

$$\sigma_1 = \tau \tag{7.58}$$

Therefore

$$\varepsilon_1 = \frac{\sigma_1(1+\nu)}{E} \tag{7.59}$$

and

$$\varepsilon_2 = \frac{-\sigma_1(1+\nu)}{E} \qquad (7.60)$$

Substituting equations (7.58)–(7.60) into (7.57), the following is obtained:

$$\sigma_1 = 2G\,\frac{\sigma_1}{E}\,(1+\nu)$$

Therefore

$$G = \frac{E}{2(1+\nu)} = \text{modulus of rigidity}$$

From Section 2.81,

$$\text{Volumetric strain} = \varepsilon_v = \varepsilon_x + \varepsilon_y + \varepsilon_z$$

But for an elemental cube under a hydrostatic stress σ,

$$\varepsilon_x = \varepsilon_y = \varepsilon_z = \sigma(1-2\nu)/E$$

Therefore

$$\varepsilon_v = 3\sigma(1-2\nu)/E \qquad (7.61)$$

Now,

$$\frac{\text{Volumetric stress } (\sigma)}{\text{Volumetric strain } (\varepsilon_v)} = \text{bulk modulus } (K)$$

Therefore

$$\varepsilon_v = \sigma/K \qquad (7.62)$$

Equating (7.61) and (7.62),

$$K = \frac{E}{3(1-2\nu)} \qquad (7.63)$$

EXAMPLE 7.5

A solid circular-section rotating shaft, of diameter 0.2 m, is subjected to combined bending, torsion and axial load, where the maximum direct stresses due to bending occur on the top and bottom surfaces of the shaft.

If a shear pair is attached to the shaft, then the strains recorded from this pair are as follows:

$$\left.\begin{array}{l}\varepsilon_1^T = 200 \times 10^{-6}\\ \varepsilon_2^T = \;\;80 \times 10^{-6}\end{array}\right\}\text{when the shear pair is at the top}$$

$$\left.\begin{aligned}\varepsilon_1^B &= 100 \times 10^{-6}\\ \varepsilon_2^B &= -20 \times 10^{-6}\end{aligned}\right\}\text{when the shear pair is at the bottom}$$

$E = 2 \times 10^{11}$ N/m²

$\nu = 0.3$

Using the above information, determine the applied bending moment, torque and axial load.

SOLUTION

Let

ε_{1T}^T and ε_{2T}^T = strain in gauges 1 and 2 due to T at the top

ε_{1T}^B and ε_{2T}^B = strain in gauges 1 and 2 due to T at the bottom

ε_{1M}^T and ε_{2M}^T = strain in gauges 1 and 2 due to M at the top

ε_{1M}^B and ε_{2M}^B = strain in gauges 1 and 2 due to M at the bottom

ε_{1D}^T and ε_{2D}^T = strain in gauges 1 and 2 due to the axial load at the top

ε_{1D}^B and ε_{2D}^B = strain in gauges 1 and 2 due to the axial load at the bottom

By inspection it can be deduced that

$$\varepsilon_{1T}^T = -\varepsilon_{2T}^T \quad \varepsilon_{1T}^B = -\varepsilon_{2T}^B \quad \varepsilon_{1T}^T = \varepsilon_{1T}^B \quad \varepsilon_{2T}^T = \varepsilon_{2T}^B$$

$$\varepsilon_{1M}^T = \varepsilon_{2M}^T \quad \varepsilon_{1M}^B = \varepsilon_{2M}^B \quad \varepsilon_{1M}^T = -\varepsilon_{1M}^B \quad \varepsilon_{2M}^T = -\varepsilon_{2M}^B$$

$$\varepsilon_{1D}^T = \varepsilon_{2D}^T = \varepsilon_{1D}^B = \varepsilon_{2D}^B$$

Hence, *at the top,*

$$\varepsilon_1^T = \varepsilon_{1T}^T + \varepsilon_{1M}^T + \varepsilon_{1D}^T \tag{7.64}$$

$$\varepsilon_2^T = -\varepsilon_{1T}^T + \varepsilon_{1M}^T + \varepsilon_{1D}^T \tag{7.65}$$

and *at the bottom,*

$$\varepsilon_1^B = \varepsilon_{1T}^T - \varepsilon_{1M}^T + \varepsilon_{1D}^T \tag{7.66}$$

$$\varepsilon_2^B = -\varepsilon_{1T}^T - \varepsilon_{1M}^T + \varepsilon_{1D}^T \tag{7.67}$$

To obtain T

Taking equation (7.65) from equation (7.64), or equation (7.67) from equation (7.66),

$$\gamma = \varepsilon_1^T - \varepsilon_2^T = 200 \times 10^{-6} - 80 \times 100^{-6}$$

$$\underline{\gamma = 120 \times 10^{-6}}$$

Now,

$$G = \frac{E}{2(1+\nu)} = \frac{2 \times 10^{11}}{2.6}$$

$$\underline{G = 7.69 \times 10^{10}\ \text{N/m}^2}$$

Therefore

$$\underline{\tau = G\gamma = 9.228 \text{ MN/m}^2}$$

Now,

$$J = \frac{\pi \times 0.2^4}{32} = 1.571 \times 10^{-4} \text{ m}^4$$

and

$$\underline{T = \frac{\tau J}{r} = 14.497 \text{ kN m}}$$

To obtain M

Take equation (7.66) from equation (7.64), or equation (7.67) from equation (7.65), to give

$$2\varepsilon_{1M}^{M} = \varepsilon_1^T - \varepsilon_1^B = \varepsilon_2^T - \varepsilon_2^B$$

or

$$\underline{\varepsilon_{1M}^{T} = 50 \times 10^{-6}}$$

Let σ_b' be the stress due to bending, acting along the strain gauges, as shown in Figure 7.28.

Consider the equilibrium of the triangle in Figure 7.29:

$$\sigma_b' \times \sqrt{2} = \frac{\sigma_b}{2}$$

Therefore

$$\underline{\sigma_b' = \sigma_b/2} \tag{7.68}$$

where σ_b is the bending stress in the x direction due to M.

From equation (7.40),

$$\varepsilon_{1M}^{M} = \frac{1}{E}(\sigma_b' - \nu\sigma_b')$$

$$= \frac{\sigma_b'}{E}(1 - \nu) = \frac{\sigma_b(1 - \nu)}{2E} = 50 \times 10^{-6} \tag{7.69}$$

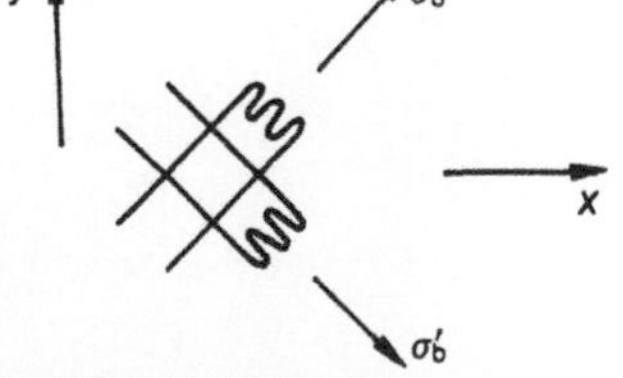

Figure 7.28

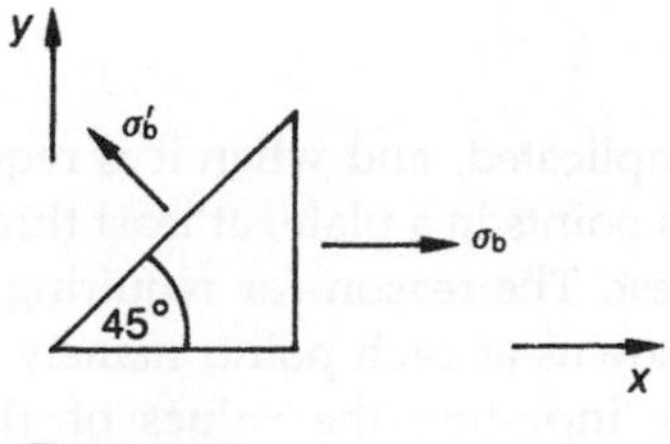

Figure 7.29

Therefore

$$\sigma_b = \pm\frac{2 \times 2 \times 10^{11} \times 50 \times 10^{-6}}{0.7} = +28.57 \text{ MN/m}^2$$

Now,

$$M = \frac{\sigma}{y} \times I$$

$$= \frac{28.57 \times 7.855 \times 10^{-5}}{0.1}$$

$$M = 22.44 \text{ kN m}$$

To obtain direct load

$$\varepsilon_1^T + \varepsilon_2^T = 2\varepsilon_{1M}^T + 2\varepsilon_{1D}^T \quad (7.70)$$

and

$$\varepsilon_1^B + \varepsilon_2^B = -2\varepsilon_{1M}^T + 2\varepsilon_{1D}^T \quad (7.71)$$

Adding equations (7.70) and (7.71),

$$\varepsilon_1^T + \varepsilon_2^T + \varepsilon_1^B + \varepsilon_2^B = 4\varepsilon_{1D}^T$$

Therefore

$$\varepsilon_{1D}^T = \frac{280 \times 10^{-6} + 80 \times 10^{-6}}{4} = 90 \times 10^{-6} \quad (7.72)$$

In a manner similar to that adopted for the derivation of equations (7.68) and (7.69),

$$\varepsilon_{1D}^T = \frac{1}{2E}(\sigma_d - \nu\sigma_d) \quad (7.73)$$

where σ_d is the stress due to the axial load.

From equations (7.72) and (7.73),

$$\sigma_d = 51.43 \text{ MN/m}^2$$

Therefore

$$\text{Axial load} = 51.43 \times \frac{\pi \times 0.2^2}{4} = 1.62 \text{ MN (tensile)}$$

7.12 Strain rosettes

In practice, strain systems are usually very complicated, and when it is required to determine experimentally the stresses at various points in a plate, at least three strain gauges have to be used at each point of interest. The reason for requiring at least three strain gauges is that there are three unknowns at each point, namely the two principal strains and their direction. Thus, by inputting the values of the three strains, together with their 'positions', into equation (7.26), three simultaneous equations will result, the solution of which will yield the two principal strains, ε_1 and ε_2, and their 'direction' θ.

To illustrate the method of determining ε_1, ε_2 and θ, consider the strains of Figure 7.30.

Let

$$\left.\begin{aligned} \varepsilon_1 &= \text{maximum principal strain} \\ \varepsilon_2 &= \text{minimum principal strain} \\ \theta &= \text{angle between } \varepsilon_1 \text{ and } \varepsilon_\theta \end{aligned}\right\} \text{Unknowns to be determined}$$

$$\left.\begin{aligned} \varepsilon_\theta &= \text{direct strain at an angle } \theta \text{ from } \varepsilon_1 \\ \varepsilon_\alpha &= \text{direct strain at an angle } \alpha \text{ from } \varepsilon_\theta \\ \varepsilon_\beta &= \text{direct strain at an angle } \beta \text{ from } \varepsilon_\alpha \end{aligned}\right\} \begin{gathered}\text{experimentally} \\ \text{measured strains}\end{gathered} \tag{7.74}$$

Substituting each measured value of equation (7.74), in turn, into equation (7.26), together with its 'direction' the following three simultaneous equations are obtained:

$$\varepsilon_\theta = \tfrac{1}{2}(\varepsilon_1 + \varepsilon_2) + \tfrac{1}{2}(\varepsilon_1 - \varepsilon_2)\cos 2\theta \tag{7.75}$$

$$\varepsilon_\alpha = \tfrac{1}{2}(\varepsilon_1 + \varepsilon_2) + \tfrac{1}{2}(\varepsilon_1 - \varepsilon_2)\cos[2(\theta + \alpha)] \tag{7.76}$$

$$\varepsilon_\beta = \tfrac{1}{2}(\varepsilon_1 + \varepsilon_2) + \tfrac{1}{2}(\varepsilon_1 - \varepsilon_2)\cos[2(\theta + \alpha + \beta)] \tag{7.77}$$

From equations (7.75)–(7.77), it can be seen that the only unknowns are ε_1, ε_2 and θ, which can be readily determined.

For mathematical convenience, manufacturers of strain gauges normally supply

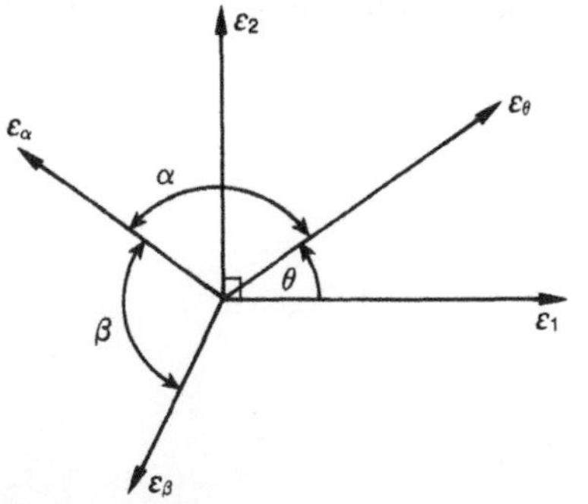

Figure 7.30

strain rosettes, where the angles α and β have values of 45° or 60° or 120°. (The strain gauge technique is described in greater detail in Chapter 16.)

If, however, the experimentalist has difficulty in attaching a standard rosette to an awkward zone, then he or she can attach three linear strain gauges at this point, and choose suitable values for α and β. In such cases, it will be necessary to use equations (7.75)–(7.77) to determine the required unknowns.

7.12.1 *The 45° rectangular rosette (see Figure 7.31)*

Substituting α and β into equations (7.75)–(7.77), the three simultaneous equations (7.78)–(7.80) are obtained:

$$\varepsilon_\theta = \tfrac{1}{2}(\varepsilon_1 + \varepsilon_2) + \tfrac{1}{2}(\varepsilon_1 - \varepsilon_2)\cos 2\theta \tag{7.78}$$

$$\varepsilon_\alpha = \tfrac{1}{2}(\varepsilon_1 + \varepsilon_2) - \tfrac{1}{2}(\varepsilon_1 - \varepsilon_2)\sin 2\theta \tag{7.79}$$

$$\varepsilon_\beta = \tfrac{1}{2}(\varepsilon_1 + \varepsilon_2) - \tfrac{1}{2}(\varepsilon_1 - \varepsilon_2)\cos 2\theta \tag{7.80}$$

Adding equations (7.78)–(7.80),

$$\varepsilon_\theta + \varepsilon_\beta = \varepsilon_1 + \varepsilon_2 \tag{7.81}$$

Equation (7.81) is known as the *first invariant of strain*, which states that the sum of mutually perpendicular direct strains at a point is constant.

From equations (7.79) and (7.80),

$$-\tfrac{1}{2}(\varepsilon_1 - \varepsilon_2)\sin 2\theta = \varepsilon_\alpha - \tfrac{1}{2}\varepsilon_1 - \tfrac{1}{2}\varepsilon_2 \tag{7.82}$$

$$-\tfrac{1}{2}(\varepsilon_1 - \varepsilon_2)\cos 2\theta = \varepsilon_\beta - \tfrac{1}{2}\varepsilon_1 - \tfrac{1}{2}\varepsilon_2 \tag{7.83}$$

Substituting equation (7.81) into (7.82) and (7.83), and then dividing equation (7.82) by equation (7.83),

$$\tan 2\theta = \frac{\varepsilon_\alpha - \frac{1}{2}\varepsilon_\theta - \frac{1}{2}\varepsilon_\beta}{\varepsilon_\beta - \frac{1}{2}\varepsilon_\theta - \frac{1}{2}\varepsilon_\beta}$$

$$\underline{\tan 2\theta = \frac{(\varepsilon_\theta - 2\varepsilon_\alpha + \varepsilon_\beta)}{(\varepsilon_\theta - \varepsilon_\beta)}} \tag{7.84}$$

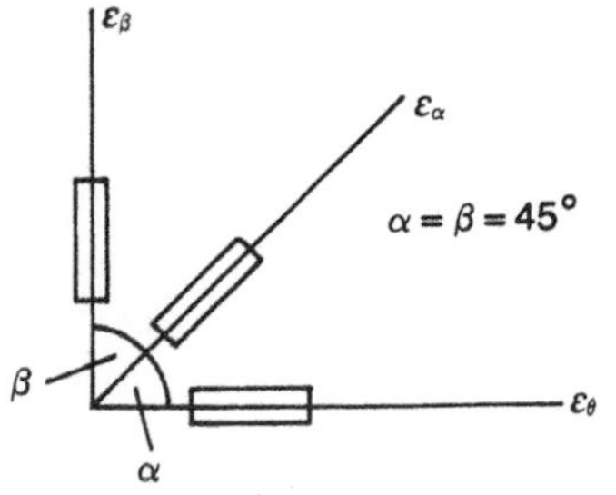

Figure 7.31 45° rectangular rosette

7.12.2 To determine ε_1 and ε_2

Taking equation (7.80) from equation (7.78),

$$\varepsilon_\theta - \varepsilon_\beta = (\varepsilon_1 - \varepsilon_2)\cos 2\theta \tag{7.85}$$

Hence, from equations (7.81) and (7.85),

$$\varepsilon_1 = \frac{\varepsilon_\theta + \varepsilon_\beta}{2} + \frac{\varepsilon_\theta - \varepsilon_\beta}{2\cos 2\theta} \tag{7.86}$$

$$\varepsilon_2 = \frac{\varepsilon_\theta + \varepsilon_\beta}{2} - \frac{\varepsilon_\theta - \varepsilon_\beta}{2\cos 2\theta} \tag{7.87}$$

Equation (7.84) can be represented by the mathematical triangle of Figure 7.32 where

$$\begin{aligned}\text{Hypotenuse} &= \sqrt{\varepsilon_\theta^2 + 4\varepsilon_\alpha^2 + \varepsilon_\beta^2 - 4\varepsilon_\theta\varepsilon_\alpha - 4\varepsilon_\alpha\varepsilon_\beta + 2\varepsilon_\theta\varepsilon_\beta + \varepsilon_\theta^2 + \varepsilon_\beta^2 - 2\varepsilon_\theta\varepsilon_\beta} \\ &= \sqrt{2}\sqrt{(\varepsilon_\theta - \varepsilon_\alpha)^2 + (\varepsilon_\beta - \varepsilon_\alpha)^2}\end{aligned} \tag{7.88}$$

$$\cos 2\theta = \frac{\varepsilon_\theta - \varepsilon_\beta}{\sqrt{2}\sqrt{(\varepsilon_\theta - \varepsilon_\alpha)^2 + (\varepsilon_\beta - \varepsilon_\alpha)^2}} \tag{7.89}$$

$$\sin 2\theta = \frac{\varepsilon_\theta - 2\varepsilon_\alpha + \varepsilon_\beta}{\sqrt{2}\sqrt{(\varepsilon_\theta - \varepsilon_\alpha)^2 + (\varepsilon_\beta - \varepsilon_\alpha)^2}} \tag{7.90}$$

Substituting equations (7.89) and (7.90) into equations (7.86) and (7.87),

$$\varepsilon_1, \varepsilon_2 = \tfrac{1}{2}(\varepsilon_\theta + \varepsilon_\beta) \pm \frac{\sqrt{2}}{2}\sqrt{(\varepsilon_\theta - \varepsilon_\alpha)^2 + (\varepsilon_\beta - \varepsilon_\alpha)^2} \tag{7.91}$$

7.12.3 The 120° equiangular rosette (see Figure 7.33)

For greater precision, 120° equiangular rosettes are preferred to 45° rectangular rosettes.

Substituting α and β into equations (7.75)–(7.77), the simultaneous equations of

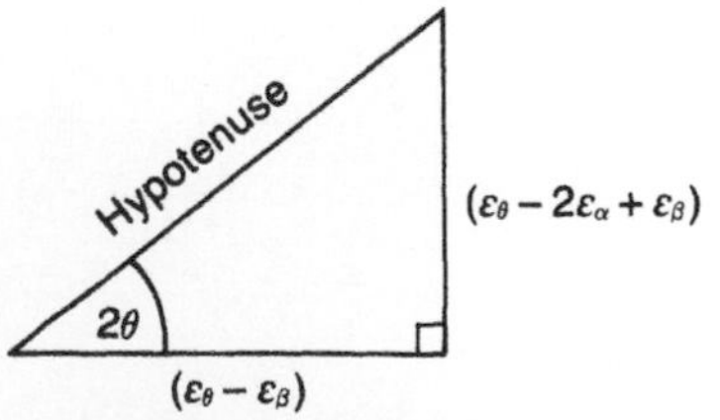

Figure 7.32 Mathematical triangle

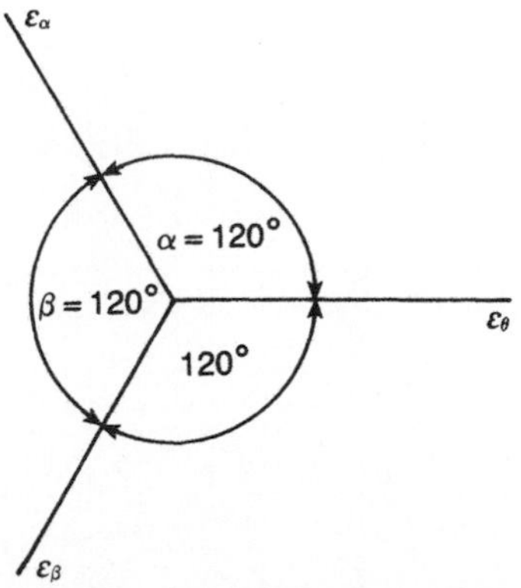

Figure 7.33 120° equiangular rosette

(7.92)–(7.94) are obtained:

$$\varepsilon_\theta = \tfrac{1}{2}(\varepsilon_1 + \varepsilon_2) + \tfrac{1}{2}(\varepsilon_1 - \varepsilon_2)\cos 2\theta \tag{7.92}$$

$$\varepsilon_\alpha = \tfrac{1}{2}(\varepsilon_1 + \varepsilon_2) + \tfrac{1}{2}(\varepsilon_1 - \varepsilon_2)\cos[2(\theta + 120°)] \tag{7.93}$$

$$\varepsilon_\beta = \tfrac{1}{2}(\varepsilon_1 + \varepsilon_2) + \tfrac{1}{2}(\varepsilon_1 - \varepsilon_2)\cos[2(\theta + 240°)] \tag{7.94}$$

Equations (7.93) and (7.94) can be rewritten in the forms

$$\varepsilon_\alpha = \tfrac{1}{2}(\varepsilon_1 + \varepsilon_2) + \tfrac{1}{2}(\varepsilon_1 - \varepsilon_2)(-\cos 2\theta \cos 60° + \sin 2\theta \sin 60°) \tag{7.95}$$

$$\varepsilon_\beta = \tfrac{1}{2}(\varepsilon_1 + \varepsilon_2) + \tfrac{1}{2}(\varepsilon_1 - \varepsilon_2)(-\cos 2\theta \cos 60° - \sin 2\theta \sin 60°) \tag{7.96}$$

Adding together equations (7.92), (7.95) and (7.96),

$$\varepsilon_\theta + \varepsilon_\alpha + \varepsilon_\beta = \tfrac{3}{2}(\varepsilon_1 + \varepsilon_2)$$

$$\varepsilon_1 + \varepsilon_2 = \tfrac{2}{3}(\varepsilon_\theta + \varepsilon_\alpha + \varepsilon_\beta) \tag{7.97}$$

Taking equation (7.96) from (7.95),

$$\varepsilon_\alpha - \varepsilon_\beta = (\varepsilon_1 + \varepsilon_2)\sin 2\theta \sin 60° \tag{7.98}$$

Taking equation (7.96) from (7.92),

$$\varepsilon_\theta - \varepsilon_\beta = \tfrac{1}{2}(\varepsilon_1 + \varepsilon_2)(\tfrac{3}{2}\cos 2\theta + \sin 2\theta \sin 60°) \tag{7.99}$$

Dividing equation (7.99) by (7.98),

$$\frac{\varepsilon_\theta - \varepsilon_\beta}{\varepsilon_\alpha - \varepsilon_\beta} = \frac{1}{2}\left(\frac{3}{2}\,\frac{\cot 2\theta}{\sin 60°} + 1\right)$$

$$\frac{2(\varepsilon_\theta - \varepsilon_\beta)}{\varepsilon_\alpha - \varepsilon_\beta} = \frac{3}{2}\,\frac{2}{\sqrt{3}}\cot 2\theta + 1$$

Therefore

$$\sqrt{3}\cot 2\theta = \frac{2(\varepsilon_\theta - \varepsilon_\beta)}{\varepsilon_\alpha - \varepsilon_\beta} - \frac{(\varepsilon_\alpha - \varepsilon_\beta)}{\varepsilon_\alpha - \varepsilon_\beta}$$

or

$$\tan 2\theta = \frac{\sqrt{3}(\varepsilon_\alpha - \varepsilon_\beta)}{(2\varepsilon_\theta - \varepsilon_\beta - \varepsilon_\alpha)} \tag{7.100}$$

7.12.4 *To determine ε_1 and ε_2*

Equation (7.100) can be represented by the mathematical triangle of Figure 7.34. From this figure,

$$\cos 2\theta = \frac{2\varepsilon_\theta - \varepsilon_\alpha - \varepsilon_\beta}{\sqrt{2}\sqrt{(\varepsilon_\theta - \varepsilon_\alpha)^2 + (\varepsilon_\alpha - \varepsilon_\beta)^2 + (\varepsilon_\theta - \varepsilon_\beta)^2}} \tag{7.101}$$

$$\sin 2\theta = \frac{\sqrt{3}(\varepsilon_\alpha - \varepsilon_\beta)}{\sqrt{2}\sqrt{(\varepsilon_\theta - \varepsilon_\alpha)^2 + (\varepsilon_\alpha - \varepsilon_\beta)^2 + (\varepsilon_\theta - \varepsilon_\alpha)^2}} \tag{7.102}$$

Substituting equation (7.102) into (7.98),

$$\varepsilon_1 - \varepsilon_2 = \frac{2\sqrt{2}}{3}\sqrt{(\varepsilon_\theta - \varepsilon_\alpha)^2 + (\varepsilon_\alpha - \varepsilon_\beta)^2 + (\varepsilon_\theta - \varepsilon_\beta)^2} \tag{7.103}$$

Adding equations (7.97)–(7.103),

$$\varepsilon_1 = \tfrac{1}{3}(\varepsilon_\theta + \varepsilon_\alpha + \varepsilon_\beta) + \frac{\sqrt{2}}{3}\sqrt{(\varepsilon_\theta - \varepsilon_\alpha)^2 + (\varepsilon_\alpha - \varepsilon_\beta)^2 + (\varepsilon_\theta - \varepsilon_\beta)^2} \tag{7.104}$$

Similarly,

$$\varepsilon_2 = \tfrac{1}{3}(\varepsilon_\theta + \varepsilon_\alpha + \varepsilon_\beta) - \frac{\sqrt{2}}{3}\sqrt{(\varepsilon_\theta - \varepsilon_\alpha)^2 + (\varepsilon_\alpha - \varepsilon_\beta)^2 + (\varepsilon_\theta - \varepsilon_\beta)^2} \tag{7.105}$$

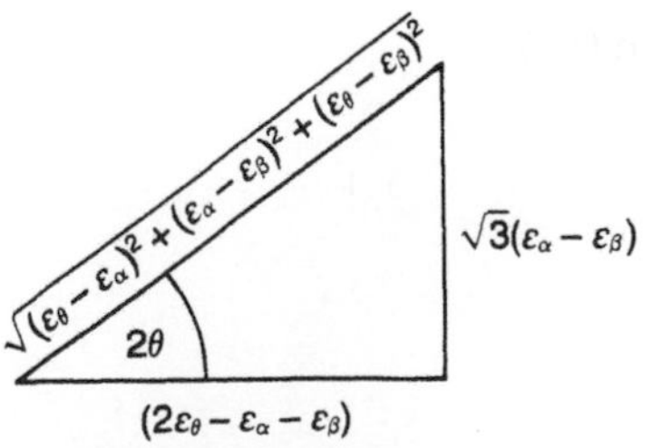

Figure 7.34 Mathematical triangle

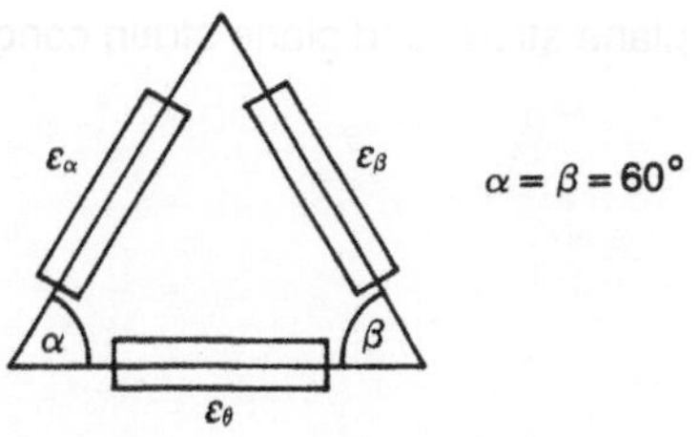

Figure 7.35 60° delta rosette

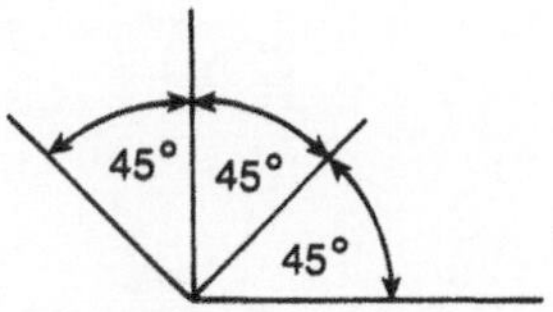

Figure 7.36 Four-gauge 45° fan rosette

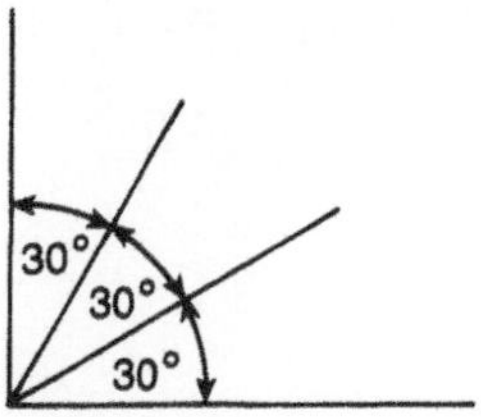

Figure 7.37 T-delta rosette

7.12.5 Other types of strain gauge rosette

These include the 60° delta of Figure 7.35, the four-gauge 45° fan of Figure 7.36 and the four-gauge T-delta of Figure 7.37.

The advantages of using four-gauge rosettes are that they can lead to greater precision, particularly if one of the strains is zero.

EXAMPLE 7.6

A 120° equiangular rosette records the following values of strain at a point:

$$\varepsilon_\theta = 300 \times 10^{-6}$$

$$\varepsilon_\alpha = -100 \times 10^{-6}$$

$$\varepsilon_\beta = 150 \times 10^{-6}$$

Determine the directions and magnitudes of the values of principal stresses, together with the maximum shear stress at this point, for plane stress and plane strain conditions.

$$E = 2 \times 10^{11}\ \text{N/m}^2 \quad \nu = 0.3$$

SOLUTION

To determine θ

From equation (7.100),

$$\tan 2\theta = \frac{\sqrt{3}(\varepsilon_\alpha - \varepsilon_\beta)}{(2\varepsilon_\theta - \varepsilon_\beta - \varepsilon_\alpha)}$$

$$= \frac{\sqrt{3} \times (-250)}{(600 - 150 + 100)} = -1.270$$

$$\underline{\theta = -19.11°}$$

For the directions of σ_1 and σ_2, see Figure 7.38

From equation (7.104),

$$\varepsilon_1 = 10^{-6}(116.7 + 0.471 \times 495)$$

$$\underline{\varepsilon_1 = 350 \times 10^{-6}}$$

$$\underline{\varepsilon_2 = -116.4 \times 10^{-6}}$$

To determine σ_1, σ_2 and $\hat{\tau}$ for plane stress

From equations (7.42) and (7.43),

$$\underline{\sigma_1 = \frac{2 \times 10^{11}}{0.91}(350 - 34.92) \times 10^{-6} = 69.25\ \text{MN/m}^2}$$

$$\underline{\sigma_2 = \frac{2 \times 10^{11}}{0.91}(-116.4 + 105) \times 10^{-6} = -2.51\ \text{MN/m}^2}$$

$$\hat{\tau} = \frac{\sigma_1 - \sigma_2}{2} = \underline{35.88\ \text{MN/m}^2}$$

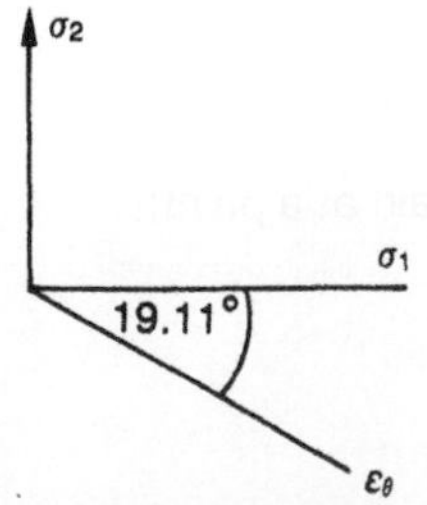

Figure 7.38 Directions of σ_1 and σ_2

To determine σ_1, σ_2 and $\hat{\tau}$ for plane strain

From equations (7.51) and (7.52),

$$\underline{\sigma_1 = 80.8 \text{ MN/m}^2} \quad \underline{\sigma_2 = 9.05 \text{ MN/m}^2}$$

$$\underline{\hat{\tau} = 35.88 \text{ MN/m}^2}$$

7.13 Computer program for principal stresses and strains

Listing 7.1 gives a computer program in BASIC, for determining principal stresses and strains from co-ordinate stresses or co-ordinate strains.

The *input* for determining principal stresses from co-ordinate stresses is as follows:

Type in σ_x (direct stress in the x direction)

Type in σ_y (direct stress in the y direction)

Type in τ_{xy} (shear stress in the x–y plane)

Typical input and output values are given in Listings 7.2 and 7.3 for Examples 7.1 and 7.2.

The input for determining principal stresses and strains from co-ordinate strains is as follows:

Type in ε_x (direct strain in x direction)

Type in ε_y (direct strain in y direction)

Type in γ_{xy} (shear strain in the x–y plane)

Type in E (elastic modulus)

Type in ν (Poisson's ratio)

NB When this part of the program is being used, the condition of *plane stress* is assumed.

7.14 The constitutive laws for a lamina of a composite in global co-ordinates

To obtain the constitutive laws, in global co-ordinates, for a lamina of a composite, consider a lamina with orthogonal properties as shown in Figure 7.39.

In the figure, the local axes of the lamina are x and y and the global axes are X and Y, where the local x axis lies at an angle θ to the global X axis. The global X and Y axes are also known as reference axes.

Listing 7.1 **Computer program for calculating σ_1, σ_2, etc., from either the co-ordinate stresses or the co-ordinate strains**

```
100  REMark principal stresses & strains
110  CLS
120  PRINT:PRINT"principal stresses & strains"
130  PRINT:PRINT"copyright of C.T.F. Ross":PRINT
140  PRINT"if inputting CO-ORDINATE STRESSES type 1; else if inputting CO-ORDINATE STRAINS,
     type 0":PRINT
150  INPUT STR
160  IF str=1 OR str=0 THEN GO TO 180
170  PRINT:PRINT"incorrect data":PRINT:GO to 140
180  IF str=0 THEN GO TO 300
190  PRINT:PRINT"stress in x direction=";:INPUT sigmax
200  PRINT"stress in y direction=;:INPUT sigmay
210  PRINT"shear stress in x-y plane=";:INPUT tauxy
220  const=.5 * SQRT((sigmax-sigmay)^2+4 * tauxy^2)
230  sigma1=.5 * (sigmax+sigmay)+const
240  sigma2=.5 * (sigmax+sigmay)-const
250  theta=.5 * ATAN(2 * tauxy/ (sigmax-sigmay))
260  GO TO 600
300  PRINT:PRINT"strain in x direction=";:INPUT ex
310  PRINT"strain in y direction=";:INPUT ey
320  PRINT"shear in x-y plane=;:INPUT gxy
330  const=.5 * SQRT((ex-ey)^2+gxy^2)
340  e1=.5 * (ex+ey)+const
350  e2=.5 * (ex+ey)-const
360  theta=.5 * ATAN(gxy/ (ex-ey))
370  PRINT:PRINT"elastic modulus=";:INPUT e
380  PRINT"poisson's ratio=";:INPUT nu
600  PI= 3.14159
610  IF str=1 THEN PRINT:PRINT:PRINT,"principal stresses":PRINT
620  IF STR=0 THEN PRINT:PRINT:PRINT,"principal stresses & principal strain s":PRINT:GO to 800
630  PRINT"stress in x direction=";sigmax
640  PRINT"stress in y direction=";sigmay
650  PRINT"shear in x-y plane=";tauxy
660  PRINT:PRINT
670  PRINT"maximum principal stress=";sigma1
680  PRINT"minimum principal stress=";sigma2
690  PRINT"theta=";theta * 180/PI;" degrees"
695  PRINT"maximum shear stress="; (sigma1-sigma2) /2
700  GO TO 1000
800  PRINT"strain in x direction=";ex
810  PRINT"strain in y direction=";ey
820  PRINT"shear strain in x-y plane=";gxy
830  PRINT"elastic modulus=";e
840  PRINT"poisson's ratio";nu
850  PRINT:PRINT
860  PRINT"maximum principal strain=";e1
870  PRINT"maximum principal strain=";e2
875  PRINT"theta=";theta * 180/PI;" degrees"
880  PRINT"maximum shear strain=";e1-e2
890  sigma1=e * (e1+nu * e2) / (1-nu^2)
900  sigma2=e * (e2+nu*e1) / (1-nu^2)
910  PRINT"maximum principal stress=";sigma1
920  PRINT"minimum principal stress=";sigma2
930  PRINT"maximum shear stress="; (sigma1-sigma2) /2
1000 PRINT:PRINT:PRINT:PRINT
1020 STOP
```

Listing 7.2 **Computer output for Example 7.1**

```
principal stresses

stress in x direction=100
stress in y direction=50
shear in x-y plane=40

maximum principal stress=122.1699
minimum principal stress=27.83009
theta=28.99731 degrees
maximum shear stress=47.16991
```

Listing 7.3 **Computer output for Example 7.2**

```
principal stresses

stress in x direction=40
stress in y direction=-80
shear in x-y plane=-50

maximum principal stress=58.1025
minimum principal stress=-98.1025
theta=-19.90279 degrees
maximum shear stress=78.1025
```

From equations (2.19) and (2.20) the relationship between the local strains and stresses is given by

$$\begin{Bmatrix} \varepsilon_x \\ \varepsilon_y \\ \gamma_{xy} \end{Bmatrix} = \begin{bmatrix} S_{11} & S_{12} & S_{13} \\ S_{21} & S_{22} & S_{23} \\ S_{31} & S_{32} & S_{33} \end{bmatrix} \begin{Bmatrix} \sigma_x \\ \sigma_y \\ \tau_{xy} \end{Bmatrix} \tag{7.106}$$

$$\{\varepsilon_{xy}\} = [S]\{\sigma_{xy}\} \tag{7.107}$$

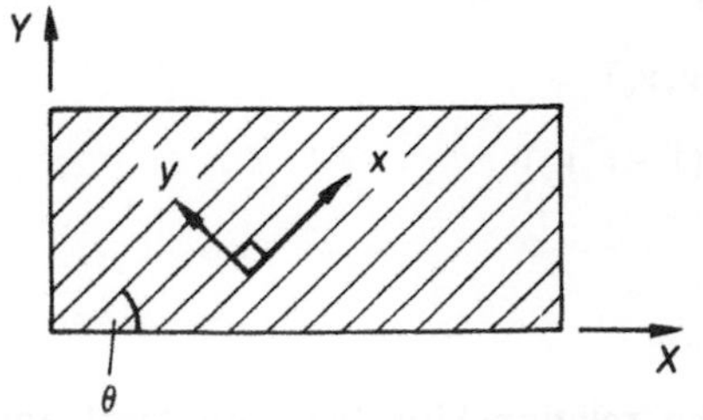

Figure 7.39 A lamina from a composite

where $[S]$ is a matrix of compliance functions for the lamina, where for plane stress,

$$S_{11} = \frac{1}{E_x} \quad S_{12} = S_{21} = \frac{-\nu_x}{E_x} = \frac{-\nu_y}{E_y}$$

$$S_{22} = \frac{1}{E_y} \quad S_{33} = G_{xy} \quad S_{13} = S_{31} = S_{23} = S_{32} = 0$$

and for plane strain,

$$S_{11} = (1 - \nu_{xz}\,\nu_{zx})/E_x$$

$$S_{12} = -(\nu_{xy} + \nu_{xz}\,\nu_{zy})/E_x$$

$$S_{22} = (1 - \nu_{yz}\,\nu_{zy})/E_y$$

$$S_{33} = 1/G_{xy}$$

$$S_{13} = S_{23} = 0$$

where

$$\nu_{yx} = \nu_y = \nu_{xy}\,E_y/E_x \quad \nu_x = \nu_{xy}$$

$$\nu_{zx} = \nu_{xz}\,E_z/E_x \quad \nu_{zy} = \nu_{yz}\,E_z/E_y$$

$$\{\varepsilon_{xy}\} = \begin{Bmatrix} \varepsilon_x \\ \varepsilon_y \\ \gamma_{xy} \end{Bmatrix} \quad \{\sigma_{xy}\} = \begin{Bmatrix} \sigma_x \\ \sigma_y \\ \sigma_{xy} \end{Bmatrix}$$

Similarly, from equation (2.21), the relationship between the local stresses and strains is given by

$$\begin{Bmatrix} \sigma_x \\ \sigma_y \\ \tau_{xy} \end{Bmatrix} = \begin{bmatrix} k_{11} & k_{12} & k_{13} \\ k_{21} & k_{22} & k_{23} \\ k_{31} & k_{32} & k_{33} \end{bmatrix} \begin{Bmatrix} \varepsilon_x \\ \varepsilon_y \\ \gamma_{xy} \end{Bmatrix} \tag{7.108}$$

or

$$\{\sigma_{xy}\} = [k]\{\varepsilon_{xy}\} \tag{7.109}$$

where $[k] = [S^{-1}]$ is a material matrix for plane stress, and

$$k_{11} = E_x/(1 - \nu_x\nu_y) \quad k_{22} = E_y/(1 - \nu_x\nu_y)$$

$$k_{12} = k_{21} = \nu_x E_y/(1 - \nu_x\nu_y) = \nu_y E_x/(1 - \nu_x\nu_y)$$

$$k_{33} = 0$$

$$k_{13} = k_{31} = k_{23} = k_{32} = 0$$

Now from equations (7.1) and (7.2) the relationships between local and global stresses are given by

$$\sigma_x = \sigma_X \cos^2\theta + \sigma_Y \sin^2\theta + 2\tau_{XY} \sin\theta \cos\theta$$

$$\sigma_y = \sigma_{X+90°} = \sigma_X \sin^2\theta + \sigma_Y \cos^2\theta - 2\tau_{XY} \sin\theta \cos\theta$$

$$\tau_{xy} = -\sigma_X \sin\theta \cos\theta + \sigma_Y \sin\theta \cos\theta + \tau_{XY} (\cos^2\theta - \sin^2\theta)$$

where σ_x, σ_y and τ_{xy} are local stresses and σ_X, σ_Y and τ_{XY} are global or reference stresses, or in matrix form

$$\begin{Bmatrix} \sigma_x \\ \sigma_y \\ \tau_{xy} \end{Bmatrix} = \begin{bmatrix} C^2 & S^2 & 2SC \\ S^2 & C^2 & -2SC \\ -SC & SC & (C^2 - S^2) \end{bmatrix} \begin{Bmatrix} \sigma_X \\ \sigma_Y \\ \tau_{XY} \end{Bmatrix} \tag{7.110}$$

where $S = \sin\theta$ and $C = \cos\theta$

$$\{\sigma_{xy}\} = [DC]\{\sigma_{XY}\} \tag{7.111}$$

or

$$\{\sigma_{XY}\} = [DC]^{-1}\{\sigma_{xy}\} \tag{7.112}$$

where

$$[DC] = \begin{bmatrix} C^2 & S^2 & 2SC \\ S^2 & C^2 & -2SC \\ -SC & SC & (C^2 - S^2) \end{bmatrix} \text{ and } [DC^{-1}] = \begin{bmatrix} C^2 & S^2 & -2SC \\ S^2 & C^2 & 2SC \\ SC & -SC & (C^2 - S^2) \end{bmatrix} \tag{7.113}$$

Similarly,

$$\{\varepsilon_{xy}\} = [DC_1]\{\varepsilon_{XY}\} \text{ and } [DC_1] = \begin{bmatrix} C^2 & S^2 & SC \\ S^2 & C^2 & -SC \\ -2SC & 2SC & (C^2 - S^2) \end{bmatrix} \tag{7.114}$$

Now from equation (7.109)

$$\{\sigma_{xy}\} = [k]\{\varepsilon_{xy}\} \tag{7.115}$$

Substituting equation (7.114) into equation (7.115),

$$\{\sigma_{xy}\} = [k][DC_1]\{\varepsilon_{XY}\} \tag{7.116}$$

Substituting equation (7.116) into equation (7.112),

$$\{\sigma_{XY}\} = [DC^{-1}][k][DC_1]\{\varepsilon_{XY}\} \tag{7.117}$$

Equation (7.117) is the equivalent global stress–strain relationship for the lamina of Figure 7.39, in terms of the global or reference areas X and Y.

Equation (7.117) can be rewritten as

$$\{\sigma_{XY}\} = [k']\{\varepsilon_{XY}\} \tag{7.118}$$

where $[k'] = [DC^{-1}][k][DC_1]$ is a stiffness matrix

$$[k'] = \begin{bmatrix} k'_{11} & k'_{12} & k'_{13} \\ k'_{21} & k'_{22} & k'_{23} \\ k'_{31} & k'_{32} & k'_{33} \end{bmatrix} \tag{7.119}$$

and

$$k'_{11} = \frac{1}{\gamma}\left[E_x\cos^4\theta + E_y\sin^4\theta + (2\nu_xE_y + 4\gamma G)\cos^2\theta\sin^2\theta\right]$$

$$k'_{12} = k'_{21} = \frac{1}{\gamma}\left[\nu_xE_y(\cos^4\theta + \sin^4\theta) + (E_x + E_y - 4\gamma G)\cos^2\theta\sin^2\theta\right]$$

$$k'_{13} = k'_{31} = \frac{1}{\gamma}\left[\cos^3\theta\sin\theta(E_x - \nu_xE_y - 2\gamma G) - \cos\theta\sin^3\theta(E_y - \nu_xE_y - 2\gamma G)\right]$$

$$k'_{22} = \frac{1}{\gamma}\left[E_y\cos^4\theta + E_x\sin^4\theta + \sin^2\theta\cos^2\theta(2\nu_xE_y + 4\gamma G)\right]$$

$$k'_{23} = k'_{32} = \frac{1}{\gamma}\left[\cos\theta\sin^3\theta\left(E_x - \nu_x E_y - 2\gamma G\right) - \cos^3\theta\sin\theta\left(E_y - \nu_x E_y - 2\gamma G\right)\right]$$

$$k'_{33} = \frac{1}{\gamma}\left[\sin^2\theta\cos^2\theta(E_x + E_y - 2\nu_xE_y - 2\gamma G) + \gamma G(\cos^4\theta + \sin^4\theta)\right]$$

where

$$\gamma = (1 - \nu_x\nu_y)$$

Similarly, to obtain the global strains of the lamina of Figure 7.39 in terms of the global stresses, consider equation (7.106), i.e.

$$\{\varepsilon_{xy}\} = [S]\{\sigma_{xy}\} \tag{7.120}$$

Substituting equation (7.111) into equation (7.120),

$$\{\varepsilon_{xy}\} = [S][DC]\{\sigma_{XY}\} \tag{7.121}$$

Equating (7.114) and (7.121),

$$\{\varepsilon_{XY}\} = [DC_1^{-1}][S][DC]\{\sigma_{XY}\} = [S']\{\sigma_{XY}\} \tag{7.122}$$

where $[S'] = [DC_1^{-1}][S][DC]$ is an overall compliance matrix

$$[S'] = \begin{bmatrix} S'_{11} & S'_{12} & S'_{13} \\ S'_{21} & S'_{22} & S'_{23} \\ S'_{31} & S'_{32} & S'_{33} \end{bmatrix}$$

and

$$S'_{11} = S_{11}\cos^4\theta + S_{22}\sin^4\theta + (2S_{12} + S_{33})\cos^2\theta\sin^2\theta$$

$$S'_{12} = S'_{21} = (S_{11} + S_{22} - S_{33})\cos^2\theta\sin^2\theta + S_{12}(\cos^4\theta\sin^4\theta)$$

$$S'_{13} = S'_{31} = (2S_{22} - 2S_{12} - S_{33})\cos^3\theta\sin\theta - (2S_{22} - 2S_{12} - S_{33})\sin^3\theta\cos\theta$$

$$S'_{22} = S_{11}\sin^4\theta + S_{22}\cos^4\theta + (2S_{12} + S_{33})\cos^2\theta\sin^2\theta$$

$$S'_{23} = S'_{32} = (2S_{11} - 2S_{12} - S_{33})\cos\theta\sin^3\theta - (2S_{22} - 2S_{12} - S_{33})\sin\theta\cos^3\theta$$

$$S'_{33} = 4(S_{11} - 2S_{12} + S_{33})\cos^2\theta\sin^2\theta + S_{33}(\cos^2\theta - \sin^2\theta)^2$$

EXAMPLES FOR PRACTICE 7

1. Prove that the following relationships apply to the 60° delta rosette of Figure 7.35:

$$\tan 2\theta = \frac{\sqrt{3}(\varepsilon_\beta - \varepsilon_a)}{(2\varepsilon_\theta - \varepsilon_a - \varepsilon_\beta)}$$

$$\varepsilon_1, \varepsilon_2 = \tfrac{1}{3}(\varepsilon_\theta + \varepsilon_a + \varepsilon_\beta)$$

$$\pm\frac{\sqrt{2}}{3}\sqrt{(\varepsilon_\theta - \varepsilon_a)^2 + (\varepsilon_a - \varepsilon_\beta)^2 + (\varepsilon_\theta - \varepsilon_\beta)^2}$$

2. At a certain point A in a piece of material, the magnitudes of the direct stresses are -10 MN/m², 30 MN/m² and 40 MN/m², as shown in Figure 7.40.

Determine the magnitude and direction of the principal stresses and the maximum shear stress. (Portsmouth, March 1982).

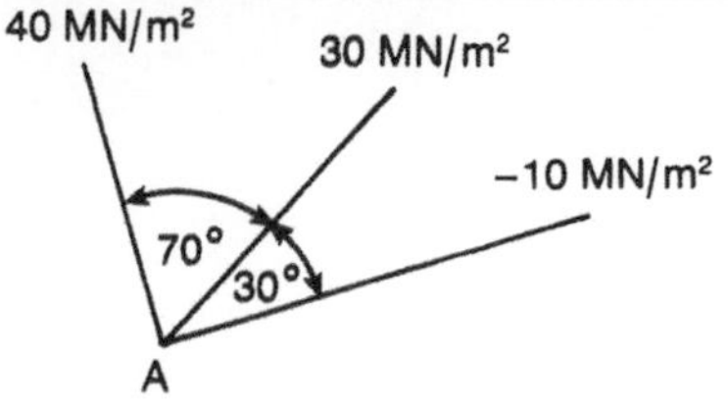

Figure 7.40

$\{\sigma_1 = 62.67$ MN/m², $\sigma_2 = -21.11$ MN/m², $\hat{\tau} = 41.89$ MN/m²$\}$

3. The web of a rolled steel joist has three linear strain gauges attached to a point A and indicating strains as shown in Figure 7.41.

Determine the magnitude and direction of the principal stresses, and the value of the maximum shear stress at this point, assuming the following to apply:

Elastic modulus = 2×10^{11} N/m²

Poisson's ratio = 0.3

(Portsmouth, 1982)

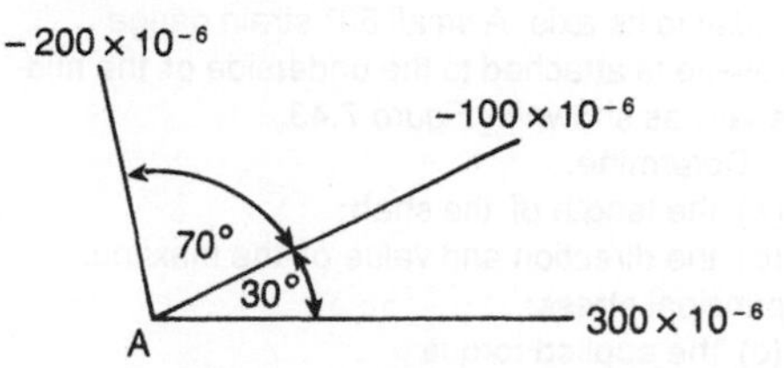

Figure 7.41

$\{\sigma_1 = 62.2$ MN/m², $\sigma_2 = -66.7$ MN/m², $-21.35°$, $\tau = 64.45$ MN/m²$\}$

4. A solid stainless steel propeller shaft of a power boat is of diameter 2 cm. A 45° rectangular rosette is attached to the shaft, where the central gauge (Gauge No. 2) is parallel to the axis of the shaft.

Assuming bending and thermal stresses are negligible, determine the thrust and torque that the shaft is subjected to, given that the recorded strains are as follows:

$\varepsilon_1 = -300 \times 10^{-6}$

$\varepsilon_2 = -142.9 \times 10^{-6}$ (Gauge No. 2)

$\varepsilon_3 = 200 \times 10^{-6}$

$E = 2 \times 10^{11}$ N/m²

$\nu = 0.3$

{−8.98 kN, 60.4 MN m}

5. At a point in a two-dimensional stress system, the known stresses are as shown in Figure 7.42.

Determine the principal stresses σ_y, α and $\hat{\tau}$.

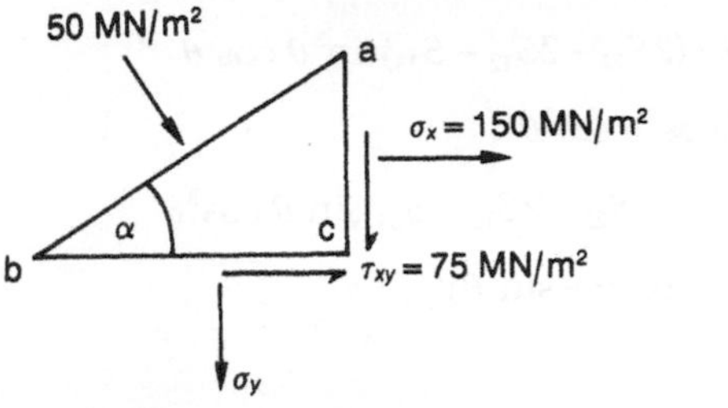

Figure 7.42

{$-20.56°$, $\sigma_1 = 178$ MN/m^2, $\sigma_2 = -50$ MN/m^2, $\sigma_y = 21.9$ MN/m^2}

6. A solid circular-section steel shaft, of 0.03 m diameter, is simply supported at its ends and is subjected to a torque and a load that is radial to its axis. A small 60° strain gauge rosette is attached to the underside of the mid-span, as shown in Figure 7.43.

Determine:

(a) the length of the shaft;

(b) the direction and value of the maximum principal stress;

(c) the applied torque

$E = 2.1 \times 10^{11}$ N/m^2 $\quad \nu = 0.28 \quad \rho = 7860$ kg/m^3

$g = 9.81$ m/s^2

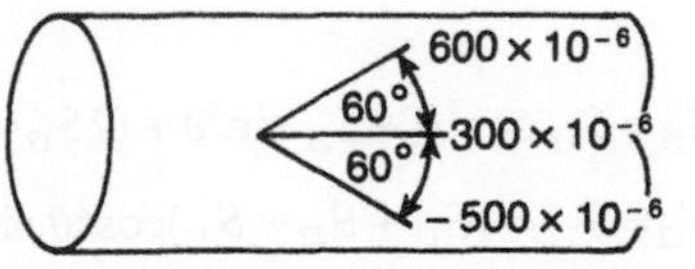

Figure 7.43

{(a) 5.08 m; (b) 37.65° anticlockwise from the middle gauge, 146.5 MN/m^2 – 68.9 MN/m^2; 552.3 N m}

7. A solid cylindrical aluminium-alloy shaft of 0.03 m diameter is subjected to a combined radial and axial load and a torque. A 120° strain gauge rosette is attached to the shaft and records the strains shown in Figure 7.44.

Determine the values of the axial load and torque, given the following:

$E = 1 \times 10^{11}$ N/m^2 $\quad \nu = 0.32$

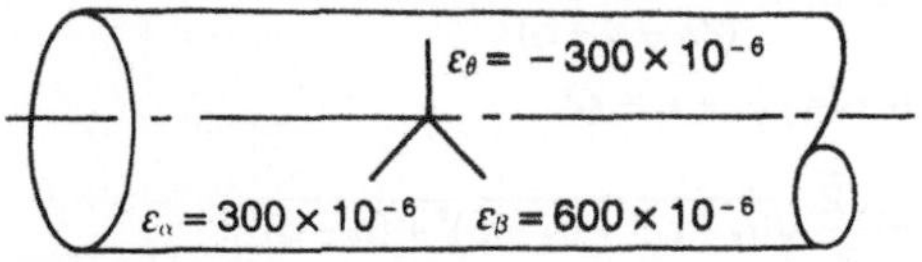

Figure 7.44

{15.4 kN, 208.7 N m}

8 Membrane theory for thin-walled circular cylinders and spheres

8.1 Introduction

The use of thin-walled circular cylinders and spheres for containing gases or liquids under pressure is a popular industrial requirement. This is partly because of the nature of fluid pressure and partly because such loads can be most efficiently resisted by in-plane membrane stresses, acting in curved shells.

In general, thin-walled shells have a very small bending resistance in comparison with their ability to resist loads in membrane tension. Although thin-walled shells also have a relatively large capability to resist pressure loads in membrane compression, the possibility arises that in such cases, failure can take place owing to structural instability (buckling) at stresses which may be a small fraction of that to cause yield, as shown in Figure 8.1.

Thin-walled circular cylinders and spheres appear in many forms, including submarine pressure hulls, the legs of offshore drilling rigs, containment vessels for nuclear reactors, boilers, condensers, storage tanks, gas holders, pipes, pumps and many other different types of pressure vessel.

Ideally, from a structural viewpoint, the perfect vessel to withstand uniform internal pressure is a thin-walled spherical shell, but such a shape may not necessarily be the most suitable from other considerations. For example, a submarine pressure hull, in the form of a spherical shell, is not a suitable shape for hydrodynamic purposes, nor for containing large quantities of equipment nor large numbers of personnel, and in any case, docking a spherically shaped vessel may present problems.

Furthermore, pressure vessels of spherical shape may present difficulties in housing or storage, or in transport, particularly if the pressure vessel is being carried on the back of a lorry ('truck' in the USA).

Another consideration in deciding whether the pressure vessel should be cylindrical or spherical is from the point of view of its cost of manufacture. For

Figure 8.1 Buckled forms of thin-walled cylinders under uniform external pressure

example, although a spherical pressure vessel may be more structurally efficient than a similar cylindrical pressure vessel, the manufacture of the former may be considerably more difficult than the latter, so that additional labour costs of constructing a spherical pressure vessel may be much greater than any material savings that may be gained, especially as extruded cylindrical tubes can often be purchased 'off the shelf'.

Circular cylindrical shells are usually blocked off by domes, but can be blocked off by circular plates; however, if circular plates are used, their thickness is relatively large in comparison with the thickness of the shell dome ends.

8.2 Circular cylindrical shells under uniform internal pressure

The two major methods of failure of a circular cylindrical shell under uniform pressure are as follows.

8.2.1 Failure due to circumferential or hoop stress (σ_H)

If failure is due to the hoop stress, then fracture occurs along a longitudinal seam, as shown in Figure 8.2.

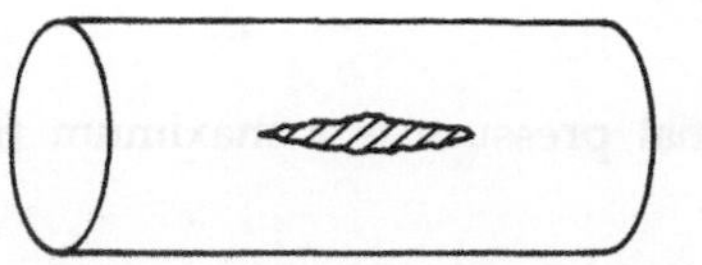

Figure 8.2 Fracture due to hoop stress

Consider the circular cylinder of Figure 8.3, which may be assumed to split in half, along two longitudinal seams, owing to the hoop stress, σ_H. To determine σ_H, in terms of the applied pressure and the geometrical properties of the cylinder, consider the equilibrium of one half of the cylindrical shell, as shown in Figure 8.4.

Resolving vertically,

$$\sigma_H \times t \times 2 \times L = \int_0^{\pi} P \times R \times d\theta \times \sin\theta \times L$$

or

$$\sigma_H \times t \times 2 = PR[-\cos\theta]_0^{\pi}$$
$$= 2PR$$

Therefore

$$\sigma_H = \frac{PR}{t} \tag{8.1}$$

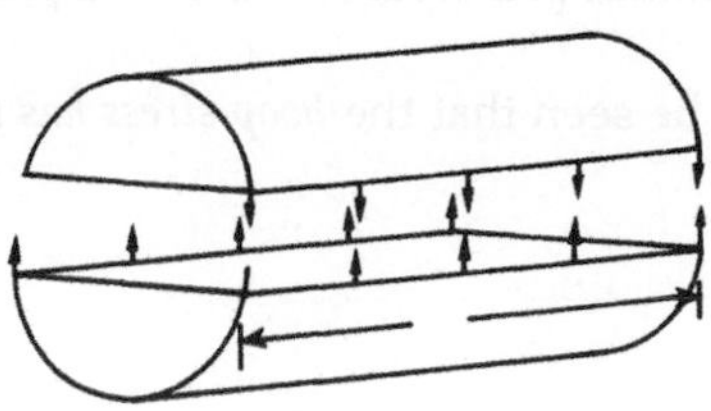

Figure 8.3 Failure due to hoop stress

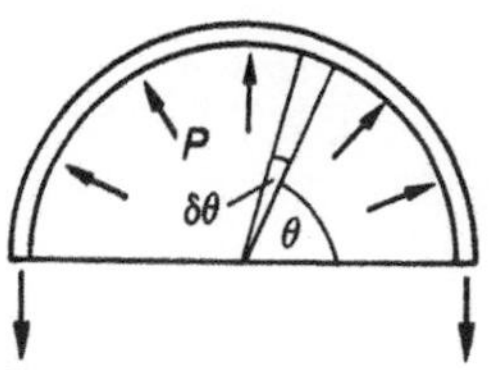

Figure 8.4 Equilibrium of circular cylinder

where

σ_H = hoop stress, which under internal pressure is a maximum principal stress
P = internal pressure
R = internal radius
t = wall thickness of cylinder
L = length of cylinder

If η_L is the structural efficiency of a longitudinal joint of the cylinder ($\eta_L \leq 1$), then

$$\sigma_H = \frac{PR}{\eta_L t} \tag{8.2}$$

8.2.2 Failure due to longitudinal stress (σ_L)

If failure is due to the longitudinal stress, then fracture will occur along a circumferential seam, as shown by Figure 8.5.

Consider the circular cylinder of Figure 8.6, which may be assumed to split in two along a circumferential seam, owing to the longitudinal stress, σ_L.

Resolving horizontally,

$$P \times \pi R^2 = \sigma_L \times 2\pi Rt$$

Therefore

$$\sigma_L = \frac{PR}{2t} \tag{8.3}$$

where σ_L is the longitudinal stress or the minimum principal stress, if the pressure is internal.

Comparing equations (8.1) and (8.3), it can be seen that the *hoop stress has twice the magnitude of the longitudinal stress.*

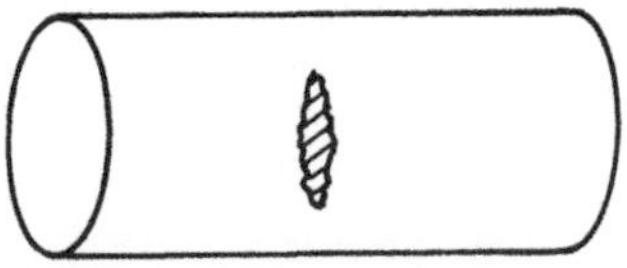

Figure 8.5 Fracture due to longitudinal stress

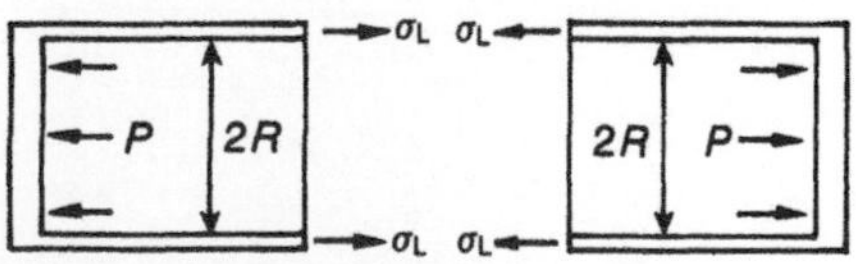

Figure 8.6 Failure along a circumferential seam

If η_c is the structural efficiency of a circumferential joint of the cylinder ($\eta_c \leqslant 1$), then

$$\sigma_L = \frac{PR}{2\eta_c t} \tag{8.4}$$

For edge effects, see Section 8.4.

EXAMPLE 8.1

Two identical extruded tubes of internal radius 5 m and of wall thickness 2 cm are joined together along a circumferential seam to form the main body of a pressure vessel. Assuming that the ends of the vessel are blocked off by very thick inextensible plates, determine the maximum stress in the vessel when it is subjected to an internal pressure of 0.4 MPa.

It may be assumed that the joint efficiency equals 52%.

SOLUTION

Now as the tubes are extruded, there is no longitudinal joint, so that

$$\underline{\eta_L = 1}$$

Therefore, from equation (8.2),

$$\sigma_H = \text{hoop stress} = \frac{0.4 \times 10^6 \times 5}{2 \times 10^{-2}}$$

$$\underline{\sigma_H = 100 \text{ MN/m}^2}$$

Now,

$$\eta_c = 0.52$$

Therefore, from equation (8.4),

$$\sigma_L = \text{longitudinal stress} = \frac{0.4 \times 10^6 \times 5}{2 \times 0.52 \times 2 \times 10^{-2}}$$

$$\underline{\sigma_L = 96.2 \text{ MN/m}^2}$$

i.e.

$$\underline{\text{Maximum stress} = \sigma_H = 100 \text{ MN/m}^2}$$

EXAMPLE 8.2

A circular cylindrical pressure vessel of internal diameter 10 m is blocked off at its ends by thick plates, and it is to be designed to sustain an internal pressure of 1 MPa.

Assuming the following to apply, determine a suitable value for the wall thickness of the cylinder:

Longitudinal joint efficiency = 98%
Circumferential joint efficiency = 46%
Maximum permissible stress = 100 MN/m^2

SOLUTION

Consideration of hoop stress

From equation (8.2),

$$\sigma_H = \frac{PR}{\eta_L t}$$

Therefore

$$t = \frac{PR}{\eta_L \sigma_H} = \frac{1 \times 5}{0.98 \times 100}$$

$$\underline{t = 0.051 \text{ m}}$$

Consideration of longitudinal stress

From equation (8.4),

$$\sigma_L = \frac{PR}{2\eta_c t}$$

or

$$t = \frac{PR}{2\eta_c \sigma_L}$$

$$= \frac{1 \times 5}{1 \times 0.46 \times 100}$$

$$\underline{t = 0.0544 \text{ m}}$$

i.e.

Design wall thickness = 5.44 cm

8.3 Thin-walled spherical shells under uniform internal pressure

Under uniform internal pressure, a thin-walled spherical shell of constant thickness will have a constant membrane tensile stress, where all such stresses will be principal stresses.

Let σ be the membrane stress in the spherical shell. For such structures, fracture will occur along a diameter, as shown in Figure 8.7.

To determine the membrane principal stress σ, in terms of the applied pressure P

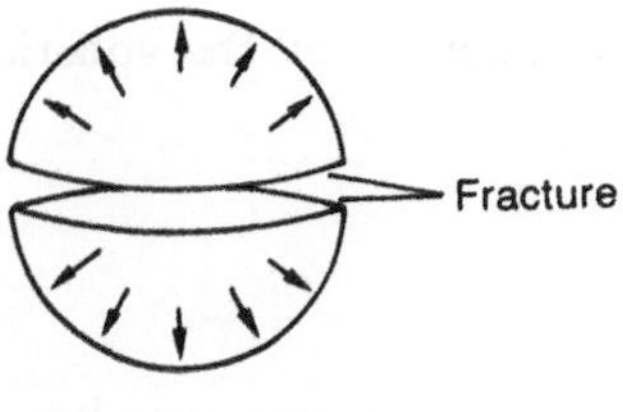

Figure 8.7 Fracture of spherical shell

and the geometry of the spherical shell, consider the equilibrium of one half of the spherical shell, as shown in Figure 8.8.

Resolving vertically,

$$\sigma \times 2\pi Rt = \int_0^{\pi/2} P \times 2\pi b \times R\,d\theta \times \sin\theta$$

but

$$b = R\cos\theta$$

Therefore

$$\sigma \times 2\pi Rt = 2\pi R^2 P \int_0^{\pi/2} \cos\theta \sin\theta\,d\theta$$

or

$$\sigma t = -PR \int_0^{\pi/2} \cos\theta\,d(\cos\theta)$$

$$= PR\left[-\frac{\cos^2\theta}{2}\right]_0^{\pi/2}$$

$$= \frac{PR}{2}$$

Therefore

$$\sigma = \frac{PR}{2t} \tag{8.5}$$

From equations (8.1) and (8.5), it can be seen that the maximum principal stress in a spherical shell has half the value of the maximum principal stress in a circular cylinder of the same radius.

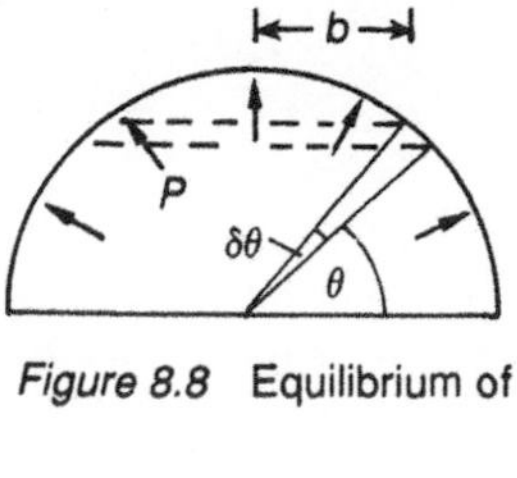

Figure 8.8 Equilibrium of hemispherical shell

If η is the structural efficiency of a joint on a diameter of the spherical shell ($\eta \leq 1$), then

$$\sigma = \frac{PR}{2\eta t} \tag{8.6}$$

EXAMPLE 8.3

A thin-walled spherical vessel, of internal diameter 10 m, is to be designed to withstand an internal pressure of 1 MN/m^2.

Assuming the following apply, determine a suitable value of the wall thickness:

Joint efficiency = 75%
Maximum permissible stress = 100 MN/m^2

SOLUTION

From equation (8.6),

$$\sigma = \frac{PR}{2\eta t}$$

or

$$t = \frac{PR}{2\eta\sigma} = \frac{1 \times 5}{2 \times 0.75 \times 100} = 0.033 \text{ m}$$

$$\underline{t = 3.3 \text{ cm}}$$

EXAMPLE 8.4

A submarine pressure hull of external diameter 10 m may be assumed to be composed of a long cylindrical shell, blocked off by two hemispherical shell domes. Neglecting buckling due to external pressure, and the effects of discontinuity at the intersections between the domes and the cylinder, determine suitable thicknesses for the cylindrical shell body and the hemispherical dome ends.

The following may be assumed:

Maximum permissible stress = 200 MN/m^2
Diving depth of submarine = 250 m
Density of sea water = 1020 kg/m^3
Acceleration due to gravity g = 9.81 m/s^2
Longitudinal joint efficiency = 90%
Circumferential joint efficiency = 70%

SOLUTION

Cylindrical body

From equation (8.2),

$$t = \frac{PR}{\eta_L \times \sigma_H}$$

but

$$P = 250\ \text{m} \times 1020\ \frac{\text{kg}}{\text{m}^3} \times 9.81\ \frac{\text{m}}{\text{s}^2}$$

$$= 2.5\ \text{MPa}$$

Therefore

$$t = \frac{2.5 \times 10^6 \times 5}{0.9 \times 200 \times 10^6}$$

$$\underline{t = 0.0694\ \text{m} = 6.94\ \text{cm}}$$

Dome ends

From equation (8.6),

$$t = \frac{PR}{2\eta\sigma}$$

$$= \frac{2.5 \times 10^6 \times 5}{2 \times 0.7 \times 200 \times 10^6}$$

$$\underline{t = 0.0446\ \text{m} \times 4.46\ \text{cm}}$$

i.e.

Wall thickness of cylindrical body = 6.94 cm
Wall thickness of dome ends = 4.46 cm

Figure 8.9 Axisymmetric buckling of an oblate dome under uniform external pressure

Figure 8.10 Lobar buckling of a hemispherical or prolate dome under uniform external pressure

NB From the above calculations, it can be seen that the required wall thickness of a submarine pressure hull increases roughly in proportion to its diving depth. Thus, to ensure that the submarine has a sufficient reserve buoyancy for a given diameter, it is necessary to restrict its diving depth or to use a material of construction which has a better strength to weight ratio. It should also be noted that under external pressure, the cylindrical section of a submarine pressure hull can buckle at a pressure which may be a fraction of that to cause yield, as shown in Figure 8.1. In a similar manner, the dome ends can also buckle at a pressure which may be a small fraction of that to cause yield, as shown in Figure 8.9 and 8.10.

8.3.1 *Liquid required to raise the pressure inside a circular cylinder (based on small-deflection elastic theory)*

In the case of a circular cylinder, assuming that it is just filled with liquid, the additional liquid that will be required to raise the internal pressure will be partly as a result of the swelling of the structure and partly as a result of the compression of the liquid itself, as shown by the following components:

(a) Longitudinal extension of the cylinder.
(b) Radial extension of the cylinder.
(c) Compressibility of the liquid.

Consider a thin-walled circular cylinder, blocked off at its ends by thick inextensible plates, and just filled with the liquid.

The calculation for the additional liquid to raise the internal pressure to P will be composed of the following three components.

8.3.2 *Longitudinal extension of the circular cylinder*

Let

u = the longitudinal movement of one end of the cylinder, relative to the other, as shown in Figure 8.11

$$\varepsilon_L = \text{longitudinal strain in cylinder} = u/L \tag{8.7}$$

The change of volume of the cylinder due to u is given by

$$\underline{\delta V_1 = \pi R^2 u = \pi R^2 \varepsilon_L L} \tag{8.8}$$

8.3.3 *Radial extension of cylinder*

Let

w = radial deflection of cylinder, as shown in Figure 8.12

$$\varepsilon_H = \text{hoop strain} = \frac{2\pi(R+w) - 2\pi R}{2\pi R} = \frac{w}{R} \tag{8.9}$$

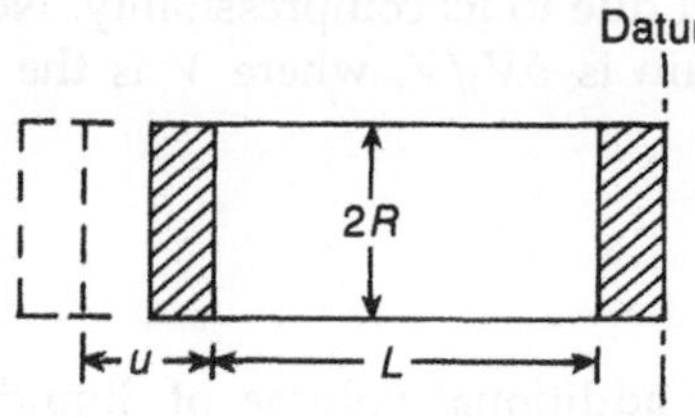

Figure 8.11 Longitudinal movement of the cylinder

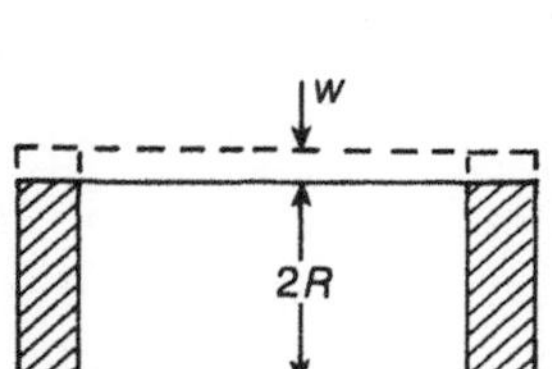

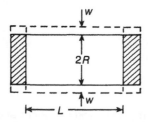

Figure 8.12 Radial deflection of the cylinder

The change of volume of the cylinder due to w is given by

$$\delta V_2 = 2\pi RLw = 2\pi R^2 L\varepsilon_H \tag{8.10}$$

Alternatively, from Section 2.8.1, the total volumetric strain of the circular cylindrical shell is given by

$$\frac{\delta V}{V} = (\varepsilon_x + \varepsilon_y + \varepsilon_z)$$

or

$$\frac{\delta V_s}{V} = (\varepsilon_L + \varepsilon_H + \varepsilon_H)$$

or

$$\begin{aligned}\delta V_s &= \pi R^2 L(\varepsilon_L + 2\varepsilon_H)\\ &= \delta V_1 + \delta V_2 \text{ (as obtained above)}\end{aligned}$$

which is the change in volume due to swelling of the shell.

8.3.4 *Compressibility of liquid*

The liquid will compress owing to the raising of its pressure by P. From Section 2.8.1,

$$\frac{\text{Volumetric stress}}{\text{Volumetric strain}} = K = \text{bulk modulus}$$

Let δV_3 be the change in volume of the liquid due to its compressibility. Now, P is the volumetric stress and the volumetric strain is $\delta V_3/V$, where V is the internal volume of the cylinder. Therefore,

$$\delta V_3 = \frac{PV}{K} \tag{8.11}$$

From equations (8.8), (8.10) and (8.11), the additional volume of liquid that is required to be pumped in to raise the internal pressure of the cylinder by P is

$$\delta V = \delta V_1 + \delta V_2 + \delta V_3$$

or

$$\delta V = \pi R^2 L\varepsilon_L + 2\pi R^2 L\varepsilon_H + PV/K \tag{8.12}$$

EXAMPLE 8.5

A thin-walled circular cylinder of internal diameter 10 m, wall thickness 2 cm and length 5 m is just filled with water. Determine the additional water that is required to be pumped into the vessel to raise its pressure by 0.5 MPa.

$$E = 2 \times 10^{11}\ \text{N/m}^2 \quad \nu = 0.3$$
$$K = 2 \times 10^{9}\ \text{N/m}^2$$

SOLUTION

From equation (8.8),

$$\delta V_1 = \pi R^2 L \varepsilon_L$$

but

$$\varepsilon_L = \frac{1}{E}(\sigma_L - \nu\sigma_H)$$
$$= \frac{PR}{Et}\left(\tfrac{1}{2} - \nu\right)$$
$$= \frac{0.5 \times 10^6 \times 5 \times 0.2}{2 \times 10^{11} \times 2 \times 10^{-2}}$$

or

$$\underline{\varepsilon_L = 1.25 \times 10^{-4}}$$

Therefore

$$\delta V_1 = \pi \times 25 \times 5 \times 1.25 \times 10^{-4}$$
$$\underline{\delta V_1 = 0.049\ \text{m}^3} \tag{8.13}$$

From equation (8.10),

$$\underline{\delta V_2 = 2\pi R^2 L \varepsilon_H}$$

but

$$\varepsilon_H = \frac{1}{E}(\sigma_H - \nu\sigma_L)$$
$$= \frac{PR}{Et}(1 - \nu/2)$$
$$= \frac{0.5 \times 10^6 \times 5 \times 0.85}{2 \times 10^{11} \times 2 \times 10^{-2}}$$

or

$$\underline{\varepsilon_H = 5.313 \times 10^{-4}}$$

Therefore

$$\delta V_2 = 2\pi \times 25 \times 5 \times 5.313 \times 10^{-4}$$
$$\underline{\delta V_2 = 0.417\ \text{m}^3} \tag{8.14}$$

From equation (8.11),

$$\delta V_3 = \frac{0.5 \times 10^6 \times \pi R^2 L}{K}$$

$$= \frac{0.5 \times 10^6 \times \pi \times 25 \times 5}{2 \times 10^9}$$

$$\underline{\delta V_3 = 0.098 \text{ m}^3} \tag{8.15}$$

From equations (8.13)–(8.15), the additional volume of water required to be pumped into the vessel to raise its internal pressure by 0.5 MPa is

$$\delta V = \delta V_1 + \delta V_2 + \delta V_3$$

$$= 0.049 + 0.417 + 0.098$$

$$\underline{\delta V = 0.564 \text{ m}^3} \tag{8.16}$$

From equation (8.16), it can be seen that the bulk of the additional water required to raise the pressure was because of the radial expansion of the cylinder.

8.3.5 *Additional liquid required to raise the internal pressure of a thin-walled spherical shell (based on small-deflection elastic theory)*

Assume that the spherical shell is just filled with liquid, and let w be the radial deflection of the sphere due to an internal pressure increase of P, as shown in Figure 8.13.

The additional liquid that is required to be pumped into the vessel will be because of the following:

(a) Swelling of the structure due to the application of P.
(b) Compressibility of the liquid itself.

If ε is the membrane strain due to P, then

$$\varepsilon = \frac{2\pi(R + w) - \pi R}{2\pi R} = \frac{w}{R} \tag{8.17}$$

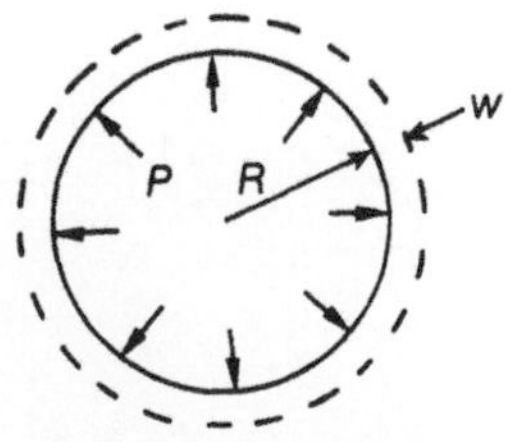

Figure 8.13 Deflected form of spherical shell

but

$$\varepsilon = \frac{1}{E}(\sigma - \nu\sigma) = \frac{PR}{2tE}(1 - \nu) \tag{8.18}$$

The change in volume due to swelling of the shell is

$$\delta V_s = 4\pi R^2 w = 4\pi R^3 \varepsilon$$

$$= 4\pi R^3 \frac{PR}{2tE}(1 - \nu)$$

or

$$\delta V_s = \frac{2\pi P R^4 (1 - \nu)}{Et} \tag{8.19}$$

Alternatively, from section 2.8.1, the total volumetric strain due to swelling of the shell is

$$\frac{\delta V}{V} = (\varepsilon_x + \varepsilon_y + \varepsilon_z)$$

or

$$\frac{\delta V_s}{V} = (\varepsilon + \varepsilon + \varepsilon)$$

or

$$V_s = \tfrac{4}{3}\pi R^3 3\varepsilon = 4\pi R^3 \varepsilon \text{ (as obtained above)}$$

The compressibility of the liquid can be calculated from

$$\delta V_L = \frac{PV}{K} \tag{8.20}$$

where

δV_L = change in the volume of the liquid due to compression
V = internal volume of sphere = $\tfrac{4}{3}\pi R^3$

EXAMPLE 8.6

A thin walled spherical shell, of diameter 10 m and a wall thickness 2 cm, is just filled with water. Determine the additional water that is required to be pumped into the vessel to raise its internal pressure by 0.5 MPa.

$E = 2 \times 10^{11}$ N/m^2 $\quad \nu = 0.3$
$K = 2 \times 10^{9}$ N/m^2

SOLUTION

From equation (8.19), the increase in volume of the spherical shell due to its swelling under pressure is

$$\delta V_s = \frac{2\pi \times 0.5 \times 10^6 \times 5^4(0.7)}{2 \times 10^{11} \times 2 \times 10^{-2}}$$

$$\underline{\delta V_s = 0.344 \text{ m}^3} \tag{8.21}$$

From equation (8.20), the additional volume of water to be pumped in, owing to the compressibility of the water, is

$$\delta V_L = 0.5 \times 10^6 \times \tfrac{4}{3} \times \pi \times 5^3 \times \frac{1}{2 \times 10^9}$$

$$= \underline{0.131 \text{ m}^3} \tag{8.22}$$

That is, the total additional quantity of water that is required to be pumped in to raise the pressure is

$$\delta V = \delta V_s + \delta V_L$$

$$\underline{\delta V = 0.475 \text{ m}^3}$$

From equations (8.21) and (8.22), it can be seen that for this spherical shell, the bulk of the additional liquid that was required to raise the pressure by 0.5 MPa was due to the swelling of the shell.

8.4 Bending stresses in circular cylinders under uniform pressure

The theory presented in this chapter neglects the effect of bending stresses at the edges of the cylinder, where the vessel may be firmly clamped, as shown in Figure 8.14.

Figure 8.15 shows the theoretical and experimental values [12, 13] for the radial deflection of a thin-walled steel cylinder clamped at its ends. The theory was based on the solution of a fourth-order shell differential equation, which is beyond the scope of this book.

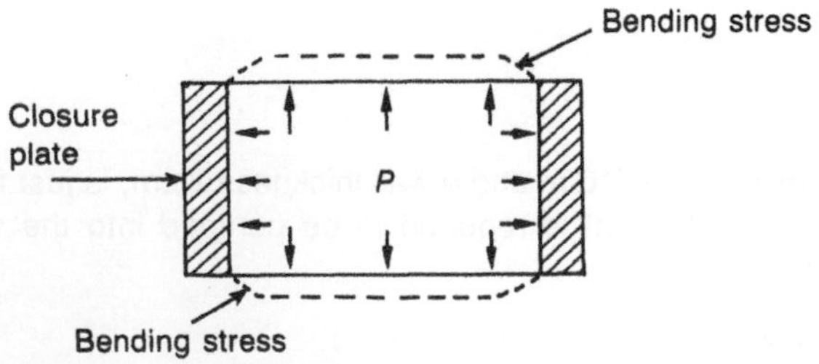

Figure 8.14 Deflected form of cylinder under internal pressure

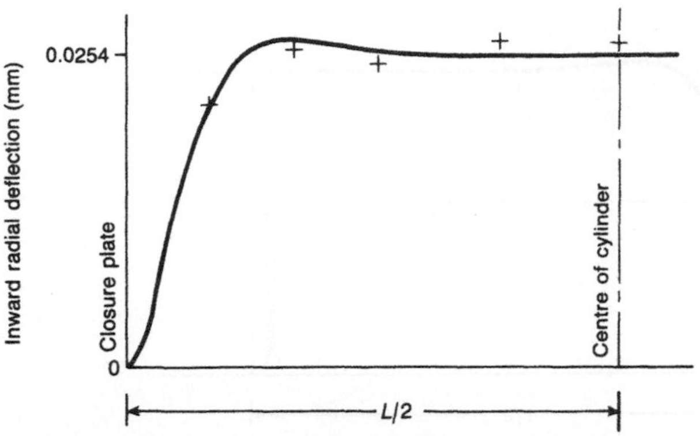

Figure 8.15 Theoretical and experimental values for the radial deflection of model No. 7 at 0.6895 MPa

The vessel was firmly clamped at its ends, but free to move longitudinally, and it had the following properties:

Internal diameter	= 26.04 cm
Wall thickness	= 0.206 cm
Length of cylindrical shell	= L = 25.4 cm
External pressure	= 0.6895 MPa
E	= 1.93×10^{11} N/m^2
ν	= 0.3 (assumed)
Initial out-of-roundness	= 0.0102 cm

Plots of the theoretical and experimental stresses are shown in Figures 8.16–8.19 from one end of the vessel (closure plate) to its mid-span.

From Figures 8.15–8.19, it can be seen that the effects of bending are very localized.

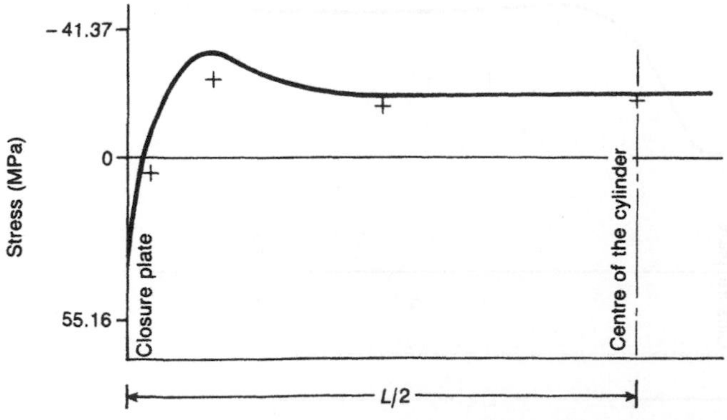

Figure 8.16 Longitudinal stress of the outermost fiber at 0.6895 MPa (model No. 7)

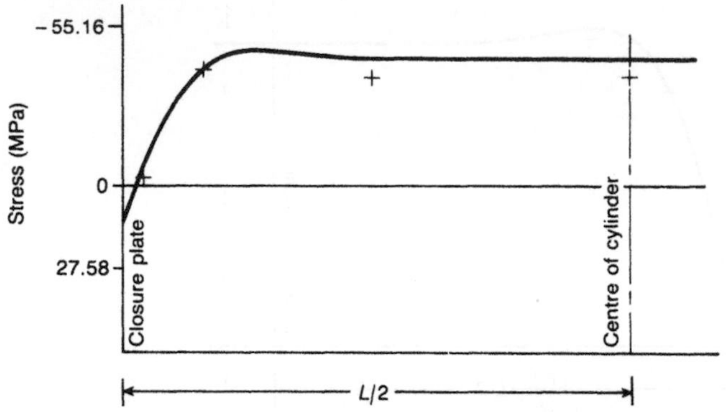

Figure 8.17 Circumferential stress of the outermost fiber at 0.6895 MPa (model No. 7)

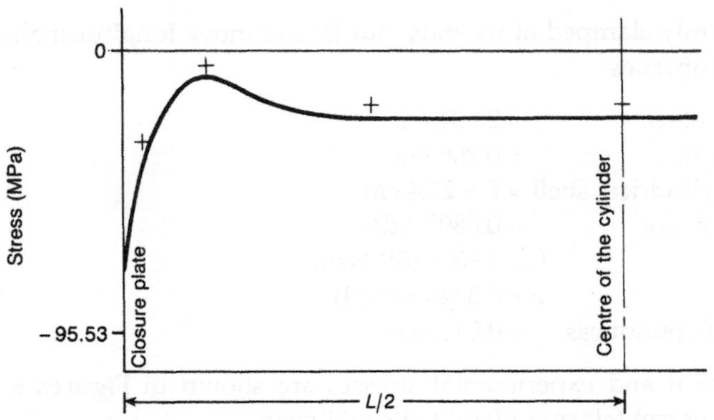

Figure 8.18 Longitudinal stress of the innermost fiber at 0.6895 MPa (model No. 7)

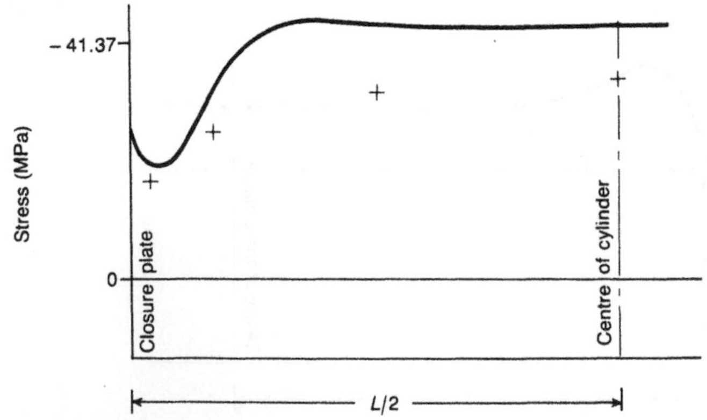

Figure 8.19 Circumferential stress of the innermost fiber at 0.6895 MPa (model No. 7)

8.5 Circular cylindrical shell with hemispherical ends

It is quite common to seal the ends of a circular cylindrical pressure vessel with hemispherical end caps as shown in Figure 8.20.

When such a vessel is subjected to uniform internal pressure, bending stresses can occur at the joints, because if both the circular cylindrical shell and the hemispherical end caps are made from the same material and of the same wall thickness, the circular cylinder will tend to have a larger radial deflection than the hemispherical shell; this will cause bending stresses at the joints. To overcome or reduce the bending stresses at the joints, the wall thickness of the hemispherical shell must be less than the wall thickness of the circular cylindrical shell. A theory for this will now be produced.

Let

w_c = radial deflection of the circular cylinder at the joint due to an internal pressure P

w_s = radial deflection of the hemispherical shell at the joint due to an internal pressure P

R = radius of the circular cylinder

t_c = wall thickness of the circular cylinder

t_s = wall thickness of the hemispherical shell.

Let ε_H be the hoop strain in the cylinder, which from equation (8.9) is

$$\varepsilon_H = \frac{w_c}{R} = \frac{1}{E}(\sigma_H - \nu\sigma_L)$$

or

$$w_c = \frac{R}{E}\,\frac{PR}{t_c}\left(1 - \frac{\nu}{2}\right)$$

$$\frac{PR^2}{t_c E}\left(1 - \frac{\nu}{2}\right) \tag{8.23}$$

Let ε be the circumferential strain in the hemispherical shell equal to the meridional

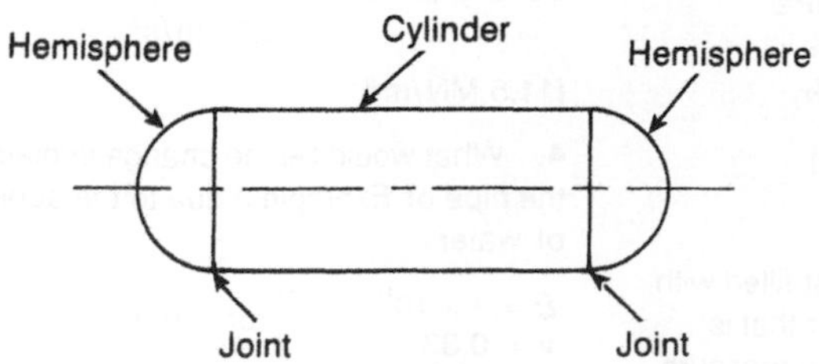

Figure 8.20 Typical pressure vessel

strain in the hemispherical shell, which from equation (8.17) is given by

$$\varepsilon = \frac{w_s}{R} = \frac{1}{E}(\sigma - \nu\sigma)$$

$$= \frac{PR}{2t_sE}(1-\nu)$$

or

$$w_s = \frac{PR^2}{2t_sE}(1-\nu) \tag{8.24}$$

For no bending to occur at the joint between the circular cylinder and the hemispherical shell, $w_c = w_s$, or equations (8.23) and (8.24) are equal:

$$\frac{PR^2}{t_cE}\left(1-\frac{\nu}{2}\right) = \frac{PR^2}{2t_sE}(1-\nu)$$

Therefore

$$\frac{t_s}{t_c} = \frac{(1-\nu)}{2(1-\nu/2)}$$

If $\nu = 0.3$, then

$$t_s = 0.412t_c$$

EXAMPLES FOR PRACTICE 8

1. A boiler, which may be assumed to be composed of a thin-walled cylindrical shell body of internal diameter 4 m, is blocked off by two thin-walled hemispherical dome ends. Neglecting the effects of discontinuity at the intersection between the dome and cylinder, determine suitable thicknesses for the cylindrical shell body and the hemispherical dome ends.

The following may be assumed:

Maximum permissible stress = 100 MN/m^2
Design pressure = 1 MPa
Longitudinal joint efficiency = 75%
Circumferential joint efficiency = 50%

{Cylinder $t = 2.67$ cm; dome $t = 2$ cm}

2. If the vessel of Example 1 is just filled with water, determine the additional water that is required to be pumped in, to raise the pressure by 1 MPa.

The following may be assumed to apply:

Length of cylindrical portion of vessel = 6 m

$E = 2\times 10^{11}$ N/m^2 $\quad \nu = 0.3 \quad K = 2\times 10^9$ N/m^2

{0.12 m^3}

3. A copper pipe of internal diameter 1.25 cm and wall thickness 0.16 cm is to transport water from a tank that is situated 30 m above it.

Determine the maximum stress in the pipe, given the following:

Density of water = 1000 kg/m^3
$g = 9.81$ m/s^2

{11.5 MN/m^2}

4. What would be the change in diameter of the pipe of Example 3 due to the applied head of water?

$E = 1\times 10^{11}$ N/m^2, $\nu = 0.33$ } for copper

{1.2 μm}

5. A thin-walled spherical pressure vessel of 1 m internal diameter is fed by a pipe of internal diameter 3 cm and wall thickness 0.16 cm.

Assuming that the material of construction of the spherical pressure vessel has a yield stress of 0.7 of that of the pipe, determine the wall thickness of the spherical shell.

{3.8 cm}

6. A spherical pressure vessel of internal diameter 2 m is constructed by bolting together two hemispherical domes with flanges.

Assuming that the number of bolts used to join the two hemispheres together is 12, determine the wall thickness of the dome and the diameter of the bolts, given the following:

Maximum applied pressure	= 0.7 MPa
Permissible stress in spherical shell	= 50 MPa
Permissible stress in bolts	= 200 MPa

{$t = 0.7$ cm, $d = 3.4$ cm}

7. A thin-walled circular cylinder, blocked off by inextensible end places, contains a liquid under zero gauge pressure. Show that the additional liquid that is required to be pumped into the vessel, to raise its internal gauge pressure by P, is the same under the following two conditions:

(a) when axial movement of the cylinder is completely free;
(b) when the vessel is totally restrained from axial movement.

It may be assumed that Poisson's ratio (ν) for the cylinder material is 0.25.

9

Energy methods

9.1 Introduction

Energy methods in structural mechanics are some of the most useful methods of theoretical analysis, as they lend themselves to computer solutions of complex structural problems [5,14], and they can also be extended to computer solutions of many other problems in engineering science [15–22].

There are many energy theorems and principles [14], but only the most popular methods will be considered here, and applications of some of these will be made to practical problems.

9.2 The method of minimum potential (Rayleigh–Ritz)

This states that to satisfy the elasticity and equilibrium equations of an elastic body, the derivative of the total potential, with respect to the displacements, must be zero i.e.

$$\frac{\partial \pi_p}{\partial u} = 0 \tag{9.1}$$

where

π_p = total potential
u = displacement

9.3 The principle of virtual work

This states that if an elastic body under a system of external forces is given a small virtual displacement, then the increase in external virtual work done to the body is equal to the increase in internal virtual strain energy stored in the body.

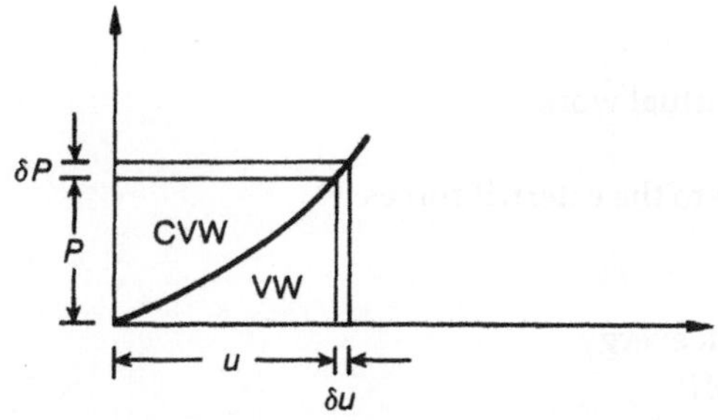

Figure 9.1 Force–displacement relationship

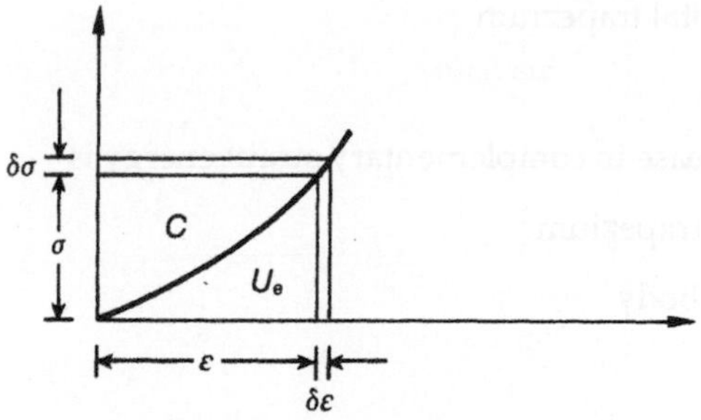

Figure 9.2 Stress–strain relationship

Consider Figure 9.1 and 9.2, where

δu = a small virtual displacement
$\delta\varepsilon$ = a small virtual strain, as a result of δu
P = external force
σ = stress due to P

From Figure 9.1, the increase in virtual work is

$$\mathrm{d}(\mathrm{VW}) = \text{area of vertical trapezium} \tag{9.2}$$

and from Figure 9.2, the increase in virtual strain energy is

$$\mathrm{d}(U_e) = \text{area of vertical trapezium} \tag{9.3}$$

Equating (9.2) and (9.3)

$$\underline{P \times \delta u = \sigma \times \delta\varepsilon \times \text{volume of the body}} \tag{9.4}$$

9.4 The principle of complementary virtual work

This theorem states that if an elastic body under a system of external forces is subjected to a small virtual force, then the increase in complementary virtual work is equal to the increase in complementary strain energy.

From Figure 9.1,

CVW = complementary virtual work
δP = virtual force
u = displacement due to the external forces

and from Figure 9.2,

C = complementary strain energy
$\delta\sigma$ = virtual stress due to δP
ε = strain due to the external forces

Hence, from Figure 9.1, the increase in complementary virtual work is

$$d(\text{CVW}) = \text{area of horizontal trapezium}$$

$$= \delta P \times u \tag{9.5}$$

Similarly, from Figure 9.2, the increase in complementary strain energy is

$$d(C) = \text{area of horizontal trapezium}$$

$$= \delta\sigma \times \varepsilon \times \text{volume of body} \tag{9.6}$$

Equating (9.5) and (9.6),

$$\underline{\delta P \times u = \delta\sigma \times \varepsilon \times \text{volume of the body}} \tag{9.7}$$

9.5 Castigliano's first theorem

This is really an extension to the principle of complementary virtual work, and applies to bodies that behave in a *linear elastic* manner.

Now when a body is linear elastic,

$$\delta(C) = \delta(U_e) = \delta P \times u$$

so that, in the limit,

$$\frac{\partial U_e}{\partial P} = u \tag{9.8}$$

9.6 Castigliano's second theorem

This theorem is an extension of Castigliano's first theorem, and it is particularly suitable for analyzing statically indeterminate frameworks.

Castigliano's second theorem simply states that for a framework with redundant members or forces,

$$\frac{\partial U_e}{\partial R} = \lambda \tag{9.9}$$

where

U_e = the strain energy of the whole frame
λ = initial lack of fit of the members of the framework
R = the force in any redundant member, to be determined

If there is no initial lack of fit (i.e. the members of the framework have been made precisely), then

$$\frac{\partial U_e}{\partial R} = 0$$

For most *pin-jointed frames*, the loads are axial; hence the strain energy is given by

$$U = \sum_{i=1}^{N} \frac{P_i^2 l_i}{2A_i E_i}$$

$$\frac{\partial U}{\partial R} = \frac{\partial U}{\partial P_i} \frac{\partial P_i}{\partial R}$$

Therefore

$$\frac{\partial U}{\partial R} = \sum_{i=1}^{N} \frac{P_i l_i}{A_i E_i} \frac{\partial P_i}{\partial R} = \text{initial lack of fit}$$

$$= 0 \text{ (for most undergraduate problems)} \tag{9.10}$$

Thus, by applying equation (9.10) to each redundant member or 'force', in turn, the required number of simultaneous equations is obtained, and hence the unknown redundant 'forces' can be determined. Once the redundant 'forces' are known, the other 'forces' can be determined through statics.

9.7 Strain energy stored in a rod under axial loading

Consider the load–displacement relationship of a uniform-section rod, as shown in Figure 9.3.

Now when the rod is displaced by u, the force required to achieve this displacement is P. Furthermore, as the average force during this load–displacement

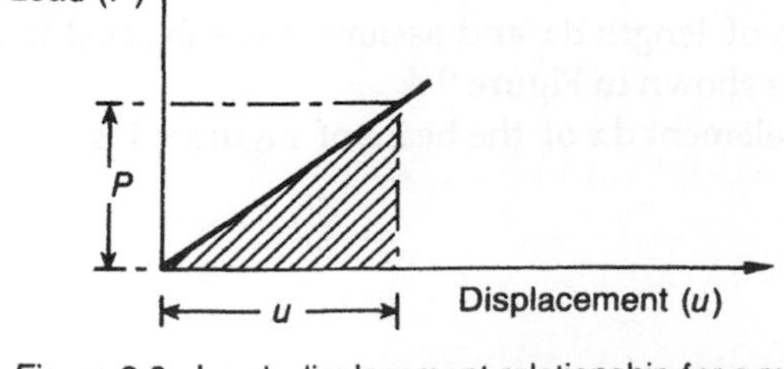

Figure 9.3 Load–displacement relationship for a rod

relationship is $P/2$, then

$$\text{Work done} = \text{the shaded area of Figure 9.3}$$
$$= \frac{P}{2} \times u \qquad (9.11)$$

However, this work done will be stored by the rod in strain energy U_e, so that

$$U_e = \tfrac{1}{2} P \times u \qquad (9.12)$$

but

$$P = \sigma \times A \qquad (9.13)$$

and

$$u = \varepsilon \times l = \sigma \times l/E \qquad (9.14)$$

where

σ = stress due to P
ε = strain due to P
A = cross-sectional area of rod
l = length of rod

Substituting equations (9.13) and (9.14) into (9.12),

$$U_e = \frac{\sigma^2}{2E} \times \text{volume of the rod} \qquad (9.15)$$

For an element of the rod of length dx,

$$d(U_e) = \frac{\sigma^2}{2E} \times A \times dx \qquad (9.16)$$

where $d(U_e)$ is the strain energy in the rod element.

9.8 Strain energy stored in a beam subjected to couples of magnitude *M* at its ends

Consider an element of the beam of length dx and assume it is subjected to two end couples, each of magnitude M, as shown in Figure 9.4.

The work done in bending the element dx of the beam of Figure 9.4 is

$$WD = \tfrac{1}{2} M\theta$$

but

$$\theta = dx/R$$

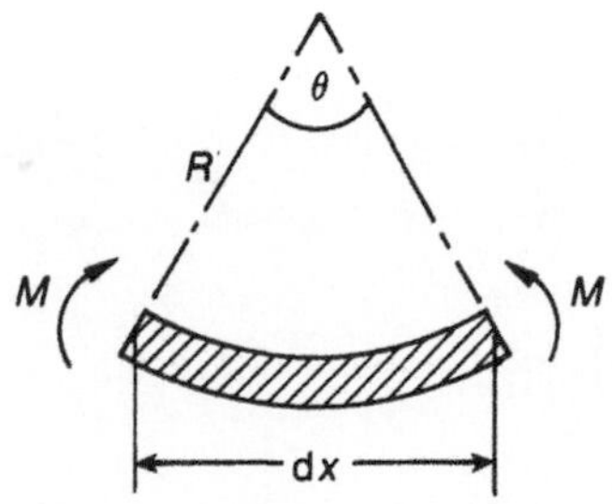

Figure 9.4 Element of beam in pure bending

Therefore

$$WD = \tfrac{1}{2} M \times dx/R \tag{9.17}$$

Now from equation (4.1),

$$\frac{1}{R} = \frac{M}{EI} \tag{9.18}$$

Substituting equation (9.18) into (9.17),

$$\underline{WD = d(U_b) = \frac{M^2}{2EI} \times dx} \tag{9.19}$$

where $d(U_b)$ is the bending strain energy in the beam element.

9.9 Strain energy due to a torque *T* stored in a uniform circular-section shaft

Let

θ = angle of twist of one end of the shaft relative to the other
T = applied torque

Then

$$WD = U_T = \tfrac{1}{2} T \times \theta \tag{9.20}$$

However, from equation (6.5),

$$\theta = T \times dx/(G \times J) \tag{9.21}$$

Therefore, substituting equation (9.21) into (9.20),

$$U_T = \frac{T^2 \times dx}{GJ} \tag{9.22}$$

where U_T is the strain energy due to torsion in an element dx.

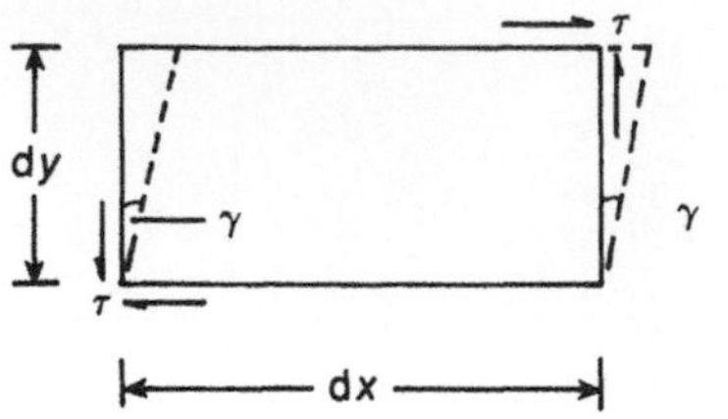

Figure 9.5 Rectangular element in pure shear

9.9.1 *Shear strain energy due to a system of complementary shear stresses τ*

Consider the rectangular element of Figure 9.5, which is in a state of pure shear. From Figure 9.5, it can be seen that the WD by the shear stresses τ, in changing the shape of the element, is

$$\mathrm{WD} = U_s = \frac{\tau}{2} \times (t \times \mathrm{d}x) \times (\mathrm{d}y \times \gamma)$$

but

$$\gamma = \tau / G$$

Therefore

$$U_s = \frac{\tau^2}{2G} \times \text{volume} \tag{9.23}$$

where

U_s = shear strain energy
volume = $t \times \mathrm{d}x \times \mathrm{d}y$
t = *thickness of plate*

EXAMPLE 9.1

A rod consists of three elements, each of length 0.5 m, joined firmly together. If this rod is subjected to an axial tensile force of 1 MN, determine the total extension of the rod, using strain energy principles and given that the cross-sectional areas of the elements are

Section 1: A_1 = 5E-3 m^2
Section 2: A_2 = 3E-3 m^2
Section 3: A_3 = 1E-2 m^2
Elastic modulus = 1×10^{11} N/m^2

SOLUTION

Section 1: $\sigma_1 = \dfrac{1\text{ MN}}{5\text{E-}3\text{ m}^2} = \underline{200\text{ MN/m}^2}$

Section 2: $\sigma_2 = \dfrac{1\text{ MN}}{3\text{E-}3\text{ m}^2} = \underline{333.3\text{ MN/m}^2}$

Section 3: $\sigma_3 = \dfrac{1\text{ MN}}{1\text{E-}2\text{ m}^2} = \underline{100\text{ MN/m}^2}$

Let δ be the total deflection of the rod, so that

$$WD = \tfrac{1}{2} \times 1\text{ MN} \times \delta \tag{9.24}$$

From equation (9.15),

$$U_e = \frac{\sigma_1^2}{2E} \times A_1 \times l + \frac{\sigma_2^2}{2E} \times A_2 \times l + \frac{\sigma_3^2}{2E} \times A_3 \times l \tag{9.25}$$

Equating (9.24) and (9.25),

$$0.5\delta = \frac{0.5}{2\text{E}5}(200^2 \times 5\text{E-}3 + 333.3^2 \times 3\text{E-}3 + 100^2 \times 1\text{E-}2)$$

or

$$\delta = \frac{1}{2\text{E}5}(200 + 333.3 + 100)$$

$$\underline{\delta = 3.17\text{ mm}}$$

EXAMPLE 9.2

Determine the maximum deflection of the end-loaded cantilever of Figure 9.6 using strain energy principles.

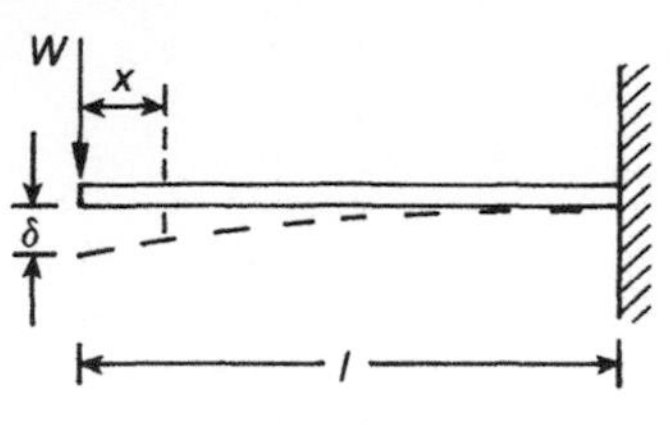

Figure 9.6 End-loaded cantilever

SOLUTION

Let

I = second moment of area of the beam section, about a horizontal axis

$$\text{WD by the load} = \tfrac{1}{2} W \times \delta \tag{9.26}$$

U_b = bending strain energy in the beam

$$= \int_0^l \frac{M^2}{2EI}\,dx$$

but

$$M = -Wx$$

Therefore

$$U_b = \frac{W}{2EI}\int_0^l x^2\, dx$$

$$= \frac{W^2 l^3}{6EI} \quad (9.27)$$

Equation (9.26) and (9.27),

$$\delta = \frac{Wl^3}{3EI} \text{ (as required)}$$

EXAMPLE 9.3

Determine the deflection at mid-span for the centrally loaded beam of Figure 9.7, which is simply supported at its ends.

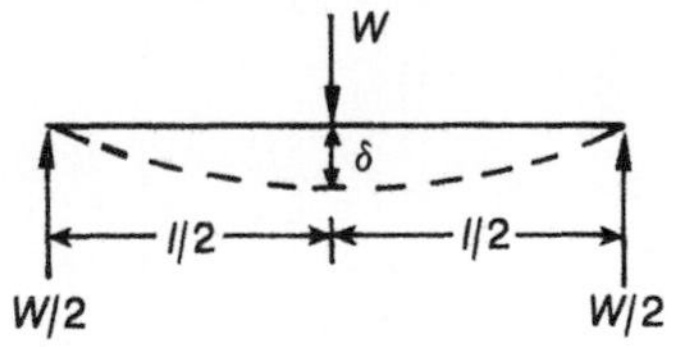

Figure 9.7 Centrally loaded beam

SOLUTION

$$\text{WD by load} = \tfrac{1}{2} W\delta \quad (9.28)$$

$$U_b = \int \frac{M^2}{2EI}\, dx$$

$$= \frac{1}{2EI}\int_0^{l/2} \left(\frac{Wx}{2}\right)^2 dx \times 2 \quad (9.29)$$

$$= \frac{W^2}{4EI}\left[\frac{x^3}{3}\right]_0^{l/2}$$

or

$$U_b = \frac{W^2 l^3}{96EI} \quad (9.30)$$

Equating (9.28) and (9.30),

$$\delta = \frac{Wl^3}{48EI} \text{ (as required)}$$

NB It should be noted that in equation (9.29), the upper limit of the integral was $l/2$. It was necessary to integrate in this manner because the expression for the bending moment, namely $M = Wx/2$, applied only between $x = 0$ and $x = l/2$.

9.10 Deflection of thin curved beams

In this theory, it will be assumed that the effects of shear strain energy and axial strain energy are negligible, and that all the work done by external loads on these beams is in fact absorbed by them in the form of bending strain energy.

Castigliano's theorems will be used in this section, and the method will be demonstrated by applying it to a number of examples.

Castigliano's first theorem requires a *load to act in the direction of the required displacement.* When the value of such a displacement is to be determined, at a point where no load points in the same direction, it will be necessary to assume an imaginary load to act in the direction of the required displacement, and, later, to set this imaginary load to zero.

EXAMPLE 9.4

A thin curved beam is in the form of a quadrant of a circle, as shown in Figure 9.8. One end of this quandrant is firmly fixed to a solid base, whilst the other end is subjected to a point load W, acting vertically downwards. Determine expressions for the vertical and horizontal displacements at the free end.

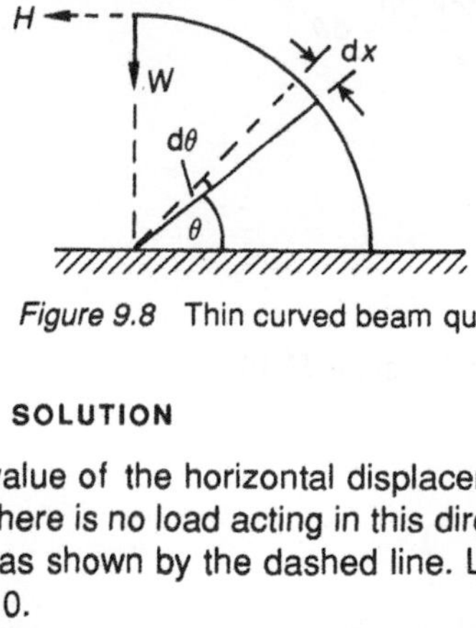

Figure 9.8 Thin curved beam quadrant

SOLUTION

As the value of the horizontal displacement is required to be determined at the free end, and as there is no load acting in this direction, it will be necessary to assume *an imaginary load H*, as shown by the dashed line. Later in the calculation, it will be necessary to note that $H = 0$.

Consider an element dx at any given angle θ from the base. At this point the bending moment is

$$M = -WR\cos\theta - HR(1 - \sin\theta)$$

NB The *sign convention* for bending moment is not important when using Castigliano's first theorem, providing that moments increasing curvature are of opposite sign to moments decreasing curvature.

Let

δV = downward vertical displacement at the free end
δH = horizontal deflection (to the left) at the free end

From equation (9.19), the bending strain energy of the element dx is

$$\mathrm{d}(U_b) = \frac{M^2}{2EI}\,\mathrm{d}x$$

or

$$U_b = \int \frac{M^2}{2EI}\,\mathrm{d}x$$

From Castigliano's first theorem,

$$\delta V = \frac{\partial U_b}{\partial W} = \frac{\partial U_b}{\partial M}\,\frac{\partial M}{W} = \frac{1}{EI}\int M\,\frac{\partial M}{\partial W}\,\mathrm{d}x \qquad (9.31)$$

Therefore

$$\delta V = \frac{1}{EI}\int_0^{\pi/2} [WR\cos\theta + HR(1-\sin\theta)]\,R\cos\theta\,R\,\mathrm{d}\theta$$

but

$$H = 0$$

Therefore

$$\delta V = \frac{WR^3}{EI}\int_0^{\pi/2}\cos^2\theta\,\mathrm{d}\theta = \frac{WR^3}{EI}\int_0^{\pi/2}\frac{(1+\cos 2\theta)}{2}\,\mathrm{d}\theta$$

$$\underline{\delta V = \pi WR^3/4EI} \qquad (9.32)$$

Similarly

$$\delta H = \frac{\partial U_b}{\partial H} = \frac{\partial U_b}{\partial M}\,\frac{\partial M}{\partial H} = \frac{1}{EI}\int M\,\frac{\partial M}{\partial H}\,\mathrm{d}x$$

$$= \frac{1}{EI}\int_0^{\pi/2} [WR\cos\theta + HR(1-\sin\theta)]R(1-\sin\theta)R\,\mathrm{d}\theta$$

but

$$\underline{H = 0}$$

Therefore

$$\delta H = \frac{WR^3}{EI}\int_0^{\pi/2}(\cos\theta - \sin\theta\cos\theta)\,\mathrm{d}\theta$$

$$= \frac{WR^3}{EI}\left[\sin\theta - \frac{\sin^2\theta}{2}\right]_0^{\pi/2}$$

$$\underline{\delta H = WR^3/2EI} \qquad (9.33)$$

Thus, it can be seen that, owing to W, there is a horizontal displacement in addition to a vertical displacement. The value of the resultant displacement can be obtained by considerations of Pythagoras's theorem, together with elementary trigonometry.

EXAMPLE 9.5

A thin curved beam, which is in the form of a semi-circle, is firmly fixed at its base and is subjected to point loads at its free end, as shown in Figure 9.9. Determine expressions for the vertical and horizontal displacements at its free end.

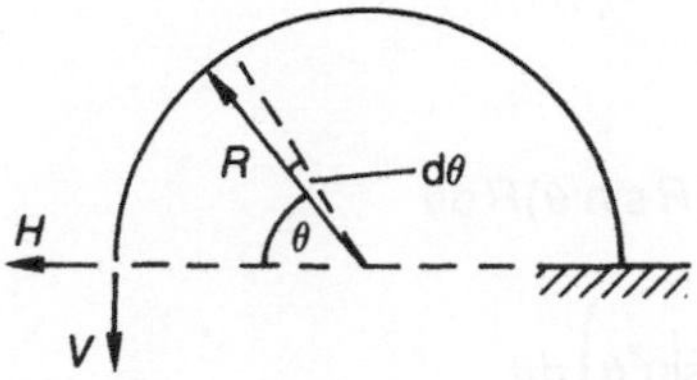

Figure 9.9 Thin curved beam (semi-circle)

SOLUTION

Let

δV = deflection in the direction of V
δH = deflection in the direction of H

At any angle θ, the bending moment acting on an element of the beam, of length $R\,d\theta$, is

$$M = VR(1 - \cos\theta) - HR\sin\theta$$

Now

$$\delta V = \frac{1}{EI}\int M\frac{\partial M}{\partial V}R\,d\theta$$

and

$$\delta H = \frac{1}{EI}\int M\frac{M}{H}R\,d\theta$$

Therefore

$$\delta V = \frac{1}{EI}\int [VR(1 - \cos\theta) - HR\sin\theta]R(1 - \cos\theta)R\,d\theta$$

$$= \frac{R^3}{EI}\int [V(1 - 2\cos\theta + \cos^2\theta) - H(\sin\theta - \sin\theta\cos\theta)]\,d\theta$$

$$= \frac{R^3}{EI}\int V\left(1 - 2\cos\theta + \frac{(1 + \cos 2\theta)}{2}\right) - H(\sin\theta - \sin\theta\cos\theta)\,d\theta$$

$$= \frac{R^3}{EI}\left[V\left(\theta - 2\sin\theta + \frac{\theta}{2} + \frac{\sin 2\theta}{4}\right) - H\left(-\cos\theta - \frac{\sin^2\theta}{2}\right)\right]_0^{\pi}$$

$$= \frac{R^3}{EI}\left\{V\left[\pi - 0 + \frac{\pi}{2} + 0\right] - 0 - H[1 - 0] + H[-1 - 0]\right\}$$

$$\delta V = \frac{R^3}{EI}\left(\frac{3\pi}{2}V - 2H\right) \tag{9.34}$$

$$\delta H = \frac{1}{EI}\int M\frac{\partial M}{\partial H}R\,d\theta$$

$$= \frac{1}{EI}\int [VR(1 - \cos\theta) - HR\sin\theta](-R\sin\theta)R\,d\theta$$

$$= \frac{R^3}{EI}\left(\int(-\sin\theta + \sin\theta\cos\theta)V + H\sin^2\theta\right)d\theta$$

$$= \frac{R^3}{EI}\left\{\left[\cos\theta + \frac{\sin^2\theta}{2}\right]V + \int H\frac{(1 - \cos 2\theta)}{2}\,d\theta\right\}$$

$$= \frac{R^3}{EI}\left\{\left[\cos\theta + \frac{\sin^2\theta}{2}\right]V + \frac{H}{2}\left[\theta - \frac{\sin 2\theta}{2}\right]\right\}_0^{\pi}$$

$$= \frac{R^3}{EI}[(-1 + 0) - (1 + 0)]V + \frac{H}{2}[(\pi - 0) - (0 - 0)]$$

$$\delta H = \frac{R^3}{EI}\left(-2V + \frac{\pi H}{2}\right) \tag{9.35}$$

EXAMPLE 9.6

Determine expressions for the deflections at the free end of the thin curved beam shown in Figure 9.10.

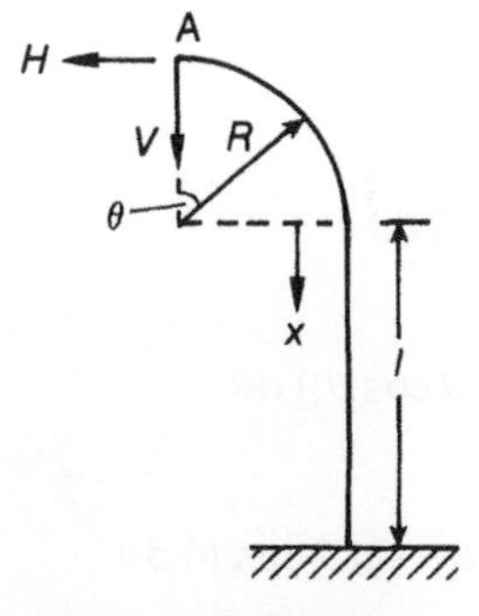

Figure 9.10 Thin beam (part curved and part straight)

SOLUTION

For convenience, this beam will be considered in two sections, i.e. a curved section and a straight section.

Let

δV = deflection at the free end in the direction of V

$= \delta V_c + \delta V_s$

δH = deflection at free end in the direction of H

$= \delta H_c + \delta H_s$

where

δV_c = component of deflection at the free end in the direction of V due to the curved section

δH_c = component of deflection at the free end in the direction of H due to the curved section

δV_s = component of deflection at the free end in the direction of V due to the straight section

δH_s = component of deflection at the free end in the direction of H due to the straight section

As the problem is a linear elastic one, it is perfectly acceptable to add together the appropriate components of δV and δH.

Curved section

At any angle θ, the bending moment is

$$M = VR\sin\theta + HR(1 - \cos\theta)$$

and

$$\delta V_c = \frac{1}{EI}\int_0^{\pi/2} M \frac{\partial M}{\partial V} R\,d\theta$$

$$= \frac{1}{EI}\int_0^{\pi/2} [VR\sin\theta + HR(1 - \cos\theta)]R\sin\theta\, R\,d\theta$$

$$= \frac{R^3}{EI}\int_0^{\pi/2} [V\sin^2\theta + H(\sin\theta - \sin\theta\cos\theta)]\,d\theta$$

$$= \frac{R^3}{EI}\int \left(V\frac{(1 - \cos 2\theta)}{2} + H(\sin\theta - \sin\theta\cos\theta)\right)d\theta$$

$$= \frac{R^3}{EI}\left[\frac{V}{2}\left(\theta - \frac{\sin 2\theta}{2}\right) + H\left(-\cos\theta - \frac{\sin^2\theta}{2}\right)\right]_0^{\pi/2}$$

$$\underline{\delta V_c = \frac{R^3}{EI}\left[\left(\frac{\pi V}{4}\right) + \frac{H}{2}\right]} \tag{9.36}$$

Similarly

$$\delta H_c = \frac{1}{EI}\int_0^{\pi/2} M \frac{\partial M}{\partial H} R\,d\theta$$

$$= \frac{1}{EI}\int [VR\sin\theta + HR(1-\cos\theta)]R(1-\cos\theta)R\,d\theta$$

$$= \frac{R^3}{EI}\int [V\sin\theta(1-\cos\theta) + H(1-\cos\theta)^2]\,d\theta$$

$$= \frac{R^3}{EI}\int [V(\sin\theta - \sin\theta\cos\theta) + H(1 - 2\cos\theta + \cos^2\theta)]\,d\theta$$

$$= \frac{R^3}{EI}\left[V\left(-\cos\theta - \frac{\sin^2\theta}{2}\right) + \int H\left(1 - 2\cos\theta + \frac{(1-\cos 2\theta)}{2}\right)d\theta\right]_0^{\pi/2}$$

$$= \frac{R^3}{EI}\left[V\left(-\cos\theta - \frac{\sin^2\theta}{2}\right) + H\left(\theta - 2\sin\theta + \frac{\theta}{2} - \frac{\sin 2\theta}{4}\right)\right]_0^{\pi/2}$$

$$\delta H_c = \frac{R^3}{EI}\left[\frac{V}{2} + H\left(\frac{3\pi}{4} - 2\right)\right] \tag{9.37}$$

Straight section

At any distance x, the bending moment is

$$M = VR + H(R + x)$$

Now,

$$\delta V_s = \frac{1}{EI}\int_0^l M\frac{\partial M}{\partial V}\,dx$$

$$= \frac{1}{EI}\int_0^l [VR + H(R+x)]R\,dx = \frac{R}{EI}\left[VRx + HRx + \frac{Hx^2}{2}\right]_0^l$$

$$\delta V_s = \frac{R}{EI}\left[VRl + H\left(Rl + \frac{l^2}{2}\right)\right] \tag{9.38}$$

Similarly,

$$\delta H_s = \frac{1}{EI}\int_0^l [VR + H(R+x)](R+x)\,dx$$

$$= \frac{1}{EI}\left[VR^2x + \frac{VRx^2}{2} + H\left(R^2x + Rx^2 + \frac{x^3}{3}\right)\right]_0^l$$

Therefore

$$\delta H_s = \frac{1}{EI}\left[VRl\left(R + \frac{l}{2}\right) + Hl\left(R^2 + Rl + \frac{l^2}{3}\right)\right] \tag{9.39}$$

From equations (9.36)–(9.39),

$$\delta V = \frac{1}{EI}\left[R^3\left(\frac{\pi V}{4} + \frac{H}{2}\right) + R[VRl + Hl(R + l/2)]\right] \tag{9.40}$$

$$\delta H = \frac{1}{EI}\left\{R^3\left[\frac{V}{2} + H\left(\frac{3\pi}{4} - 2\right)\right] + VRl\left(R + \frac{l}{2}\right) + Hl\left(R^2 + Rl + \frac{l^2}{3}\right)\right\} \tag{9.41}$$

EXAMPLE 9.7

Determine the deflection at the free end of the thin curved beam shown in Figure 9.11, which has an out-of-plane concentrated load applied to its free end.

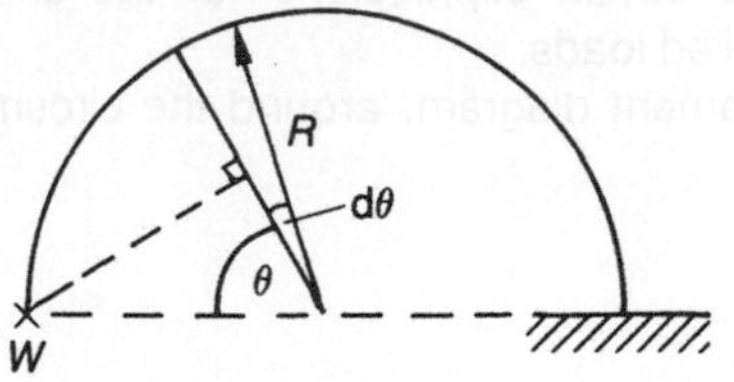

Figure 9.11 Thin curved beam under an out-of-plane load

SOLUTION

In this case, the beam is subjected to both bending and torsion; hence, it will be necessary to consider bending strain energy in addition to torsional strain energy.

At any angle θ, the element $R\,d\theta$ of the beam is subjected to a bending moment M and a torque T, which are evaluated as follows:

$$M = WR\sin\theta$$

$$T = WR(1 - \cos\theta)$$

From equations (9.19) and (9.22), the total strain energy, U, is given by

$$U = \int \frac{M^2}{2EI}\,dx + \int \frac{T^2}{2GJ}\,dx$$

Let δW be the out-of-plane deflection under the load W, acting at the free end. Now,

$$\delta W = \frac{1}{EI}\int M\frac{\partial M}{\partial W}R\,d\theta + \frac{1}{GJ}\int T\frac{\partial T}{\partial W}R\,d\theta$$

$$= \frac{R}{EI}\int WR\sin\theta R\sin d\theta + \frac{R}{GJ}\int [WR(1 - \cos\theta)]R(1 - \cos\theta)\,d\theta$$

$$= \frac{WR^3}{EI}\int_0^{\pi}\frac{(1 - \cos 2\theta)}{2}\,d\theta + \frac{WR^3}{GJ}\int_0^{\pi}(1 - 2\cos\theta + \cos^2\theta)\,d\theta$$

$$= WR^3 \int \left\{ \frac{1}{2}\left[\theta - \frac{\sin 2\theta}{2}\right]\frac{1}{EI} + \frac{1}{GJ}\left[\theta - 2\sin\theta + \frac{1}{2}\left(\theta + \frac{\sin 2\theta}{2}\right)\right]\right\}_0^{\pi}$$

$$= WR^3 \left\{ \frac{1}{2EI}[\pi] + \frac{1}{GJ}\left[\left(\pi - 0 + \frac{\pi}{2} + 0\right) - [0]\right]\right\}$$

$$\delta W = WR^3\left(\frac{\pi}{2EI} + \frac{3\pi}{2GJ}\right) \tag{9.42}$$

EXAMPLE 9.8

Determine expressions for the values of the bending moments at the points A and B of the thin ring shown in Figure 9.12, and also obtain expressions for the changes in diameter of the ring in the directions of the applied loads.

Hence, or otherwise, sketch the bending moment diagram, around the circumference of the ring, when $H=0$.

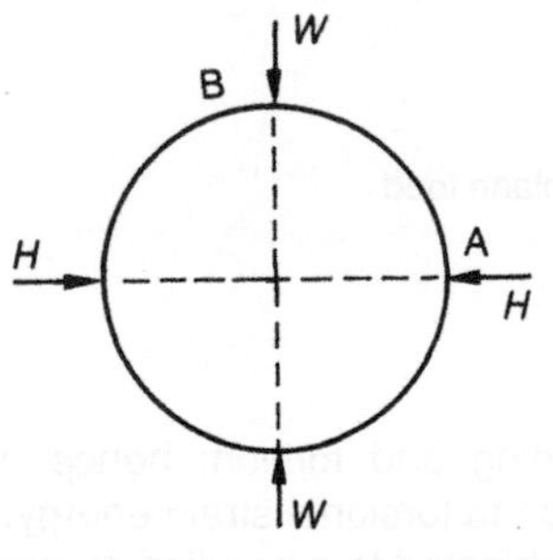

Figure 9.12 Thin ring, under diametral loads

SOLUTION

Because of symmetry, it is only necessary to consider a quadrant of the ring, as shown in Figure 9.13.

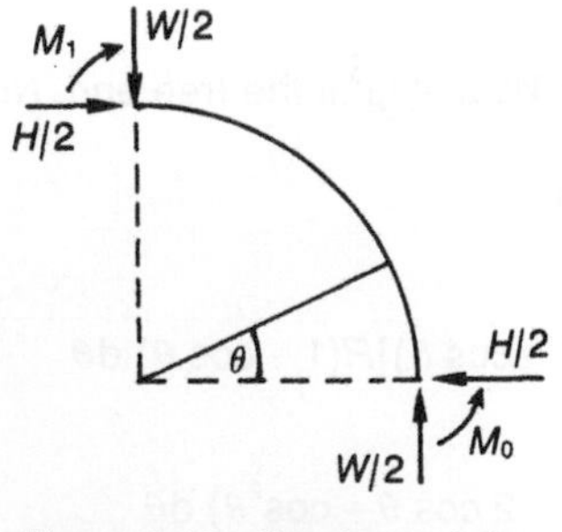

Figure 9.13 Quadrant of ring

Let

M_0 = unknown bending moment at A, which can be assumed to be the redundancy

M_1 = unknown bending moment at B, which can be obtained from M_0, together with considerations of elementary statics

As this problem is statically indeterminate, it will be necessary to use Castigliano's second theorem to determine the redundancy M_0.

At angle θ, the bending moment acting on an element $R\,d\theta$ is

$$M = M_0 + \frac{WR}{2}(1-\cos\theta) - \frac{HR}{2}\sin\theta \tag{9.43}$$

***To find* M_0**

From Castigliano's second theorem, $\partial U_b/\partial M_0$, the initial lack of fit, is zero (in this case, as the ring is assumed to be geometrically perfect) or

$$\frac{\partial U_b}{\partial M_0} = \frac{\partial U_b}{\partial M}\frac{\partial M}{\partial M_0}$$

i.e.

$$0 = \frac{1}{EI}\int M\frac{M}{M_0}R\,d\theta$$

$$0 = \frac{4}{EI}\int_0^{\pi/2}\left(M_0 + \frac{WR}{2}(1-\cos\theta) - \frac{HR}{2}\sin\theta\right)R\,d\theta$$

$$0 = \frac{4}{EI}\int_0^{\pi/2}\left(M_0 + \frac{WR}{2} - \frac{WR\cos\theta}{2} - \frac{HR\sin\theta}{2}\right)R\,d\theta$$

$$0 = \left[M_0\theta + \frac{WR}{2}\theta - \frac{WR}{2}\sin\theta + \frac{HR}{2}\cos\theta\right]_0^{\pi/2}$$

$$0 = M_0\frac{\pi}{2} + WR\frac{\pi}{4} - \frac{WR}{2} + 0 + 0 - \frac{HR}{2}$$

$$\underline{M_0 = (W+H)\frac{R}{\pi} - \frac{WR}{2}} \tag{9.44}$$

An expression for M_1 can be determined from equation (9.43) by setting $\theta = 90°$ and by the use of equation (9.44), i.e.

$$M_1 = M_0 + \frac{WR}{2}\left(1-\cos\frac{\pi}{2}\right) - \frac{HR}{2}\sin\frac{\pi}{2}$$

or

$$\underline{M_1 = \frac{WR}{\pi} + HR\left(\frac{1}{\pi} - \frac{1}{2}\right)} \tag{9.45}$$

To find the *change in the diameter* of the ring in the direction of W,

$$\delta V = \frac{1}{EI}\int M\frac{\partial M}{\partial W}\,dx$$

$$= \frac{4}{EI}\int_0^{\pi/2}\left(M_0 + \frac{WR}{2}(1-\cos\theta) - \frac{HR}{2}\sin\theta\right)\frac{R}{2}(1-\cos\theta)\,R\,d\theta$$

$$= \frac{4R^3}{EI}\int_0^{\pi/2}\left(\frac{W}{\pi} + \frac{H}{\pi} - \frac{W}{2} + \frac{W}{2} - \frac{W}{2}\cos\theta - \frac{H}{2}\sin\theta\right)\tfrac{1}{2}(1-\cos\theta)\,d\theta$$

$$= \frac{2R^3}{EI}\left[\frac{(W+H)}{\pi}(\theta - \sin\theta) - \frac{W}{2}\sin\theta + \frac{H}{2}\cos\theta\right.$$

$$\left. + \int\left(\frac{W}{2}\cos^2\theta + \frac{H}{2}\sin\theta\cos\theta\right)d\theta\right]_0^{\pi/2}$$

$$= \frac{2R}{EI}\left[\frac{(W+H)}{\pi}\left(\frac{\pi}{2} - 1\right) - \frac{W}{2} - \frac{H}{2}\right.$$

$$\left. + \int_0^{\pi/2}\left(\frac{W}{2}\frac{(1+\cos 2\theta)}{2} + \frac{H}{2}\sin\theta\cos\theta\right)d\theta\right]$$

$$= \frac{2R^3}{EI}\left\{\frac{(W+H)}{\pi}\left(\frac{\pi}{2} - 1\right) - \frac{W}{2} - \frac{H}{2} + \frac{W}{4}\left[\theta + \frac{\sin 2\theta}{2}\right]_0^{\pi/2} + \frac{H}{4}\left[\sin^2\theta\right]_0^{\pi/2}\right\}$$

$$= \frac{2R^3}{EI}\left(-\frac{(W+H)}{\pi} + \frac{W\pi}{8} + 0 + \frac{H}{4}\right)$$

$$\delta V = \frac{2R^3}{EI}\left[W\left(\frac{\pi}{8} - \frac{1}{\pi}\right) + H\left(\frac{1}{4} - \frac{1}{\pi}\right)\right] \tag{9.46}$$

By a similar process, the change in diameter of the ring in the direction of H can be found:

$$\delta H = \frac{R^3}{EI}\left[H\left(\frac{\pi}{8} - \frac{1}{\pi}\right) + W\left(\frac{1}{4} - \frac{1}{\pi}\right)\right] \tag{9.47}$$

NB Changes in the diameters and $2\times\delta V$ and $2\times\delta H$, respectively.

To determine the bending moment diagram, when H = 0.

From equation (9.43),

$$M = \left(\frac{WR}{\pi} - \frac{WR}{2}\right) + \frac{WR}{2}(1-\cos\theta)$$

$$M = \frac{WR}{\pi} - \frac{WR}{2}\cos\theta$$

@ $\theta = \dfrac{\pi}{2}$

$$M_1 = \frac{WR}{\pi}$$

@ $\theta = 0$

$$M_0 = \left(\frac{WR}{\pi} - \frac{WR}{2}\right)$$

The bending moment diagram is shown in Figure 9.14.

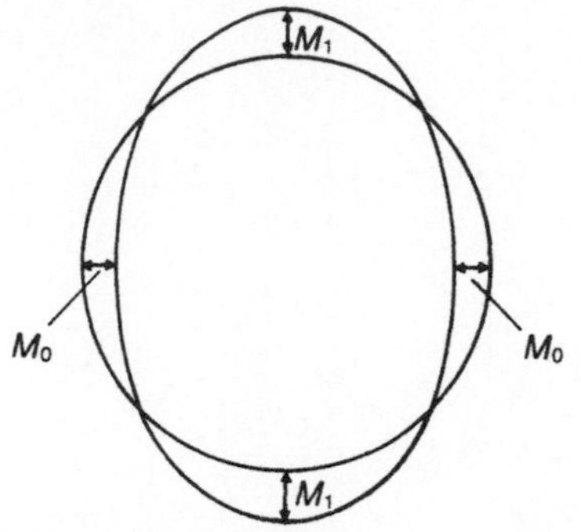

Figure 9.14 Bending moment distribution, when $H = 0$

9.11 Suddenly applied and impact loads

When structures are subjected to suddenly applied or impact loads, the resulting stresses can be considerably greater than those which will occur if these same loads are gradually applied.

Using a theoretical solution, it will be proven later that if a stationary load were suddenly applied to a structure, at zero velocity, then the stresses set up in the structure would have twice the magnitude that they would have had if the load were gradually applied.

The importance of stresses due to impact will be demonstrated by a number of worked examples, but prior to this, it will be necessary to make a few definitions.

9.12 Resilience

This is a term that is often used when considering elastic structures under impact. The resilience of an elastic body is a measurement of the amount of elastic strain energy that the body will store under a given load.

EXAMPLE 9.9

A vertical rod, of cross-sectional area A, is secured firmly at its top end and has an inextensible collar of negligible mass attached firmly to its bottom end, as shown in Figure 9.15. If a mass of magnitude M is dropped onto the collar from a distance h above it, determine the maximum values of stress and deflection that will occur owing to this impact. Neglect energy and other losses, and assume that the mass is in the form of an annular ring.

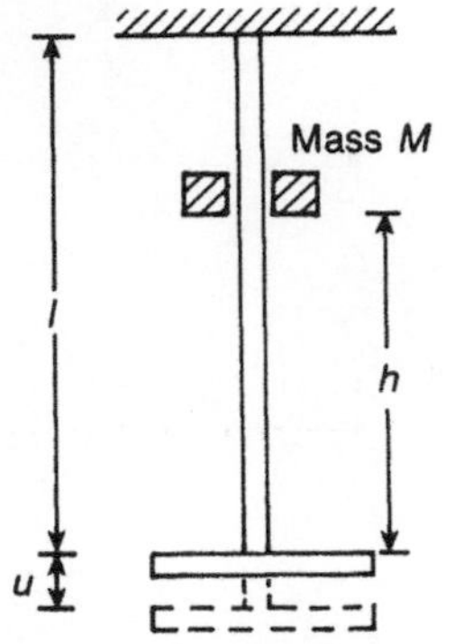

Figure 9.15 Rod under impact

SOLUTION

Energy and friction losses are neglected, so that the rod is assumed to absorb all the energy on impact. From the point of view of the structural analyst, this assumption is reasonable, as it simplifies the solution and the errors are on the so-called 'safe side'.

Let u be the maximum deflection of the rod on impact. Then

$$\text{PE of mass} = Mg(h + u)$$

$$= \text{strain energy stored in the rod}$$

i.e.

$$Mg(h + u) = \frac{\sigma^2}{2E} \times \text{volume of rod} \tag{9.48}$$

where

σ = the maximum stress in the rod due to impact
PE = potential energy

but

$$u = \varepsilon l = \sigma l / E$$

which on substitution into equation (9.48) gives

$$Mg(h + \sigma l/E) = \frac{\sigma^2}{2E} \times Al$$

or

$$\sigma^2 - \frac{2Mg\sigma}{A} - \frac{2MghE}{Al} = 0$$

Therefore

$$\sigma = \frac{Mg}{A} + \sqrt{\frac{M^2g^2}{A^2} + \frac{2MghE}{Al}} \tag{9.49}$$

and

$$u = \sigma l / E \tag{9.50}$$

9.12.1 *Suddenly applied load*

If the mass M in Figure 9.15 where *just above* the top surface of the collar and *suddenly released*, so that $h = 0$ in equation (9.49) then

$$\sigma = \frac{2Mg}{A}$$

That is, *a suddenly applied load*, at zero velocity, *induces twice the stress* of a gradually applied load of the same magnitude.

If, in Example 9.9, the following applied, then the maximum stress on impact and, also, the maximum stress if the same load were gradually applied to the top of the collar can be determined:

$$M = 1 \text{ kg} \quad A = 400 \text{ mm}^2 \quad l = 1 \text{ m} \quad h = 0.5 \text{ m}$$

$$E = 2 \times 10^{11} \text{ N/m}^2 \qquad g = 9.81 \text{ m/s}^2$$

From equation (9.49) the maximum stress on impact is

$$\sigma = \frac{1 \times 9.81}{400\text{E-6}} + \sqrt{\frac{1^2 \times 9.81^2}{(400\text{E-6})^2} + \frac{2 \times 1 \times 9.81 \times 0.5 \times 2\text{E11}}{400\text{E-6} \times 1}}$$

$$\sigma = 0.025 + 70.04 = 70.07 \text{ MN/m}^2$$

If the load were gradually applied to the top of the collar,

Maximum static stress = 0.025 mN/m^2

i.e. *the impact stress is about 2800 times greater than the static stress.*

EXAMPLE 9.10

A simply supported beam of uniform section is subjected to an impact load at mid-span, as shown in Figure 9.16. Determine expressions for the maximum bending moment and deflection of the beam.

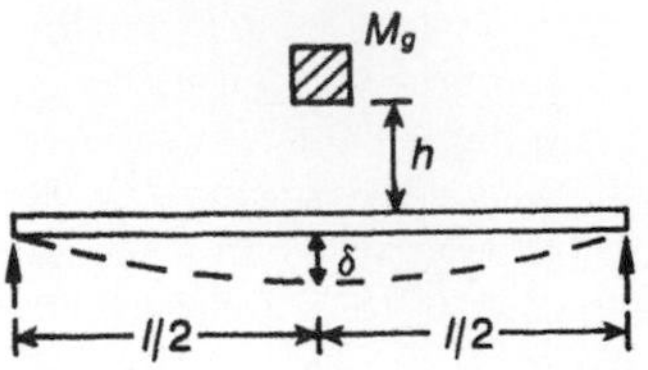

Figure 9.16 Beam under impact

SOLUTION

Let

δ = the maximum central deflection of the beam due to impact

W_e = equivalent static load to cause the deflection δ

$$\text{PE of load} = Mg(h + \delta) \tag{9.51}$$

From equation (9.30),

$$U_b = \frac{W_e^2 l^3}{96EI} \tag{9.52}$$

Equating (9.51) and (9.52),

$$Mg(h + \delta) = \frac{W_e^2 l^3}{96EI} \tag{9.53}$$

but from Example 9.3,

$$\delta = \frac{W_e l^3}{48EI}$$

which on substitution into equation (9.53) results in

$$-Mgh - \frac{Mgl^3 W_e}{48EI} + \frac{l^3}{96EI} W_e^2 = 0$$

or

$$W_e^2 - 2MgW_e - \frac{96EIMgh}{l^3} = 0$$

Therefore

$$\underline{W_e = Mg + \sqrt{M^2g^2 + 96EIMgh/l^3}} \tag{9.54}$$

If Mg were suddenly applied, at zero velocity, so that $h = 0$ in equation (9.54), then

$$\underline{W_e = 2Mg}$$

i.e. once again, it can be seen that a suddenly applied load, at zero velocity, has twice the value of a gradually applied load.

NB $\quad \delta = \dfrac{W_e l^3}{48EI}$

EXAMPLE 9.11

A uniform-section beam, which is initially horizontal, is simply supported at one end, and is supported at the other end by an elastic wire, as shown in Figure 9.17. Assuming that a mass of magnitude 2 kg, which is situated at a height of 0.2 m above the beam, is dropped onto the mid-span of the beam, determine the central deflection of the beam and the maximum stress induced in the wire. The beam and wire may be assumed to have negligible masses, and the following may also be assumed to apply:

E = 2E11 N/m^2 (for both beam and wire)
I = second moment of area of the beam section about a horizontal plane = 2E-8 m^4
A = cross-sectional area of wire = 1.3E-6 m^2
L = length of beam = 2 m
l = length of wire = 0.9 m
h = 0.4 m

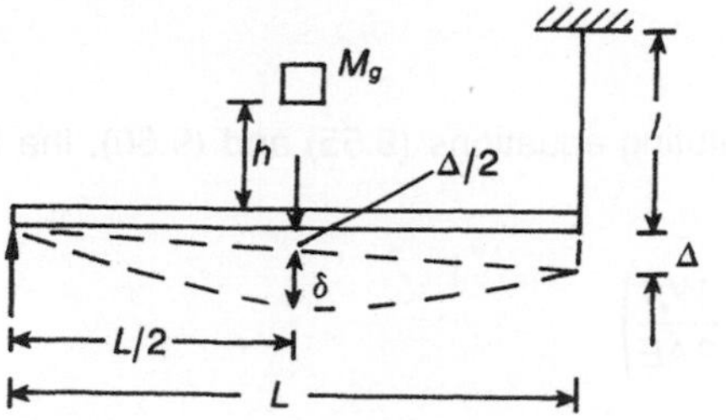

Figure 9.17 Impact on a beam

SOLUTION

Let

W_e = equivalent static load to cause the deflections δ and Δ
Δ = deflection of wire due to impact
δ = maximum central deflection of beam due to flexure alone, on impact

i.e. the maximum central deflection of the beam due to impact is $\delta + \Delta/2$.

From equilibrium considerations, the maximum force in the wire is $W_e/2$ or

$$\Delta = \frac{W_e l}{2AE} \tag{9.55}$$

Now, from equation (9.30),

$$U_b = \frac{W_e^2 L^3}{96EI} \tag{9.56}$$

From equation (9.15), the strain energy in the wire is

$$U_e = \frac{\sigma^2}{2E} \times Al$$

but

$$\sigma = E\varepsilon = \frac{E\Delta}{l} = \frac{W_e}{2A}$$

or

$$U_e = \frac{W_e^2 \times l}{8AE} \tag{9.57}$$

so that the total strain energy is

$$U = \frac{W_e L^3}{96EI} + \frac{W_e^2 l}{8AE} \tag{9.58}$$

$$\text{PE of mass} = Mg(h + \delta + \Delta/2) \tag{9.59}$$

but

$$\delta = \frac{W_e L^3}{48EI} \tag{9.60}$$

Equating (9.58) and (9.59), and by substituting equations (9.55) and (9.60), the following relationship is obtained:

$$\frac{W_e^2 L^3}{96EI} + \frac{W_e^2 l}{8AE} = Mg\left(h + \frac{W_e L^3}{48EI} + \frac{W_e l}{2AE}\right) \tag{9.61}$$

Substituting the appropriate values into equation (9.61),

$$\frac{W_e^2 \times 8}{96 \times 2\text{E}11 \times 2\text{E-}8} + \frac{W_e^2 \times 0.9}{8 \times 1.3\text{E-}6 \times 2\text{E}11}$$

$$= 2 \times 9.81\left(0.2 + \frac{W_e \times 8}{48 \times 2\text{E}11 \times 2\text{E-}8} + \frac{W_e \times 0.9}{2 \times 1.3\text{E-}6 \times 2\text{E}11}\right)$$

or

$$W_e^2(2.083\text{E-}5 + 4.327\text{E-}7) = 3.924 + W_e(8.175\text{E-}4 + 3.396\text{E-}5)$$

Therefore

$$2.12\text{E-}5\, W_e^2 - 8.515\text{E-}4\, W_e - 3.924 = 0$$

$$W_e = \frac{8.515\text{E-}4 + \sqrt{(8.515\text{E-}4)^2 + 4 \times 2.127\text{E-}5 \times 3.924}}{2 \times 2.127\text{E-}5}$$

$$= 20 + 430$$

$$\underline{W_e = 450\ \text{N}}$$

and

<u>Maximum stress in wire due to impact = 346.1 MN/m^2</u>

EXAMPLE 9.12

A concrete pillar of length 4 m has a cross-sectional area of 0.25 m², where 10% is composed of steel reinforcement and the remainder of concrete. If a mass of 10 kg is dropped onto the top of the concrete pillar, from a height of 0.12 m above it, determine the maximum stresses in the steel and the concrete due to impact.

E_s = elastic modulus of steel $= 2 \times 10^{11}$ N/m²

E_c = elastic modulus of concrete $= 1.5 \times 10^{10}$ N/m²

$g = 9.81$ m/s²

SOLUTION

Let

δ = the maximum deflection of the column under impact, as shown in Figure 9.18

$$\text{Work done, WD} = Mg(h+\delta) \quad (9.62)$$

$$\text{Strain energy} = \frac{\sigma_c^2}{2E_c} \times 0.225 \times 4 + \sigma_s^2 \times \frac{0.025 \times 4}{2E_s} \quad (9.63)$$

where

σ_c = maximum stress in the concrete due to impact
$= E_c\varepsilon_c$
σ_s = maximum stress in the steel due to impact
$= E_s\varepsilon_s$
ε_c and ε_s = maximum strain in concrete and steel, respectively

Equating (9.62) and (9.63),

$$Mg(h+\delta) = 3\text{E-}11\,\varepsilon_c^2 + 2.5\text{E-}13\,\varepsilon_s^2$$

$$98.1(0.12 + \varepsilon_c \times 4) = 3\text{E-}11\,\sigma_c^2 + 2.5\text{E-}13\,\sigma_s^2$$

$$98.1\left(0.12 + \frac{\sigma_c}{E_c} \times 4\right) = 3\text{E-}11\,\sigma_c^2 + 2.5\text{E-}13\,\sigma_s^2$$

$$11.772 + 2.616\text{E-}8\,\sigma_c = 3\text{E-}11\,\sigma_c^2 + 2.5\text{E-}13\,\sigma_s^2 \quad (9.64)$$

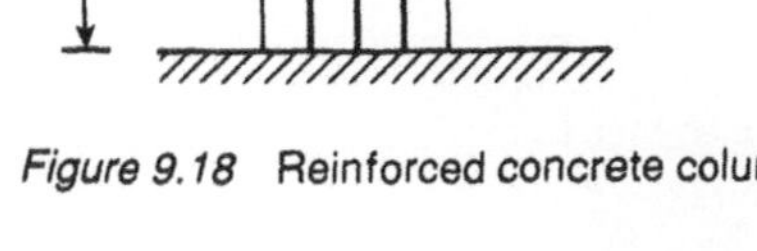

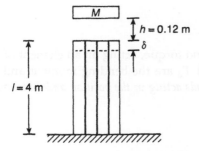

Figure 9.18 Reinforced concrete column

but

$$\varepsilon_c = \varepsilon_s$$

and

$$\frac{\sigma_c}{E_c} = \frac{\sigma_s}{E_s}$$

or

$$\sigma_s = \sigma_c E_s / E_c = 13.33\sigma_c \tag{9.65}$$

Substituting equation (9.65) into (9.64),

$$7.444\text{E-}11\sigma_c^2 - 2.616\text{E-}8\sigma_c - 11.772 = 0$$

Therefore

$$\sigma_c = \frac{2.616\text{E-}8 + \sqrt{6.844\text{E-}16 + 3.505\text{E-}9}}{1.4888\text{E-}10}$$

$$\underline{\sigma_c = 0.398\ \text{MN/m}^2} \tag{9.66}$$

and from equation (9.65),

$$\underline{\sigma_s = 5.303\ \text{MN/m}^2}$$

9.13 Unit load method

This is similar to equations of the form of (9.31), and it is applicable to linear elastic structures. Using this method, the displacement at any point in a structure can be obtained from equations such as (9.67) and (9.68):

$$\delta = \int \frac{M \times M_u}{EI}\, dx \tag{9.67}$$

and

$$\delta = \int \frac{T \times T_u}{GJ}\, dx \tag{9.68}$$

where M and T are the bending moment and torque, acting on an element of length dx, due to the external loads, and M_u and T_u are the bending moment and torque acting on the same element, due to *unit loads acting in the position and direction of the required displacements.*

EXAMPLE 9.13

Apply the unit load method to Example 9.4.

SOLUTION

From section 9.10

$$M = -WR\cos\theta - HR(1 - \sin\theta) \tag{9.69}$$

To determine δV, assume $W = 1$ and $H = 0$. Hence,

$$\underline{M_u = -R\cos\theta} \tag{9.70}$$

Substituting (9.69) and (9.70) into (9.67),

$$\delta V = \frac{1}{EI}\int_0^{\pi/2} [WR\cos\theta + HR(1 - \sin\theta)]R\cos\theta R\,d\theta$$

which is identical to δV in Example 9.4.

Similarly, *if* δH *is required*, assume $W = 0$ and $H = 1$, so that

$$\underline{M_u = -R(1 - \sin\theta)} \tag{9.71}$$

Substituting equations (9.69) and (9.71) into (9.67),

$$\delta H = \frac{1}{EI}\int_0^{\pi/2} [WR\cos\theta + HR(1 - \sin\theta)]R(1 - \sin\theta)R\,d\theta$$

which is identical to δH in Example 9.4.

9.14 Plastic collapse of beams

The design of structures on elastic theory is somewhat illogical, as it does not include the effects of residual stresses due to manufacture. For structures made from materials which have a definite yield point, a better estimate of the collapse load of the structure, whether or not it has residual stresses, can be obtained through the plastic theory. This theory is based on the fact that the structure will cease to carry

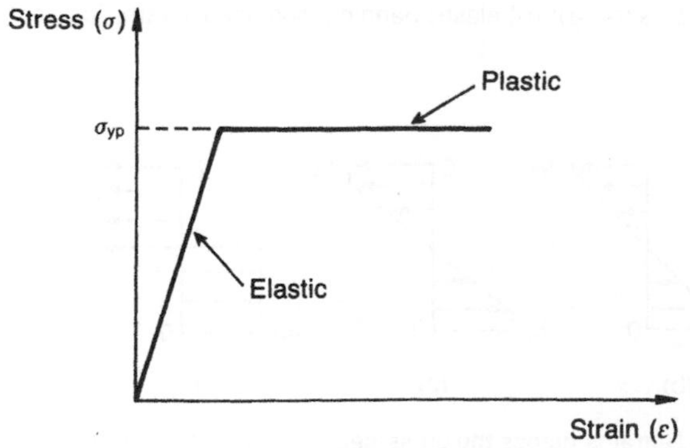

Figure 9.19 Ideally elastic–plastic material

load when it becomes a mechanism, the 'hinges' of the mechanism being plastic hinges. In the plastic hinge theory, the material is assumed to be ideally elastic–plastic, as shown in Figure 9.19.

To demonstrate the plastic theory, consider the encastré beam of Figure 9.20(a).

Within the elastic limit, the bending moment diagram will be as in Figure 9.20(b), and the stress diagram across a cross-section will be as in Figure 9.21(b). If the load W is increased, the beam will yield at B, assuming that $a > b$; the stress diagram will now take the form of Figure 9.21(c). That is, although yield has taken place, the beam will not fail, because the cross-section at B still has the ability to resist deformation. Further increase of the load W may cause other sections to become elastic–plastic, especially at A and C, and the plastic penetration at B will become even deeper.

Eventually at a certain load, the section at B becomes fully plastic, as shown in Figure 9.21(d). When this section at B becomes fully plastic, this part of the beam

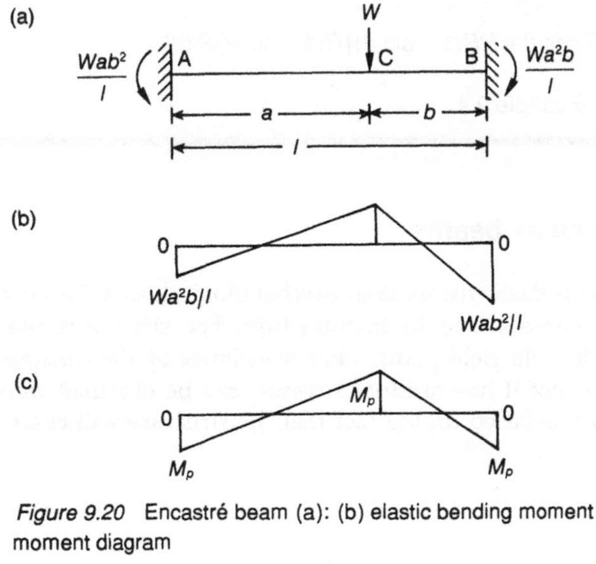

Figure 9.20 Encastré beam (a): (b) elastic bending moment diagram; (c) plastic bending moment diagram

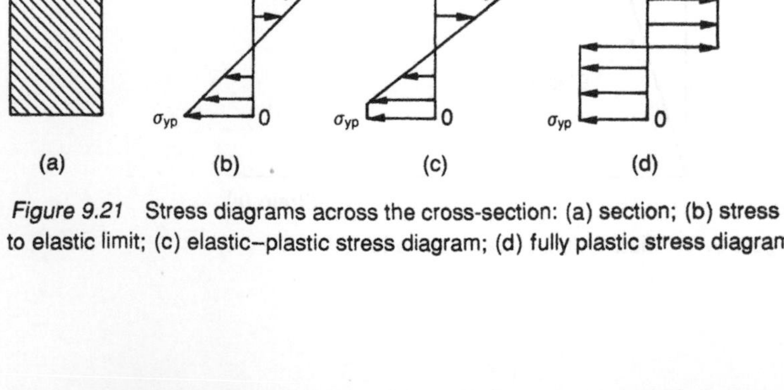

Figure 9.21 Stress diagrams across the cross-section: (a) section; (b) stress diagram up to elastic limit; (c) elastic–plastic stress diagram; (d) fully plastic stress diagram

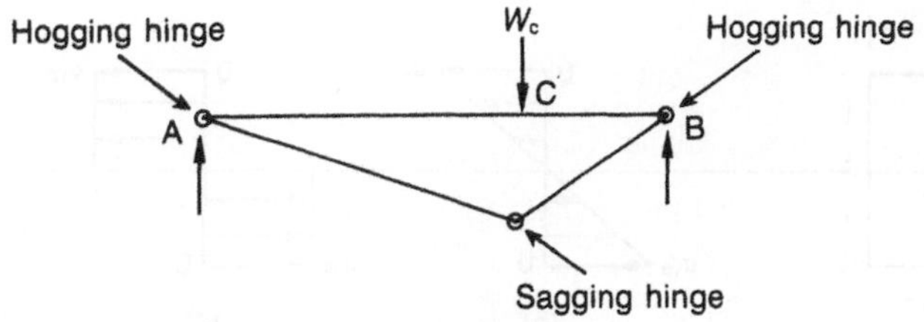

Figure 9.22 Plastic hinges in the beam

cannot withstand further load, as a plastic hinge has formed here; the beam simply rotates at B and transforms the bending resistance of the beam, due to an increasing load, to sections A and C. Further increase of load will eventually cause fully plastic hinges to occur at A and C. The plastic bending moment diagram is shown in Figure 9.20(c). When all three plastic hinges have occurred at A, B and C, the beam cannot resist further load, and the slightest increase in load will cause catastrophic deformation, the beam turning into a mechanism, the hinges of the mechanism being plastic hinges, as shown in Figure 9.22.

9.14.1 *Plastic neutral axis*

The plastic neutral axis lies at the centre of area of the section or the central axis, because the stress distribution is constant across the beam's section.

From Section 3.2.2, it can be seen that to determine the horizontal central axis of a beam, the area below the central axis must be equal to the area above the central axis.

9.14.2 *Load factor (λ)*

The safety factor for the plastic hinge theory is called the load factor, where

$$\text{load factor} = \lambda = \frac{\text{plastic collapse load}}{\text{working load}}$$

9.14.3 *Shape factor (S)*

The shape factor is a measure of the geometry of the beam's section, where

$$\text{shape factor} = S = \frac{M_P}{M_{yp}}$$

where

M_p = plastic moment of resistance of the section
M_{yp} = elastic moment of resistance of the section up to the yield point

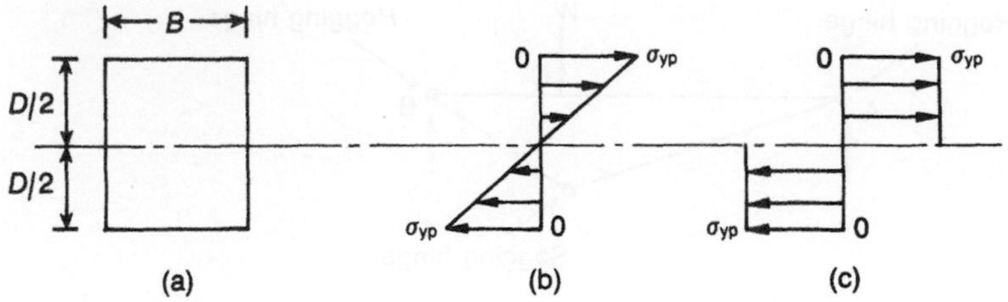

Figure 9.23 Stress diagrams: (a) cross-section; (b) stress diagram up to yield point; (c) fully plastic stress diagram

9.14.4 Shape factor for a rectangular section

Consider the rectangular section of Figure 9.23(a).

To determine M_{yp}, we can use the elastic theory of equation (4.1), i.e.

$$\frac{\sigma}{y} = \frac{M}{I}$$

$$\frac{\sigma_{yp}}{D/2} = \frac{M_{yp}}{BD^3} \times 12$$

Therefore

$$M_{yp} = \sigma_{yp} BD^2/6 \tag{9.72}$$

which is the *elastic moment of resistance of the section* up to the yield point.

From the stress diagram of Figure 9.23(c)

$$M_p = \text{plastic moment of resistance}$$

$$= \sigma_{yp} \times B \times \frac{D}{2} \times \frac{D}{4} \times 2 \tag{9.73}$$

$$M_p = \sigma_{yp} BD^2/4$$

Thus the shape factor is

$$S = \frac{M_p}{M_{yp}}$$

$$= \frac{\sigma_{yp} \times BD^2}{4} \; \frac{6}{\sigma_{yp} \times BD^2}$$

$$= 1.5$$

For a rolled steel joist $S = 1.14$ and for a circular cross-section $S = 1.7$.

EXAMPLE 9.14

Using the plastic hinge theory, design a suitable sectional modulus for the encastré beam of Figure 9.24, given that

$$\lambda = 3 \quad S = 1.14 \quad \text{and} \quad \sigma_{yp} = 300 \text{ MPa}$$

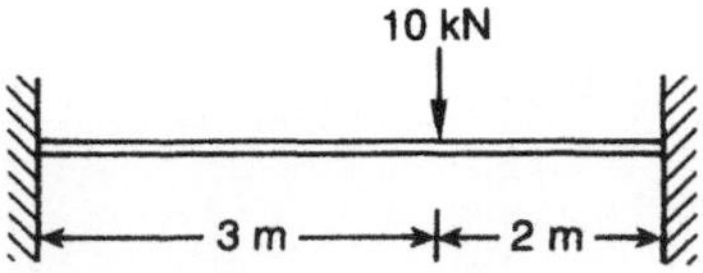

Figure 9.24 Encastré beam

SOLUTION

The beam will collapse plastically as shown in Figure 9.25.

When the beam collapses, it can be assumed that the beam element remains straight, as shown in Figure 9.25. The reason for the assumption is that, although the beam will be curved, as the bending moments at A, B and C cannot change, the curvature of the beam will remain essentially unchanged during failure.

From small-deflection theory,

$$3\theta = 2\alpha$$

Therefore

$$\alpha = 3\theta/2 \quad (9.74)$$

Also,

$$\beta = \alpha + \theta = 3\theta/2 + \theta$$

$$\underline{\beta = 5\theta/2} \quad (9.75)$$

From the principle of virtual work,

Virtual work done by the load = virtual work done by the hinges during rotation

i.e.

$$10 \times 3\theta = M_p\theta + M_p\beta + M_p\alpha$$

or

$$30\theta = M_p\theta + 5M_p\theta/2 + 3M_p\theta/2$$

i.e.

$$M_p = 30/5$$

$$M_p = 6 \text{ kN m}$$

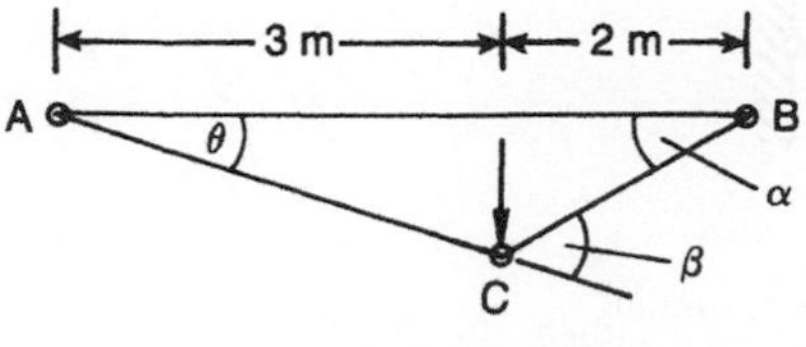

Figure 9.25 Beam mechanism

Design $M_p = M_p \times \lambda = 18$ kN m. Therefore

$$M_{yp} = \frac{M_p}{S} = \frac{18}{1.14} = 15.79 \text{ kN m}$$

and

$$Z = \frac{M_{yp}}{\sigma_{yp}} = \frac{15.79 \times 10^3}{300 \times 10^6}$$

Z = sectional modulus = 5.263×10^{-5} m^3

EXAMPLE 9.15

Using the plastic hinge theory, determine a suitable sectional modulus for the propped cantilever of Figure 9.26, given that

$$\lambda = 3 \quad S = 1.14 \quad \text{and} \quad \sigma_{yp} = 300 \text{ MPa}$$

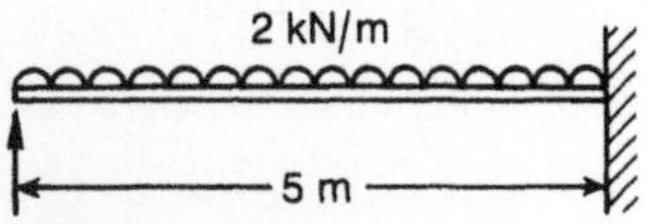

Figure 9.26 Propped cantilever

SOLUTION

In this case it will be necessary for two plastic hinges to occur, as the beam is statically indeterminate to the first degree.

The collapse mechanism is shown in Figure 9.27 where it can be seen that the sagging hinge is assumed to occur at a distance X to the right of mid-span.

Assuming θ is small,

$$(2.5 + X)\theta = (2.5 - X)\alpha$$

or

$$\alpha = \frac{(2.5 + X)}{(2.5 - X)}\theta \tag{9.76}$$

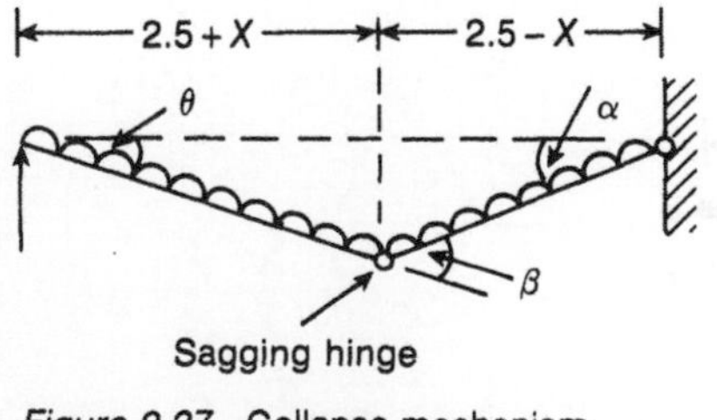

Figure 9.27 Collapse mechanism

Now,

$$\beta = \alpha + \theta = \frac{(2.5 + X)}{(2.5 - X)}\,\theta + \theta$$

$$= \frac{(2.5 + X + 2.5 - X)}{(2.5 - X)}\,\theta$$

$$\beta = \frac{5}{(2.5 - X)}\,\theta \tag{9.77}$$

Virtual work done by falling load =
virtual work done by plastic hinges resisting rotation

$$(2.5 + X) \times (2.5 + X)\theta + (2.5 - X) \times (2.5 - X)\alpha = M_p\beta + M_p\alpha$$

Therefore

$$(2.5 + X)^2\theta + (2.5 - X)(2.5 + X)\theta = M_p \times \frac{5}{(2.5 - X)}\,\theta + \frac{M_p(2.5 + X)}{(2.5 - X)}\,\theta$$

or

$$M_p\,\frac{(5 + 2.5 + X)}{(2.5 - X)} = (2.5 + X)^2 + (2.5 - X)(2.5 + X)$$

$$M_p = (2.5 + X)^2 \times \frac{(2.5 - X)}{(7.5 + X)} + \frac{(2.5 - X)(2.5 + X)(2.5 - X)}{(7.5 + X)}$$

$$= \frac{(6.25 + 5X + X^2)(2.5 - X) + (2.5 - X)(6.25 - X^2)}{(7.5 + X)}$$

$$= \frac{15.625 + 12.5X - 6.25X - 5X^2 + 15.625 - 6.25X}{(7.5 + X)}$$

$$M_p = \frac{31.25 - 5X^2}{(7.5 + X)} \tag{9.78}$$

For maximum M_p,

$$\frac{dM_p}{dX} = 0$$

$$M_p = \frac{u}{v}$$

$$\frac{dM_p}{dX} = \frac{v\,du/dX - u\,dv/dX}{v^2}$$

or

$$\frac{dM_p}{dX} = \frac{(7.5 + X)(-10X) - (31.25 - 5X^2)}{v^2}\,(1) = 0$$

i.e.

$$(7.5+X)(-10X)-31.2175X^2=0$$
$$-75X-10X^2-31.25+5X^2=0$$

or

$$-75X-5X^2-31.25=0$$
$$5X^2+75X+31.25=0$$

Thus

$$X=\frac{-75\pm\sqrt{5625-625}}{10}$$
$$=\frac{-75\pm70.711}{10}$$
$$\underline{X=-0.429\text{ m}} \tag{9.79}$$

Substituting equation (9.79) into (9.78),

$$M_p=\frac{31.25-5\times0.429^2}{7.5-0.429}=4.289\text{ kN m}$$

Design $M_p=\lambda\times M_p=7.892$ kN m. Therefore

$$M_{yp}=\frac{M_p}{S}=\frac{7.892}{1.14}$$
$$=6.923$$

and

$$Z=\frac{M_{yp}}{\sigma_{yp}}=\frac{6.923\times1000}{300\times10^6}$$
$$Z=2.308\times10^{-5}\text{ m}^3$$

9.15 Residual stresses in beams

In this section the theory will be based on the material behaving ideally elastic–plastic, as shown in Figure 9.19.

The following example will demonstrate how to calculate the residual stresses that remain when a load is removed, after this load has caused the beam to become partially plastic.

EXAMPLE 9.16

A steel beam of constant rectangular section, of depth 0.2 m and width 0.1 m, is subjected to four-point loading, as shown in Figure 9.28. Determine:

(a) the depth of plastic penetration at mid-span;

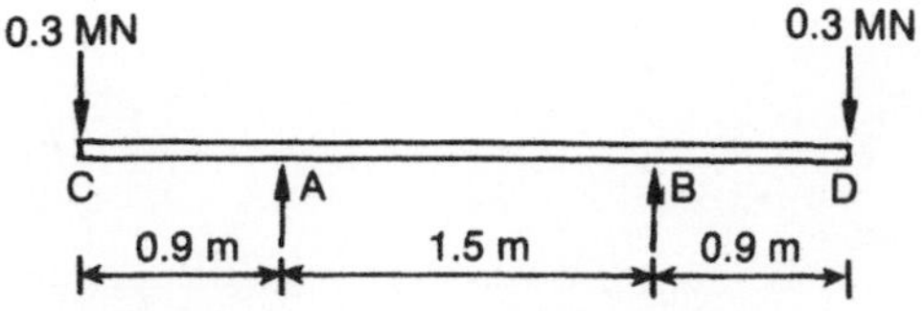

Figure 9.28 Beam under four-point loading

(b) the central deflection; and
(c) the length of the beam over which yield takes place.

If the above load is removed:

(d) determine the residual central deflection; and
(e) plot the residual stress distribution at mid-span.

The following may be assumed to apply:

$\sigma_{yp} = 350$ MN/m^2 $\quad E = 2E11$ N/m^2

SOLUTION

(a) To determine the depth of plastic penetration

At mid-span and throughout the length AB, the bending moment is

$$M = 0.3 \text{ MN} \times 0.9 \text{ m} = \underline{0.27 \text{ MN m}}$$

Owing to this bending moment, the stress distribution will be as in Figure 9.29.
Let

M_e = moment of resistance of the elastic portion of the beam's section
M_p' = moment of resistance of the plastic portion of the beam's section
I_e = second moment of area of the elastic portion of the beam's section
$= B \times (2H)^3/12 = 0.6667 BH^3$

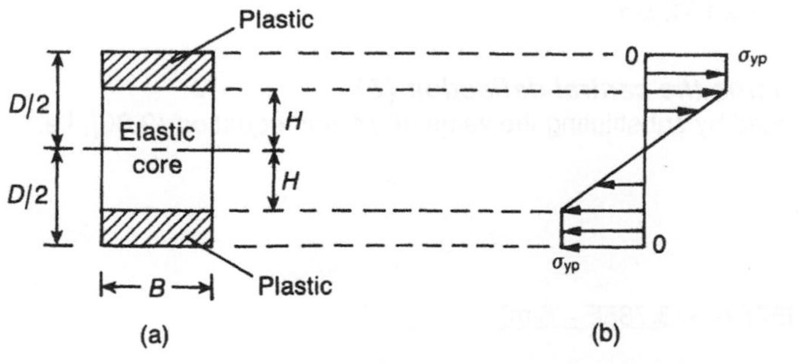

Figure 9.29 Elastic–plastic deformation of beam section: (a) section; (b) stress distribution

From elementary elastic theory [1],

$$\frac{\sigma}{y} = \frac{M}{I} = \frac{E}{R} \qquad (9.80)$$

where

σ = stress at any distance y from the neutral axis (NA)
M = bending moment
I = second moment of area about NA
R = radius of curvature of NA

Applying equation (9.80) to the elastic portion of the beam's section,

$$M_e = \frac{\sigma_{yp} \times I_e}{H}$$

$$= \frac{350 \times 0.6667 \times 0.1 \times H^3}{H}$$

$$\underline{M_e = 23.33H^2} \qquad (9.81)$$

From Figure 9.29,

$$M_p' = 2 \times \sigma_{yp} \times B \times (D/2 - H) \times (D/2 + H)/2$$

$$= 35 \times (0.1 - H) \times (0.1 + H)$$

$$M_p' = 0.35 - 35\ H^2 \qquad (9.82)$$

Now

$$M = M_e + M_p'$$

or

$$0.27 = 23.33H^2 + 0.35 - 35\ H^2$$

$$11.667H^2 = 0.08$$

$$\underline{H = 0.0828\ \text{m}} \qquad (9.83)$$

Therefore the *depth of plastic penetration* at the top and the bottom of the beam is

$$0.1 - 0.0828 = \underline{1.72\ \text{cm}}$$

(b) To determine the central deflection (δ)

This can be calculated by substituting the value of M_e into equation (9.80), i.e.

$$\frac{M_e}{I_e} = \frac{E}{R}$$

Now,

$$I_e = 0.6667BH^3 = \underline{3.785\text{E} - 5\ \text{m}^4}$$

and

$$M_e = 23.33H^2 = \underline{0.16\ \text{MN m}}$$

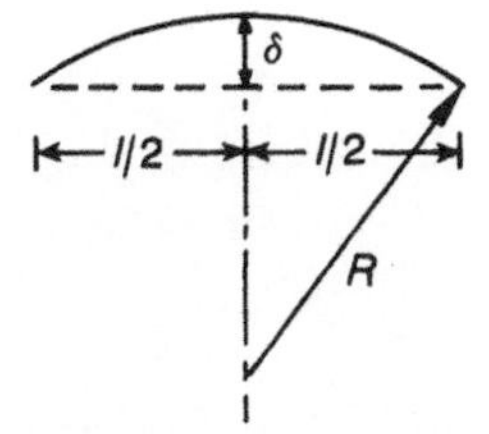

l = length between supports

Figure 9.30 Deflected form of beam

Therefore

$$R = \frac{EI_e}{M_e} = \frac{2E11 \times 3.785E\text{-}5}{0.16E6}$$

$$\underline{R = 47.31\ \text{m}}$$

The central deflection (δ) can be calculated from the properties of a circle, as shown in Figure 9.30.

Thus

$$\delta(2R - \delta) = (l/2)^2 \qquad (9.84)$$

$$\delta(94.62 - \delta) = 0.75^2$$

$$-\delta^2 + 94.62\delta - 0.5626 = 0$$

$$\delta = \frac{94.62 - \sqrt{94.62^2 - 4 \times 0.5626}}{2}$$

$$\underline{\delta = 5.945E\text{-}3\ \text{m} = 5.945\ \text{mm}}$$

(c) To determine the length of the beam over which yield takes place

Let

$$M_{yp} = \text{moment required just to cause first yield}$$

$$= \frac{\sigma_{yp} \times I}{(D/2)}$$

$$= \frac{\sigma_{yp} \times BD^2}{6} = \frac{350 \times 0.1 \times 0.2^2}{6}$$

$$\underline{M_{yp} = 0.233\ \text{MN m}}$$

From Figure 9.31,

$$Wl' = M_{yp}$$

or

$$0.3l' = 0.2333$$

Therefore

$$\underline{l' = 0.778\ \text{m}}$$

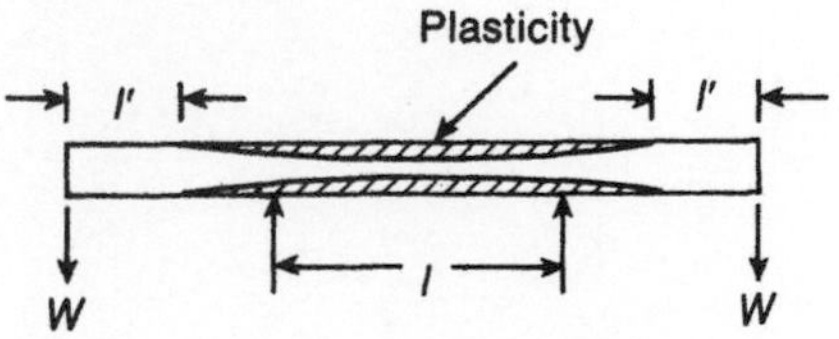

Figure 9.31 Plastic region on beam

Therefore the length of the beam over which plasticity occurs is

$$1.8 + 1.5 - 0.778 \times 2 = \underline{1.744\text{ m}}$$

(d) To determine the residual central deflection (δ_R)

On unloading the beam, it behaves elastically, as shown by Figure 9.32, i.e.

$$R = \frac{EI}{M} = \frac{2\text{E}11 \times 0.1 \times 0.2^3}{12 \times 0.27\text{E}6}$$

$$\underline{R = 49.383\text{ m}}$$

From equation (9.84)

$$\delta_1 (98.765 - \delta_1) = 0.75^2$$

or

$$-\delta_1^2 + 98.765\delta_1 - 0.5625 = 0$$

Therefore

$$\delta_1 = \frac{98.765 - \sqrt{98.765^2 - 4 \times 0.5625}}{2}$$

$$= 5.48\text{E-3 m}$$

Therefore the residual deflection is

$$\underline{\delta_R = 5.95 - 5.48 = 0.47\text{ mm}}$$

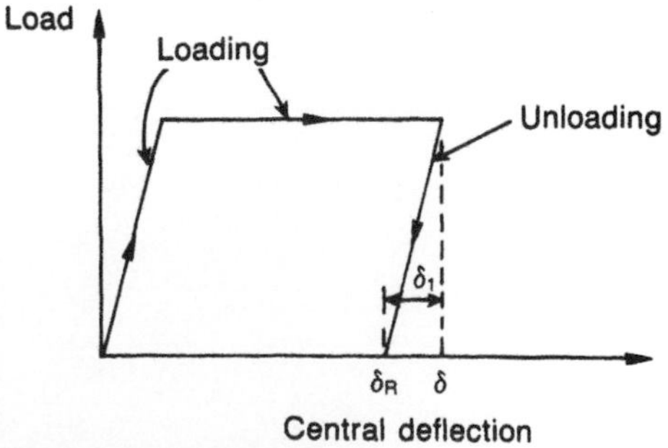

Figure 9.32 Load–deflection relationship

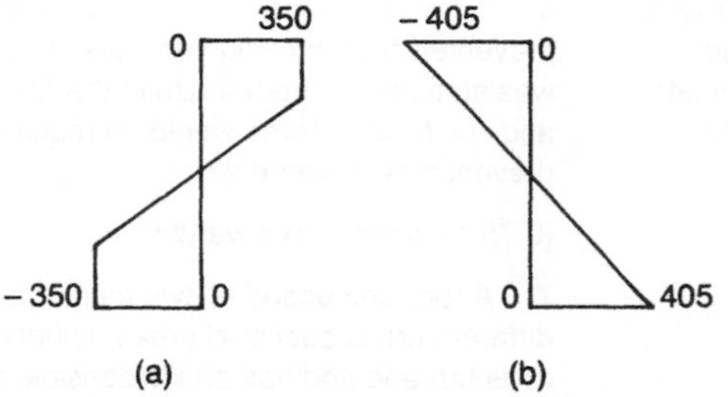

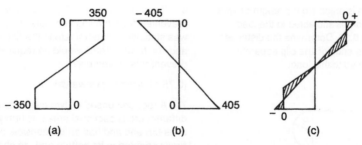

Figure 9.33 Stress distribution across section: (a) elastic–plastic stress distribution, on loading; (b) elastic stress distribution, on unloading; (c) residual stress distribution (shown shaded)

(e) To determine the residual stress distribution

This can be obtained by superimposing the elastic–plastic stress distribution, on loading, with the elastic stress distribution, on unloading, as shown in Figures 9.33(a)–(c).

EXAMPLES FOR PRACTICE 9

1. Determine expressions for the horizontal and vertical deflections of the free end of the thin curved beam shown in Figures 9.34(a)–(c).

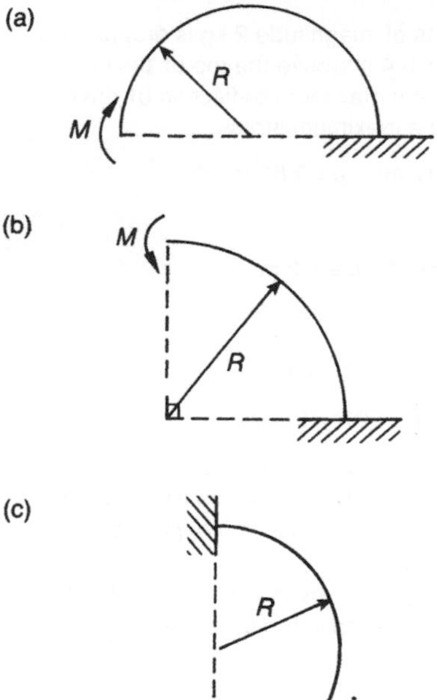

Figure 9.34

{(a) Horizontal deflection = $2MR^2/EI$ to the left, vertical deflection = $\pi MR^2/EI$ upwards; (b) horizontal deflection = $MR^2(\pi/2-1)/EI$ to the left, vertical deflection = MR^2/EI downwards; (c) horizontal deflection = $\delta H = R^3/EI((3\pi/2)H + 2V)$ to the left, vertical deflection = $\delta V = R^3(2H + \pi V/2)/EI$ upwards}

2. A thin curved beam consists of a length AB, which is of semi-circular form, and a length BC, which is a quadrant, as shown in Figure 9.35.

Determine the vertical deflection at A due to a downward load W, applied at this point. The end C is firmly fixed, and the beam's cross-section may be assumed to be uniform. (Portsmouth, March 1982)

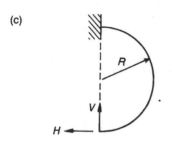

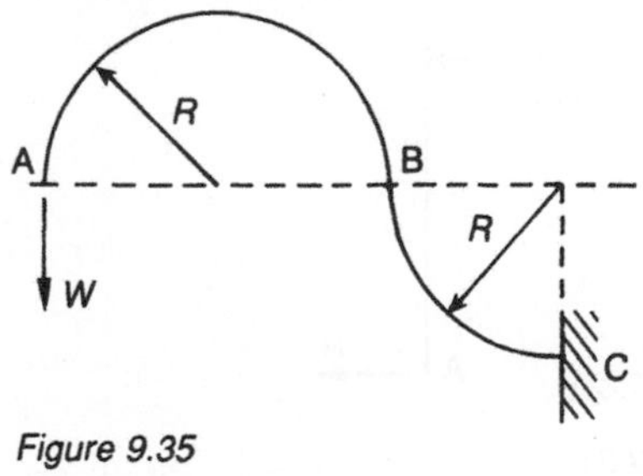

Figure 9.35

$$\left\{\delta V = \frac{WR^3}{EI}\left(\frac{25\pi}{4} - 6\right)\right\}$$

3. A clip, which is made from a length of wire of uniform section, is subjected to the load shown in Figure 9.36. Determine the distance by which the free ends of the clip separate, when it is subjected to this load.

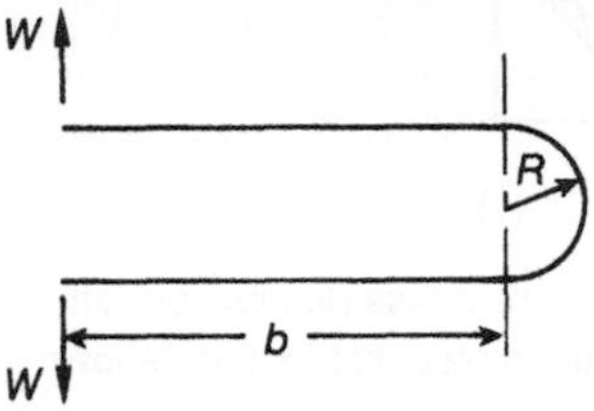

Figure 9.36

$$\left\{\frac{2W}{EI}\left[\frac{b^3}{3}+R\left(\frac{b^2\pi}{2}+\frac{R^2\pi}{4}+2bR\right)\right]\right\}$$

4. Determine the vertical and horizontal displacements at the free end of the rigid-jointed frame shown in Figure 9.37.

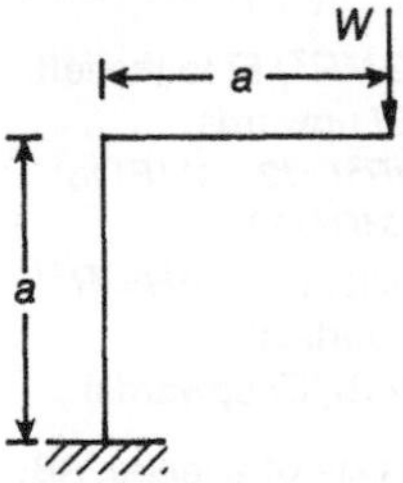

Figure 9.37

{$\delta V = 4Wa^3/3EI$ downwards, $\delta H = Wa^3/2EI$ to the right}

5. Determine the horizontal deflection at the free end A of the rigid-jointed frame of Figure 9.38.

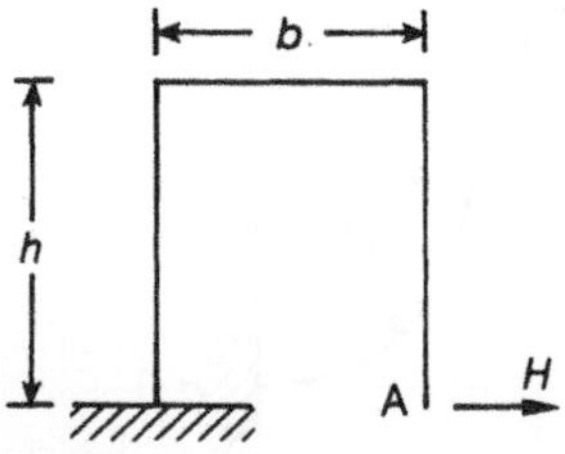

Figure 9.38

$$\left\{\delta H = \frac{Hh^2}{EI}\left(\frac{2h}{3}+b\right)\right\}$$

6. If the framework of Figure 9.38 were prevented from moving vertically at A, so that it was statically indeterminate to the first degree and $h = b$, what force would be required to prevent this movement?

{0.75 H, acting downwards}

7. A rod, composed of two elements of different cross-sectional areas, is firmly fixed at its top end and has an inextensible collar, firmly secured to its bottom end, as shown in Figure 9.39.

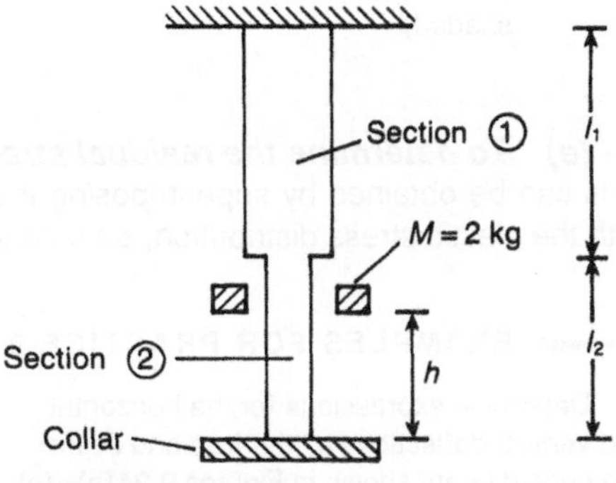

Figure 9.39

If a mass of magnitude 2 kg is dropped from a height of 0.4 m above the top of the collar, determine the maximum deflection of the rod, and also the maximum stress.

$E = 2E11\ N/m^2 \quad g = 9.81\ m/s^2$

Section ①

$A_1 = 500\ mm^2 \quad l_1 = 1.2\ m$

Section ②

$A_2 = 300\ mm^2 \quad l_2 = 0.8\ m$

{0.63 mm, 83 MN m²}

8. A reinforced concrete pillar, of length 3 m, is fixed firmly at its base and is free at the top, onto which a 20 kg mass is dropped, as shown in Figure 9.40.

Assuming the following apply, determine the maximum stresses in the steel and the concrete:

Steel

A_s = cross-sectional area of steel reinforcement = 0.01 m²
E_s = elastic modulus in steel = 2E11 N/m²

Concrete

A_c = cross-sectional area of concrete reinforcement = 0.2 m^2
E_c = elastic modulus in concrete = 1.4E10 N/m^2

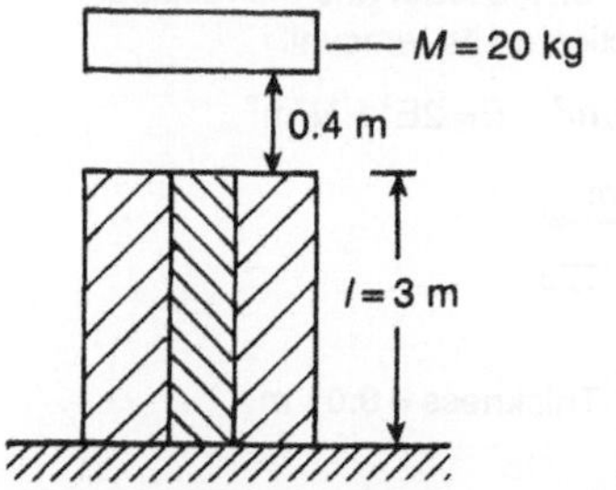

Figure 9.40 Reinforced concrete pillar

{σ(concrete) = −1.46 MN/m^2, σ(steel) = −20.89 MN/m^2}

9. An initially horizontal beam, of length 2 m, is supported at its ends by two wires, one made from aluminium alloy and the other from steel, as shown in Figure 9.41. If a mass of 203.87 kg is dropped a distance of 0.1 m above the mid-span of the beam, determine the deflection fo the beam under this load. The following may be assumed to apply:

Steel

E_s = 2E11 N/m^2
l_s = 2 m
A_s = 2E-4 m^2

Aluminium alloy

$E_{Al} = 7 \times 10^{10}$ N/m^2
l_{Al} = 1 m
A_{Al} = 4E-4 m^2
EI (for beam) = 2E6 N/m^2

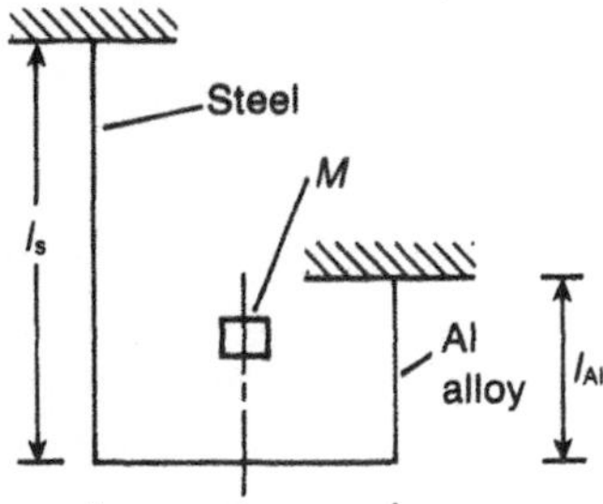

Figure 9.41

{δ = 6.49 mm, σ_s = 154.98 MN/m^2, σ_{Al} = 77.48 MN/m^2}

10. Determine the shape factors for the cross-sections shown in Figures 9.42(a)–(c).

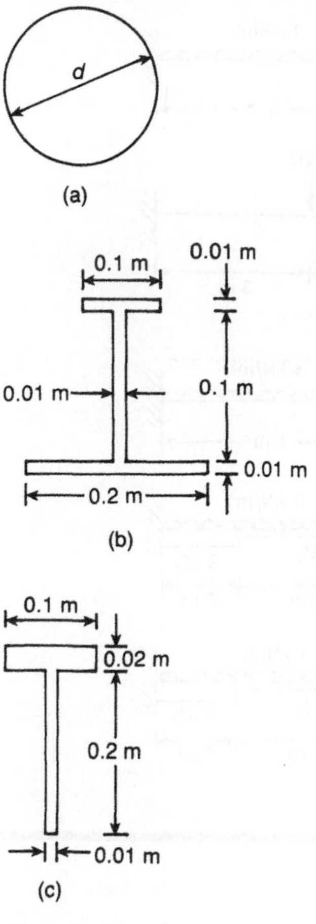

Figure 9.42 Cross sections: (a) solid circular section; (b) RSJ; (c) tee beam

{(a) 1.7; (b) 1.327; (c) 1.811}

11. Using the plastic hinge theory, determine suitable sectional moduli for the beams of Figure 9.43, given that

$\lambda = 3$, $S = 1.14$ and $\sigma_{yp} = 300$ MPa

{(a) 2.632×10^{-4}; (b) 1.096×10^{-4}; (c) 4.511×10^{-5}; (d) 1.563×10^{-5}; (e) 1.469×10^{-5}; (f) 1.108×10^{-5}; all in m^3}

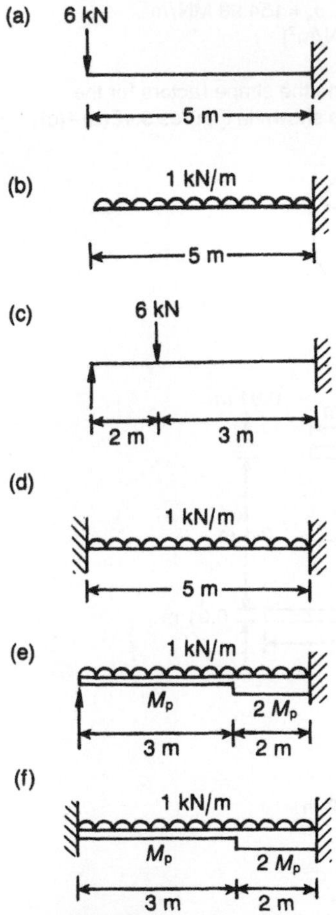

Figure 9.43

12. A uniform-section tee beam, with the sectional properties shown in Figure 9.44(a), is subjected to four-point bending, as shown in Figure 9.44(b). Determine the central deflection on application of the load, and the residual central deflection on its removal.

$\sigma_{yp} = 300\ \mathrm{MN/m^2}$ $E = 2\mathrm{E}11\ \mathrm{N/m^2}$

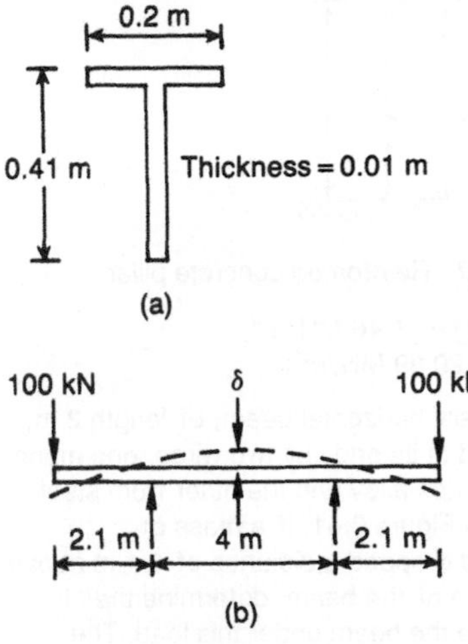

Figure 9.44 Beam under four-point loading

{0.025 m, 0.0055 m}

10

Theories of elastic failure

10.1 Introduction

There are many theories as to which stress or strain, or any combination of these, causes the onset of yield, but in this chapter we will restrict ourselves to a consideration of the five major theories, namely:

(a) Maximum principal stress theory.
(b) Maximum principal strain theory.
(c) Total strain energy theory.
(d) Maximum shear stress theory.
(e) Shear strain energy theory.

The combination of stresses and strains that causes the onset of yield is of much importance in stress analysis, as the onset of yield is very often related to the ultimate failure of the structure.

For convenience, the five major theories are related to a triaxial principal stress system, where

$$\sigma_1 > \sigma_2 > \sigma_3$$

σ_1 = maximum principal stress

σ_3 = minimum principal stress

σ_2 = minimax principal stress

The reason for choosing the triaxial principal stress system to investigate yield criteria is that such a system describes the complete stress situation at a point, without involving the complexities caused by the shear stresses on six planes, which would have resulted if a different three-dimensional co-ordinate system were adopted. The five major theories of elastic failure, together with some of the reasons why they have gained popularity, are described below.

10.2 Maximum principal stress theory (Rankine)

This theory states that yield will occur when

$$\underline{\sigma_1 = \sigma_{yp}} \tag{10.1}$$

or, if the material is in compression, when

$$\underline{\sigma_3 = \sigma_{ypc}} \tag{10.2}$$

where

σ_{yp} = yield stress in tension, obtained in simple uniaxial tensile test

σ_{ypc} = yield stress in compression, obtained in the simple uniaxial compression test

10.3 Maximum principal strain theory (St Venant)

Experimental tests have revealed that comparison between the maximum principal stress theory and experiment was very often found to be poor, and St Venant suggested that perhaps yield occurred owing to the maximum principal strain, rather than the maximum principal stress, as the former involved all three principal stresses and Poisson's ratio, whilst the latter did not.

The maximum principal strain theory states that yield will occur when

$$\varepsilon_1 = \frac{1}{E}\left[\sigma_1 - \nu(\sigma_2 + \sigma_3)\right] = \sigma_{yp}/E$$

or

$$\underline{\sigma_1 - \nu(\sigma_2 + \sigma_3) = \sigma_{yp}} \tag{10.3}$$

or if the material is in compression,

$$\underline{\sigma_3 - \nu(\sigma_1 + \sigma_2) = \sigma_{ypc}} \tag{10.4}$$

where ε_1 is the maximum principal strain.

In *two dimensions*, $\sigma_3 = 0$; therefore equations (10.3) and (10.4) become

$$\sigma_1 - \nu\sigma_2 = \sigma_{yp} \tag{10.5}$$

$$\sigma_2 - \nu\sigma_1 = \sigma_{ypc} \tag{10.6}$$

For convenience, for the remaining three theories, the assumption will be made that σ_{yp} is of the same magnitude as σ_{ypc}. If the magnitude of σ_{ypc} is less than that of σ_{yp}, the theories can be easily modified to take this into account.

10.4 Total strain energy theory (Beltrami and Haigh)

This theory states that elastic failure will occur when the total strain energy per unit volume, at a point, reaches the total strain energy per unit volume of a specimen made from the same material, when it is subjected to a simple uniaxial test.

Now for a three-dimensional stress system, the total strain energy per unit volume, in terms of the three principal stresses, is

$$U_T/\text{vol} = \tfrac{1}{2}\sigma_1\varepsilon_1 + \tfrac{1}{2}\sigma_2\varepsilon_2 + \tfrac{1}{2}\sigma_3\varepsilon_3$$

but from section 7.10,

$$\varepsilon_1 = \frac{1}{E}[\sigma_1 - \nu(\sigma_2 + \sigma_3)] = \text{maximum principal strain}$$

$$\varepsilon_2 = \frac{1}{E}[\sigma_2 - \nu(\sigma_1 + \sigma_3)] = \text{minimax principal strain}$$

$$\varepsilon_3 = \frac{1}{E}[\sigma_3 - \nu(\sigma_1 + \sigma_2)] = \text{minimum principal strain}$$

i.e.

$$U_T/\text{vol} = \frac{1}{2E}(\sigma_1^2 + \sigma_2^2 + \sigma_3^2) - \frac{\nu}{E}(\sigma_1\sigma_2 + \sigma_1\sigma_3 + \sigma_2\sigma_3) \tag{10.7}$$

In the simple uniaxial test,

$$\sigma_1 = \sigma_{yp} \quad \text{and} \quad \sigma_2 = \sigma_3 = 0$$

Therefore

$$U_T/\text{vol} = \sigma_{yp}^2/2E \tag{10.8}$$

Equating (10.7) and (10.8),

$$\underline{\sigma_1^2 + \sigma_2^2 + \sigma_3^2 - 2\nu(\sigma_1\sigma_2 + \sigma_1\sigma_3 + \sigma_2\sigma_3) = \sigma_{yp}^2} \tag{10.9}$$

This theory, like the principal strain theory, also involves all three principal stresses and Poisson's ratio.

In *two dimensions*, $\sigma_3 = 0$; therefore equation (10.9) becomes

$$\underline{\sigma_1^2 + \sigma_2^2 - 2\nu\sigma_1\sigma_2 = \sigma_{yp}^2} \tag{10.10}$$

10.5 Maximum shear stress theory (Tresca)

The problem with the three previous theories is that they all fail in the case of hydrostatic stress. Experiments have shown that, whether or not a solid piece of material is soft and ductile or hard and brittle when it is subjected to a large uniform external pressure, then despite the fact that the yield stress is grossly exceeded, the material does not suffer elastic breakdown in this condition. For example, lumps of chalk or similar low-strength substances can survive intact at great depths in the oceans.

In such cases, all the principal stresses are equal to the water pressure P, so that

$$\sigma_1 = \sigma_2 = \sigma_3 = -P \tag{10.11}$$

From equation (10.11), it can be seen that there are no shear stresses in a hydrostatic stress condition, and this is why low-strength materials survive intact under large values of water pressure.

An argument held against this hypothesis is that in a normal hydrostatic stress condition, the stresses are all compressive, and this is the reason why failure does not take place. However, a Russian scientist carried out tests on a piece of glass, which by a process of heating and cooling was believed to be subjected to a hydrostatic tensile stress of about 6.895E9 Pa (1E6 lbf/in^2) at a certain point in the material, but inspection revealed that there were no signs of cracking at this point.

Thus, it can be concluded that for elastic failure to take place, the material must distort (change shape), and for this to occur, it is necessary for shear stress to exist.

The maximum shear stress theory states that elastic failure will take place when the maximum shear stress at a point equals the maximum shear stress obtained in a specimen, made from the same material, in the simple uniaxial test, i.e.

$$\underline{\sigma_1 - \sigma_3 = \sigma_{yp}} \tag{10.12}$$

In two dimensions, $\sigma_3 = 0$, and equation (10.12) becomes

$$\sigma_1 - \sigma_2 = \sigma_{yp} \tag{10.13}$$

10.6 Maximum shear strain energy theory (Hencky and von Mises)

The problem with the maximum shear stress theory is that it states that

$$\underline{\tau_{yp} = 0.5\sigma_{yp}}$$

where

$$\tau_{yp} = \text{shear stress at yield}$$

However, torsional tests on mild steel specimens have found that

$$\underline{\tau_{yp} \doteqdot 0.577\sigma_{yp}}$$

and this implies that the maximum shear stress theory is not always suitable. The reason for this may be due to the fact that it ignores the effects of σ_2 and ν.

A theory, therefore, that takes into consideration all these factors, and does not fail under the hydrostatic stress condition, is the shear strain energy theory, which states that elastic failure takes place when the shear strain energy per unit volume, at a point, equals the shear strain energy per unit volume in a specimen of the same material, in the simple uniaxial test.

Now, the shear strain energy/vol is

Total strain energy/vol – hydrostatic strain energy/vol

i.e.

$$SSE = U_T - U_H$$

where

$$U_T = \text{total strain energy} = \frac{1}{2E}\left[(\sigma_1^2 + \sigma_2^2 + \sigma_3^2) - 2\nu(\sigma_1\sigma_2 + \sigma_1\sigma_3 + \sigma_2\sigma_3)\right]$$

$$\times \text{ volume (see equation (10.7))}$$

$$U_H = \text{hydrostatic strain energy} = \frac{1}{2E}(3 \times P^2 - 2\nu \times 3P^2) \times \text{volume}$$

$$P = \text{hydrostatic stress} = (\sigma_1 + \sigma_2 + \sigma_3)/3$$

and U_H is obtained by substituting P for σ_1, σ_2 and σ_3 into U_T. Therefore

$$SSE/vol = \frac{1}{2E}\left[\sigma_1^2 + \sigma_2^2 + \sigma_3^2 - 2\nu(\sigma_1\sigma_2 + \sigma_1\sigma_3 + \sigma_2\sigma_3)\right.$$

$$\left. + (2\nu - 1) \times 3 \times (\sigma_1 + \sigma_2 + \sigma_3)^2/9\right]$$

$$= \frac{(1+\nu)}{6E}\left[(\sigma_1 - \sigma_2)^2 + (\sigma_1 - \sigma_3)^2 + (\sigma_2 - \sigma_3)^2\right]$$

but

$$G = E/[2(1 + \nu)]$$

Therefore

$$\underline{SSE/vol = \frac{1}{12G}\left[(\sigma_1 - \sigma_2)^2 + (\sigma_1 - \sigma_3)^2 + (\sigma_2 - \sigma_3)^2\right]} \quad (10.14)$$

The shear strain energy/vol for a specimen in the uniaxial tensile test can be obtained from equation (10.14) by substituting

$$\sigma_1 = \sigma_{yp} \quad \text{and} \quad \sigma_2 = \sigma_3 = 0$$

i.e.

$$SSE/vol = \sigma_{yp}^2/6G \quad (10.15)$$

Equating (10.14) and (10.15), the criterion for yielding, according to the shear strain energy theory, is

$$\underline{(\sigma_1 - \sigma_2)^2 + (\sigma_1 - \sigma_3)^2 + (\sigma_2 - \sigma_3)^2 = 2\sigma_{yp}^2} \quad (10.16)$$

In *two dimensions,* $\sigma_3 = 0$, and equation (10.16) reduces to

$$\sigma_1^2 + \sigma_2^2 - \sigma_1\sigma_2 = \sigma_{yp}^2 \quad (10.17)$$

Another interpretation of equation (10.16) is that elastic failure takes place when the *von Mises stress* reaches yield, where

$$\text{von Mises stress} = \sqrt{(\sigma_1 - \sigma_2)^2 + (\sigma_1 - \sigma_3)^2 + (\sigma_2 - \sigma_3)^2}/\sqrt{2} \quad (10.18)$$

or in *two dimensions*

$$\text{von Mises stress} = \sqrt{\sigma_1^2 + \sigma_2^2 - \sigma_1\sigma_2}$$

10.7 Yield loci

In two dimensions, equations (10.1), (10.2), (10.5), (10.6), (10.10), (10.13) and (10.17) can be expressed graphically, as shown in Figure 10.1.

The figures are obtained by plotting the above equations, using σ_1 as the horizontal co-ordinate and σ_2 as the vertical co-ordinate, and the interpretation of each figure is that yield will not occur according to the theory under consideration if

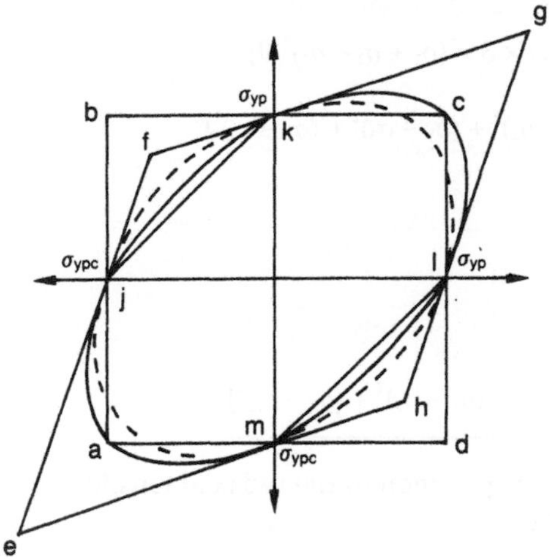

Figure	Theory enclosed by figure
abcd	Maximum principal stress theory
efgh	Maximum principal strain theory
ajkclm	Maximum shear stress theory
– – – ellipse	Total strain energy theory
_____ ellipse	Distortion of shear strain energy theory

Figure 10.1 Yield loci for a two-dimensional stress system

the point described by the values of σ_1 and σ_2 does not fall outside the appropriate figure.

EXAMPLE 10.1

A long thin-walled cylinder of wall thickness 2 cm and internal diameter 10 m is subjected to a uniform internal pressure. Determine the pressure that will cause yield, based on the five major theories of elastic failure, and given the following:

$$\sigma_{yp} = \sigma_{ypc} = 300 \text{ MN/m}^2$$

$$\nu = 0.3$$

SOLUTION

From Chapter 8

$$\sigma_1 = \text{hoop stress} = \frac{PR}{t} = \frac{P \times 5}{2\text{E-}2} = 250P$$

$$\sigma_2 = \text{longitudinal stress} = \frac{PR}{2t} = 125P$$

$$\sigma_3 = \text{radial stress on inside surface of cylinder wall} = -P$$

where

R = internal radius of cylinder
t = wall thickness
P = internal pressure – to be determined

Maximum principal stress theory

$$\sigma_1 = \sigma_{yp}$$

$$250P = 300 \text{ MPa}$$

$$\underline{P = 1.2 \text{ MPa}}$$

Maximum principal strain theory

$$\sigma_1 - \nu(\sigma_2 + \sigma_3) = \sigma_{yp}$$

$$250P - 0.3(125P - P) = 300$$

Therefore

$$P = \frac{300}{212.8} = \underline{1.41 \text{ MPa}}$$

Total strain energy theory

$$\sigma_1^2 + \sigma_2^2 + \sigma_3^2 - 2\nu(\sigma_1\sigma_2 + \sigma_1\sigma_3 + \sigma_2\sigma_3) = \sigma_{yp}^2$$

$$[62\,500 + 15\,625 + 1 - 0.6(31\,250 - 250 - 125)]P^2 = 90\,000$$

Therefore

$$P^2 = 90\,000/59\,601$$

or

$$\underline{P = 1.229 \text{ MPa}}$$

Maximum shear stress theory

$$\sigma_1 - \sigma_3 = \sigma_{yp}$$

or

$$251\ P = 300$$

$$\underline{P = 1.195 \text{ MPa}}$$

Shear strain energy theory

$$(\sigma_1 - \sigma_2)^2 + (\sigma_1 - \sigma_3)^2 + (\sigma_2 - \sigma_3)^2 = 2\sigma_{yp}^2$$

$$(15\,625 + 63\,001 + 15\,876)P^2 = 18\,000$$

Therefore

$$\underline{P = 1.38 \text{ MPa}}$$

From the calculations in this section, it can be seen that the maximum principal strain theory is the most optimistic, and the maximum shear stress theory is the most pessimistic. As expected, the shear strain energy theory is more optimistic than the maximum shear stress theory.

EXAMPLE 10.2

A solid circular-section shaft of diameter 0.1 m is subjected to a bending moment of 15 kN m. Determine the required torque to cause yield, based on the five major theories of elastic failure, given the following:

$$\sigma_{yp} = 300 \text{ MN/m}^2 \quad \nu = 0.3$$

SOLUTION

This is a two-dimensional system of stress; hence, it is necessary to use equations (10.1), (10.5), (10.10), (10.13) and (10.17).

Now,

$$I = \frac{\pi \times 0.1^4}{64} = 4.909\text{E-6 m}^4$$

$$\bar{y} = 0.05 \text{ m}$$

Therefore

$$\sigma_x = \frac{M\bar{y}}{I} = \text{maximum stress in axial direction}$$

$$= \frac{15E3 \times 0.05}{4.909E\text{-}6}$$

$$\underline{\sigma_x = 152.8 \text{ MPa}}$$

By inspection,

$$\underline{\sigma_y = 0}$$

Now from Chapter 6,

$$T = \frac{\tau_{xy} \times J}{r} = \frac{\tau_{xy} \times 9.818E\text{-}6}{0.05}$$

Therefore

$$\underline{T = 1.964E\text{-}4\tau_{xy}}$$

Also, from Chapter 7,

$$\sigma_1 = \frac{(\sigma_x + \sigma_y)}{2} + \tfrac{1}{2}\sqrt{(\sigma_x - \sigma_y)^2 + 4\tau_{xy}^2}$$

$$= \frac{152.8}{2} + \tfrac{1}{2}\sqrt{152.8^2 + 4\tau_{xy}^2}$$

Therefore

$$\underline{\sigma_1 = 76.4 + k}$$

and

$$\underline{\sigma_2 = 76.4 - k}$$

where

$$\underline{k = \tfrac{1}{2}\sqrt{152.8^2 + 4\tau_{xy}^2}}$$

Maximum principal stress theory

$$\sigma_1 = \sigma_{yp}$$

or

$$\underline{76.4 + k = 300}$$

Therefore

$$k = 223.6 = \tfrac{1}{2}\sqrt{152.8^2 + 4\tau_{xy}^2}$$

$$447.2^2 = 152.8^2 + 4\tau_{xy}^2$$

$$\underline{\tau_{xy} = 210.1 \text{ MPa}}$$

but

$$T = 1.964\text{E-}4\tau_{xy}$$

Therefore

$$\underline{T = 41.3\ \text{kN m}}$$

Maximum principal strain theory

$$\sigma_1 - \nu\sigma_2 = \sigma_{yp}$$

$$76.4 + k - 0.3 \times 76.4 + 0.3k = 300$$

or

$$1.3k = 246.52$$

or

$$k = 189.6 = \tfrac{1}{2}\sqrt{152.8^2 + 4\tau_{xy}^2}$$

Therefore

$$\underline{\tau_{xy} = 173.5\ \text{MPa}}$$

and

$$\underline{T = 34.08\ \text{kN m}}$$

Total strain energy theory

$$\sigma_1^2 + \sigma_2^2 - 2\nu\sigma_1\sigma_2 = \sigma_{yp}^2$$

$$(76.4 + k)^2 + (76.4 - k)^2 - 0.6 \times (76.4 + k) \times (76.4 - k)$$

$$= 90\,000$$

or

$$k = 177.4 = \tfrac{1}{2}\sqrt{152.8^2 + 4\tau_{xy}^2}$$

Therefore

$$\underline{\tau_{xy} = 160.1}$$

and

$$\underline{T = 31.45\ \text{kN m}}$$

Maximum shear stress theory

$$\sigma_1 - \sigma_2 = \sigma_{yp}$$

$$76.4 + k - 76.4 + k = 300$$

or

$$k = 150 = \tfrac{1}{2}\sqrt{152.8^2 + 4\tau_{xy}^2}$$

Therefore

$$\underline{\tau_{xy} = 129.1\ \text{MPa}}$$

and

$$\underline{T = 25.35\ \text{kN m}}$$

Shear strain energy theory

$$\sigma_1^2 + \sigma_2^2 - \sigma_1\sigma_2 = \sigma_{yp}^2$$

$$(76.4 + k)^2 + (76.4 - k)^2 - (76.4 + k) \times (76.4 - k) = 300^2$$

or

$$k = 167.5 = \tfrac{1}{2}\sqrt{152.8^2 + 4\tau_{xy}^2}$$

Therefore

$$\underline{\tau_{xy} = 149\ \text{MPa}}$$

and

$$\underline{T = 29.27\ \text{kN m}}$$

The calculations here have shown that for this example, the maximum principal stress theory is the most optimistic, and the maximum shear stress theory is the most pessimistic, where the ratio of the former to the latter is about 1.6 : 1.

10.8 Conclusions

The examples have shown that the predictions from various yield criteria can be very different. Furthermore, from the heuristic arguments of Section 10.5, it would appear that the only two theories that do not fail the hydrostatic stress condition are the maximum shear stress theory and the shear strain energy theory, and in any case, when materials such as mild steel are tested to destruction in tension, the characteristic 'cup and cone' failure mode (Figure 10.2) indicates the importance that shear stress plays in elastic failure. Because of this, many structural designers often

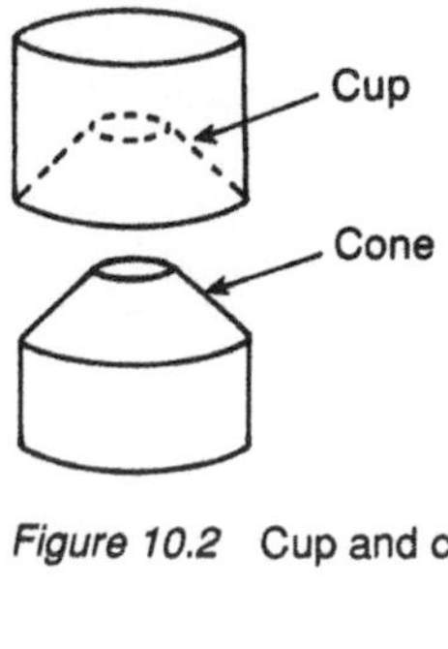

Figure 10.2 Cup and cone failure, indicating the importance of shear stress

prefer to use the maximum shear stress and the shear strain energy theories when designing structures involving two- and three-dimensional stress systems, where the former often lends itself to neat mathematical computations.

Finally, it can be concluded from the calculations for both examples in this chapter that in certain two- and three-dimensional stress systems, some of the theories of elastic failure can be dangerous when applied to practical cases.

EXAMPLES FOR PRACTICE 10

1. A submarine pressure hull, which may be assumed to be a long thin-walled circular cylinder, of external diameter 10 m and wall thickness 5 cm, is constructed from high tensile steel.

Assuming that buckling does not occur, determine the maximum permissible diving depths that the submarine can achieve, without suffering elastic failure, based on the five major theories of yield and given the following:

$\sigma_{yp} = -\sigma_{ypc} = 400$ MN/m² $\quad \nu = 0.3$

Density of water = 1020 kg/m²

$g = 9.81$ m/s²

{400 m, 470 m, 412.3 m, 404 m, 466.6 m}

2. A circular-section torsion specimen, of diameter 2 cm, yields under a pure torque of 0.25 kN m. What is the shear stress due to yield? What is the yield stress according to (a) Tresca, (b) Hencky–von Mises? What is the ratio τ_{yp}/σ_{yp} according to these two theories?

{159.1 MPa; (a) 318.3 MPa; (b) 275.6 MPa; 0.5; 0.577}

3. A shaft of diameter 0.1 m is found to yield under a torque of 30 kN m.

Determine the pure bending moment that will cause a similar shaft, with no torque applied to it, to yield, assuming that the Tresca theory applies. What would be the bending moment to cause yield if the Hencky–von Mises theory applied?

{30 kN m, 30 kN m}

11 Thick cylinders and spheres

11.1 Introduction

If the thickness to radius ratio of a shell exceeds 1:30, the theory for the thin shell starts to break down. The reason for this is that for thicker shells, the radius of the shell changes appreciably over its thickness, so that the membrane strain can no longer be assumed to be constant over the thickness. Thick shells are of great importance in ocean engineering, civil engineering, nuclear engineering, etc.

In this chapter, in addition to considering thick shells under pressure, considerations will also be made of the plastic collapse of thick cylinders and discs. The collapse of discs and rings due to high-speed rotation will also be discussed.

11.2 Derivation of the hoop and radial stress equations for a thick-walled cylinder

The following convention will be used, where all the stresses and strains are assumed to be positive if they are tensile. At any radius r (see Figure 11.1),

σ_θ = hoop stress

σ_r = radial stress

σ_x = longitudinal stress

ε_θ = hoop strain

ε_r = radial strain

ε_x = longitudinal strain (assumed to be constant)

w = radial deflection

From Figure 11.2, it can be seen that at any radius r,

$$\varepsilon_\theta = \frac{2\pi(r+w) - 2\pi r}{2\pi r}$$

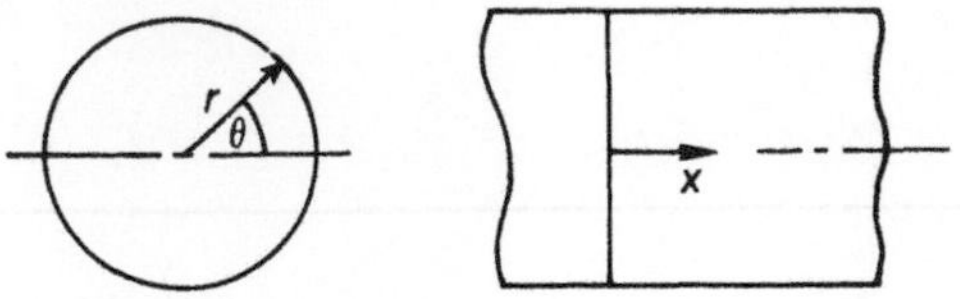

Figure 11.1 Thick cylinder

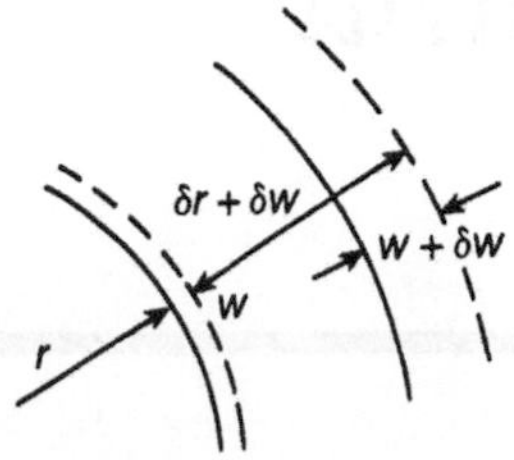

Figure 11.2 Deformation at any radius *r*

or

$$\varepsilon_\theta = w/r \tag{11.1}$$

Similarly,

$$\varepsilon_r = \frac{\delta w}{\delta r} = \frac{dw}{dr} \tag{11.2}$$

From the standard stress–strain relationships (see Chapter 2),

$$E\varepsilon_x = \sigma_x - \nu\sigma_\theta - \nu\sigma_r = \text{a constant}$$

$$E\varepsilon_\theta = \frac{Ew}{r} = \sigma_\theta - \nu\sigma_x - \nu\sigma_r \tag{11.3}$$

$$E\varepsilon_r = E\,\frac{dw}{dr} = \sigma_r - \nu\sigma_\theta - \nu\sigma_x \tag{11.4}$$

Multiplying (11.3) by *r*,

$$Ew = \sigma_\theta \times r - \nu\sigma_x \times r - \nu\sigma_r \times r \tag{11.5}$$

and differentiating (11.5) w.r.t. *r*,

$$E\,\frac{dw}{dr} = \sigma_\theta - \nu\sigma_x - \nu\sigma_r + r\left(\frac{d\sigma_\theta}{dr} - \nu\,\frac{d\sigma_x}{dr} - \nu\,\frac{d\sigma_r}{dr}\right) \tag{11.6}$$

Subtracting (11.4) from (11.6),

$$(\sigma_\theta - \sigma_r)(1 + \nu) + r\,\frac{d\sigma_\theta}{dr} - \nu r\,\frac{d\sigma_x}{dr} - \nu r\,\frac{d\sigma_r}{dr} = 0 \tag{11.7}$$

Since ε_x is constant

$$\sigma_x - \nu\sigma_\theta - \nu\sigma_r = \text{constant} \tag{11.8}$$

Differentiating (11.8) w.r.t. r,

$$\frac{d\sigma_x}{dr} - \nu\frac{d\sigma_\theta}{dr} - \nu\frac{d\sigma_r}{dr} = 0$$

or

$$\frac{d\sigma_x}{dr} = \nu\left(\frac{d\sigma_\theta}{dr} + \frac{d\sigma_r}{dr}\right) \tag{11.9}$$

Substituting (11.9) into (11.7),

$$(\sigma_\theta - \sigma_r)(1+\nu) + r(1-\nu^2)\frac{d\sigma_\theta}{dr} - \nu r(1+\nu)\frac{d\sigma_r}{dr} = 0 \tag{11.10}$$

and dividing through by $(1+\nu)$,

$$\sigma_\theta - \sigma_r + r(1-\nu)\frac{d\sigma_\theta}{dr} - \nu r\frac{d\sigma_r}{dr} = 0 \tag{11.11}$$

Considering now the equilibrium of an element of the shell, as shown in Figure 11.3,

$$2\sigma_\theta\,\delta r \sin\left(\frac{\delta\theta}{2}\right) + \sigma_r r\,\delta\theta - (\sigma_r + \delta\sigma_r)(r+\delta r)\delta\theta = 0$$

In the limit,

$$\sigma_\theta - \sigma_r - r\frac{d\sigma_r}{dr} = 0 \tag{11.12}$$

Subtracting (11.11) from (11.12),

$$\frac{d\sigma_\theta}{dr} + \frac{d\sigma_r}{dr} = 0 \tag{11.13}$$

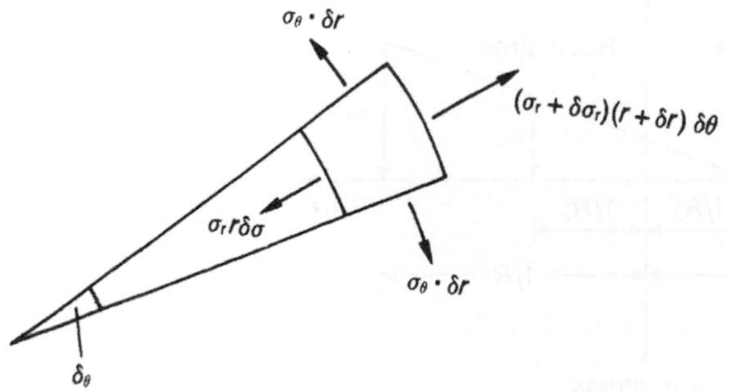

Figure 11.3 Shell element

Therefore

$$\sigma_\theta + \sigma_r = \text{constant} = 2A \tag{11.14}$$

Subtracting (11.12) from (11.14),

$$2\sigma_r + r\frac{d\sigma_r}{dr} = 2A$$

or

$$\frac{1}{r}\frac{d(\sigma_r r^2)}{dr} = 2A$$

$$\frac{d(\sigma_r r^2)}{dr} = 2Ar$$

Integrating,

$$\sigma_r r^2 = Ar^2 - B$$

$$\sigma_r = A - \frac{B}{r^2} \tag{11.15}$$

From (11.14),

$$\sigma_\theta = A + \frac{B}{r^2} \tag{11.16}$$

11.3 Lamé line

Equations (11.15) and (11.16) can be represented by a single straight line if they are plotted against $1/r^2$, as shown in Figure 11.4, where σ_r is on the left of the diagram and σ_θ is on the right.

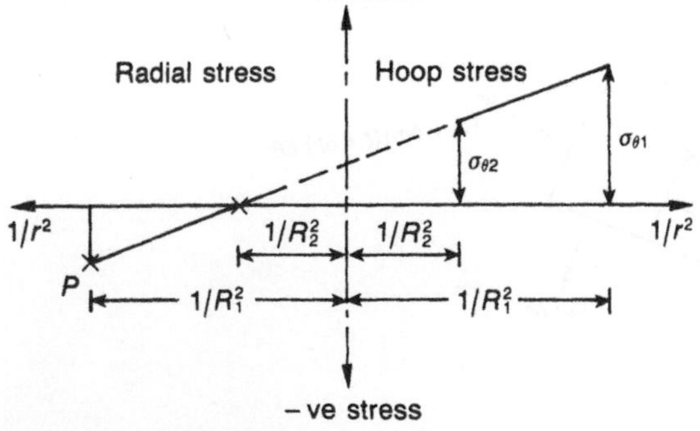

Figure 11.4 Lamé line for the case of internal pressure

In Figure 11.4, we see the case of a thick-walled cylinder of internal radius R_1 and external radius R_2, subjected to an internal pressure P. Now two points are known on this straight line: the radial stresses on the internal and external surfaces, which are $-P$ and zero, respectively (shown by a 'X'). Hence, the straight line can be drawn, and any stress calculated throughout the thickness of the walls by equating similar triangles.

In Figure 11.4,

$\sigma_{\theta 1}$ = internal hoop stress, which can be seen to be a maximum stress

$\sigma_{\theta 2}$ = external hoop stress

Furthermore, from Figure 11.4, it can be seen that both $\sigma_{\theta 1}$ and $\sigma_{\theta 2}$ are tensile, and that σ_r is compressive.

To calculate $\sigma_{\theta 1}$ and $\sigma_{\theta 2}$

Equating similar triangles in Figure 11.4,

$$\frac{\sigma_{\theta 1}}{(1/R_1^2 + 1/R_2^2)} = \frac{P}{(1/R_1^2 - 1/R_2^2)}$$

or

$$\sigma_{\theta 1} = \frac{P(1/R_1^2 + 1/R_2^2)}{(1/R_1^2 - 1/R_2^2)} \times \frac{R_1^2 R_2^2}{R_1^2 R_2^2}$$

$$\sigma_{\theta 1} = \frac{P(R_1^2 + R_2^2)}{(R_2^2 - R_1^2)}$$

Similarly,

$$\frac{\sigma_{\theta 2}}{(1/R_2^2 + 1/R_2^2)} = \frac{P}{(1/R_1^2 - 1/R_2^2)}$$

or

$$\sigma_{\theta 2} = \frac{P(1/R_2^2) \times 2}{(1/R_1^2 - 1/R_2^2)} \times \frac{R_1^2 R_2^2}{R_1^2 R_2^2}$$

$$\sigma_{\theta 2} = \frac{2PR_1^2}{(R_2^2 - R_1^2)}$$

EXAMPLE 11.1

A thick-walled cylinder of internal diameter 0.1 m is subjected to an internal pressure of 50 MPa. If the maximum permissible stress in the cylinder is limited to 200 MPa, determine the minimum possible external diameter.

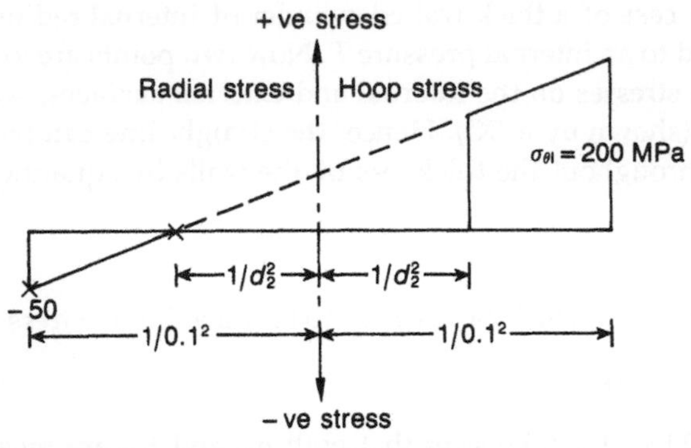

Figure 11.5 Lamé line for thick cylinder

SOLUTION

Let d_2 be the external diameter – to be determined

The Lamé line for this case would appear as shown in Figure 11.5, where it can be seen that the Lamé line is obtained by knowing two values of radial stress. These are the radial stresses on the internal and external surfaces of the cylinder, which are −50 MPa and zero, respectively. However, as d_2 is unknown, a third point is required, which in this case is the maximum stress (i.e. $\sigma_{\theta I}$); this is 200 MPa.

Equating similar triangles in Figure 11.5,

$$\frac{50}{(1/0.1^2 - 1/d_2^2)} = \frac{200}{(1/0.1^2 + 1/d_2^2)}$$

or,

$$\frac{(1/0.1^2 + 1/d_2^2)}{(1/0.1^2 - 1/d_2^2)} \times \left(\frac{0.1^2 \times d_2^2}{0.1^2 \times d_2^2}\right) = 4$$

$$\left(\frac{d_2^2 + 0.1^2}{d_2^2 - 0.1^2}\right) = 4$$

$$d_2^2 + 0.1^2 = 4d_2^2 - 4 \times 0.1^2$$

$$3d_2^2 = 0.1^2 \times 5$$

$$\underline{d_2 = 0.129 \text{ m}}$$

EXAMPLE 11.2

If the vessel of Example 11.1 is subjected to an external pressure of 50 MPa, determine the maximum value of stress that would occur in this vessel.

SOLUTION

The Lamé line for this case is as shown in Figure 11.6, where it can be seen that two

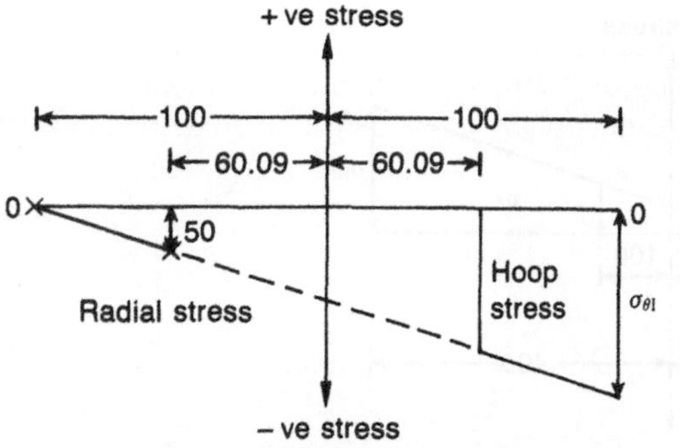

Figure 11.6 Lamé line for external pressure case

values of radial stress are known. These are the radial stresses on the internal and external surfaces, which are zero and −50 MPa, respectively. Let $\sigma_{\theta I}$ = hoop stress on the internal surface, which from Figure 11.6 can be seen to have the largest magnitude.

By equating similar triangles,

$$\frac{\sigma_{\theta I}}{100+100} = \frac{-50}{100-60.09}$$

$$\sigma_{\theta I} = \frac{-50 \times 200}{39.91}$$

$$\underline{\sigma_{\theta I} = -250.6 \text{ MPa}}$$

NB It should be noted in Examples 11.1 and 11.2 that the maximum stress for both cases was the internal hoop stress.

EXAMPLE 11.3

A steel ring of external diameter 10 cm and internal diameter 5 cm is to be shrunk into a solid steel shaft of diameter 5 cm, where all the dimensions are nominal. If the interference fit at the common surface between the ring and the shaft is 0.01 cm, based on a diameter, determine the maximum stress in the material.

$E = 2\text{E}11 \text{ N/m}^2 \quad \nu = 0.3$

SOLUTION

Consider first the steel ring

The Lamé line for the steel ring will be as shown in Figure 11.7, where the radial stress on its outer surface will be zero, and that on its internal surface will be P_c (the radial pressure at the common surface between the steel ring and the shaft).

Let

$\sigma_{\theta IR}$ = hoop stress (maximum stress) on the internal surface of the ring

σ_{rIR} = radial stress on the internal surface of the ring

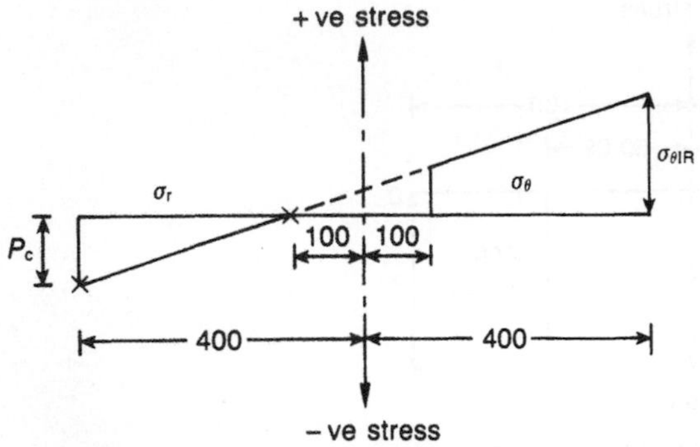

Figure 11.7 Lamé line for steel ring

Equating similar triangles,

$$\frac{\sigma_{\theta IR}}{400 + 100} = \frac{P_c}{400 - 100}$$

Therefore

$$\underline{\sigma_{\theta IR} = 1.667P_c} \tag{11.17}$$

Consider now the shaft

For this case, the Lamé line will be horizontal, because if it had any slope at all, the stresses at the centre would be infinite for a finite value of P_c, which is impossible.

From Figure 11.8, it can be seen that for a solid shaft, under an axisymmetric external pressure of P_c,

$$\sigma_r = \sigma_\theta = -P_c \text{ (everywhere)} \tag{11.18}$$

Let

w_R = increase in the radius of the ring at its inner surface

w_s = increase in the radius of the shaft at its outer surface

Now, applying the expression

$$E\varepsilon_\theta = \frac{w}{r} = \sigma_\theta - \nu\sigma_r - \nu\sigma_x$$

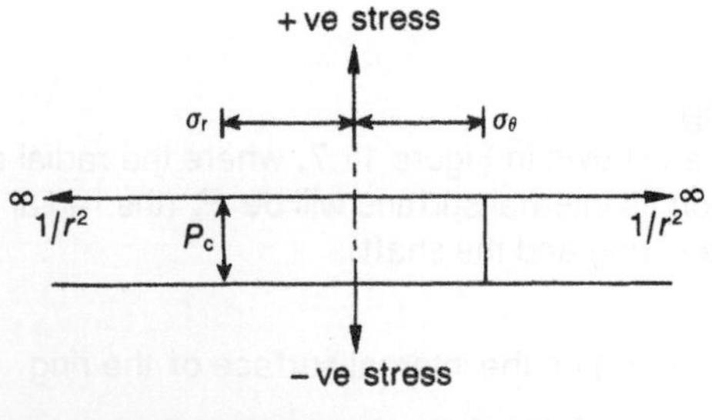

Figure 11.8 Lamé line for a solid shaft

to the inner surface of the ring,

$$\frac{Ew_R}{2.5\text{E-}2} = \sigma_{\theta IR} - \sigma_{rIR}$$

but

$$\sigma_{rIR} = -P_c$$

Therefore

$$\frac{2\text{E}11 \times w_R}{2.5\text{E-}2} = 1.667P_c + 0.3P_c$$

$$\underline{w_R = 2.459\text{E-}13P_c} \qquad (11.19)$$

Similarly for the shaft,

$$\frac{Ew_s}{2.5\text{E-}2} = \sigma_{\theta s} - \nu\sigma_{rs}$$

but

$$\sigma_{\theta s} = \sigma_{rs} = -P_c$$

Therefore

$$\frac{2\text{E}11 \times w_s}{2.5\text{E-}2} = -P_c(1-\nu)$$

or

$$\underline{w_s = -8.75\text{E-}14P_c} \qquad (11.20)$$

Now the interference fit on the diameters is 0.01E-2m. Therefore the interference fit on the radii is 5E-5 m = $w_R - w_s$ or

$$(2.459\text{E-}13 + 8.75\text{E-}14)P_c = 5\text{E-}5$$

$$\underline{P_c = 149.97 \text{ MPa}}$$

From (11.17), the maximum stress is

$$\sigma_{\theta IR} = 1.667 \times P_c$$

$$\underline{\sigma_{\theta IR} = 250 \text{ MPa}}$$

11.4 Compound cylinders

A cylinder made from two different materials is sometimes found useful in engineering when one material is suitable for resisting corrosion in a certain environment, but because this material is expensive or weak, another material is used to strengthen it.

EXAMPLE 11.4

An aluminium-alloy disc of constant thickness, and of internal and external radii R_1 and R_2, respectively, is shrunk onto a solid steel shaft of external radius $R_1 + \delta$. Show that the maximum stress ($\hat{\sigma}$) in the disc is given by

$$\hat{\sigma} = \delta \Bigg/ \left\{ R_1 \left[\left(\frac{1-\nu_s}{E_s} + \frac{\nu_a}{E_a} \right) \left(\frac{R_2^2 - R_1^2}{R_1^2 + R_2^2} \right) + \frac{1}{E_a} \right] \right\}$$

where

E_s = elastic modulus of steel
E_a = elastic modulus of aluminium alloy
ν_s = Poisson's ratio for steel
ν_a = Poisson's ratio for aluminium alloy

SOLUTION

let P_c be the radial pressure at the common surface – to be determined. Now the radial stress for the disc on the external surface is zero, and the radial stress on the internal surface of the disc is $-P_c$; hence, the Lamé line will take the form shown in Figure 11.9.

Equating similar triangles in Figure 11.9,

$$\frac{P_c}{(1/R_1^2 - 1/R_2^2)} = \frac{\hat{\sigma}}{(1/R_1^2 + 1/R_2^2)}$$

Therefore

$$P_c = \hat{\sigma} \times \frac{(R_2^2 - R_1^2)}{(R_2^2 + R_1^2)} \tag{11.21}$$

Now for the steel shaft, the Lamé line must be horizontal, otherwise the stresses at the centre of the shaft will be infinite for a finite value of P_c, i.e.

$\sigma_{\theta s}$ = hoop stress in the shaft
= radial stress in the shaft (σ_{rs})
= $-P_c$ (everywhere in the shaft)

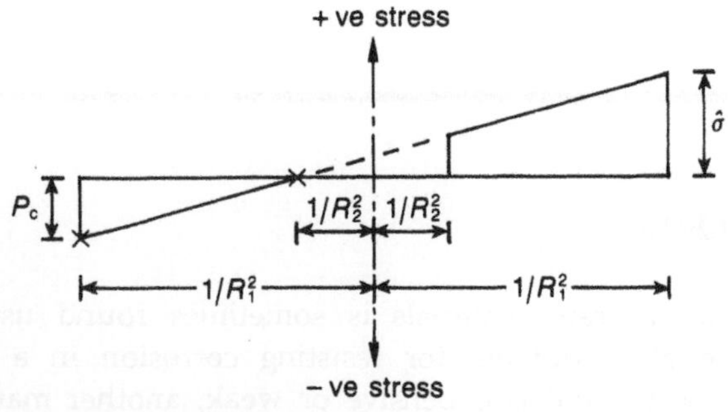

Figure 11.9 Lamé line for the disc

Let

w_a = increase in radius of the aluminium disc at its internal surface

w_s = increase in radius of the steel shaft on its external surface

Applying the expression

$$E\varepsilon_\theta = \frac{Ew}{r} = \sigma_\theta - \nu\sigma_r - \nu\sigma_x$$

to the internal surface of the aluminium disc,

$$\frac{E_a w_a}{R_1} = \hat{\sigma} + \nu_a P_c$$

or

$$w_a = \frac{R_1}{E_a}(\hat{\sigma} + \nu_a P_c) \tag{11.22}$$

Similarly, for the steel shaft,

$$\frac{E_s w_s}{R_1} = -P_c + \nu_s P_c$$

or

$$w_s = \frac{-R_1}{E_s} \times P_c(1 - \nu_s) \tag{11.23}$$

but

$$\delta = w_a - w_s$$

$$= \frac{R_1}{E_a}(\hat{\sigma} + \nu_a P_c) + \frac{R_1}{E_s} P_c(1 - \nu_s) \tag{11.24}$$

Substituting (11.21) into (11.24),

$$\underline{\hat{\sigma} = \delta \Bigg/ \left\{ R_1 \left[\left(\frac{1 - \nu_s}{E_s} + \frac{\nu_a}{E_a} \right) \left(\frac{R_2^2 - R_1^2}{R_1^2 + R_2^2} \right) + \frac{1}{E_a} \right] \right\}}$$

EXAMPLE 11.5

A steel cylinder with external and internal diameters of 10 cm and 8 cm, respectively, is shrunk onto a aluminium-alloy cylinder with internal and external diameters of 5 cm and 8 cm, respectively, where all the dimensions are nominal.

Find the radial pressure at the common surface due to shrinkage alone, so that when there is an internal pressure of 150 MPa, the maximum hoop in the inner cylinder is 110 MPa. Determine, also, the maximum hoop stress in the outer cylinder, and plot the distributions across the sections.

For steel

$E_s = 2E11\ N/m^2 \quad \nu_s = 0.3$

For aluminium alloy

$E_a = 6.7E10\ N/m^2 \quad \nu_a = 0.32$

SOLUTION

Let

P_c^s = the radial pressure at the common surface due to shrinkage alone

σ_θ^s = the hoop stress due to shrinkage alone

σ_θ^p = the hoop stress due to pressure alone

$\sigma_{\theta,10s}$ = hoop stress in the steel at its 10 cm diameter

$\sigma_{\theta,8s}$ = hoop stress in the steel at its 8 cm diameter

$\sigma_{r,10s}$ = radial stress in the steel at its 10 cm diameter

$\sigma_{r,8s}$ = radial stress in the steel at its 8 cm diameter

$\sigma_{\theta,8a}$ = hoop stress in the aluminium alloy at its 8 cm diameter

$\sigma_{r,8a}$ = radial stress in the aluminium alloy at its 8 cm diameter

$\sigma_{\theta,5a}$ = hoop stress in the aluminium alloy at its 5 cm diameter

$\sigma_{r,5a}$ = radial stress in the aluminium alloy at its 5 cm diameter

***Consider first the stress due to* shrinkage alone**

For the aluminium-alloy tube, the Lamé line due to shrinkage alone is shown in Figure 11.10.

Equating similar triangles in Figure 11.10,

$$\frac{\sigma_{\theta,5a}^s}{400+400} = \frac{-P_c^s}{400-156.3}$$

Therefore

$$\underline{\sigma_{\theta,5a}^s = -3.282P_c^s} \tag{11.25}$$

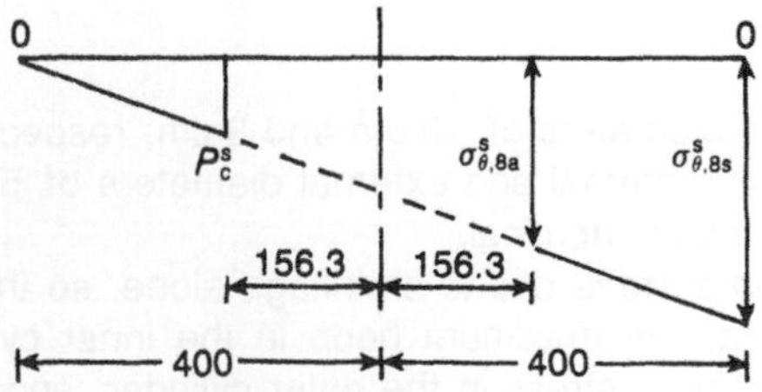

Figure 11.10 Lamé line for aluminium-alloy tube

Similarly,

$$\frac{\sigma^s_{\theta,8a}}{400+156.3} = \frac{-P^s_c}{400-156.3}$$

Therefore

$$\underline{\sigma^s_{\theta,8a} = -2.282P^s_c} \tag{11.26}$$

For the steel tube, the Lamé line due to shrinkage alone will be as shown in Figure 11.11.

Equating similar triangles in Figure 11.11,

$$\frac{\sigma^s_{\theta,8s}}{156.25+100} = \frac{P^s_c}{156.25-100}$$

$$\underline{\sigma^s_{\theta,8s} = 4.556P^s_c} \tag{11.27}$$

Consider now the stresses due to pressure alone

Let

P = internal pressure

P^p_c = pressure at the common surface due to pressure alone

For the aluminium-alloy tube, the Lamé line due to pressure alone will be as shown in Figure 11.12.

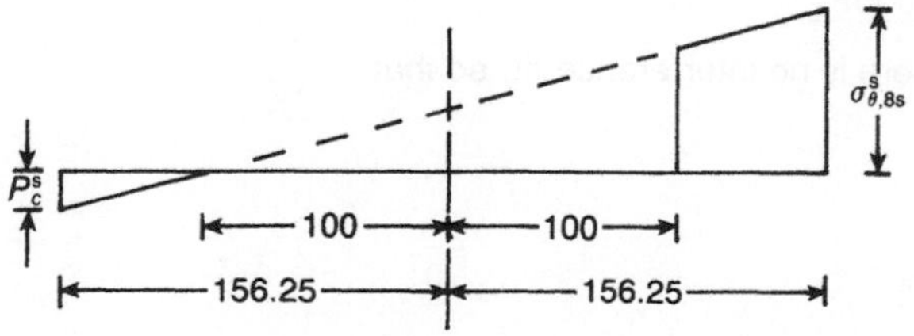

Figure 11.11 Lamé line for steel tube due to shrinkage

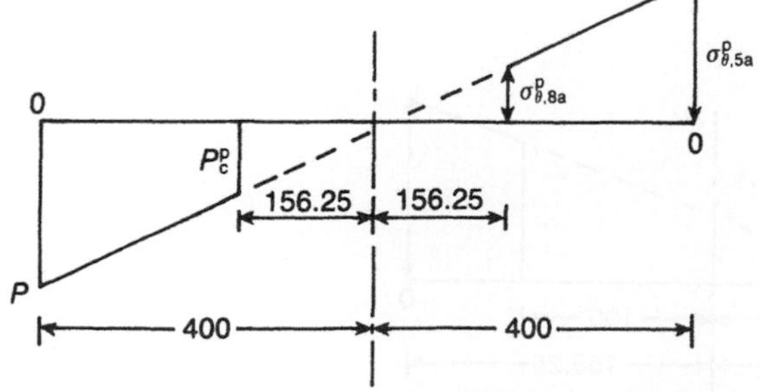

Figure 11.12 Lamé line in aluminium alloy due to pressure alone

Equating similar triangles in Figure 11.12,

$$\frac{P - P_c^p}{400 - 156.25} = \frac{\sigma_{\theta,8a}^p + P}{400 + 156.25}$$

or

$$\frac{150 - P_c^p}{243.75} = \frac{\sigma_{\theta,8a}^p + 150}{556.25}$$

$$\underline{\sigma_{\theta,8a}^p = 192.3 - 2.282P_c^p} \qquad (11.28)$$

Similarly,

$$\frac{P - P_c^p}{243.75} = \frac{\sigma_{\theta,5a}^p + P}{800}$$

$$\frac{150 - P_c^p}{243.75} = \frac{\sigma_{\theta,5a}^p + 150}{800}$$

Therefore

$$\underline{\sigma_{\theta,5a}^p = 342.5 - 3.282P_c^p} \qquad (11.29)$$

For the steel tube due to pressure alone, the Lamé line is as shown in Figure 11.13.
Equating similar triangles in Figure 11.13,

$$\frac{\sigma_{\theta,8s}^p}{156.25 + 100} = \frac{P_c^p}{156.25 - 100}$$

$$\underline{\sigma_{\theta,8s}^p = 4.556P_c^p} \qquad (11.30)$$

Owing to pressure alone, there is no interference fit, so that

$$w_s^p = w_a^p$$

Now,

$$\frac{E_s \times w_s^p}{4E\text{-}2} = \sigma_{\theta,8s}^p + \nu_s P_c^p$$

Therefore

$$w_s^p = \frac{4E\text{-}2}{2E11}(4.556 + 0.3)P_c^p$$

$$\underline{w_s^p = 9.712E\text{-}13P_c^p} \qquad (11.31)$$

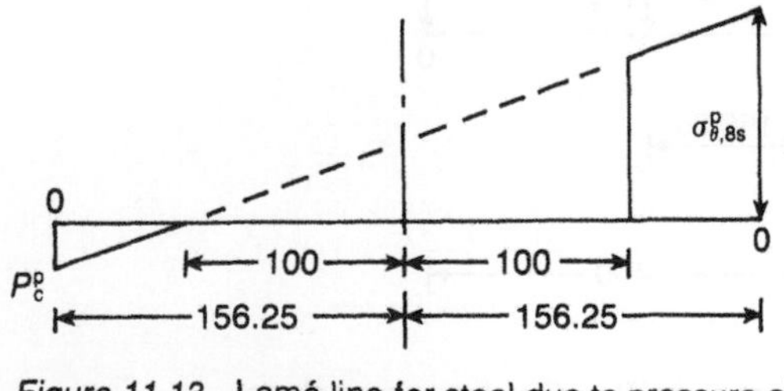

Figure 11.13 Lamé line for steel due to pressure alone

Similarly,

$$\frac{E_a \times w_a^p}{4E\text{-}2} = \sigma_{\theta,8a}^p + \nu_a P_c^p$$

Therefore

$$w_a^p = \frac{4E\text{-}2}{6.7E10}\,(\sigma_{\theta,8a}^p + \nu_a P_c^p)$$

$$= \frac{4E\text{-}2}{6.7E10}\,[192.3 + (-2.282 + 0.32)P_c^p]$$

$$\underline{w_a^p = 1.148E\text{-}10 - 1.171E\text{-}12P_c^p} \tag{11.32}$$

Equating (11.31) and (11.32),

$$9.712E\text{-}13P_c^p = 1.148E\text{-}10 - 1.171E\text{-}12P_c^p$$

Therefore

$$P_c^p = 53.59 \text{ MPa} \tag{11.33}$$

Substituting (11.33) into (11.28) and (11.29),

$$\sigma_{\theta,5a}^p = 342.5 - 3.282 \times 53.59 = \underline{166.6 \text{ MPa}}$$

$$\sigma_{\theta,8a}^p = 192.3 - 2.282 \times 53.59 = \underline{70 \text{ MPa}}$$

Now the maximum hoop stress in the inner tube lies either on its outer surface or on its inner surface, so that

$$\text{either} \quad \sigma_{\theta,8a}^p + \sigma_{\theta,8a}^s = 110 \tag{11.34}$$

$$\text{or} \quad \sigma_{\theta,5a}^p + \sigma_{\theta,5a}^s = 110 \tag{11.35}$$

From (11.34),

$$P_c^s = -17.5 \text{ MPa}$$

and from (11.35)

$$P_c^s = 17.27 \text{ MPa}$$

i.e. the required

$$\underline{P_c^s = 17.27 \text{ MPa}}$$

Hence, owing to the *combined effects of pressure and shrinkage*,

$$\underline{P_c = P_c^p + P_c^s = 17.27 + 53.59 = 70.86 \text{ MPa}}$$

The resultant hoop stress in the steel tube on its 8 cm diameter is

$$\underline{\sigma_{\theta,8s} = 4.556 \times (P_c^s + P_c^p) = 322.8 \text{ MPa}}$$

Similarly, the resultant hoop stress in the aluminium-alloy tube on its 8 cm diameter is

$$\sigma_{\theta,8a} = 192.3 - 2.282\,(P_c^p + P_c^s) = \underline{30.6 \text{ MPa}}$$

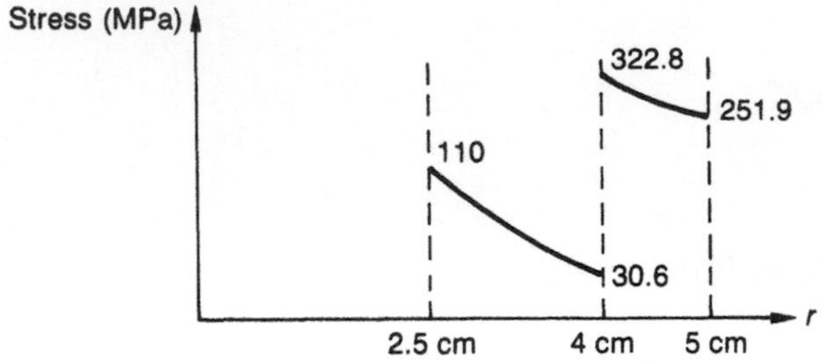

Figure 11.14 Stress distribution across the compound cylinder

and,

$$\sigma_{\theta,5a} = 342.5 - 3.282(P_c^p + P_c^s) = \underline{110\ \text{MPa}}$$

The hoop stress distribution through the two walls is shown in Figure 11.14.

11.5 Plastic yielding of thick tubes

The following assumptions are made in this theory:

1. The tube is constructed from an ideally elastic–plastic material.
2. The longitudinal stress is the 'minimax' stress.
3. Yield occurs according to Tresca's criterion.

For this case, the equilibrium considerations of (11.12) apply i.e.

$$\sigma_\theta - \sigma_r - r\frac{d\sigma_r}{dr} = 0 \tag{11.36}$$

Now, Tresca's criterion is that

$$\sigma_\theta - \sigma_r = \sigma_{yp}$$

$$\sigma_\theta = \sigma_{yp} + \sigma_r \tag{11.37}$$

Substituting (11.37) into (11.36),

$$\sigma_{yp} + \sigma_r - \sigma_r - r\frac{d\sigma_r}{dr} = 0$$

$$d\sigma_r = \sigma_{yp}\frac{dr}{r}$$

$$\sigma_r = \sigma_{yp}\ln r + C \tag{11.38}$$

For the case of a partially plastic cylinder, as shown in Figure 11.15,

$$@\ r = r_2 \quad \sigma_r = -P_2$$

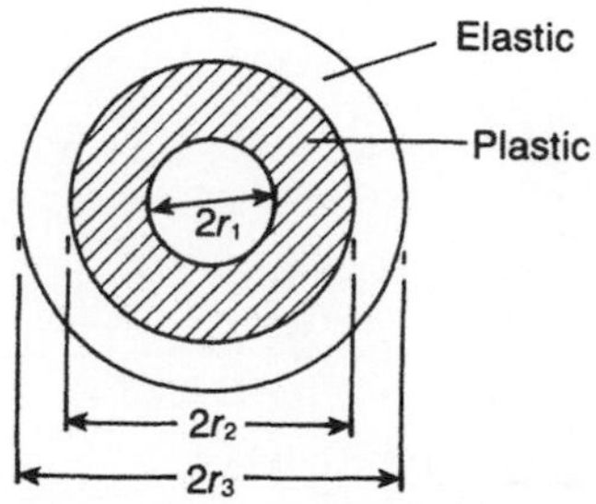

Figure 11.15 Partially plastic cylinder

Substituting the above boundary value into (11.38)

$$-P_2 = \sigma_{yp} \ln r_2 = C$$

Therefore

$$C = -\sigma_{yp} \ln r_2 - P_2$$

and

$$\underline{\sigma_r = \sigma_{yp} \ln(r/r_2) - P_2} \qquad (11.39)$$

Similarly, from (11.37),

$$\underline{\sigma_\theta = \sigma_{yp}\{1 + \ln(r/r_2)\} - P_2} \qquad (11.40)$$

where

r_1 = internal radius

r_2 = outer radius of plastic section of cylinder

r_3 = external radius

P_1 = internal pressure

P_2 = radial pressure at outer radius of plastic zone

The tube can be assumed to behave as a compound cylinder, with the internal portion behaving plastically and the external portion elastically.

The Lamé line for the elastic portion of the cylinder is shown in Figure 11.16.

In Figure 11.16, $\hat{\sigma}_{\theta e}$ is the elastic hoop stress at $r = r_2$, so that according to Tresca's criterion, on this radius,

$$\sigma_{yp} = \hat{\sigma}_{\theta e} + P_2 \qquad (11.41)$$

From Figure 11.16

$$\frac{P_2}{(1/r_2^2 - 1/r_3^2)} = \frac{\hat{\sigma}_{\theta e}}{(1/r_2^2 + 1/r_3^2)}$$

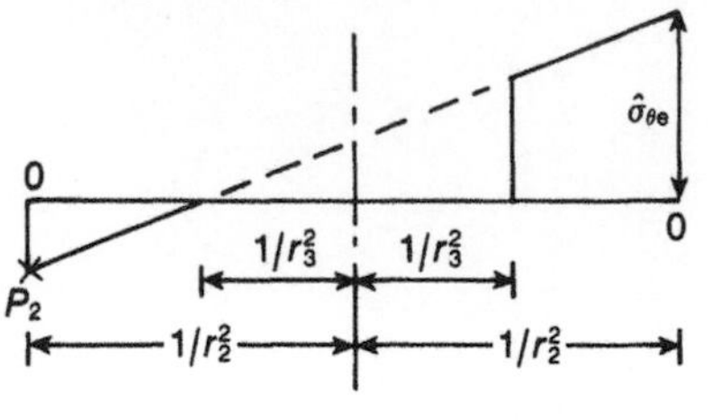

Figure 11.16 Lamé line for elastic zone

Therefore

$$\hat{\sigma}_{\theta e} = \frac{P_2(r_3^2 + r_2^2)}{(r_3^2 - r_2^2)} \tag{11.42}$$

Substituting (11.42) into (11.41),

$$P_2 = \sigma_{yp}(r_3^2 - r_2^2)/2r_3^2 \tag{11.43}$$

Consider now the portion of the cylinder that is plastic.

Substituting (11.43) into (11.39) and (11.40), the stress distributions in the plastic zone are given by

$$\underline{\sigma_r = -\sigma_{yp}\left[\ln\left(\frac{r_2}{r}\right) + \frac{(r_3^2 - r_2^2)}{2r_3^2}\right]} \tag{11.44}$$

$$\underline{\sigma_\theta = \sigma_{yp}\left[\frac{(r_3^2 + r_2^2)}{2r_3^2} - \ln\left(\frac{r_2}{r}\right)\right]} \tag{11.45}$$

To find the *pressure just to cause yield*, put

$$\sigma_r = -P_1 \quad @ \quad r = r_1$$

where P_1 is the internal pressure that causes the onset of yield. Therefore

$$\underline{P_1 = \sigma_{yp}\left[\ln\left(\frac{r_2}{r_1}\right) + \left(\frac{r_3^3 - r_2^2}{2r_3^2}\right)\right]} \tag{11.46}$$

but if yield is only on the inside surface,

$$r_1 = r_2$$

in (11.46), so that

$$\underline{P_1 = \sigma_{yp}[(r_3^2 - r_1^2)/2r_3^2]} \tag{11.47}$$

To determine the *plastic collapse pressure* P_p, put $r_2 = r_3$ in (11.46). Therefore

$$\underline{P_p = \sigma_{yp}\ln(r_3/r_1)} \tag{11.48}$$

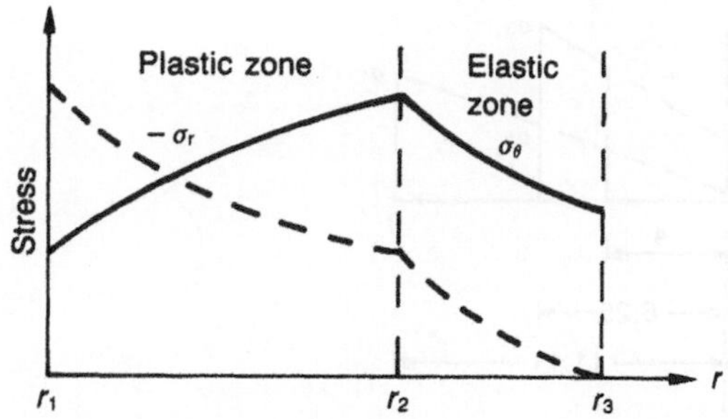

Figure 11.17 Stress distributions in a partially plastic cylinder

To determine the hoop stress distribution in the plastic zone, $\sigma_{\theta p}$, it must be remembered that

$$\sigma_{yp} = \sigma_\theta - \sigma_r$$

Therefore

$$\underline{\sigma_{\theta p} = \sigma_{yp}[1 + \ln(r_3/r_1)]} \tag{11.49}$$

Plots of the stress distributions in a partially plastic cylinder, under internal pressure, are shown in Figure 11.17.

EXAMPLE 11.6

A high tensile steel cylinder of 0.5 m outer diameter and 0.4 m inner diameter is shrunk onto a mild-steel cylinder of 0.3 m bore. If the interference fit is such that when the internal pressure is 100 MN/m^2, the inner face of the inner cylinder is on the point of yielding, determine the internal pressure which will cause plastic penetration through half the thickness of the inner cylinder. The material of the outer cylinder may be assumed to be of a higher quality, so that it does not yield, and the inner cylinder material is perfectly elastic–plastic, yielding at a constant shear stress of 140 MN/m^2. Both materials may be assumed to have the same elastic modulus and Poisson's ratio. (Portsmouth, 1983)

SOLUTION

The Lamé line for the compound cylinder at the onset of yield is shown in Figure 11.18.
In Figure 11.18,

σ_1 = hoop stress on inner surface of inner cylinder

σ_2 = hoop stress on outer surface of inner cylinder

σ_3 = hoop stress on inner surface of outer cylinder

As yield occurs on the inner surface of the inner cylinder,

$$\frac{\sigma_1 - (-100)}{2} = \tau_{yp} = 140$$

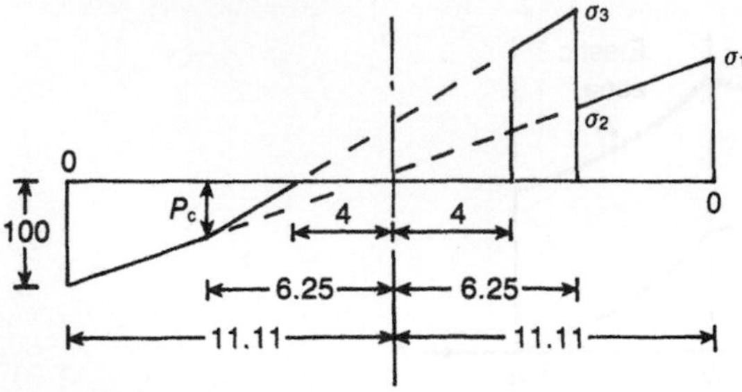

Figure 11.18 Lamé line for compound cylinder

Therefore

$$\sigma_1 = 180 \text{ MPa}$$

where τ_{yp} is the yield stress in shear.

Equating similar triangles in Figure 11.18,

$$\frac{\sigma_1 + 100}{11.11 + 11.11} = \frac{100 - P_c}{11.11 - 6.25}$$

or

$$\frac{280}{22.22} = \frac{100 - P_c}{4.86}$$

Therefore

$$\underline{P_c = 38.76 \text{ MPa}}$$

Similarly from Figure 11.18,

$$\frac{\sigma_2 + 100}{11.11 + 6.25} = \frac{280}{22.22}$$

Therefore

$$\underline{\sigma_2 = 118.8 \text{ MPa}} \qquad (11.50)$$

Also from Figure 11.18,

$$\frac{\sigma_3}{6.25 + 4} = \frac{P_c}{6.25 - 4}$$

but

$$P_c = 38.76$$

Therefore

$$\underline{\sigma_3 = 176.6 \text{ MPa}} \qquad (11.51)$$

Consider, now, plastic penetration of the inner cylinder to a diameter of 0.35 m. The

Lamé line in the elastic zones will be as shown in Figure 11.19. From this figure,

$$\frac{\sigma_6 + P_3}{2} = 140$$

Therefore

$$\sigma_6 = 280 - P_3 \tag{11.52}$$

Similarly,

$$\frac{P_3 - P_2}{8.16 - 6.25} = \frac{\sigma_6 + P_3}{8.16 \times 2} = \frac{280}{16.32}$$

Therefore

$$P_3 - P_2 = \frac{280}{16.32} \times 1.91 = 32.77$$

or

$$P_2 = P_3 - 32.77 \tag{11.53}$$

Also from Figure 11.19,

$$\frac{\sigma_4}{(6.25 + 4)} = \frac{P_2}{6.25 - 4}$$

or

$$\underline{\sigma_4 = 4.56 P_2} \tag{11.54}$$

Substituting (11.53) into (11.54),

$$\underline{\sigma_4 = 4.56 P_3 - 149.3} \tag{11.55}$$

and

$$\frac{\sigma_5 + P_3}{8.16 + 6.25} = \frac{280}{16.32}$$

Therefore

$$\underline{\sigma_5 = 247.2 - P_3} \tag{11.56}$$

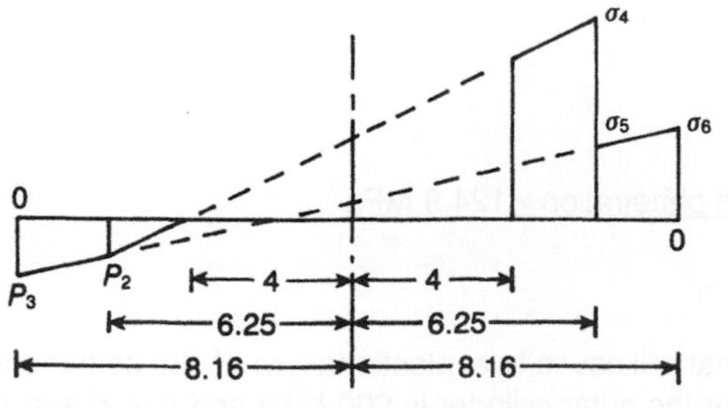

Figure 11.19 Lamé line in elastic zones

Consider strains during the additional pressurization. Now,

$$w = \frac{r}{E}(\sigma_\theta - \nu\sigma_r)$$

which will be the same for both cylinders at the common surface, i.e.

$$\frac{1}{E}[(\sigma_5 - \sigma_2) - \nu(P_2 - P_c)] = \left(\frac{1}{E}(\sigma_4 - \sigma_3) - \nu(P_2 - P_c)\right)$$

or

$$\sigma_5 - \sigma_2 = \sigma_4 - \sigma_3 \tag{11.57}$$

Substituting (11.50), (11.51), (11.55) and (11.56) into (11.57),

$$247.2 - P_3 - 118.8 = 4.56P_3 - 149.3 - 176.6$$

or

$$5.56P_3 = 454.3$$

Therefore

$$\underline{P_3 = 81.71 \text{ MPa}}$$

Consider the *yielded portion*. Now,

$$\sigma_r = \sigma_{yp}\ln r + C$$

$$\sigma_{yp} = 280 \text{ MPa}$$

and

$$@\ r = 0.175 \quad \sigma_r = -P_3 = -81.71$$

Therefore

$$\underline{C = 406.32}$$

$$@\ r = 0.15 \quad \sigma_r = -P$$

Therefore

$$-P = 280\ln(0.15) + 406.3$$

$$= -531.2 + 406.3$$

$$\underline{P = 124.9 \text{ MPa}}$$

i.e.

$$\underline{\text{Pressure to cause plastic penetration} = 124.9 \text{ MPa}}$$

EXAMPLE 11.7

Determine the internal pressure that will cause total plastic failure of the compound vessel of Example 11.6, given that σ_{yp} for the outer cylinder is 600 MPa and that E and ν are the same for both cylinders.

SOLUTION

Now,

$$P_p = \sigma_{yp} \ln\left(\frac{r_3}{r_1}\right)$$

$$= 280 \ln\left(\frac{0.2}{0.15}\right) + 600 \ln\left(\frac{0.25}{0.2}\right)$$

$$= 80.55 + 133.89$$

Therefore

$$\underline{P_p = 214.44 \text{ MPa}}$$

i.e.

<u>Plastic collapse pressure = 214.44 MPa</u>

11.6 Thick spherical shells

Consider a hemispherical element of a thick spherical shell at any radius r under a compressive radial stress P, as shown in Figure 11.20.

Let w be the radial deflection at any radius r so that

Hoop strain = w/r

and

Radial strain = dw/dr

From three-dimensional stress–strain relationships,

$$E\frac{w}{r} = \sigma - \nu\sigma + \nu P \tag{11.58}$$

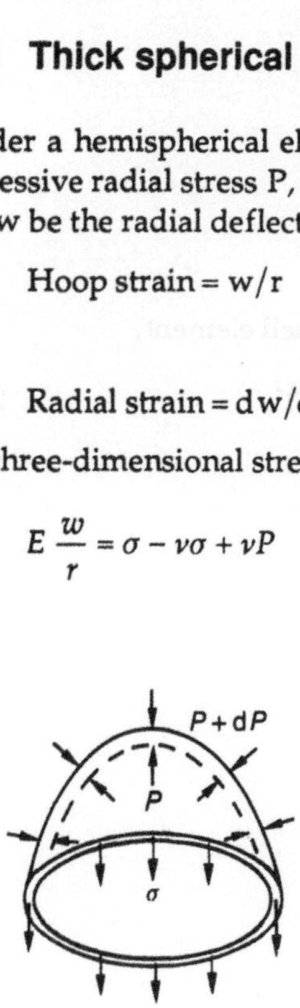

Figure 11.20 Hemispherical shell element

and

$$E\frac{dw}{dr} = -P - \nu\sigma - \nu\sigma$$

$$= -P - 2\nu\sigma \tag{11.59}$$

Multiplying (11.58) by r,

$$Ew = \sigma r - \nu\sigma r + \nu Pr$$

and differentiating w.r.t. r,

$$E\frac{dw}{dr} = \sigma + r\frac{d\sigma}{dr} - \nu\sigma - \nu r\frac{d\sigma}{dr} + \nu P + \nu r\frac{dP}{dr}$$

$$= (1-\nu)\left(\sigma - r\frac{d\sigma}{dr}\right) + \nu\left(P + r\frac{dP}{dr}\right) \tag{11.60}$$

Equating (11.59) and (11.60),

$$-P - 2\nu\sigma = (1-\nu)\left(\sigma - r\frac{d\sigma}{dr}\right) + \nu\left(P + r\frac{dP}{dr}\right)$$

or

$$(1+\nu)(\sigma + P) + r(1-\nu)\frac{d\sigma}{dr} + \nu r\frac{dP}{dr} = 0 \tag{11.61}$$

Considering now the equilibrium of the hemispherical shell element,

$$\sigma \times 2\pi r \times dr = P \times \pi r^2 - (P + dP) \times \pi \times (r + dr)^2 \tag{11.62}$$

Neglecting higher-order terms, (11.62) becomes

$$\sigma + P = (-r/2)\frac{dP}{dr} \tag{11.63}$$

Substituting (11.63) into (11.61),

$$-(r/2)\frac{dP}{dr}(1+\nu) + r(1-\nu)\frac{d\sigma}{dr} + \nu r\frac{dP}{dr} = 0$$

or

$$\frac{d\sigma}{dr} - \frac{1}{2}\frac{dP}{dr} = 0 \tag{11.64}$$

which on integrating becomes

$$\sigma - P/2 = A \tag{11.65}$$

Substituting (11.65) into (11.63),

$$3P/2 + A = (-r/2)\,\frac{dP}{dr}$$

or

$$-\frac{1}{r^2}\,\frac{d(Pr^3)}{dr} = 2A$$

or

$$\frac{d(Pr^3)}{dr} = -2Ar^2$$

which on integrating becomes

$$P \times r^3 = -2Ar^3 + B$$

or

$$\underline{P = -2A/3 + B/r^3} \tag{11.66}$$

and

$$\underline{\sigma = 2A/3 + B/2r^3} \tag{11.67}$$

11.7 Rotating discs

These are a common feature in engineering, which from time to time suffer failure due to high-speed rotation.

In this section, equations will be obtained for calculating hoop and radial stresses, and the theory will be extended for calculating angular velocities to cause plastic collapse of rotating discs and rings.

Consider a uniform thickness disc, of density ρ, rotating at a constant angular velocity ω.

From Section 11.2,

$$E\,\frac{dw}{dr} = \sigma_r - \nu\sigma_\theta \tag{11.68}$$

and

$$E\,\frac{w}{r} = \sigma_\theta - \nu\sigma_r \tag{11.69}$$

or

$$Ew = \sigma_\theta \times r + \nu\sigma_r \times r \tag{11.70}$$

Differentiating (11.70) w.r.t. r,

$$E\,\frac{dw}{dr} = \sigma_\theta + r\,\frac{d\sigma_\theta}{dr} - \nu\sigma_r - \nu r\,\frac{d\sigma_r}{dr} \tag{11.71}$$

Equating (11.68) and (11.71),

$$(\sigma_\theta - \sigma_r)(1+\nu) + r\frac{d\sigma_\theta}{dr} - \nu r\frac{d\sigma_r}{dr} = 0 \qquad (11.72)$$

Considering equilibrium of an element of the disc, as shown in Figure 11.21,

$$2\sigma_\theta \times dr \times \sin\left(\frac{d\theta}{2}\right) + \sigma_r \times r \times d\theta$$

$$-(\sigma_r + d\sigma_r)(r+dr)\,d\theta = \rho \times \omega^2 \times r^2 \times dr \times d\theta$$

In the limit, this reduces to

$$\sigma_\theta - \sigma_r - r\frac{d\sigma_r}{dr} = \rho\omega^2 r^2 \qquad (11.73)$$

Substituting (11.73) into (11.72),

$$\left(r\frac{d\sigma_r}{dr} + \rho\omega^2 r^2\right)(1+\nu) + r\frac{d\sigma_\theta}{dr} - \nu r\frac{d\sigma_r}{dr} = 0$$

or

$$\frac{d\sigma_\theta}{dr} + \frac{d\sigma_r}{dr} = -\rho\omega^2 r^2(1+\nu)$$

which on integrating becomes

$$\sigma_\theta + \sigma_r = -(\rho\omega^2 r^2/2)(1+\nu) + 2A \qquad (11.74)$$

Subtracting (11.73) from (11.74)

$$2\sigma_r + r\frac{d\sigma_r}{dr} = -(\rho\omega^2 r^2/2)(3+\nu) + 2A$$

or

$$\frac{1}{r}\frac{d(\sigma_r \times r^2)}{dr} = -\frac{\rho\omega^2 r^2(3+\nu)}{2} + 2A$$

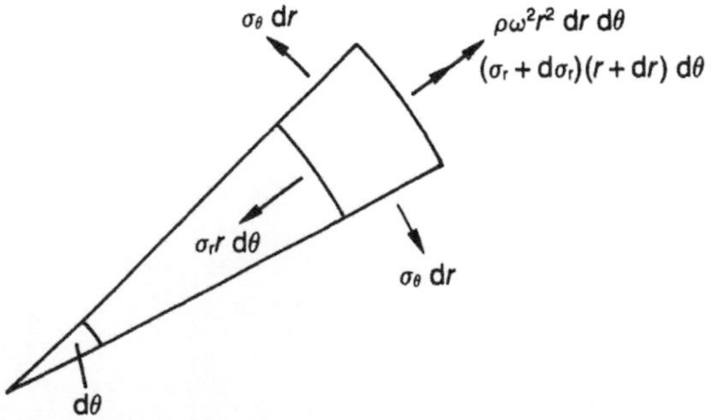

Figure 11.21 Element of disc

which on integrating becomes

$$\sigma_r \times r^2 = -(\rho\omega^2 r^4/8)(3+\nu) + Ar^2 - B$$

or

$$\underline{\sigma_r = A - B/r^2 - (3+\nu)(\rho\omega^2 r^2/8)} \tag{11.75}$$

and

$$\underline{\sigma_\theta = A + B/r^2 - (1+3\nu)(\rho\omega^2 r^2/8)} \tag{11.76}$$

EXAMPLE 11.8

Obtain an expression for the radial variation in the thickness of a disc, so that it will be of constant strength when it is rotated at an angular velocity ω.

SOLUTION

Let

t_0 = thickness at centre
t = thickness at a radius r
$t + dt$ = thickness at a radius $r + dr$
σ = stress = constant (everywhere)

Consider the equilibrium of an element of this disc at any radius r, as shown in Figure 11.22.

Resolving radially

$$2\sigma \times t \times dr \sin\left(\frac{d\theta}{2}\right) + \sigma tr\, d\theta = \sigma(r+dr)(t+dt)\, d\theta + \rho\omega^2 r^2 t\, d\theta\, dr$$

In the limit, this becomes

$$\sigma t\, dt = \sigma r\, dt + \sigma t\, dr + \rho\omega^2 rt\, dr$$

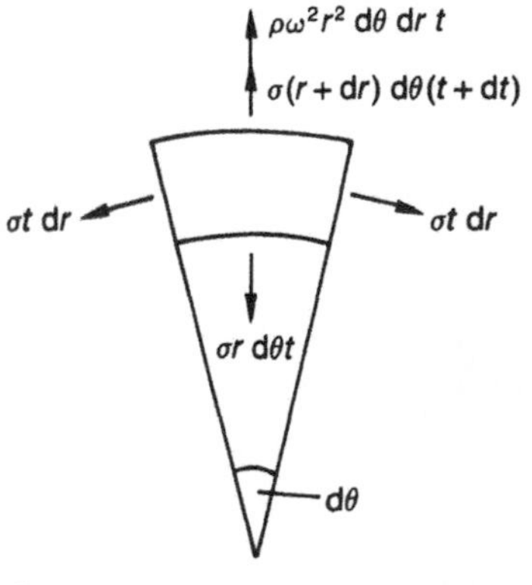

Figure 11.22 Element of constant strength disc

or

$$\frac{dt}{dr} = -\rho\omega^2 rt/\sigma$$

which on integrating becomes
$= -\rho\omega^2 r^2 t/2\sigma + \ln C$

or

$$t = Ce^{(-\rho\omega^2 r^2/2\sigma)}$$

Now, @ $r = 0$ $t = t_0 = C$; therefore

$$\underline{t = t_0 e^{(-\rho\omega^2 r^2/2\sigma)}} \tag{11.77}$$

11.8 Plastic collapse of discs

Assume that $\sigma_\theta > \sigma_r$, and that plastic collapse occurs when

$$\sigma_\theta = \sigma_{yp}$$

Let R be the external radius of the disc. From *equilibrium considerations*

$$\sigma_{yp} - \sigma_r - r\frac{d\sigma_r}{dr} = \rho\omega^2 r^2$$

or

$$\int r\, d\sigma_r = \int (\sigma_{yp} - \sigma_r - \rho\omega^2 r^2)\, dr$$

Integrating the left-hand side by parts,

$$r\sigma_r - \int \sigma_r\, dr = \sigma_{yp} r - \int \sigma_r\, dr - \rho\omega^2 r^3/3 + A$$

Therefore

$$\underline{\sigma_r = \sigma_{yp} - \rho\omega^2 r^2/3 + A/r} \tag{11.78}$$

For a *solid disc*, @ $r = 0$, $\sigma_r \neq \infty$; therefore

$$\underline{A = 0}$$

and

$$\sigma_r = \sigma_{yp} - \rho\omega^2 r^2/3$$

@ $r = R$ $\sigma_r = 0$; therefore

$$0 = \sigma_{yp} - \rho\omega^2 R^2/3$$

and

$$\omega = \frac{1}{R}\sqrt{\frac{3\sigma_{yp}}{\rho}} \tag{11.79}$$

where ω is the angular velocity of the disc, which causes plastic collapse.

For an *annular disc* of internal radius R_1 and external radius R_2, suitable boundary values for (11.78) are as follows

$$@\ r = R_1 \quad \sigma_r = 0$$

Therefore

$$A = (\rho\omega^2 R_1^2/3 - \sigma_{yp})R_1$$

and

$$\sigma_r = \sigma_{yp} - \rho\omega^2 r^2/3 + (\rho\omega^2 R_1^2/3 - \sigma_{yp})(R_1/r) \tag{11.80}$$

$$@\ r = R_2 \quad \sigma_r = 0$$

Therefore

$$0 = \sigma_{yp} - \rho\omega^2 R_2^2/3 + (\rho\omega^2 R_1^2/3 - \sigma_{yp})(R_1/R_2)$$

i.e.

$$\omega = \sqrt{\left(\frac{3\sigma_{yp}}{\rho}\right)\frac{(R_2 - R_1)}{(R_2^3 - R_1^3)}} \tag{11.81}$$

11.9 Rotating rings

Consider the equilibrium of the semi-circular ring element shown in Figure 11.23.

Let a be the cross-sectional area of the ring. Then, *resolving vertically,*

$$\sigma_\theta \times a \times 2 = \int_0^\pi \rho\omega^2 r^2 a \, d\theta \sin\theta$$

$$= \rho\omega^2 r^2 a[-\cos\theta]_0^\pi$$

$$= 2\rho\omega^2 r^2 a$$

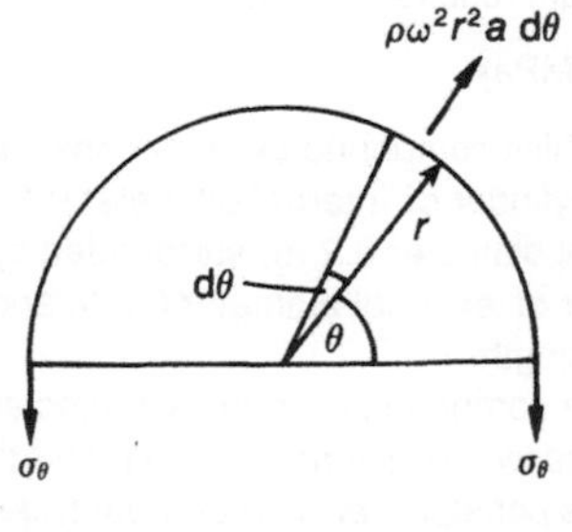

Figure 11.23 Ring element

Therefore

$$\sigma_\theta = \rho\omega^2 r^2$$

@ collapse, $\sigma_\theta = \sigma_{yp}$; therefore

$$\omega = \frac{1}{r}\sqrt{\frac{\sigma_{yp}}{\rho}} \tag{11.82}$$

where ω is the angular velocity required to fracture the ring.

EXAMPLES FOR PRACTICE 11

1. Determine the maximum permissible internal pressure that a thick-walled cylinder of internal diameter 0.2 m and wall thickness 0.1 m can be subjected to, if the maximum permissible stress in this vessel is not to exceed 250 MPa.

{150 MPa}

2. Determine the maximum permissible internal pressure that a thick-walled cylinder of internal diameter 0.2 m and wall thickness 0.1 m can be subjected to, if the cylinder is also subjected to an external pressure of 20 MPa. $\sigma_{yp} = 300$ MPa.

{212 MPa}

3. A steel ring of 9 cm external diameter and 5 cm internal diameter is to be shrunk into a solid bronze shaft, where the interference fit is 0.005E-2 m, based on the diameter.

Determine the maximum tensile stress that is set up in the material given that:

For steel

$E_s = 2\text{E}11\ \text{N/m}^2 \quad \nu_s = 0.3$

For bronze

$E_b = 1\text{E}11\ \text{N/m}^2 \quad \nu_b = 0.35$

{$P_c = 57.3$ MPa, $\hat{\sigma}_\theta = 108.3$ MPa}

4. A compound cylinder is manufactured by shrinking a steel cylinder of external diameter 22 cm and internal diameter 18 cm onto another steel cylinder of internal diameter 14 cm, the dimensions being nominal. If the maximum tensile stress in the outer cylinder is 100 MPa, determine the radial compressive stress at the common surface and the interference fit at the common diameter. Determine, also the maximum stress in the inner cylinder.

$E_s = 2\text{E}11\ \text{N/m}^2 \quad \nu_s = 0.3$

{19.8 MPa, $\delta = 0.16$ mm, -100 MPa}

5. If the inner cylinder of Example 4 were made from bronze, what would be the value of δ?

For bronze

$E_b = 1\text{E}11\ \text{N/m}^2 \quad \nu_b = 0.4$

{$\delta = 0.226$ mm}

6. If the compound cylinder of Example 4 were subjected to an internal pressure of 50 MPa, what would be the value of the maximum resultant stress?

{227.9 MPa}

7. If the compound cylinder of Example 5 were subjected to an internal pressure of 50 MPa, what would be the value of the maximum resultant stress?

{241.1 MPa}

8. A thick compound cylinder consists of a brass cylinder of internal diameter 0.1 m and external diameter 0.2 m, surrounded by a steel cylinder of external diameter 0.3 m and of the same length.

If the compound cylinder is subjected to a compressive axial load of 5 MN, and the axial strain is constant for both cylinders, determine the pressure at the common surface and the longitudinal stresses in the two cylinders due to this load.

The following assumptions may be made:

(a) σ_{Ls} = a longitudinal stress in steel cylinder = a constant

(b) σ_{LB} = longitudinal stress in brass cylinder = a constant

For steel

$E_s = 2 \times 10^{11}$ N/m² $\quad \nu_s = 0.3$

For brass

$E_B = 1 \times 10^{11}$ N/m² $\quad \nu_B = 0.4$

Portsmouth, 1984

{2.2 MPa; σ_{LS} = −96.52 MPa; σ_{LB} = −51.14 MPa}

12

The buckling of struts

12.1 Introduction

A *strut*, in its most usual form, can be described as a column under axial compression, as shown in Figure 12.1.

From Figure 12.1, it can be seen that despite the fact that struts are loaded axially, they fail owing to bending moments caused by lateral movement of the struts. The reason for this is that under increasing axial compression, the bending resistance of the strut decreases until a point is reached where the bending stiffness of the strut is so small that the slightest offset of load, or geometrical imperfection of the strut, causes catastrophic failure (or *instability*).

Similarly for a beam (or a length of rubber) under increasing axial tension, its bending stiffness increases until a point is reached where failure occurs. Beams or rods under tension are called *beam-ties* or *ties*, respectively, and beams under compression are often called *beam-columns*; in this text, considerations will only be made of the latter.

Struts appear in many and various forms, from the pillars in the hold of a ship to

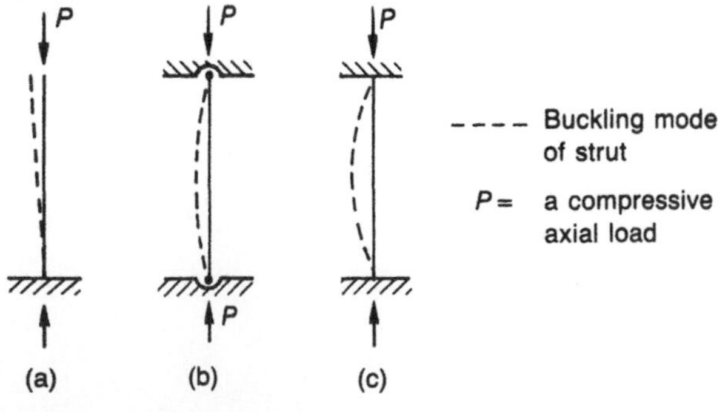

Figure 12.1 Some axially loaded struts: (a) one end free and the other fixed; (b) both ends pinned; (c) both ends clamped

the forks of a bicycle, and from pillars supporting the roofs of ancient monuments to those supporting the roofs of modern football stadiums. For some struts, the design against failure is one of guarding against instability, whilst for others, it is one of determining the stresses due to the combined action of axial and lateral loading.

12.2 Axially loaded struts

Initially 'straight' struts, which are subjected to axial compression but without additional lateral loading, as shown in Figure 12.1, are classified under the following three broad headings:

Very short struts, as shown in Figure 12.2, which fail owing to excessive stress values. For such cases, instability is not of any importance.

Very long and slender struts, which fail owing to elastic instability. For such cases, the calculation of stresses is not important, and the only material properties of interest are elastic modulus and in some cases Poisson's ratio.

Intermediate struts, whose slenderness is somewhere between the previous two extremes. Most struts tend to fall into this category, where both the elastic properties of the material and its failure stress are important. Another feature of much importance for this class of strut is its initial geometrical imperfections.

The initial geometrical imperfections of a strut can cause it to fail at a buckling load which may be a small fraction of the elastic buckling load, and the difference between these two buckling loads is sometimes called *elastic knockdown*.

The buckling behaviour of intermediate struts is sometimes termed *inelastic instability*.

12.3 Elastic instability of very long slender struts

This theory is due to Euler, and it breaks down the following classes of strut:

(a) Very short struts
(b) Intermediate struts
(c) Eccentrically loaded struts
(d) Struts with initial curvature
(e) Laterally loaded struts

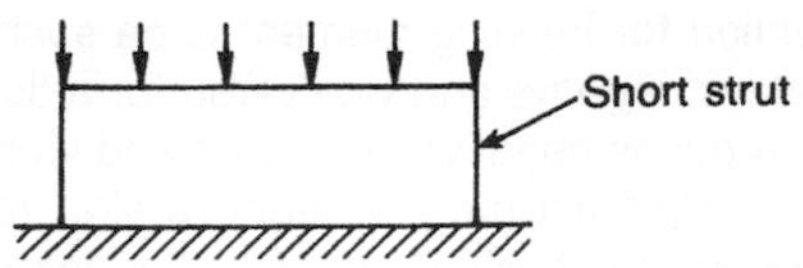

Figure 12.2 Very short strut

Despite the fact that the theory presented in this section is only applicable to very long slender struts, which are made from homogeneous, isotropic and elastic materials, the theory can be extended to cater for axially loaded intermediate struts. In addition to this, the differential equation describing the behaviour of this class of strut can be extended to deal with eccentrically loaded, initially curved and laterally loaded struts.

The Euler theory for the elastic instability of a number of initially straight struts will now be considered.

EXAMPLE 12.1

Determine the Euler buckling load for an axially loaded strut, pinned at its ends, as shown in Figure 12.3. It may be assumed that the ends of the strut are free to move axially towards each other.

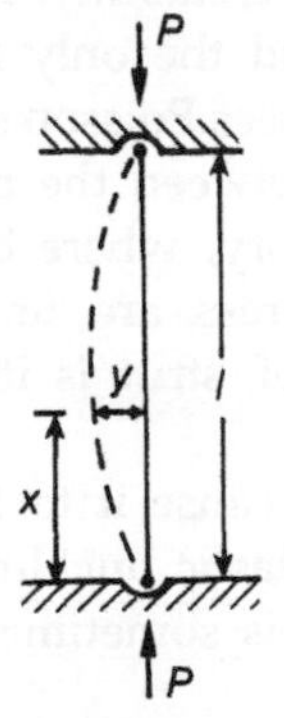

Figure 12.3 Axially loaded strut, pinned at its ends

SOLUTION

In the case of this strut, lateral movement of the strut is prevented at its ends, but the strut is free to rotate at these points (i.e. it is 'position fixed' at both ends). Experiments have shown that such struts tend to have a buckling mode, as shown by the dashed line of Figure 12.3.

Just prior to instability, let the lateral deflection of the strut be y at a distance x from its base. Now,

$$EI \frac{d^2y}{dx^2} = M = -Py \tag{12.1}$$

For struts, it is necessary for the sign convention for bending moment to be such that the *bending moment due to the product* Py *is always negative* and vice versa for a tie. That is, it is assumed that positive bending moment produces negative deflection and vice versa.

The mathematical reasons why the product Py must have a negative sign for a strut and why it must have a positive sign for a tie are dealt with in a number of texts and will not be dealt with here, as this book is concerned with applying mathematics to some problems in engineering. The student must, however, remember to define the sign convention for bending moment, based on the assumption that the product Py causes a negative bending moment for all struts.

Equation (12.1) can now be rewritten in the form

$$\frac{d^2y}{dx^2} + \frac{Py}{EI} = 0 \tag{12.2}$$

Let $\alpha^2 = P/EI$, so that (12.2) becomes

$$\frac{d^2y}{dx^2} + \alpha^2 y = 0 \tag{12.3}$$

Let

$$y = Ae^{\lambda x} \tag{12.4}$$

so that

$$\frac{d^2y}{dx^2} = \lambda^2 Ae^{\lambda x} = \lambda^2 y \tag{12.5}$$

Substituting (12.4) and (12.5) into (12.3),

$$\lambda^2 y + \alpha^2 y = 0$$

or

$$\lambda^2 + \alpha^2 = 0$$

Therefore

$$\lambda = \pm j\alpha$$

where

$$j = \sqrt{-1}$$

i.e.

$$\underline{y = A\cos \alpha x + B\sin \alpha x} \tag{12.6}$$

As there are two constants, namely A and B, it will be necessary to apply two boundary conditions to (12.6).

By inspection it can be seen that

$$@\ x = 0 \quad y = 0 \quad \text{and} \quad @\ x = l \quad y = 0 \tag{12.7}$$

From (12.7),

$$A = 0 \quad \text{and} \quad B\sin \alpha l = 0 \tag{12.8}$$

Now the condition $B = 0$ in (12.8) is not of practical interest, as the strut will not suffer lateral deflections if this were so; therefore, the only possibility for equation (12.8) to apply is that

$$\sin \alpha l = 0$$

or

$$\alpha l = 0, \pi, 2\pi, 3\pi, \text{etc.}$$

The value of αl that is of interest for the buckling of struts will be the lowest positive value, as nature has shown that the strut of Figure 12.3 will buckle by this mode, i.e.

$$\alpha l = \pi$$

or

$$\sqrt{\frac{P}{EI}}\, l = \pi$$

$$P = \frac{\pi^2 EI}{l^2} \quad (12.9)$$

Equation (12.9) is often written in the form of (12.10), where P_e, the Euler buckling load of a strut pinned at its ends, is given by

$$P_e = \frac{\pi^2 EI}{l^2} \quad (12.10)$$

EXAMPLE 12.2

Determine the Euler buckling load for an axially loaded strut, clamped at its ends, as shown in Figure 12.4. It may be assumed that the ends of the strut are free to move axially towards each other.

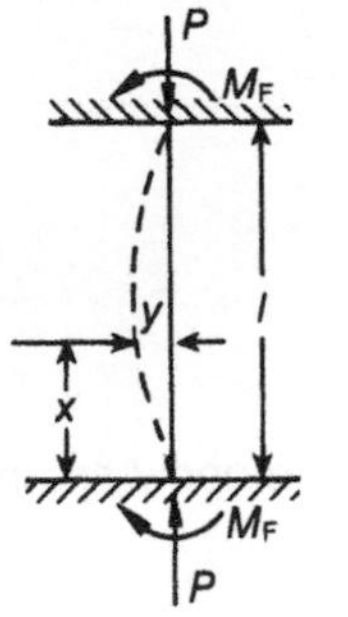

Figure 12.4 Axially loaded strut, clamped at its ends

SOLUTION

At both its ends, the strut is prevented from lateral and rotational movement, where the latter is achieved through the reaction moments M_F (i.e. it is 'direction fixed' at both ends). Experiments have shown that the strut will have the buckled form shown by the dashed line.

Just prior to buckling, let the deflection at a distance x from the base of the strut be given by y, so that

$$EI \frac{d^2y}{dx^2} = -Py + M_F \tag{12.11}$$

NB As the product Py is a negative bending moment, M_F will be a positive one.

Equation (12.11) can be rewritten in the form

$$\frac{d^2y}{dx^2} + \alpha^2 y = \frac{M_F}{EI} \tag{12.12}$$

From Example 12.1, it can be seen that the complementary function is

$$y = A\cos\alpha x + B\sin\alpha x \tag{12.13}$$

To obtain the particular integral, let

$$D = \frac{d}{dx} = \text{operator 'D'}$$

so that (12.12) becomes

$$(D^2 + \alpha^2)y = \frac{M_F}{EI}$$

or

$$y = \frac{M_F}{(D^2 + \alpha^2)EI}$$

$$= \frac{(1 + D^2/\alpha^2)^{-1}}{\alpha^2}\frac{M_F}{EI}$$

i.e.

$$y = \frac{M_F}{EI\alpha^2} \tag{12.14}$$

From equation (12.13) and (12.14), the complete solution is

$$y = A\cos\alpha x + B\sin\alpha x + \frac{M_F}{EI\alpha^2} \tag{12.15}$$

and

$$\frac{dy}{dx} = -\alpha A\sin\alpha x + \alpha B\cos\alpha x$$

Now there are three unknowns, namely A, B and M_F; hence, it will be necessary to use three boundary conditions, as follows: when

$$x = 0 \quad \frac{dy}{dx} = 0$$

Therefore

$$\underline{B = 0}$$

and when

$$\underline{x = 0 \quad y = 0}$$

Therefore

$$0 = A + \frac{M_F}{EIa^2}$$

or

$$A = \frac{-M_F}{EIa^2}$$

The third boundary condition can be when

$$\underline{x = l \quad y = 0}$$

or

$$0 = A \cos al + \frac{M_F}{EIa^2}$$

or

$$\frac{M_F}{EIa^2}(1 - \cos al) = 0$$

i.e.

$$\cos al = 1 \quad \text{or} \quad al = 0, 2\pi, 4\pi, 6\pi, \text{etc.}$$

Nature has shown that the value of interest is $al = 2\pi$ (the lowest positive root) or

$$\underline{P_e = 4\pi^2 EI/l^2} \tag{12.16}$$

where P_e is the Euler buckling load for clamped ends.

12.4 Struts with various boundary conditions

By a similar process, it can be seen that the Euler buckling load for a strut fixed at one end and free at the other, as shown in Figure 12.5, is given by

$$P_e = \frac{\pi^2 EI}{4l^2} \tag{12.17}$$

Comparing equations (12.10), (12.16) and (12.17), it can be seen that the assumed end conditions play a significant role. Indeed, the 'clamped ends' strut has an Euler buckling load four times greater than the 'pinned ends' strut and 16 times greater than the strut of Figure 12.5.

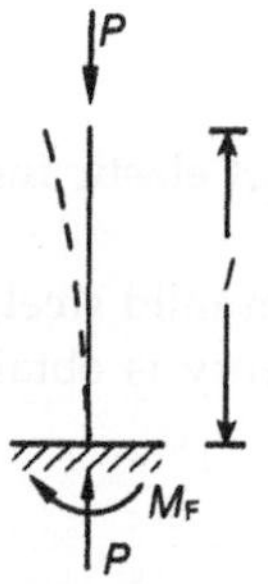

Figure 12.5 Axially loaded strut, fixed at one end and free at the other

In practice, however, it is very difficult to obtain the completely clamped condition of Figure 12.4, as there is usually some rotation due to elasticity etc. at the ends and the effect of this is to reduce drastically the predicted value for P_e from equation (12.16).

From Sections 12.3 and 12.4, it can be seen that the Euler buckling load P_{cr} can be represented by the equation

$$P_{cr} = \frac{\pi^2 EI}{L_0^2} \tag{12.18}$$

where L_0 is the effective length of the strut, dependent on the end conditions and end flexibility, as shown in Table 12.1, and l is the actual length of the strut.

The differences between the effective lengths for the Euler theory and BS 449 are because the latter allows for flexibility at the end fixings.

Table 12.1 **Effective lengths of struts (L_0)**

Type of strut	Euler	BS 449
	$L_0 = l$	$L_0 = l$
	l	l
	$0.5l$	$0.7l$
	$0.7l$	$0.85l$
	$2l$	$2l$

12.5 Limit of application of Euler theory

The Euler formulae are obviously inapplicable where they predict elastic instability loads greater than the crushing strength of the strut material.

For example, assuming that a 'pinned ends' strut is made from mild steel, with a crushing stress of 300 MN/m², the lower limit of the Euler theory is obtained, as follows:

$$P_e = \frac{\pi^2 EI}{l^2} = \frac{\pi^2 EAk^2}{l^2} \tag{12.19}$$

where

A = cross-sectional area

k = least radius of gyration

$$\text{Yield load} = 300 \frac{\text{MN}}{\text{m}^2} \times A \tag{12.20}$$

Equating (12.19) and (12.20),

$$\frac{\pi^2 EA}{(l/k)^2} = 300A \tag{12.21}$$

Let $E = 2 \times 10^{11}$ N/m² $= 2 \times 10^5$ MN/m², and substitute this into (12.21) to give

$$(l/k)^2 = \frac{\pi^2 \times 2 \times 10^5}{300} = 6579.7$$

or

$$\underline{(l/k) = 81 = \text{slenderness ratio}} \tag{12.22}$$

From (12.22), it can be seen that, for mild steel, the Euler theory is obviously inapplicable where the *slenderness ratio* is less than 80, and in practice, the theory needs correction for some values of slenderness ratios greater than this.

12.6 Rankine–Gordon formula

This formula is applicable to intermediate struts, and extends the Euler formulae to take account of the crushing stress of the strut material by a semi-empirical approach, as follows.

For a very short strut, the crushing load

$$P_c = \sigma_c A \tag{12.23}$$

where

A = cross-section

σ_c = crushing stress

Then

$$\frac{1}{P_R} = \frac{1}{P_e} + \frac{1}{P_c}$$

$$= \frac{L_0^2}{\pi^2 EI} + \frac{1}{\sigma_{yc} \times A}$$

$$= \frac{L_0^2}{\pi^2 EAk^2} + \frac{1}{\sigma_{yc} A}$$

$$= \frac{L_0^2 \sigma_{yc} + \pi^2 Ek^2}{\pi^2 EAk^2 \sigma_{yc}}$$

or

$$P_R = \frac{\pi^2 EAk^2 \sigma_{yc}}{L_0^2 \sigma_{yc} + \pi^2 Ek^2}$$

$$= \frac{\sigma_{yc}}{L_0^2 \sigma_{yc}/\pi^2 EAk^2 + \pi^2 Ek^2/\pi^2 EAk^2}$$

$$P_R = \frac{\sigma_{yc} \times A}{(\sigma_{yc}/\pi^2 E)(L_0/k)^2 + 1}$$

Let

$$a = \frac{\sigma_{yc}}{\pi^2 E} \tag{12.24}$$

Then

$$P_R = \frac{\sigma_{yc} A}{1 + a(L_0/K)^2} \tag{12.25}$$

where a is the denominator constant in the Rankine–Gordon formula, which is dependent on the boundary conditions and material properties.

A comparison of the Rankine–Gordon and Euler formulae, for geometrically perfect struts, is given in Figure 12.6. Some typical values for $1/a$ and σ_{yc} are given in Table 12.2.

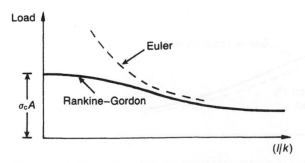

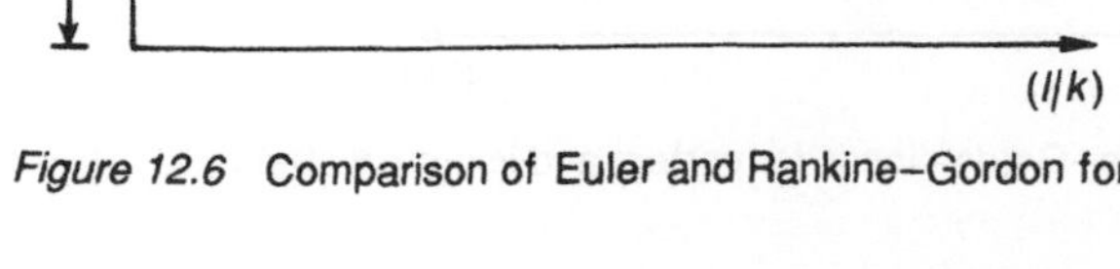

Figure 12.6 Comparison of Euler and Rankine–Gordon formulae

Table 12.2 **Rankine constants**

Material	$1/a$	σ_{yc}
Mild steel	7500	300
Wrought iron	8000	250
Cast iron	18000	560
Timber	1000	35

12.7 Effects of geometrical imperfections

For intermediate struts with geometrical imperfections, the buckling load is further decreased, as shown in Figure 12.7.

12.7.1 Johnson's parabolic formula

This is a simplified version of the Rankine–Gordon expression, and it proved to be popular in design offices prior to the invention of the hand-held calculator:

$$\underline{\text{Johnson buckling stress} = \sigma_c - b(l/k)^2} \tag{12.26}$$

where b is a constant depending on end conditions, material properties, etc.

12.8 Eccentrically loaded struts

Struts in this category, examples of which are shown in Figure 12.8, frequently occur in practice. For such problems, the main object in stress analysis is to determine the stresses due to the combined effects of bending and axial load, unlike the struts of Sections 12.3 and 12.4, where the main object was to obtain a crippling load.

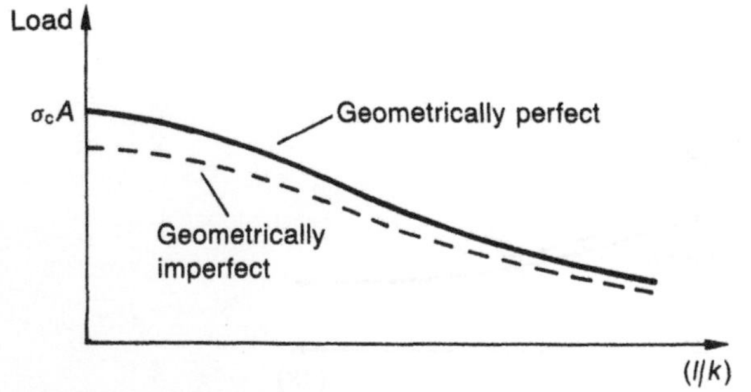

Figure 12.7 Rankine–Gordon loads for perfect and imperfect struts

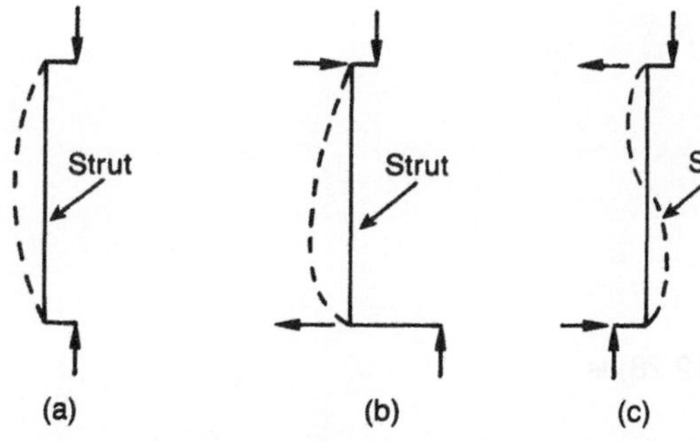

Figure 12.8 Eccentrically loaded struts: (a) eccentricity equal and on the same side of the strut; (b) eccentricity unequal and on the same side of the strut; (c) eccentricity on opposite sides of the strut

EXAMPLE 12.3

A pillar in the hold of a ship is in the form of a tube of external diameter 25 cm, internal diameter 22.5 cm and length 10 m. If the pillar is subjected to an eccentric load of 20 tonnes, as shown in Figure 12.9, calculate the maximum permissible eccentricity if the maximum permissible stress is 75 MN/m^2. It may be assumed that $E = 2 \times 10^{11}$ N/m^2.

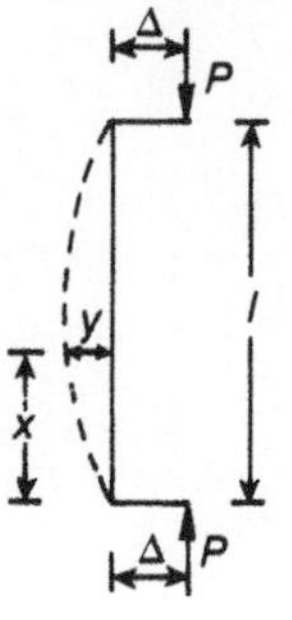

Figure 12.9 Eccentrically loaded strut

SOLUTION

At any distance x from the base,

$$EI\frac{d^2y}{dx^2} = M = -P(y + \Delta)$$

or

$$\frac{d^2y}{dx} + \frac{P}{EI}y = -\frac{P}{EI}\Delta \tag{12.27}$$

Let

$$\alpha^2 = \frac{P}{EI}$$

so that (12.27) becomes

$$\frac{d^2y}{dx^2} + \alpha^2 y = -\alpha^2 \Delta \tag{12.28}$$

The complementary function of (12.28) is

$$y = A\cos\alpha x + B\sin\alpha x \tag{12.29}$$

and the particular integral of (12.28) is

$$y = \frac{-\alpha^2\Delta}{(D^2 + \alpha^2)}$$

giving

$$y = -\Delta \tag{12.30}$$

Combining (12.29) and (12.30), the complete solution is

$$y = A\cos\alpha x + B\sin\alpha x - \Delta \tag{12.31}$$

Now there are two unknowns in (12.31), therefore, two boundary conditions will be required as follows:

$$@\ x = 0 \quad y = 0 \quad \text{and}\ @\ x = l \quad y = 0 \tag{12.32}$$

From (12.32)

$$\underline{A = \Delta} \tag{12.33}$$

and

$$0 = \Delta\cos\alpha l + B\sin\alpha l - \Delta$$

or

$$B = \frac{\Delta(1 - \cos\alpha l)}{\sin\alpha l} = \frac{\Delta 2\sin^2(\alpha l/2)}{2\sin(\alpha l/2)\cos(\alpha l/2)}$$

$$\underline{B = \Delta\tan(\alpha l/2)} \tag{12.34}$$

Substituting (12.33) and (12.34) into (12.31), the deflected form of the strut, at any distance x, is given by

$$\underline{y = \Delta[\cos\alpha x + \tan(\alpha l/2)\sin(\alpha x) - 1]} \tag{12.35}$$

The maximum deflection δ occurs at $x = l/2$, i.e.

$$\begin{aligned}\delta &= \Delta[\cos(\alpha l/2) + \tan(\alpha l/2)\sin(\alpha l/2) - 1] \\ &= \Delta\cos(\alpha l/2)[1 + \tan^2(\alpha l/2) - 1/\cos(\alpha l/2)] \\ &= \Delta\cos(\alpha l/2)[\sec^2(\alpha l/2) - 1/\cos(\alpha l/2)]\end{aligned}$$

$$\underline{\delta = \Delta[\sec(\alpha l/2) - 1]} \tag{12.36}$$

By inspection, the maximum bending moment M is given by

$$\hat{M} = P(\delta + \Delta)$$

$$\underline{\hat{M} = P\Delta \sec(\alpha l/2)} \qquad (12.37)$$

$$I = \frac{\pi}{64}[(25 \times 10^{-2})^4 - (22.5 \times 10^{-2})^4]$$

$$\underline{I = 6.594 \times 10^{-5}\ \text{m}^4}$$

$$A = \frac{\pi}{4}[(25 \times 10^{-2})^2 - (22.5 \times 10^{-2})^2]$$

$$\underline{A = 9.327 \times 10^{-3}\ \text{m}^2}$$

$$\frac{\alpha l}{2} = \sqrt{\frac{P}{EI}}\,\frac{l}{2} = \sqrt{\frac{20 \times 10^3 \times 9.81}{2 \times 10^{11} \times 6.594 \times 10^{-5}}} \times 5$$

$$\underline{\frac{\alpha l}{2} = 0.6099\ \text{rads} = 34.94°}$$

From (12.37)

$$\hat{M} = 20\,000 \times \Delta \times 1/[\cos(34.94°)] \times 9.81$$

$$\underline{\hat{M} = 239\,341\Delta}$$

Now,

$$\hat{\sigma} = \sigma(\text{direct}) \pm \sigma(\text{bending})$$

or

$$-75 \times 10^6 = \frac{-20 \times 10^3 \times 9.81}{9.327 \times 10^{-3}} - \frac{239\,341\Delta}{6.594 \times 10^{-5}} \times 12.5 \times 10^{-2}$$

$$= -2.104 \times 10^7 - 4.537 \times 10^8\Delta$$

or

$$\Delta = 0.1189\ \text{m} = \underline{11.9\ \text{cm}}$$

EXAMPLE 12.4

If the pillar of Example 12.3 were subjected to the eccentric loading of Figure 12.10, determine the maximum permissible value of Δ.

SOLUTION

The horizontal reactions R are required to achieve equilibrium, and the relationship between them and P can be obtained by taking moments, as follows:

$$Rl = P \times 2\Delta$$

or

$$R = 2P\Delta/l \qquad (12.38)$$

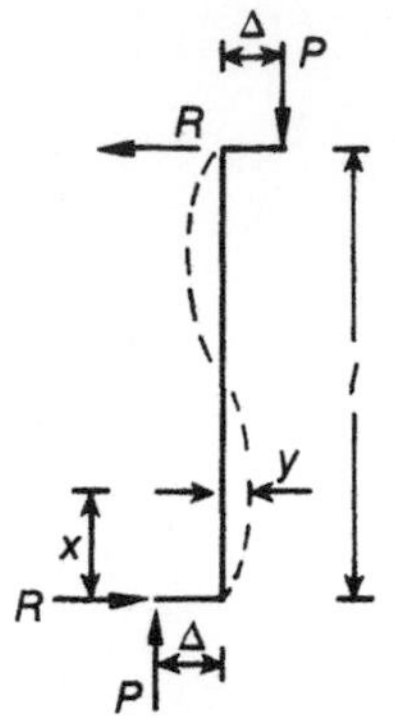

Figure 12.10 Eccentrically loaded strut

At any distance x from the base,

$$EI\frac{d^2y}{dx^2} = -P(y + \Delta) + Rx$$

or

$$\frac{d^2y}{dx^2} + \alpha^2 y = \frac{R\alpha^2 x}{P} - \alpha^2\Delta \qquad (12.39)$$

The complete solution of (12.39) is

$$\underline{y = A\cos\alpha x + B\sin\alpha x + (Rx/P) - \Delta} \qquad (12.40)$$

There are two unknowns; therefore two boundary conditions are required, as follows:

$$@\ x = 0 \quad y = 0; \quad \text{therefore } \underline{A = \Delta} \qquad (12.41)$$

$$@\ x = l/2 \quad y = 0; \quad \text{therefore } 0 = \Delta\cos(\alpha l/2) + B\sin(\alpha l/2) + (2P\Delta l/2Pl) - \Delta$$

or

$$\underline{B = -\Delta\cot(\alpha l/2)} \qquad (12.42)$$

Substituting (12.41) and (12.42) into (12.40),

$$y = \Delta[\cos\alpha x - \cot(\alpha l/2)\sin\alpha x] + \frac{2\not{P}\Delta x}{\not{P}l} - \Delta$$

$$\underline{y = \Delta[\cos\alpha x - \cot(\alpha l/2)\sin\alpha x + (2x/l) - 1]} \qquad (12.43)$$

Now,

$$M = EI\frac{d^2y}{dx^2}$$

or

$$M = EIa^2\Delta[-\cos \alpha x + \cot(\alpha l/2)\sin \alpha x]$$
$$= -P\Delta[\cos \alpha x - \cot(\alpha l/2)\sin \alpha x]$$
$$= \frac{-P\Delta[\sin(\alpha l/2)\cos \alpha x - \cos(\alpha l/2)\sin \alpha x]}{\sin(\alpha l/2)}$$

or

$$M = \frac{-P\Delta \sin(\alpha l/2 - \alpha x)}{\sin(\alpha l/2)}$$

The maximum bending moment $\hat{M}$ occurs when $\alpha l/2 - \alpha x = \pm \pi/2$, i.e.

$$\underline{\hat{M} = \pm P\Delta/[\sin(\alpha l/2)] = \pm P\Delta \operatorname{cosec}(\alpha l)}$$

Now,

$$\hat{\sigma} = \sigma(\text{direct}) \pm \sigma(\text{bending})$$

i.e.

$$-75 \times 10^6 = -21.04 \times 10^6 - \frac{20\,000 \times 9.81 \operatorname{cosec}(34.94)\Delta \times 12.5 \times 10^{-2}}{6.594 \times 10^{-5}}$$

or

$$53.96 \times 10^6 = 649.4 \times 10^6\Delta$$
$$\underline{\Delta = 0.0831 \text{ m} = 8.31 \text{ cm}}$$

12.9 Struts with initial curvature

Struts in this category are usually assumed to have an initial sinusoidal or parabolic shape of the form shown in Figure 12.11. As in Section 12.8, stresses are of more importance for this class of strut than are crippling loads.

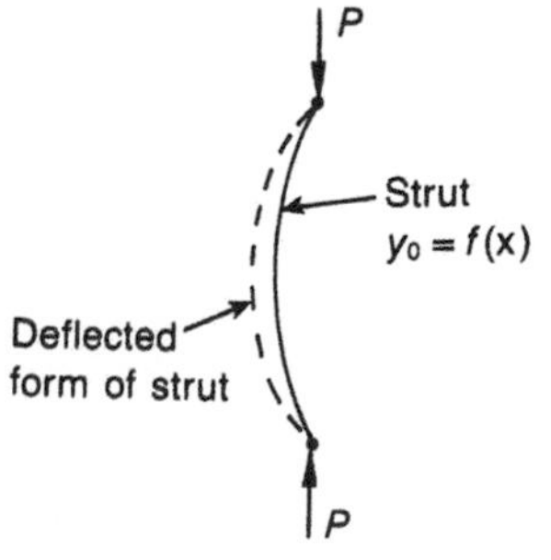

Figure 12.11 Initially curved strut

Let

$$y_0 = f(x) = \text{initial equation of strut}$$

and

$$R_0 = \text{initial radius of curvature of strut of } x$$
$$= \frac{d^2y_0}{dx^2}$$

From the relationship

$$EI\left(\frac{1}{R} - \frac{1}{R_0}\right) = M$$

$$EI\left(\frac{d^2y}{dx^2} - \frac{d^2y_0}{dx^2}\right) = M \tag{12.44}$$

Normally, solution of (12.44) is achieved by assuming the initial shape of the strut y_0 to be sinusoidal or parabolic or circular.

EXAMPLE 12.5

Determine the maximum deflection and bending moment for a strut with pinned ends, which has an initial sinusoidal curvature of the form

$$y_0 = \Delta \sin\left(\frac{\pi x}{l}\right) \tag{12.45}$$

where Δ is a small initial central deflection.

SOLUTION

Substituting (12.45) and its second derivative into (12.44), the following is obtained:

$$EI\left[\frac{d^2y}{dx^2} + \frac{\pi^2}{l^2}\Delta \sin\left(\frac{\pi x}{l}\right)\right] = -Py$$

or

$$\frac{d^2y}{dx^2} + \alpha^2 y = -\frac{\pi^2}{l^2}\Delta \sin\left(\frac{\pi x}{l}\right) \tag{12.46}$$

The complementary function of (12.46) is

$$y = A\cos \alpha x + B \sin \alpha x \tag{12.47}$$

and the particular integral is

$$y = \frac{-(\pi^2/l^2)\Delta \sin(\pi x/l)}{(D^2 + \alpha^2)}$$

$$= \frac{-(\pi^2/l^2)\Delta \sin(\pi x/l)}{[(-\pi^2/l^2) + \alpha^2]} \tag{12.48}$$

From (12.47) and (12.48), the complete solution is

$$\underline{y = A \cos \alpha x + B \sin \alpha x - \frac{(\pi^2/l^2)\Delta \sin(\pi x/l)}{[(-\pi^2/l^2) + \alpha^2]}} \tag{12.49}$$

Two suitable boundary conditions are

$$@\ x = 0 \quad y = 0; \qquad \text{therefore } \underline{A = 0}$$

$$@\ x = l/2 \quad \frac{dy}{dx} = 0; \quad \text{therefore } \underline{B = 0}$$

i.e.

$$y = \frac{(\pi^2/l^2)\Delta \sin(\pi x/l)}{[(\pi^2/l^2) - \alpha^2]} = \frac{\Delta P_e \sin(\pi x/l)}{(P_e - P)} \tag{12.50}$$

The maximum deflection δ occurs at $x = l/2$:

$$\delta = \Delta P_e/(P_e - P) \tag{12.51}$$

and

$$\hat{M} = \Delta P P_e/(P_e - P) \tag{12.52}$$

where

$$\underline{P_e = \pi^2 EI/l^2}$$

EXAMPLE 12.6

Determine the maximum permissible value of Δ for an initially curved strut of sinusoidal curvature, given the following:

(a) Sectional and material properties are as in Example 12.4
(b) Length of strut is 10 m.
(c) The strut is pinned at its ends.

SOLUTION

Now,

$$\hat{\sigma} = \sigma(\text{direct}) \pm \sigma(\text{bending})$$

$$P_e = \pi^2 EI/l^2 = 1.302 \text{ MN}$$

$$P = 20\,000 \times 9.81 = 0.196 \text{ MN}$$

Therefore

$$\hat{\sigma} = -75 \times 10^6 = -21.04 \times 10^6 - \Delta \times \frac{1.302 \times 0.196 \times 10^6 \times 12.5 \times 10^{-2}}{1.106 \times 6.594 \times 10^{-5}}$$

or

$$53.96 \times 10^6 = 437.49\Delta \times 10^6$$

or

$$\underline{\Delta = 0.123 \text{ m} = 12.3 \text{ cm}}$$

12.10 Perry–Robertson formula

From Section 12.9, it can be seen that the maximum stress for a strut with initial sinusoidal curvature is given by

$$\sigma_c = \frac{P}{A} + \frac{\Delta P P_e \bar{y}}{(P_e - P)I} \tag{12.53}$$

Putting

$$\gamma = \Delta \bar{y}/k^2 \quad \text{and} \quad \sigma_e = P_e/A \tag{12.54}$$

where k is the least radius of gyration of the strut's cross-section, and substituting (12.54) into (12.53),

$$\sigma_c = \frac{P}{A} + \frac{\gamma(P/A)\sigma_e}{(\sigma_e - P/A)}$$

Therefore

$$\sigma_c\left(\sigma_e - \frac{P}{A}\right) = \frac{P}{A}\left(\sigma_e - \frac{P}{A}\right) + \gamma\left(\frac{P}{A}\right)\sigma_e$$

or

$$\left(\frac{P}{A}\right)^2 - [\sigma_c + (\gamma + 1)\sigma_e]\frac{P}{A} + \sigma_c\sigma_e = 0$$

i.e.

$$\underline{\left(\frac{P}{A}\right) = \tfrac{1}{2}[\sigma_c + (\gamma + 1)\sigma_e] - \tfrac{1}{2}\sqrt{[\sigma_c + (\gamma + 1)\sigma_e]^2 - 4\sigma_c\sigma_e}} \tag{12.55}$$

From BS 449

$$\underline{\gamma = 0.003(L_0/k)} \tag{12.56}$$

EXAMPLE 12.7

Calculate the maximum permissible value for P for the strut of Example 12.6, using γ from BS449.

SOLUTION

$$k = \sqrt{\frac{I}{A}} = 0.084 \text{ m} \quad L_0/k = 119.05$$

Therefore

$$\underline{\gamma = 0.357}$$

From (12.55)

$$\frac{P}{A} = \frac{1}{2}\left[75 \times 10^6 + 1.357 \times \frac{1.302 \times 10^6}{9.327 \times 10^{-3}}\right]$$

$$- \tfrac{1}{2}\sqrt{\{[75 \times 10^6 + 1.357 \times 139.59 \times 10^6]^2 - 4 \times 75 \times 10^6 \times 139.59 \times 10^6\}}$$

$$= 132.7 \times 10^6 - 83.73 \times 10^6$$

$$= 48.97 \times 10^6 \text{ N/m}^2$$

i.e.

$$P = 0.456 \text{ MN} = \underline{46.56 \text{ tonnes}}$$

12.11 Dynamic instability

The theory covered in this chapter has been based on static stability, but it must be pointed out that for struts subjected to compressive periodic axial forces, there is a possibility that dynamic instability can occur when the lateral critical frequency of the beam-column is reached.

The study of dynamic stability is beyond the scope of this book, but is dealt with in much detail in reference 30.

EXAMPLES FOR PRACTICE 12

1. Determine the Euler buckling load for the axially loaded strut of Figure 12.5 by the method of Section 12.3.

$\{P_e = \pi^2 EI/4l^2\}$

2. Determine the Euler buckling load for an initially straight axially loaded strut which is pinned at one end and fixed at the other.

$\{P_e = 20.2 EI/l^2\}$

3. Find the Euler crushing load for a hollow cylindrical cast-iron column of 0.15 m external diameter and 20 mm thick, if it is 6 m long and hinged at both ends. $E = 75 \times 10^9$ N/m². Compare this load with that given by the Rankine formula using constants of 540 MN/m² and 1/1600. For what length of column would these two formulae give the same crushing load?

{363.2 kN, 386.7 kN, 4.55 m}

4. A short steel tube of 0.1 m outside diameter, when tested in compression, was found to fail under an axial load of 800 kN. A 15 m length of the same tube when tested as a pin-jointed strut failed under a load of 30 kN. Assuming that the Euler and Rankine–Gordon formulae apply to the strut, calculate

(a) the tube inner diameter;
(b) the denominator constant in the Rankine–Gordon formula.

$E = 196.5$ GN/m^2

{(a) 0.0734 m; (b) 1/9114}

5. A steel pipe of 36 mm inner diameter, 6 mm thick and 1 m long is supported so that the ends are hinged, but all expansion is prevented. The pipe is unstressed at 0 °C. Calculate the temperature at which buckling will occur.

$\sigma_c = 325$ MN/m^2 $a = 1/7500$ $E = 200$ GN/m^2
$\alpha = 11.1 \times 10^{-6}$/°C

{68.4 °C}

6. The table below shows the results of a series of buckling tests carried out on a steel tube of external diameter 35 mm and internal diameter 25 mm.

Assuming the Rankine–Gordon formula to apply, determine the numerator and denominator constants for this tube.

l(mm)	600	1000	1400	1800
P_R(kN)	150	125	110	88

{350 MN/m^2, 1/36000}

7. The result of two tests on steel struts with pinned ends were found to be:

Test number	1	2
Slenderness ratio	50	80
Average stress at failure (MN/m^2)	266.7	194.4

$A = 1$ m^2

(a) Assuming that the Rankine–Gordon formula applies to both struts, determine the numerator and denominator constants of the Rankine–Gordon formula.
(b) If a steel bar of rectangular section 0.06 m × 0.019 m and of length 0.4 m is used as a strut with both ends clamped, determine the safe load using the constants derived in (a) and employing a safety factor of 4.

{(a) 350 MN/m^2, 1/8000; (b) 342 kN}

8. A long slender strut of length L is encastré at one end and pin jointed at the other. At its pinned end, it carries an axial load P, together with a couple M. Show that the magnitude of the couple at the clamped end is given by the expression

$$M\left(\frac{\alpha L - \sin \alpha L}{\alpha L \cos \alpha L - \sin \alpha L}\right)$$

Determine the value of this couple if P is one-quarter of the Euler buckling load for this class of strut.

{−0.672 m}

9. A long strut, initially straight, securely fixed at one end and free at the other, is loaded at the free end with an eccentric load whose line of action is parallel to the original axis. Deduce an expression for the deviation of the free end from its original position.

{$\Delta(\sec \alpha L - 1)$ where $\alpha = \sqrt{P/EI}$
Δ = eccentricity}

10. A tubular steel strut of 70 mm external diameter and 50 mm internal diameter is 3.25 m long. The line of action of the compressive forces is parallel to, but eccentric from, the axis of the tube, as shown in Figure 12.12.

Find the maximum allowable eccentricity of these forces if the maximum permissible deflection (total) is not greater than 15 mm.

$E = 2 \times 10^{11}$ N/m^2 $P = 114.7$ kN

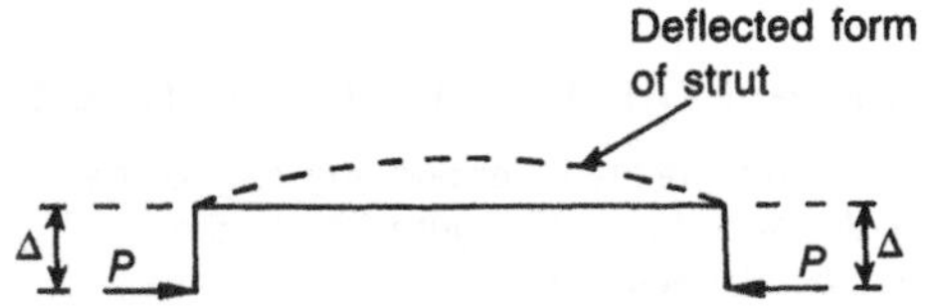

Figure 12.12

{$\Delta = 5$ mm}

11. The eccentrically loaded strut of Figure 12.13 is subjected to a compressive load P. If $EI = 20\,000$ N m^2, determine the position and value of the maximum deflection assuming the following data apply:

$P = 5000$ N $l = 3$ m $\Delta = 0.01$ m

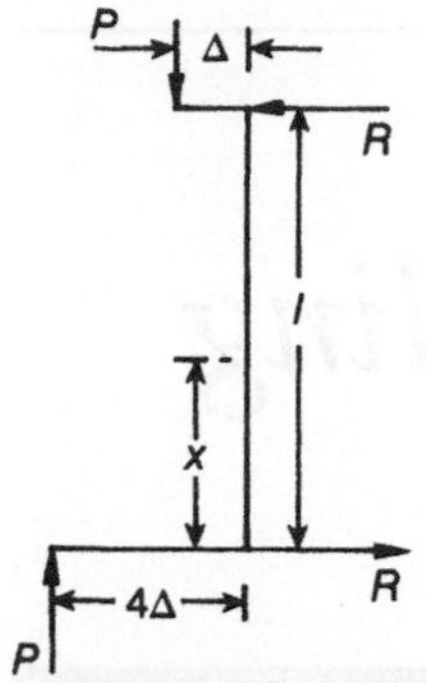

Figure 12.13

{$x = 1.38$ m, $\delta = 9.21E-3$ m}

12. Show that for the eccentrically loaded strut of Figure 12.14 the bending moment at any distance x is given by

$$M = P\Delta\left(-2\cos\alpha x + \frac{(1 + 2\cos\alpha l)}{\sin\alpha l}\sin\alpha x\right)$$

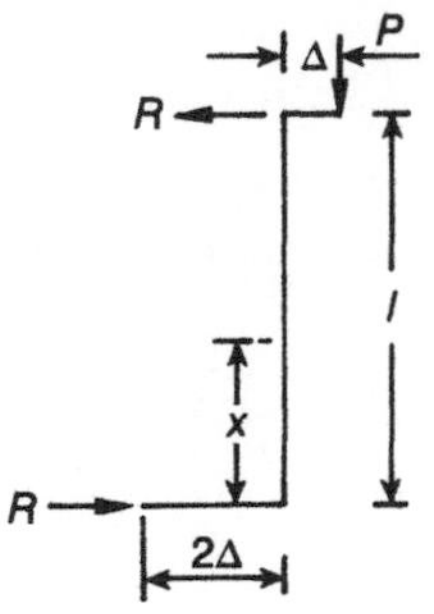

Figure 12.14

13. An initially curved strut, whose initial deflected form is small and parabolic, is symmetrical about its mid-point. If the strut is subjected to a compressive axial load P at its pinned ends, show that the maximum compressive stress is given by

$$\frac{P}{A}\left[1 + \frac{\Delta\bar{y}}{k^2}\frac{8EI}{Pl^2}\left(\sec\frac{\alpha l}{2} - 1\right)\right]$$

where

Δ = initial central deflection
k = least radius of gyration

Determine Δ for such a strut, assuming the geometrical and material properties of Example 12.6 apply.

{12.3 cm}

14. An initially curved strut, whose initial deflected form is small and circular, is subjected to a compressive axial load P at its pinned ends. Show that the total deflection y at any distance x is given by

$$y = -\frac{8\Delta}{\alpha^2 l^2}\left[\cos\alpha x + \tan(\alpha l/2)\sin\alpha x - 1\right]$$

where Δ is the initial central deflection.

Determine Δ for such a strut, assuming the geometrical and material properties of Example 12.6 apply.

{12.3 cm}

13

Unsymmetrical bending of beams

13.1 Introduction

The two most common forms of unsymmetrical (or asymmetrical) bending of straight beams are as follows:

(a) when symmetrical-section beams are subjected to unsymmetrical loads, as shown by Figure 13.1;
(b) when unsymmetrical-section beams are subjected to either symmetrical or unsymmetrical loading, as shown in Figure 13.2.

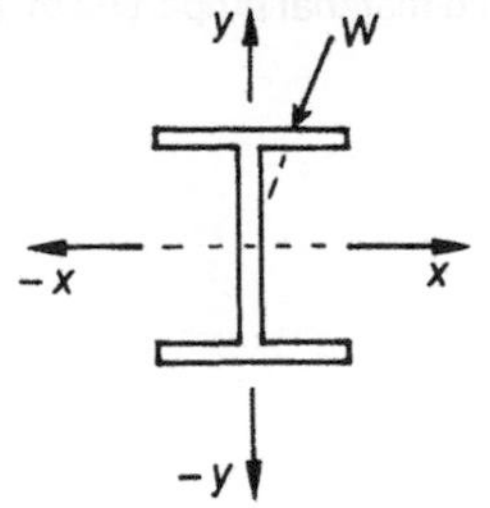

Figure 13.1 A beam of symmetrical section, subjected to an asymmetrical load

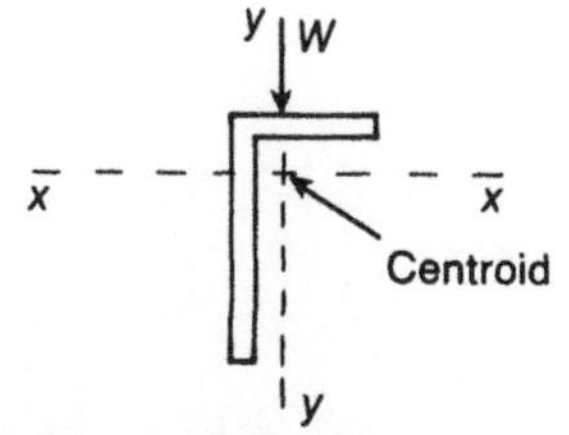

Figure 13.2 A beam of unsymmetrical section, subjected to a vertical load

13.2 Symmetrical-section beams loaded asymmetrically

In the case of the symmetrical-section beam which is loaded asymmetrically, the skew load of Figure 13.3(a) can be resolved into two components mutually perpendicular to each other and acting along the axes of symmetry, as shown in Figures 13.3(b) and (c). Assuming that the beam behaves in a linear elastic manner, the effects of bending can be considered separately, about each of the two axes of symmetry, namely xx and yy, and later the effects of each component of W can be superimposed to give the resultant stresses and deflections.

To demonstrate the process, let us assume that the beam of Figure 13.3(a) is of length l, and is simply supported at its ends, with W at mid-span.

It can readily be seen that the components of W are $W \cos \alpha$, acting along the y axis, and $W \sin \alpha$, acting along the x axis, where the former causes the beam to bend about its x–x axis, and the latter causes bending about the y–y axis.

The effect of $W \cos \alpha$ will be to cause the stress in the flange AB, namely $\sigma_{y(AB)}$, to be compressive, whilst the stress in the flange CD, namely $\sigma_{y(CD)}$, will be tensile, so that

$$\sigma_{y(AB)} = -\frac{W \cos(\alpha) l \bar{y}}{4 I_{xx}} \tag{13.1}$$

and

$$\sigma_{y(CD)} = \frac{W \cos(\alpha) l \bar{y}}{4 I_{xx}} \tag{13.2}$$

Similarly, owing to $W \sin \alpha$, the stress on the flange edges B and D, namely $\sigma_{x(BD)}$, will be compressive, whilst the stress on the flange edges A and C, namely $\sigma_{x(AC)}$, is tensile, so that

$$\sigma_{x(BD)} = -\frac{W \sin(\alpha)/\bar{x}}{4 I_{yy}} \tag{13.3}$$

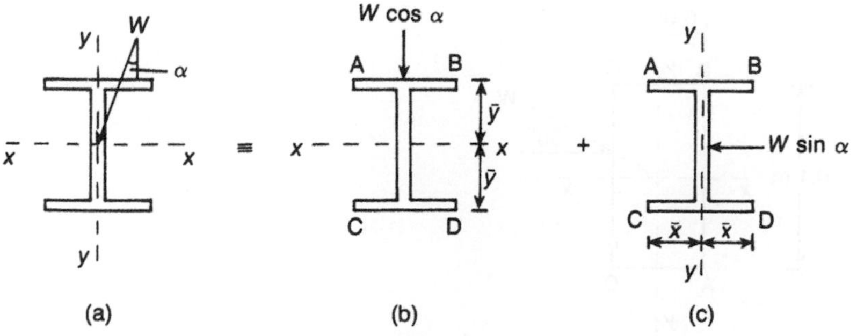

Figure 13.3 Symmetrical beam, loaded asymmetrically

and

$$\sigma_{x(AC)} = \frac{W \sin(\alpha) l \bar{x}}{4 I_{xx}} \tag{13.4}$$

The combined effects of $W \cos \alpha$ and $W \sin \alpha$ will be such that the magnitude of the maximum stresses will be largest at the points B and C, where at point B the stress is compressive, whilst at point C the stress is tensile and of the same magnitude as the stress at point B.

At the points A and D, the effects of $W \cos \alpha$ will be to cause stresses of opposite sign to the stresses caused by $W \sin \alpha$, so that the magnitude of these stresses will be less than those at the points B and C.

Thus, in general, the stress at any point in the section of the beam is given by

$$\sigma = \frac{M \cos(\alpha) y}{I_{xx}} + \frac{M \sin(\alpha) x}{I_{yy}} \tag{13.5}$$

where $\bar{x}$ and $\bar{y}$ are perpendicular distances of the outermost fibers from yy and xx, respectively, and

I_{xx} = second moment of area of section about xx

I_{yy} = second moment of area of section about yy

M is a bending moment due to W and l, and x and y are as defined in Figure 13.1.

EXAMPLE 13.1

A cantilever of length 2 m and of rectangular section 0.1 m × 0.05 m is subjected to a skew load of 5 kN, at its free end, as shown in Figure 13.4. Determine the stresses at A, B, C and D at its fixed end.

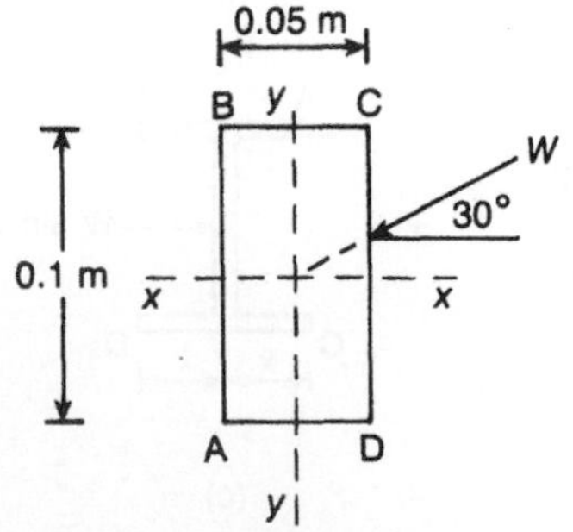

Figure 13.4 Rectangular-section beam

SOLUTION

M_y = bending moment about xx

$= W\sin 30° \times 2 = 5$ kN m

M_x = bending moment about yy

$= W\cos 30° \times 2 = 8.66$ kN m

I_{xx} = second moment of area of the cross-section about xx

$$= \frac{0.05 \times 0.1^3}{12} = 4.167\text{E-6 m}^4$$

I_{yy} = second moment of area of the cross-section about yy

$$= \frac{0.1 \times 0.05^3}{12} = 1.042\text{E-6 m}^4$$

σ_A = maximum stress at the point A

$$= -\frac{5\text{E}3 \times 0.05}{4.167\text{E-6}} - \frac{8.66\text{E}3 \times 0.025}{1.042\text{E-6}}$$

$\underline{\sigma_A = -60 - 207.9 = -267.9 \text{ MN/m}^2}$

$\underline{\sigma_B = +60 - 207.9 = -147.9 \text{ MN/m}^2}$

$\underline{\sigma_C = 60 + 207.9 = 267.9 \text{ MN/m}^2}$

$\underline{\sigma_D = -60 + 207.9 = 147.9 \text{ MN/m}^2}$

13.3 Unsymmetrical sections

To demonstrate the more complex problem of the bending of an unsymmetrical section, consider the cantilever of Figure 13.5, which is of 'Z' section.

If the symmetrical theory of bending is applied to the cantilever of Figure 13.5,

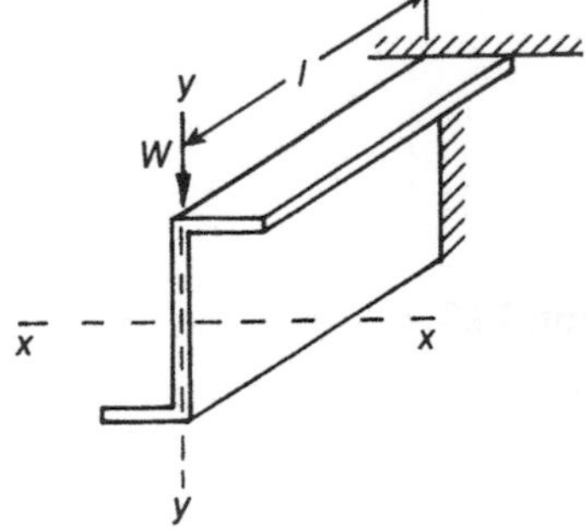

Figure 13.5 Cross-section that is unsymmetrical about *yy*

then at the built-in end, the stress in the top flange would be uniform and tensile, whilst the stress in the bottom flange would be uniform and compressive.

Thus, according to the theory of bending of symmetrical sections, the resisting couple due to the stresses would balance the bending moment about *xx* due to *W*, but if such a stress system existed, it would also cause a resisting couple about *yy*, which is impossible, as there is not applied bending moment about *yy*.

It is evident, therefore, that the theory of bending for symmetrical sections cannot be applied to unsymmetrical sections, and the mathematical explanation for this is as follows:

If

$$M_y = \text{bending moment about } yy$$

and

$$M_x = \text{bending moment about } xx$$

then according to simple bending theory

$$M_x = \sum \sigma \times y \times \delta a \tag{13.6}$$

and

$$M_y = \sum \sigma \times x \times \delta a \tag{13.7}$$

where

$$\begin{aligned} \sigma &= \text{stress due to bending} = Ey/R \\ &= \text{constant} \times y \qquad (13.8) \\ \delta a &= \text{elemental area} \end{aligned}$$

Substituting equation (13.8) into equations (13.6) and (13.7),

$$\begin{aligned} M_x &= \sum \text{constant} \times y^2 \times \delta a = \text{constant} \times \sum y^2 \times \delta a \\ &= \text{constant} \times I_{xx} \qquad (13.9) \end{aligned}$$

and

$$\begin{aligned} M_y &= \sum \text{constant} \times x \times y \times \delta a \\ &= \text{constant} \times I_{xy} \qquad (13.10) \end{aligned}$$

where

I_{xy} = the product of inertia

However, from the heuristic arguments of Section 13.3

$$M_y = 0$$

Therefore

Thus the only way that simple bending theory can be satisfied for unsymmetrical sections is for the beam to bend about those two mutually perpendicular axes where the product of inertia is zero.

It will now be shown that these two axes are in fact the *principal axes of bending* of the section, rather similar to the axes of principal stresses and principal strains, as discussed in Chapter 7.

13.4 Calculation of I_{xy}

Two elements will be considered in this section, namely a rectangle and the quadrant of a circle.

Consider a rectangle of area A in the positive quadrant of the Cartesian co-ordinate system of Figure 13.6.

Let

$\bar{h}$ = distance of the centroid of the rectangular element from Oy

$\bar{k}$ = distance of the centroid of the rectangular element from Ox

Now, from equation (13.10),

$$I_{xy} = \int xy \, da$$

which for the rectangular element for Figure 13.6 becomes

$$I_{xy} = \iint (dx \times dy)x \times y$$

$$= \int_{\bar{k}-d/2}^{\bar{k}+d/2} \int_{\bar{h}-b/2}^{\bar{h}+b/2} xy \, dx \, dy$$

$$= bd\bar{h}\bar{k}$$

Therefore

$$\underline{I_{xy} = A\bar{h}\bar{k}} \qquad (13.11)$$

where A is the area of the rectangle. Hence, for a built-up section, consisting of n

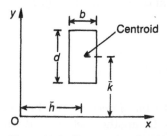

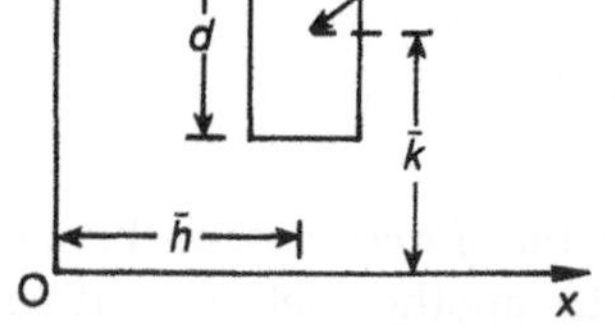

Figure 13.6 Rectangular element

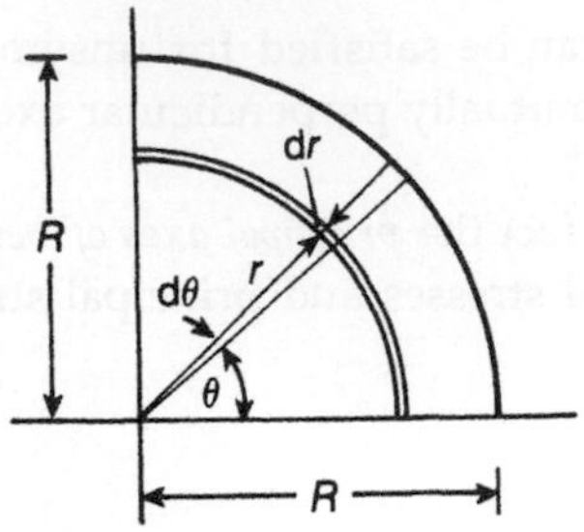

Figure 13.7 Quadrant of a circle

rectangles, equation (13.11) becomes

$$I_{xy} = \sum_{i=1}^{n} A_i \bar{h}_i \bar{k}_i \tag{13.12}$$

For the *quadrant of the circle* in Figure 13.7,

$$I_{xy} = \int xy \times da$$

$$= \int_0^{\pi/2} \int_0^R r \cos\theta \times r \sin\theta \times dr \times r\, d\theta$$

$$= \int_0^{\pi/2} \int_0^R r^3 \cos\theta \sin\theta\, dr\, d\theta$$

$$= \frac{R^4}{4} \int_0^{\pi/2} \cos\theta d(-\cos\theta)$$

$$= \frac{R^4}{4} \left[-\frac{\cos^2\theta}{2} \right]_0^{\pi/2}$$

$$\underline{I_{xy} = R^4/8}$$

13.5 Principal axes of bending

Let O*U* and O*V* be the principal axes, and O*x* and O*y* be the reference (or global) axes, as shown in Figure 13.8.

From Figure 13.8, it can be seen that

$$u = x\cos\theta + y\sin\theta \tag{13.13}$$

$$v = y\cos\theta - x\sin\theta \tag{13.14}$$

Equations (13.13) and (13.14) are in fact *co-ordinate transformations*, as these equations transform one set of orthogonal co-ordinates to another set of orthogonal co-ordinates.

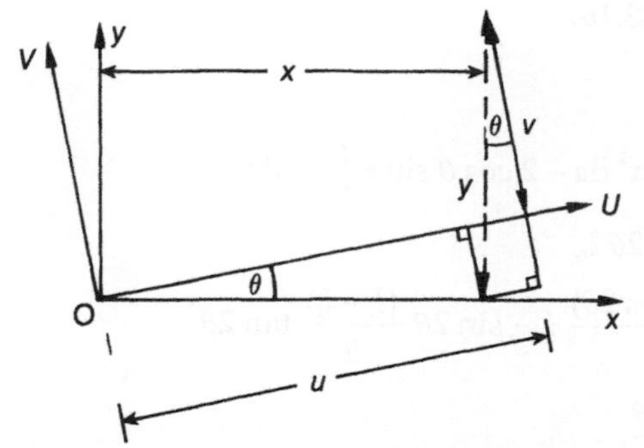

Figure 13.8 Principal and reference axes

Now to satisfy simple bending theory and equilibrium, it is necessary for the product of inertia to be zero, with reference to the principal axes of bending, namely the axes OU and OV, i.e.

$$I_{UV} = \int uv \, da = 0$$

or

$$\int (x \cos\theta + y \sin\theta) \times (y \cos\theta - x \sin\theta) \times da = 0$$

or

$$\int xy \cos^2\theta \, da - \int x^2 \cos\theta \sin\theta \, da + \int y^2 \sin\theta \cos\theta \, da - \int xy \sin^2\theta \, da$$

$$= (\cos^2\theta - \sin^2\theta) \int xy \, da + \cos\theta \sin\theta \left(\int y^2 \, da - \int x^2 \, da \right)$$

$$= \cos 2\theta I_{xy} + \tfrac{1}{2} \sin 2\theta (I_x - I_y) = 0$$

Therefore

$$\tan 2\theta = \frac{2I_{xy}}{(I_y - I_x)} \tag{13.15}$$

13.5.1 *To determine* I_U *and* I_V*, the principal second moments of area*

Now,

$$I_U = \int v^2 \, da \tag{13.16}$$

Substituting equation (13.14) into (13.16),

$$I_U = \int (y\cos\theta - x\sin\theta)^2\,da$$

$$= \cos^2\theta \int y^2\,da + \sin^2\theta \int x^2\,da - 2\cos\theta\sin\theta \int xy\,da$$

$$= \cos^2\theta\, I_x + \sin^2\theta\, I_y - \sin 2\theta\, I_{xy}$$

$$= \frac{(1+\cos 2\theta)}{2} I_x + \frac{(1-\cos 2\theta)}{2} I_y - \sin 2\theta \frac{(I_y - I_x)}{2}\tan 2\theta$$

$$\underline{I_U = \tfrac{1}{2}(I_x + I_y) + \tfrac{1}{2}(I_x - I_y)\sec 2\theta} \tag{13.17}$$

Similarly,

$$I_V = \int u^2\,da$$

$$= \cos^2\theta\, I_y + \sin^2\theta\, I_x + \sin 2\theta\, I_{xy}$$

$$\underline{I_V = \tfrac{1}{2}(I_x + I_y) - \tfrac{1}{2}(I_x - I_y)\sec 2\theta} \tag{13.18}$$

If equations (13.17) and (13.18) are added together, it can be seen that

$$I_U + I_V = I_x + I_y \tag{13.19}$$

Equation (13.19) is known as the *invariant of inertia.*

13.6 Mohr's circle of inertia

Equations (13.15), (13.17) and (13.18) can be represented by a circle of inertia, as shown by Figure 13.9, rather similar to Mohr's circles for stress and strain, as discussed in Chapter 7.

From Figure 13.9 it can readily be seen that I_U and I_V are maximum and minimum values of second moments of area, and this is why I_U and I_V are called principal second moments of area.

Plots of the variation of the radius of gyration in any direction are called *momental ellipses*, and typical momental ellipses are shown in Figure 13.10.

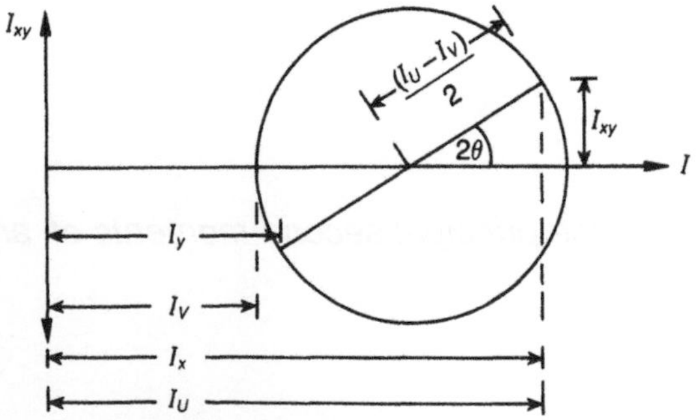

Figure 13.9 Mohr's circle of inertia

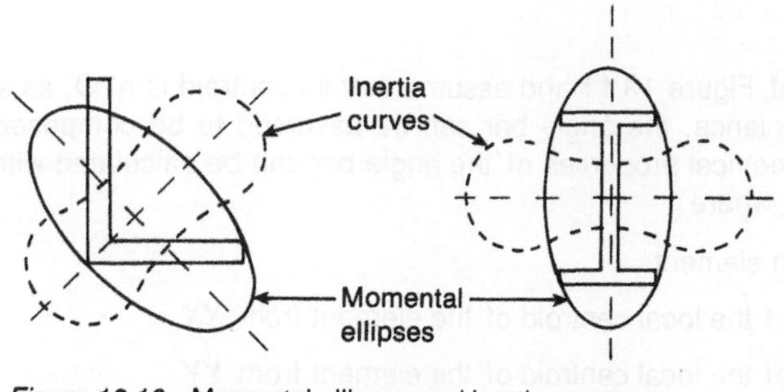

Figure 13.10 Momental ellipses and inertia curves

Structural engineers sometimes find another set of curves useful, which are known as *inertia curves*. The inertia curve of a section is obtained by plotting a radius vector equal to the second moment of area of the section at any angle θ, as shown by the dashed lines of Figure 13.10. From this Figure, it can be seen that the inertia curves are effectively plotted in a direction perpendicular to the momental ellipses.

EXAMPLE 13.2

Determine the directions and values of the principal second moments of area of the asymmetrical sections of Figures 13.11 and 13.12.

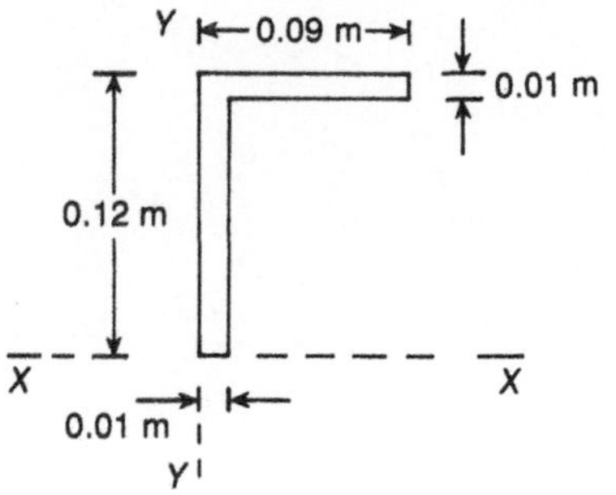

Figure 13.11

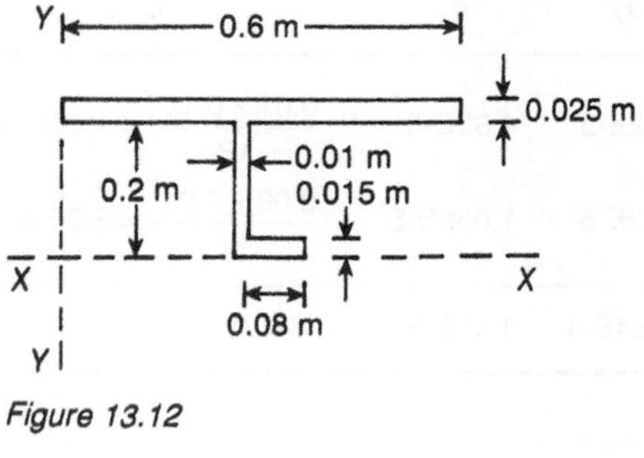

Figure 13.12

SOLUTION

Consider the angle bar of Figure 13.11 and assume that its centroid is at O, as shown in Figure 13.13. For convenience, the angle bar can be assumed to be composed of two rectangles, and the geometrical properties of the angle bar can be calculated with the aid of Tables 13.1 and 13.2, where

a = area of an element

y = distance of the local centroid of the element from XX

x = distance of the local centroid of the element from YY

i_H = second moment of area of an element about an axis passing through its local centroid and parallel to XX

i_v = second moment of area of an element about an axis passing through its local centroid and parallel to YY

ay = the product $a \times y$

ay^2 = the product $a \times y \times y$

ax = the product $a \times x$

ax^2 = the product $a \times x \times x$

Σ = summation of column, where appropriate

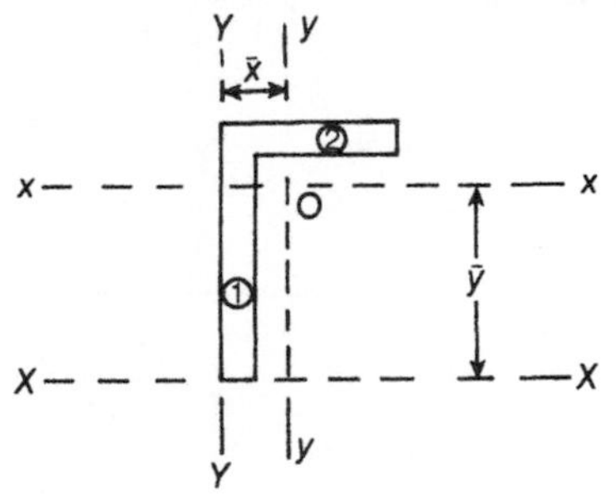

Figure 13.13

Table 13.1 **Calculation of I_{xx}**

Section	a	y	ay	ay^2	i_H
①	1.2E-3	0.06	7.2E-5	4.32E-6	$\frac{0.01 \times 0.12^3}{12} = 1.44\text{E-}6$
②	8E-4	0.115	9.2E-5	1.058E-5	$\frac{0.08 \times 0.01^3}{12} = 6.7\text{E-}9$
Σ	2E-3	–	1.64E-4	1.49E-5	1.447E-6

From Table 13.1,

$$\bar{y} = \frac{\Sigma\, ay}{\Sigma\, a} = \frac{1.64\text{E-4}}{2\text{E-3}} = \underline{0.082 \text{ m}}$$

$$I_{XX} = \sum i_H + \sum ay^2 = 1.49\text{E-5} + 1.447\text{E-6}$$

$$\underline{I_{XX} = 1.635\text{E-5 m}^4}$$

$$I_{xx} = I_{XX} - \bar{y}^2 \sum a = 1.635\text{E-5} - 0.082^2 \times 2\text{E-3}$$

$$\underline{I_{xx} = 2.902\text{E-6 m}^4} \qquad (13.20)$$

From 13.2,

$$\bar{x} = \frac{\Sigma\, ax}{\Sigma\, a} = \frac{4.6\text{E-5}}{2\text{E-3}} = \underline{0.023 \text{ m}}$$

$$I_{YY} = \sum i_V + \sum ax^2 = 4.37\text{E-7} + 2.03\text{E-6}$$

$$\underline{I_{YY} = 2.467\text{E-6 m}^4}$$

$$I_{yy} = I_{YY} - \bar{x}^2 \sum a = 2.467\text{E-6} - 0.023^2 \times 2\text{E-3}$$

$$\underline{I_{yy} = 1.409\text{E-6 m}^4} \qquad (13.21)$$

To calculate I_{xy}

From equation (13.12),

$$I_{xy} = \Sigma A\bar{h}\bar{k}$$

$$= 1.2\text{E-3} * [-(\bar{x} - 0.005)] \times [-(\bar{y} - 0.06)]$$

$$+ 8\text{E-4} \times (0.05 - \bar{x}) \times (0.115 - \bar{y})$$

$$= 1.2\text{E-3} \times 0.018 \times 0.022 + 8\text{E-4} \times 0.027 \times 0.033$$

$$\underline{I_{xy} = 1.188\text{E-6 m}^4} \qquad (13.22)$$

To calculate θ

Substituting the appropriate values from equations (13.20) to (13.22) into equation

Table 13.2 **Calculation of I_{yy}**

Section	a	x	ax	ax^2	i_V
①	1.2E-3	0.005	6E-6	3E-8	$\frac{0.12 \times 0.01^3}{12} = 1\text{E-8}$
②	8E-4	0.05	4E-5	2E-6	$\frac{0.01 \times 0.08^3}{12} = 4.27\text{E-7}$
Σ	2E-3	–	4.6E-5	2.03E-6	4.37E-7

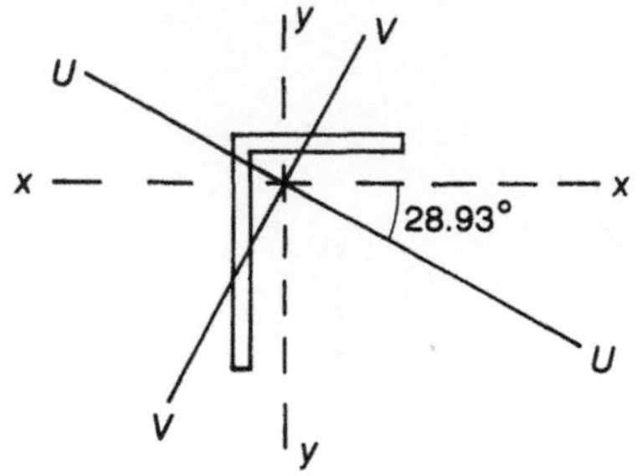

Figure 13.14 Directions of the principal axes

(13.15),

$$\tan 2\theta = \frac{2 \times 1.188\text{E-6}}{(1.409\text{E-6} - 2.902\text{E-6})}$$

$$2\theta = -57.86°$$

$$\underline{\theta = -28.93°} \quad (13.23)$$

where θ is shown in Figure 13.14,

Now,

$$I_U = \tfrac{1}{2}(I_x + I_y) + \tfrac{1}{2}(I_x - I_y)\sec 2\theta$$

and

$$I_V = \tfrac{1}{2}(I_x + I_y) - \tfrac{1}{2}(I_x - I_y)\sec 2\theta$$

Hence, by substituting the values for I_x, I_y, I_{xy} and θ into the above,

$$I_U = \tfrac{1}{2}(2.902 + 1.409)\text{E-6} + \tfrac{1}{2}(2.902 - 1.409)\text{E-6} \times \frac{1}{\cos(-57.86°)}$$

$$= 2.156\text{E-6} + 7.465\text{E-7} \times 1.88$$

$$\underline{I_U = 3.56\text{E-6 m}^4} \quad (13.24)$$

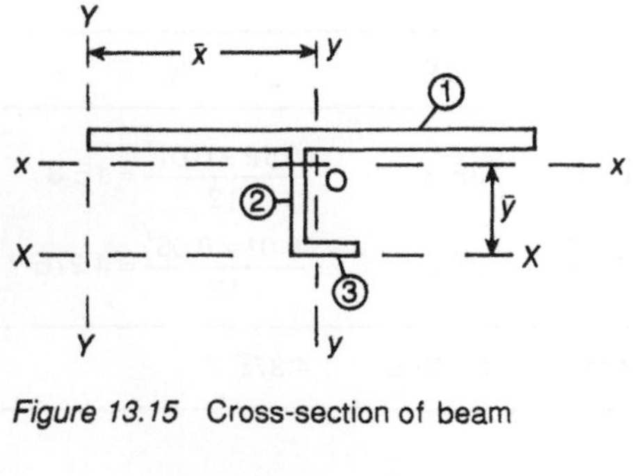

Figure 13.15 Cross-section of beam

and

$$I_V = 2.156\text{E-6} - 1.403\text{E-6}$$

$$\underline{I_V = 7.53\text{E-7 m}^4} \qquad (13.25)$$

Let the centroid of the section in Figure 13.12 be at O, as shown in Figure 13.15.

Tables 13.3 and 13.4 show the calculations for determining I_x and I_y of this section, where the symbols have the same meanings as described above.

From Table 13.3,

$$\bar{y} = \frac{\Sigma\, ax}{\Sigma\, a} = \underline{0.1866 \text{ m}}$$

$$I_{XX} = \sum i_H + \sum ay^2 = \underline{7.049\text{E-4 m}^4}$$

$$I_{xx} = I_{XX} - \bar{y}^2 \sum a = \underline{7.118\text{E-5 m}^4} \qquad (13.26)$$

From Table 13.4,

$$\bar{x} = \frac{\Sigma\, ax}{\Sigma\, a} = \underline{0.303 \text{ m}}$$

$$I_{YY} = \sum i_V + \sum ax^2 = \underline{2.124\text{E-3 m}^4}$$

$$I_{yy} = I_{YY} - \bar{x}^2 \sum a = \underline{4.531\text{E-4 m}^4} \qquad (13.27)$$

Table 13.3 **Calculation for I_x**

Section	a	y	ay	ay^2	i_H
①	0.015	0.2125	3.188E-3	6.773E-4	7.81E-7
②	2E-3	0.1	2E-4	2E-5	6.667E-6
③	1.2E-3	7.5E-3	9E-6	6.75E-8	2.25E-8
Σ	0.0182	–	3.397E-3	6.974E-4	7.471E-6

Table 13.4 **Calculation for I_y**

Section	a	x	ax	ax^2	i_V
①	0.015	0.3	4.5E-3	1.35E-3	4.5E-4
②	2E-3	0.3	6E-4	1.8E-4	1.67E-8
③	1.2E-3	0.345	4.14E-4	1.428E-4	6.4E-7
Σ	0.0182	–	5.514E-3	1.673E-3	4.51E-4

$$I_{xy} = \sum A\overline{hk}$$

$$= 0.015 \times (0.3 - \bar{x}) \times (0.2125 - \bar{y})$$

$$+2E\text{-}3 \times (0.3 - \bar{x}) \times (0.1 - \bar{y})$$

$$+1.2E\text{-}3 \times (0.345 - \bar{x}) \times (7.5E\text{-}3 - \bar{y})$$

$$= -1.166E\text{-}6 + 5.196E\text{-}7 - 9.02E\text{-}6$$

$$\underline{I_{xy} = -9.673E\text{-}6\ m^4} \qquad (13.28)$$

Thus,

$$2\theta = \tan^{-1}\left(\frac{2I_{xy}}{I_y - I_x}\right) = \tan^{-1}\left(\frac{1.935E\text{-}5}{3.819E\text{-}4}\right)$$

$$\underline{\theta = -1.45°}$$

Now,

$$I_U = \tfrac{1}{2}(I_x + I_y) + \tfrac{1}{2}(I_x - I_y)\sec 2\theta$$

$$= \tfrac{1}{2}(7.118E\text{-}5 + 4.531E\text{-}4) + \tfrac{1}{2}(7.118E\text{-}5 - 4.531E\text{-}4) \times \frac{1}{\cos(-2.9)}$$

$$\underline{I_U = 7.095E\text{-}5\ m^4}$$

Similarly,

$$\underline{I_V = 4.533E\text{-}4\ m^4}$$

13.7 Stresses in beams of asymmetrical section

Equation (13.5) can be extended to the bending of unsymmetrical sections, as shown by equation (13.29) which gives the value of bending stress at any point P in the positive quadrant for the orthogonal axes O*U* and O*V*, where θ, u and v are defined in Figure 13.16:

$$\sigma = \frac{M\cos\theta\, v}{I_{UU}} + \frac{M\sin\theta\, u}{I_{VV}} \qquad (13.29)$$

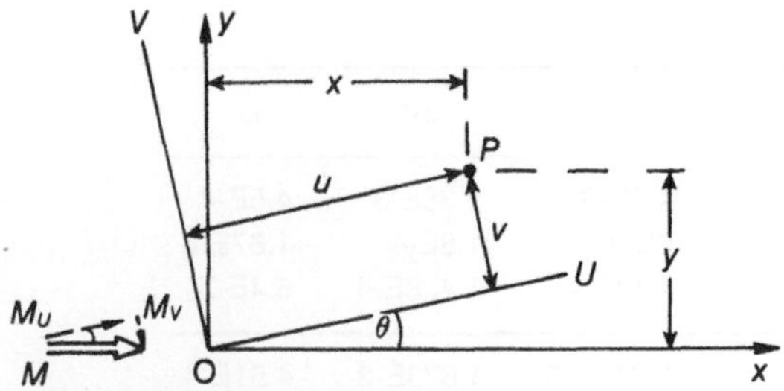

Figure 13.16 Principal axes of bending, O*u* and O*v*

From equation (13.29), it can be seen that M is due to a load that acts perpendicular to the Ox axis, where in Figure 13.16, M is shown according to the right-hand screw rule. The components of M, namely M_U and M_V, cause bending about the principal axes Ou and Ov, respectively, where

$$M_U = M\cos\theta$$

and

$$M_V = M\sin\theta$$

EXAMPLE 13.3

A cantilever of length 2 m is subjected to a concentrated load of 5 kN at its free end, as shown in Figure 13.17. Assuming that the cantilever's cross-section is as shown in Figure 13.11, determine the direction of its neutral axis and the position and magnitude of the maximum stress.

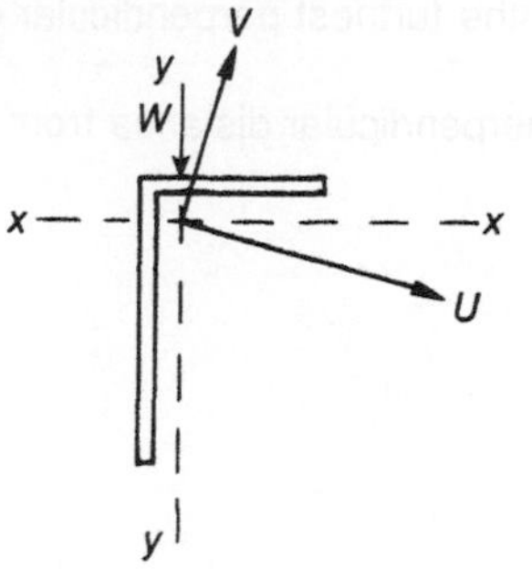

Figure 13.17 Cross-section of cantilever

SOLUTION

The maximum bending moment ($\hat{M}$) occurs at the built-in end, where

$$\hat{M} = Wl = 5\text{ kN} \times 2\text{ m} = 10\text{ kN m} \tag{13.30}$$

At the neutral axis, the bending stress is zero, i.e.

$$\frac{M\cos\theta\, v}{I_{UU}} + \frac{M\sin\theta\, u}{I_{VV}} = 0 \tag{13.31}$$

or

$$\frac{v}{u} = -\frac{I_{UU}}{I_{VV}}\tan\theta$$

$$= \tan\beta$$

where β is defined in Figure 13.18. Therefore

$$\underline{\beta = \tan^{-1}(-I_{UU} \times \tan\theta / I_{VV})} \tag{13.32}$$

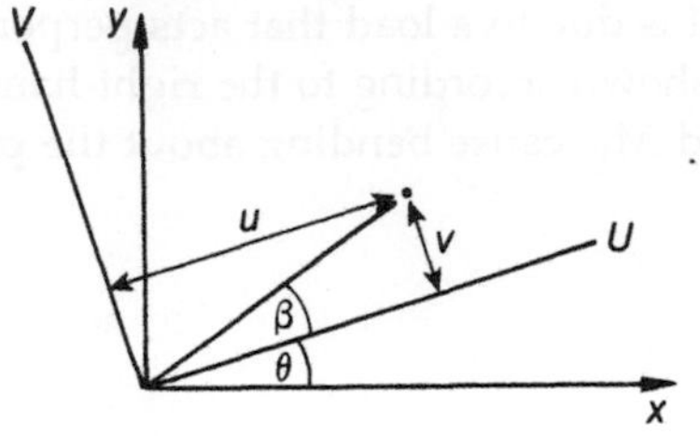

Figure 13.18 Definition of β

Substituting the appropriate values from Section 13.6 into equation (13.32),

$$\beta = \tan^{-1}[-3.56\text{E-}6 \times (-0.5527)/7.53\text{E-}7]$$

$$\underline{\beta = 69.06°}$$

The direction of the neutral axis is shown in Figure 13.19, and the largest stress due to bending will occur at a point on the section which is at the furthest perpendicular distance from the neutral axis (NA).

From Figure 13.19, it can be seen that the furthest perpendicular distance from NA is at the point B.

From equations (13.13) and (13.14),

$$u_B = x_B \cos\theta + y_B \sin\theta$$

$$= -0.013 \times 0.875 - 0.083 \times (-0.483)$$

$$\underline{u_B = 0.029 \text{ m}}$$

and

$$v_B = y_B \cos\theta - x_B \sin\theta$$

$$= -0.083 \times 0.875 - 0.013 \times 0.483$$

$$\underline{v_B = -0.0789 \text{ m}}$$

where u_B, v_B, etc., are defined in Figure 13.19.

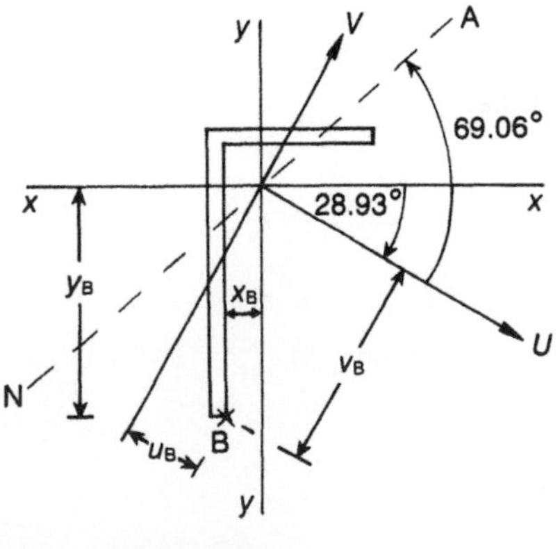

Figure 13.19 Position of neutral axis

From equation (13.27),

$$\sigma_B = \text{stress at the point B}$$

$$= \frac{10 \text{ kN m} \times 0.875 \times (-0.0789)\text{m}}{3.56\text{E-6 m}^4} + \frac{10 \text{ kN m} \times (-0.483) \times (0.029)\text{m}}{7.53\text{E-7 m}^4}$$

$$= -(193.9 + 186)\text{MN/m}^2$$

$$\underline{\sigma_B = -380 \text{ MPa(compressive)}}$$

EXAMPLE 13.4

If an encastré beam of length 2 m and with a cross-section as in Figure 13.12 is subjected to the uniformly distributed load shown in Figure 13.20, determine the position and value of the maximum stress.

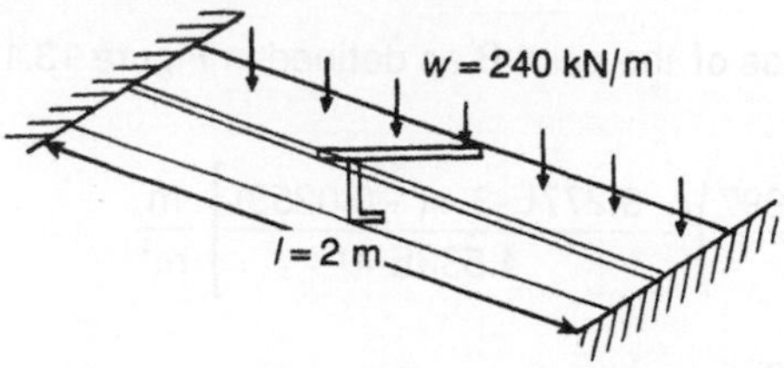

Figure 13.20 Encastré beam with a uniformly distributed load

SOLUTION

The maximum value for bending moment ($\hat{M}$) occurs at the ends, and is given by

$$\hat{M} = \frac{wl^2}{12} = \frac{240 \times 2^2}{12} = \underline{80 \text{ kN m}}$$

From equation (13.32),

$$\beta = \tan^{-1}(-I_{UU} \times \tan\theta / I_{VV})$$

$$= \tan^{-1}[-7.095\text{E-6} \times (-0.0253)/4.533\text{E-4}]$$

$$\underline{\beta = 0.23^\circ} \tag{13.33}$$

The position of the neutral axis (NA) is shown in Figure 13.21.

By inspection, it can be seen that the point of maximum stress occurs at B which is the furthest perpendicular distance from the neutral axis.

Prior to calculating σ_B, the stress at the point B, the distances u_B and v_B have to be determined, as follows:

$$u_B = x_B \cos\theta + y_B \sin\theta$$

$$= (0.295 - 0.303) \times 0.9997 + 0.1866 \times 0.0253$$

$$= -7.998\text{E-3} + 4.721\text{E-3}$$

$$\underline{u_B = -3.277\text{E-3 m}} \tag{13.34}$$

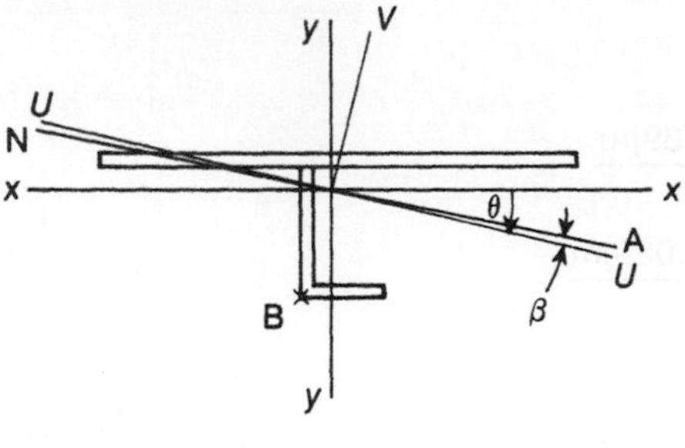

Figure 13.21 Position of neutral axis for encastré beam

$$v_B = x_B \sin\theta - y_B \cos\theta$$
$$= -8E\text{-}3 \times (-0.0253) - 0.1866 \times (0.9997)$$
$$\underline{v_B = -0.1867\ \text{m}} \qquad (13.35)$$

where, u_B, v_B, x_B and y_B are co-ordinates of the point B as defined in Figure 13.19.

From equation (13.29),

$$\sigma_B = 80\ \text{kN m}\left[\left(\frac{-0.1867 \times 0.9997}{7.095E\text{-}5}\right) - \frac{3.277E\text{-}3 \times (-0.0253)}{4.533E\text{-}4}\right]\frac{\text{m}}{\text{m}^4}$$

$$= \frac{80\ \text{kN}}{\text{m}^2} \times (-2631.3 + 0.18)$$

$$\underline{\sigma_B = -210.5\ \text{MN/m}^2\ \text{(compressive)}}$$

EXAMPLE 13.5

Determine the end deflection of the cantilever of Example 13.3, given that

$$E = 2E11\ \text{N/m}^2$$

SOLUTION

In this case, the components of load can be resolved along the two principal axes of bending, and each component of deflection can then be calculated. Hence, the resultant deflection can be obtained.

Now, from Chapter 5 the maximum deflection (δ) of an end-loaded cantilever is given by

$$\delta = \frac{Wl^3}{3EI}$$

where

W = load

l = length of cantilever

E = elastic modulus

I = second moment of area of cantilever section about its axis of bending

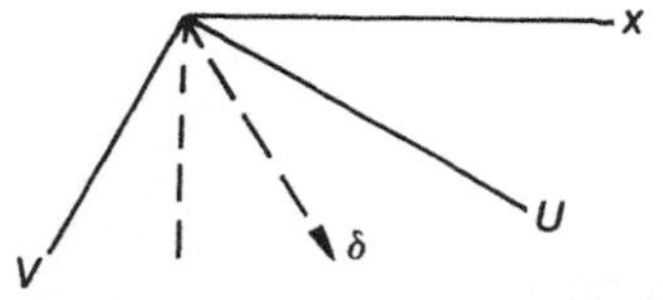

Figure 13.22

For the present problem,

δ_u = deflection under load in the u direction

$$= -\frac{W \sin \theta l^3}{3EI_V} = \frac{10 \times 10^3 \times 0.483 \times 8}{3 \times 2 \times 10^{11} \times 7.53E-7} = \underline{0.0855 \text{ m}}$$

and

δ_v = deflection under load in the v direction

$$= -\frac{W \cos \theta l^3}{3EI_U}$$

Therefore

$$\delta_V = \frac{-10 \times 10^3 \times 0.8752 \times 8}{3 \times 2 \times 10^{11} \times 3.56E\text{-}6} = \underline{-0.0328 \text{ m}}$$

These two components of deflection can be drawn as shown in Figure 13.22. Hence, from Pythagoras's theorem and from elementary trigonometry,

$$\underline{\delta = \sqrt{\delta_u^2 + \delta_v^2} = 0.0916}$$

at angle of

$$\tan^{-1}\left\{\frac{\delta_v}{\delta_u}\right. = -20.98° \text{ from the } OU \text{ axis}$$

EXAMPLE 13.6

Determine the central deflection of the encastré beam of Example 13.4, given that $E = 2 \times 10^{11}$ N/m².

SOLUTION

From Chapter 5 the maximum deflection of an encastré beam with a uniformly distributed load is given by

$$\delta = \frac{wl^4}{384EI}$$

where

w = load/unit length

l = length

E = Young's modulus

I = second moment of area

For the present problem,

$$\delta_u = \text{central deflection in the } u \text{ direction} = \frac{w \sin \theta l^4}{384EI_V}$$

$$\delta_v = \text{central deflection in the } v \text{ direction} = -\frac{w \cos \theta l^4}{384EI_U}$$

$$\delta_u = \frac{240 \times 10^3 \times (-0.0253) \times 16}{384 \times 2 \times 10^{11} \times 4.533\text{E-}4} = \underline{2.791\text{E-}3 \text{ mm}}$$

$$\delta_v = -\frac{240 \times 10^3 \times 0.9997 \times 16}{384 \times 2 \times 10^{11} \times 7.095\text{E-}5} = \underline{-0.7045 \text{ mm}}$$

The resultant deflection is

$$\underline{\delta = 0.7045 \text{ m at } 1.22°}$$

(clockwise) from the vertical.

NB The theory in this chapter does not include the additional effects of shear stresses due to torsion and bending that occur when unsymmetrical beams are loaded through their centroids, but these theories are dealt with in some detail in Chapters 6 and 14.

EXAMPLES FOR PRACTICE 13

1. Determine the direction and magnitudes of the principal second moments of area of the angle bar shown in Figure 13.23.

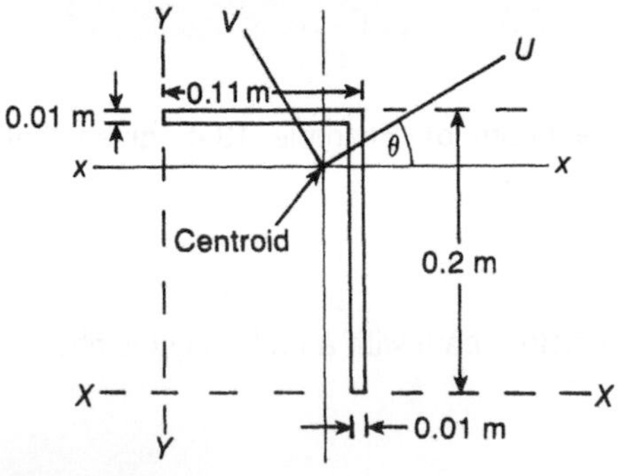

Figure 13.23

{I_{xx} = 1.27E-5 m^4, I_{yy} = 2.849E-6 m^4, I_{xy} = −3.843E-6 m^4, θ = 17.64°, I_{UU} = 1.381E-5 m^4, I_{VV} = 1.741E-6 m^4}

2. Determine the direction and magnitudes of the principal second moments of area of the section of Figure 13.24.

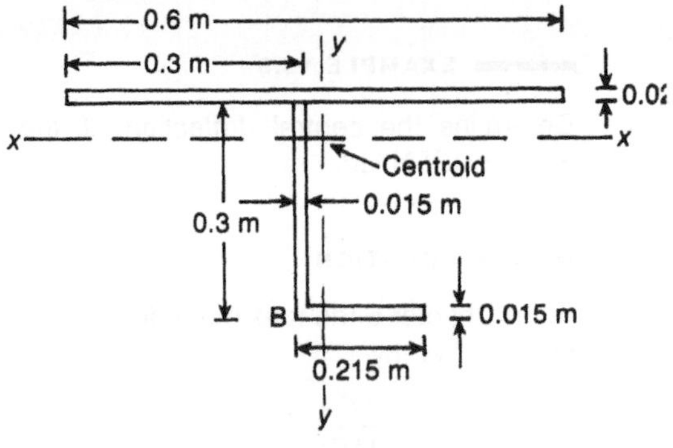

Figure 13.24

{I_{xx} = 2.844E-4 m^4, I_{yy} = 3.994E-4 m^4, I_{xy} = −7.097E-5 m^4, I_U = 1.592E-4 m^4, I_V = 5.246E-4 m^4, θ = −25.49°}

3. Determine the stresses at the corners and the maximum deflection of a cantilever of length 3 m, loaded at its free end with a concentrated load of 10 kN, as shown in Figure 13.25.

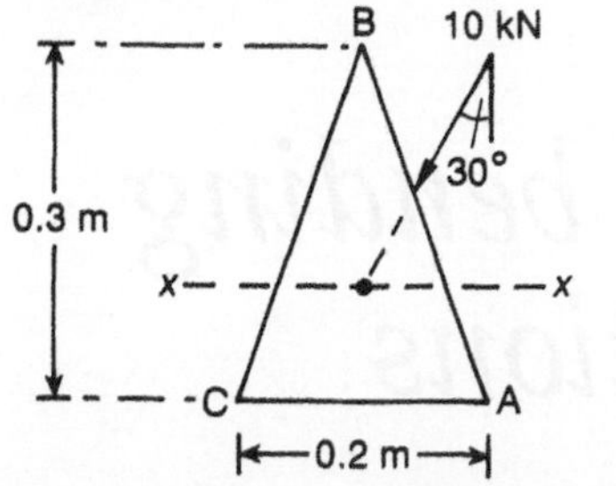

Figure 13.25 Cross-section of cantilever

{σ_A = 12.68 MPa, σ_B = 34.64 MPa, σ_C = −47.32 MPa, neutral axis is 60° clockwise from xx}

4. If a cantilever of length 3 m, and with a cross-section as shown in Figure 13.23, is subjected to a vertically applied downward load at its free end, of magnitude 4 kN, determine the position and value of the maximum bending stress.

{NA is −68.38° from UU, σ_B = −163.63 MPa}

5. A simply supported beam of length 4 m, and with a cross-section as shown in Figure 13.24, is subjected to a centrally placed concentrated load of 20 kN, acting perpendicularly to the xx axis, and through the centroid of the beam. Determine the stress at mid-span at the point B in the cross-section of the beam.

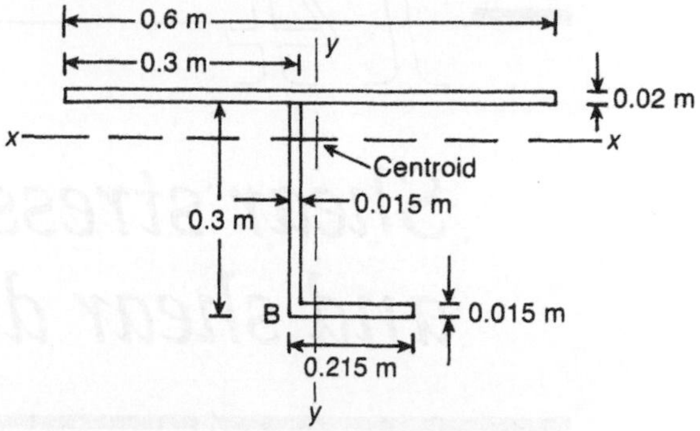

Figure 13.24

{NA is 8.23° from UU, σ_B = −19.82 MPa}

6. Determine the components of deflection under the load in the directions of the principal axes of bending for the beam of Example 4, given that

E = 2E11 N/m²

{δ_u = −0.031 m, δ_v = −0.0124 m}

7. Determine the components of the central deflection in the directions of the principal axes of bending for the beam of Example 5, given that

E = 1E11 N/m²

{δ_u = 2.19E-4 m, δ_v = −1.51E-3 m}

14

Shear stresses in bending and shear deflections

14.1 Introduction

If a horizontal beam is subjected to transversely applied vertical loads, so that the bending moment changes, then there will be vertical shearing forces at every point along the length of the beam where the bending moment changes (see Chapter 1).

Furthermore, if the cross-section of the beam is of a built-up section, such as a rolled steel joist (RSJ), or a tee beam or a channel section, then there will be horizontal shearing stresses in addition to the vertical shearing stresses, where both are caused by the same vertically applied shearing forces. Vertical shearing stresses due to bending are those that act in a vertical plane, and horizontal shearing stresses are those that act in a horizontal plane. Later in this chapter, it will be shown that for curved sections such as split tubes etc., the shearing stresses due to bending act in the planes of the curved sections. Similar arguments apply to horizontal beams subjected to laterally applied horizontal shearing forces.

To demonstrate the concept of shearing stresses due to bending, the variation of vertical shearing stresses across the section of a beam will be first considered, followed by a consideration of the variation of horizontal shearing stresses due to the same vertical shearing forces.

14.2 Vertical shearing stresses

Consider a beam subjected to a system of vertical loads, so that the bending moment changes along the length of the beam, as shown in Figure 14.1.

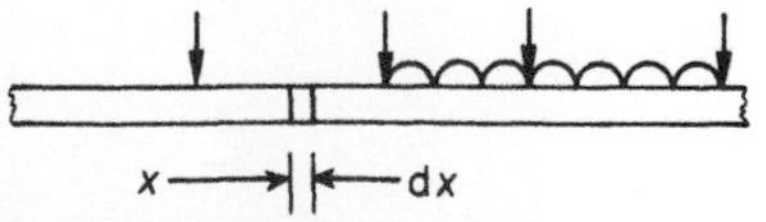

Figure 14.1 Beam subjected to vertical loads

Consider an elemental length dx of the beam, as shown in Figure 14.2(a). Consider the sub-element of Figure 14.2(b), and also the sub-sub-element in the same figure, which is shown by the heavy line. Let the stress on the left of the sub-sub-element of Figure 14.2(b) be σ, and the stress on the right of the sub-sub-element be $\sigma + d\sigma$, due to M and $M + dM$, respectively.

From Figure 14.2(b), it can be seen that there is an apparent unbalanced force in the x direction, of magnitude $d\sigma \times Z \times dy$, but as the sub-sub-element is in equilibrium, this is not possible, i.e. the only way equilibrium can be achieved is for a longitudinal horizontal stress to act tangentially along the rectangular face ABCD. It is evident that as this stress acts tangentially along the face ABCD, it must be a shearing stress.

Let τ be the shearing stress on the face ABCD. Resolving horizontally,

$$\tau \times b \times dx = \int d\sigma \times dA$$

$$= \int \frac{dM \times y \times dA}{I}$$

where

$$dA = Z \times dy$$

or

$$\tau = \frac{dM}{dx}\frac{1}{bI}\int y\,dA$$

but

$$\frac{dM}{dx} = F = \text{shearing force at } x$$

Therefore

$$\tau = \frac{F}{bI}\int y\,dA \qquad (14.1)$$

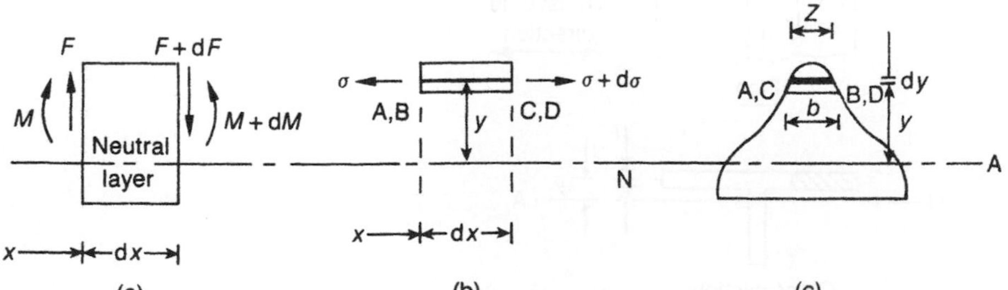

Figure 14.2 Beam element under shearing force: (a) element; (b) sub-element; (c) cross-section

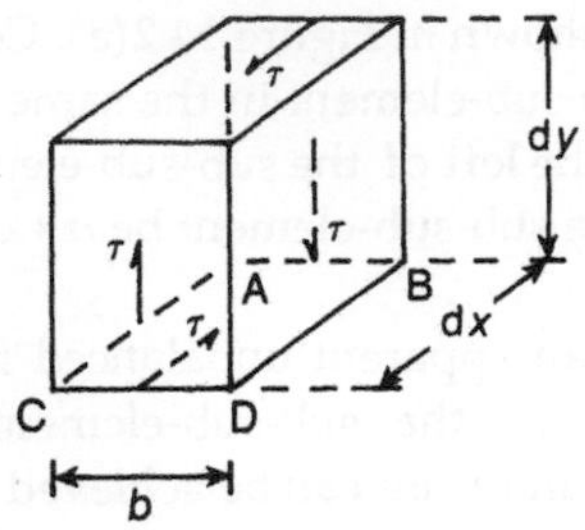

Figure 14.3 Complementary shearing stresses in a vertical plane

where, in this case, $\int y \, dA$ is the first moment of area of the section *above the plane* ABCD, and about the neutral axis, NA.

Owing to complementary shearing stresses [1], the shearing stress on the face ABCD will be accompanied by three other shearing stresses of the same magnitude, as shown in Figure 14.3.

From Figure 14.3, it can be seen that these four complementary shearing stresses act together in a vertical plane, and this is why these shearing stresses are called vertical shearing stresses.

14.3 Horizontal shearing stresses

Consider a horizontal top flange, of thickness t, on a horizontal beam under the action of vertical shearing forces, as shown in Figure 14.4.

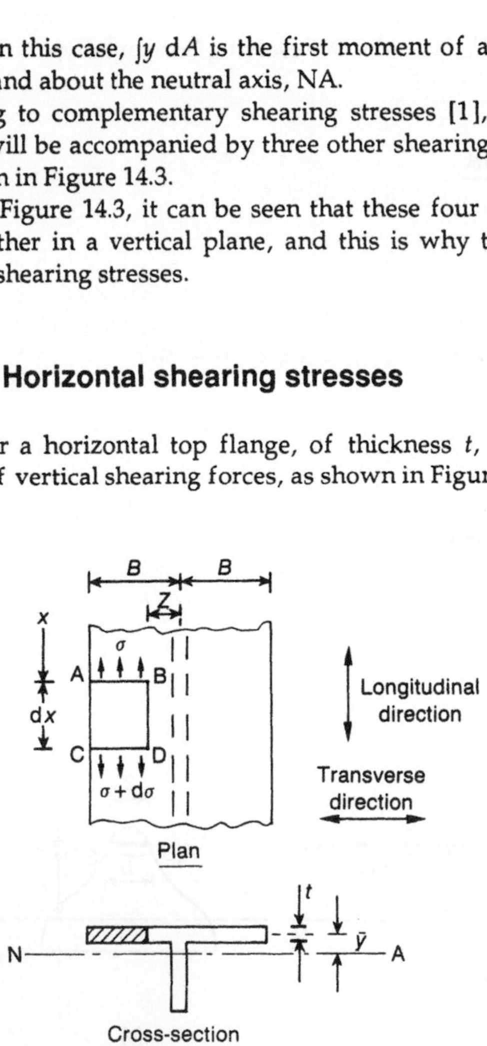

Figure 14.4 Horizontal flange on a beam

Let

$$M = \text{bending moment at AB}$$
$$M + dM = \text{bending moment at CD}$$
$$\sigma = \text{bending stress in flange at AB}$$
$$\sigma + d\sigma = \text{bending stress in flange at CD}$$

Consider the equilibrium of an element of the flange, ABCD, in the x direction. From the plan view of Figure 14.4, it can be seen that the apparent unbalanced force in the longitudinal direction on this element is $d\sigma \times t \times (B - Z)$. However, as the beam is in equilibrium, no such unbalanced force can exist, and the only way that equilibrium can be achieved in this flange element is for a horizontal shearing stress to act tangentially along BD.

Let τ be the shearing stress on the face BD. Then, considering horizontal equilibrium of the element ABCD in the x direction,

$$d\sigma \times t \times (B - Z) = \tau \times dx \times t$$

or

$$\frac{dM \times \bar{y} \times t \times (B - Z)}{I} = \tau \times dx \times t$$

or

$$\tau = \frac{dM}{dx}\frac{(B - Z)\bar{y}}{I}$$

but

$$\frac{dM}{dx} = F = \text{the vertical shearing force}$$

Therefore

$$\tau = \frac{F(B - Z)\bar{y}}{I} \qquad (14.2)$$

Equation (14.2) can be obtained directly from equation (14.1), as follows:

$$\tau = \frac{F\int y\,dA}{bI}$$

which, for the horizontal flange element ABCD of Figure 14.4, becomes

$$\tau = \frac{F}{tI} \times (B - Z) \times t \times \bar{y} = \frac{F(B - Z)\bar{y}}{I}$$

which is identical to equation (14.2).

It should be noted that, in this case, $\int y\,dA$ was the first moment of area of the flange element ABCD about NA, and b was, in fact, the flange thickness t. The

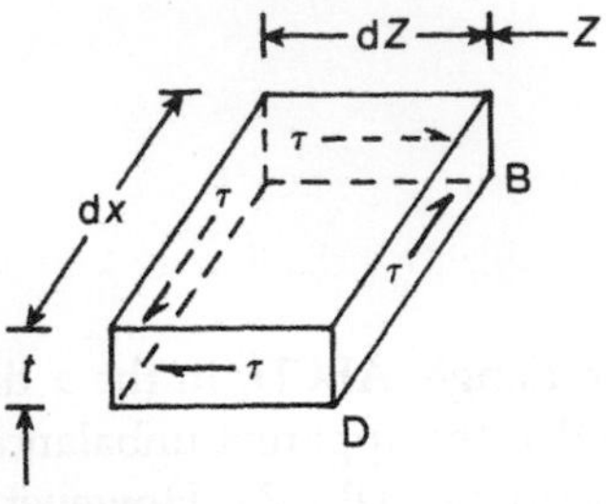

Figure 14.5 Complementary shearing stresses in a horizontal plane

reason for applying equation (14.1) in this way was because the shearing stresses in the flange act in a horizontal plane, as shown in Figure 14.5.

The horizontal shearing stress distribution of equation (14.2) can be seen to vary linearly along the width of the flange, having a maximum value at the centre of the flange, and zero values at the flange edges.

In fact, owing to the effects of complementary shearing stresses, this shearing stress will be accompanied by three other shearing stresses, where all four shearing stresses act together in a horizontal plane, as shown in Figure 14.5.

EXAMPLE 14.1

Calculate and sketch the vertical shearing stress distribution in a horizontal beam of rectangular section, subjected to a transverse vertical shearing force F, as shown in Figure 14.6.

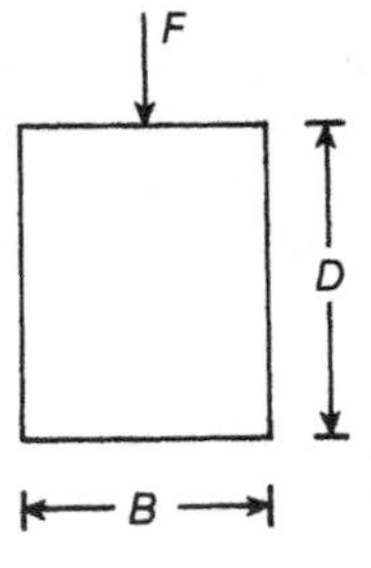

Figure 14.6 Rectangular section under a shearing force F

SOLUTION

From equation (14.1), τ the vertical shearing stress at any point y from NA (Figure 14.7), is given by

$$\tau = \frac{F\int y\,\mathrm{d}A}{bI}$$

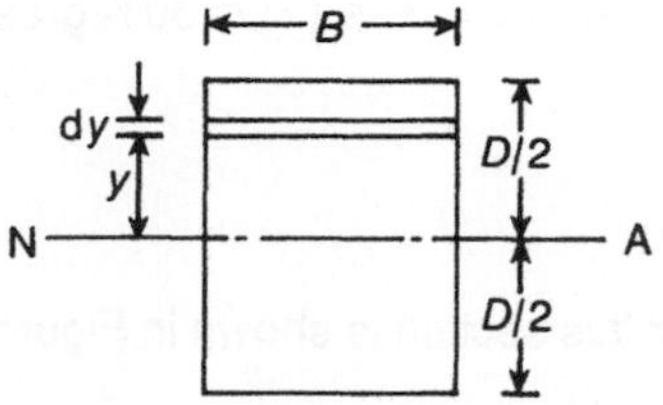

Figure 14.7 Rectangular section

Now,

$$I = BD^3/12$$

and for this case,

$$b = B$$

Therefore

$$\tau = \frac{12 \times F}{B \times BD^3} \int_y^{D/2} B \times dy \times y$$

$$= \frac{12F}{BD^3} \left[\frac{y^2}{2} \right]_y^{D/2}$$

$$\tau = \frac{6F}{BD^3} \left(\frac{D^2}{4} - y^2 \right) \tag{14.3}$$

Equation (14.3) can be seen to vary parabolically, having a maximum value at NA, and being zero at the top and bottom.

Let $\hat{\tau}$ be the maximum shearing stress. Then

$$\hat{\tau} = \frac{6F}{BD^3} \times \frac{D^2}{4}$$

$$\hat{\tau} = \frac{1.5F}{BD} \tag{14.4}$$

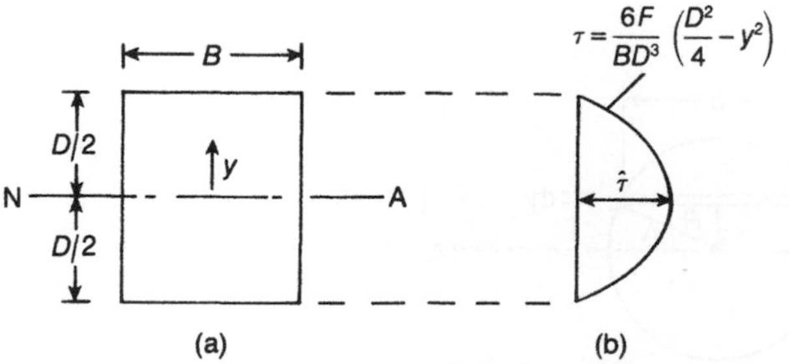

Figure 14.8 Shearing stress distribution in a rectangular section: (a) section; (b) plot of τ

Equation (14.4) shows that the maximum value of shearing stress ($\hat{\tau}$) is 50% greater than the average value (τ_{av}) where

$$\tau_{av} = \frac{F}{BD}$$

A plot of the variation of vertical shearing stress for this section is shown in Figure 14.8.

EXAMPLE 14.2

Calculate and sketch the vertical shearing stress distribution in a horizontal beam of circular section, subjected to the shearing force F, shown in Figure 14.9.

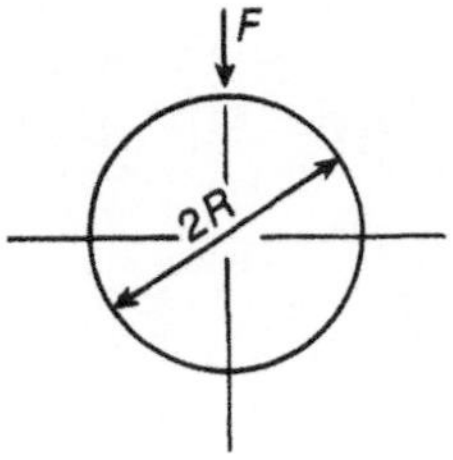

Figure 14.9 Circular section

SOLUTION

Let τ be the vertical shearing stress at any distance y from NA (Figure 14.10). Then

$$\tau = \frac{F}{bI}\int y\,dA$$

$$\tau = \frac{F}{bI}\int_y^R b\,dy\,y$$

but

$$I = \pi R^4/4$$

$$b = 2R\cos\theta$$

and

$$y = R\sin\theta$$

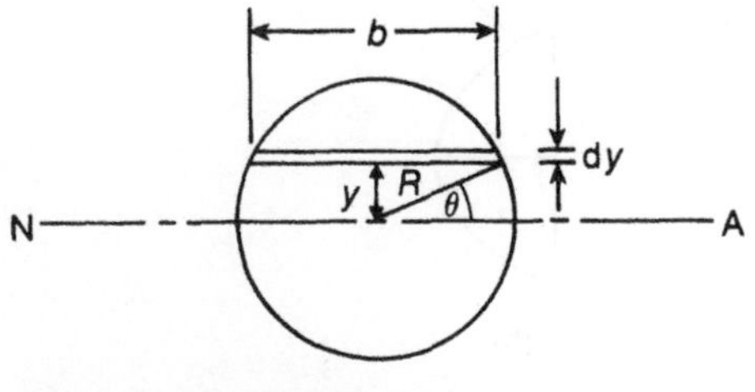

Figure 14.10 Circular section

By differentiation,

$$dy = R\cos\theta\, d\theta$$

Therefore

$$\tau = \frac{4F}{2\pi R^5 \cos\theta}\int_\theta^{\pi/2} 2R\cos\theta\, R\sin\theta\, R\cos\theta\, d\theta$$

$$= \frac{-4F}{\pi R^2 \cos\theta}\int_\theta^{\pi/2} \cos^2\theta (d\cos\theta)$$

$$= \frac{-4F}{\pi R^2 \cos\theta}\left[\frac{\cos^3\theta}{3}\right]_\theta^{\pi/2}$$

$$\tau = \frac{4F\cos^2\theta}{3\pi R^2} = \frac{4F}{3\pi R^2}(1 - \sin^2\theta)$$

but $\sin\theta = y/R$; therefore

$$\tau = \frac{4F}{3\pi R^2}\left[1 - \left(\frac{y}{R}\right)^2\right] \quad (14.5)$$

which, again, is a parabolic distribution, as shown in Figure 14.11.

Let $\hat{\tau}$ be the maximum value of shearing stress. Then

$$\hat{\tau} = \frac{4F}{3\pi R^2}$$

which can be seen to be about 33.3% larger than the average shearing stress τ_{av}, where

$$\tau_{av} = \frac{F}{\pi R^2}$$

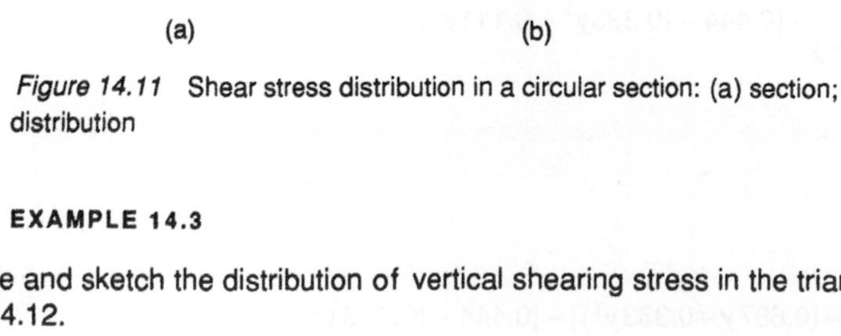

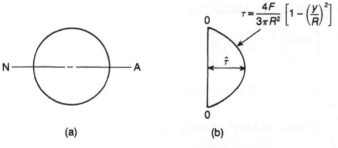

Figure 14.11 Shear stress distribution in a circular section: (a) section; (b) shear stress distribution

EXAMPLE 14.3

Calculate and sketch the distribution of vertical shearing stress in the triangular section of Figure 14.12.

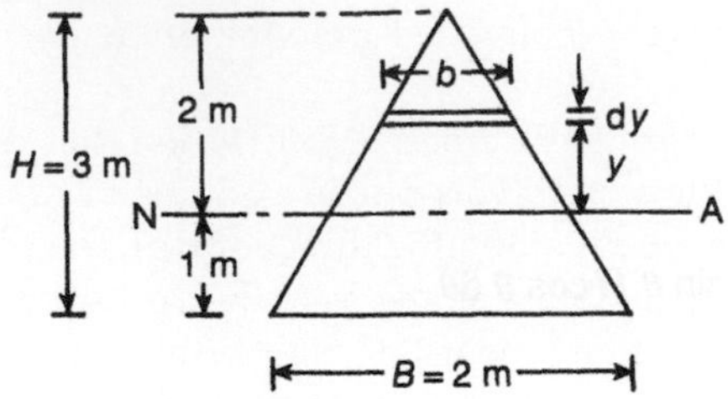

Figure 14.12 Triangular section

SOLUTION

At any distance y above NA in the triangular section of Figure 14.12,

$$\tau = \frac{F\int y\,dA}{bI}$$

but

$$I = BH^3/36 = 2\times 3^3/36 = 1.5$$

and

$$b = B\left(\frac{2}{3} - \frac{y}{H}\right) = 2(0.667 - 0.333y)$$

Therefore

$$\tau = \frac{F}{2(0.667 - 0.333y)\times 1.5}\int b\,dy\,y$$

$$= \frac{F}{(2-y)}\int 2(0.667y - 0.333y^2)\,dy$$

$$= \frac{2F}{(2-y)}\left[\frac{0.667y^2}{2} - 0.111y^3\right]_y^2$$

$$= \frac{2F}{(2-y)}\{[1.333 - 0.88889] - [0.333y^2 - 0.111y^3]\}$$

$$\tau = \frac{2F}{(2-y)}[0.444 - (0.333y^2 - 0.111y^3)] \qquad (14.6)$$

For maximum τ,

$$\frac{d\tau}{dy} = 0$$

Therefore

$$(2-y)[-(0.667y - 0.333y^2)] - [0.444 - (0.333y^2 - 0.111y^3)](-1)$$

or

$$-(2-y)(0.667y-0.333y^2)+(0.444-0.333y^2+0.111y^3)=0$$
$$-1.333y+0.667y^2+0.667y^2-0.333y^3+0.444-0.333y^2+0.111y^3=0$$
$$-0.222y^3+y^2-1.333y+0.444=0$$
$$0.222y^3-y^2+1.333y-0.444=0 \qquad (14.7)$$

Equation (14.7) has three roots, but for this case the root of interest is the lowest positive root, which can be obtained by the Newton–Raphson [31] iterative process, as follows:

$$y_{(i)}=y_{(i-1)}-\frac{f(y_{i-1})}{f'(y_{i-1})}$$

where,

y_i = ith approximation – to be determined
y_{i-1} = known approximation at the $i-1$ stage

In this case,

$$f(y)=0.222y^3-y^2+1.333y-0.444$$

and

$$f'(y)=\frac{d[f(y)]}{dy}=0.667y^2-2y+1.333$$

For the *first approximation*, let

$$y=0$$

or

$$y(0)=0$$

Therefore

$$y(1)=0+\frac{0.444}{1.333}=0.333$$

Second approximation:

$$y(2)=0.333-\frac{8.198\text{E-}3-0.111+0.444-0.444}{0.0741-0.667+1.333}$$
$$=0.333+\frac{0.103}{0.741}=0.472$$

Third approximation:

$$y(3)=0.472-\frac{0.023-0.222+0.629-0.444}{0.149-0.944+1.333}$$
$$=0.472+\frac{0.014}{0.538}=0.498$$

Fourth approximation:

$$y(4) = 0.498 - \frac{0.027 - 0.248 + 0.664 - 0.444}{0.165 - 0.996 + 1.333}$$

$$= 0.498 + \frac{1E\text{-}3}{0.502} = \underline{0.5\ \text{m}}$$

As the relative difference between y(4) and $y(3)$ is small, it can be assumed that

$$\underline{y = 0.5\ \text{m}}$$

$$\hat{\tau} = \frac{2F}{1.5}[0.444 - (0.0833 - 0.01388)] = 0.5F$$

At NA (i.e. $y = 0$),

$$\underline{\tau = 0.444F}$$

In this case, the average shear stress is

$$\tau_{av} = F/(2 \times 3/2)$$
$$= 0.333F$$

A sketch of the distribution of vertical shearing stress across the section is shown in Figure 14.13.

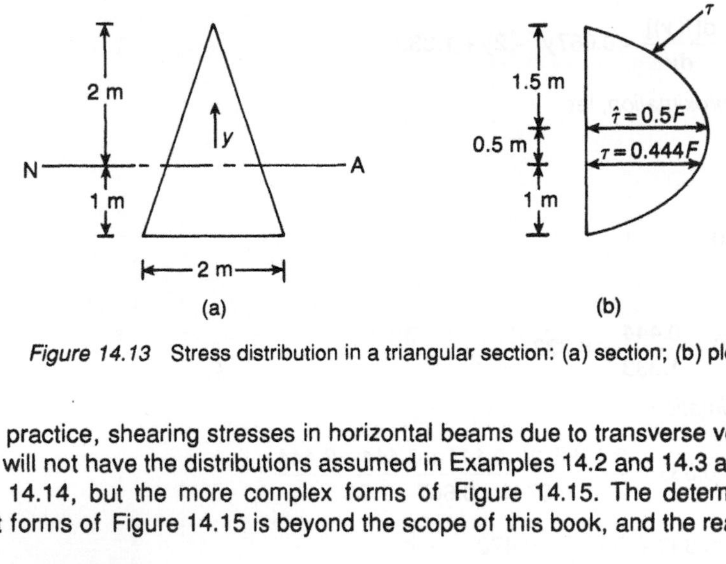

Figure 14.13 Stress distribution in a triangular section: (a) section; (b) plot of τ

NB In practice, shearing stresses in horizontal beams due to transverse vertical shearing forces will not have the distributions assumed in Examples 14.2 and 14.3 and as shown in Figure 14.14, but the more complex forms of Figure 14.15. The determination of the correct forms of Figure 14.15 is beyond the scope of this book, and the reader is referred

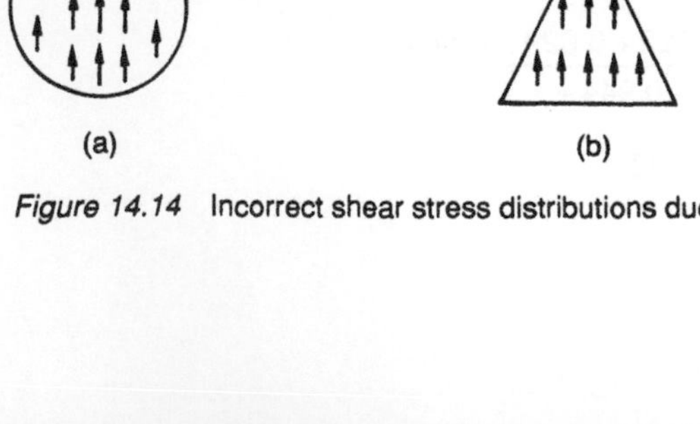

Figure 14.14 Incorrect shear stress distributions due to bending

Figure 14.15 Correct shear stress distributions due to bending

to more advanced works on this topic [31]. In both figures, the size of the arrows is related to the magnitude of the shearing stresses.

EXAMPLE 14.4

A beam of the section in Figure 14.16 is subjected to a bending moment of 0.5 MN m, so that the section bends about a horizontal plane NA. If the maximum principal stress due to this bending moment lies on the axis AA, and it is not to exceed 80% of the greatest bending stress, determine the value of the shearing force that acts on this section.

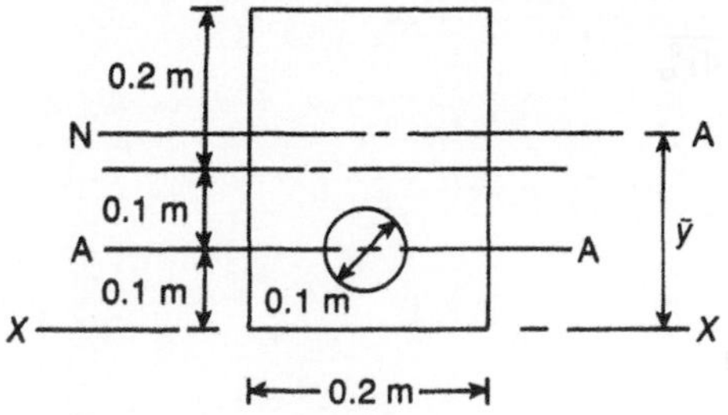

Figure 14.16 Complex section

SOLUTION

To determine $\bar{y}$, I, etc., let us use Table 14.1, where the symbols are as defined in Chapter 13.

$$\bar{y} = \frac{\sum ay}{\sum a} = 0.211 \text{ m}$$

$$I_{XX} = \sum ay^2 + \sum i_H = 4.183\text{E-3 m}^4$$

$$I = I_{XX} - \bar{y}^2 \sum a = 9.73\text{E-4 m}^4$$

$$\hat{\sigma} = \frac{0.5 \times 0.211}{9.73\text{E-4}} = 108.4 \text{ MN/m}^2$$

@

AA $\sigma_1 = 0.8 \times 108.4 =$ maximum principal stress

$\sigma_1 = 86.75 \text{ MN/m}^2$

Table 14.1

Section	a	y	ay	ay^2	i_H
①	0.08	0.2	0.016	3.2E-3	1.0667E-3
②	-7.854E-3	0.1	-7.854E-4	-7.854E-5	-4.909E-6
Σ	0.0721	—	0.0152	3.121E-3	1.062E-3

@

$$\text{AA} \quad \sigma_x = \frac{0.5 \times (0.211 - 0.1)}{9.73\text{E-}4} = 57 \text{ MN/m}^2$$

and

$$\underline{\sigma_y = 0}$$

Now from Chapter 7

$$\sigma_1 = \tfrac{1}{2}(\sigma_x + \sigma_y) + \tfrac{1}{2}\sqrt{(\sigma_x - \sigma_y)^2 + 4\tau_{xy}^2}$$

i.e.

$$86.75 = \tfrac{57}{2} + \tfrac{1}{2}\sqrt{57^2 + 4\tau_{xy}^2}$$

or

$$7525.6 = 812.3 + \tfrac{1}{4}(57^2 + 4\tau_{xy}^2)$$

or

$$4\tau_{xy}^2 = 23\,604$$

$$\underline{\tau_{xy} = 76.8 \text{ MN/m}^2} \text{ @ AA}$$

Now,

$$\tau_{xy} = \frac{F\int y\,dA}{bI}$$

which, when applied to AA, becomes

$$76.8 = \frac{F\int y\,dA}{0.1 \times 9.73\text{E-}4}$$

where

$$\int y\,dA = 0.211 \times 0.2 \times \frac{0.211}{2} - \frac{\pi \times 0.1^2}{4} \times (0.211 - 0.1)$$

$$= 4.452\text{E-}3 - 8.718\text{E-}4 = 3.58\text{E-}3$$

Therefore

$$76.8 = \frac{F \times 3.58\text{E-3}}{9.73\text{E-5}}$$

or

$$\underline{F = 2.1 \text{ MN}}$$

i.e. the *vertical shearing force* at the section is

$$\underline{F = 2.1 \text{ MN}}$$

EXAMPLE 14.5

Calculate and sketch the distribution of vertical and horizontal shearing stresses due to bending, which occur on a beam with the cross-section shown in Figure 14.17, when it is subjected to a vertical shearing force of 30 kN through its centroid.

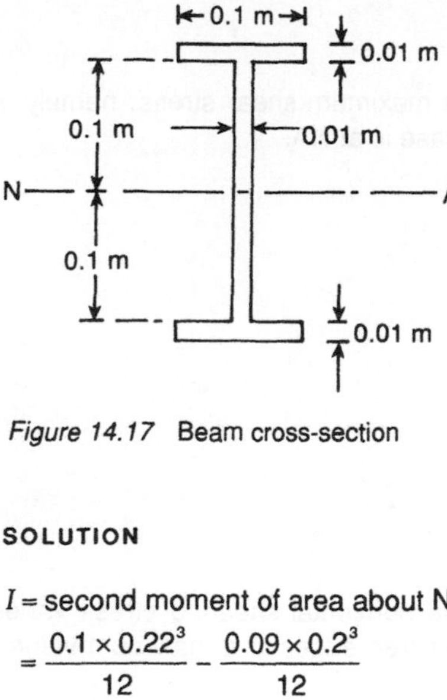

Figure 14.17 Beam cross-section

SOLUTION

I = second moment of area about NA

$$= \frac{0.1 \times 0.22^3}{12} - \frac{0.09 \times 0.2^3}{12}$$

$$\underline{I = 2.873\text{E-5 m}^4}$$

Vertical shearing stress

At the *top of the flange*,

$$\int y \, dA = 0$$

Therefore

$$\underline{\tau_1 = 0}$$

At the *bottom of the flange,*

$$\int y\,dA = 0.1 \times 0.01 \times 0.105 = 1.05\text{E-4 m}^3$$

Therefore

$$\tau_2 = \frac{30\text{E3} \times 1.05\text{E-4}}{0.1 \times 2.873\text{E-5}}$$

$$\underline{\tau_2 = 1.096 \text{ MN/m}^2}$$

At the *top of the web,*

$$\int y\,dA = 1.05\text{E-4 m}^3$$

Therefore

$$\tau_3 = \frac{30\text{E3} \times 1.05\text{E-4}}{0.01 \times 2.873\text{E-5}}$$

$$\underline{\tau_3 = 10.96 \text{ MN/m}^2}$$

From equation (14.1), it can be seen that the maximum shear stress, namely $\hat{\tau}$, occurs where $(\int y\,dA)/b$ is a maximum, which in this case is at NA.

@ NA,

$$\int y\,dA = 1.05\text{E-4} + 0.1 \times 0.01 \times 0.05$$

$$= 1.55\text{E-4}$$

Therefore

$$\hat{\tau} = \frac{30\text{E3} \times 1.55\text{E-4}}{0.01 \times 2.873\text{E-5}}$$

$$\underline{\hat{\tau} = 16.19 \text{ MN/m}^2}$$

Horizontal shear stress

From equation (14.2), it can be seen that the horizontal shearing stress varies linearly along the flanges of the RSJ, from zero at the free edges to a maximum value $\hat{\tau}_F$ at the centre:

$$\tau_F = \frac{F(B-Z)\bar{y}}{I}$$

and

$$\hat{\tau}_F = \frac{FB\bar{y}}{I} = \frac{30\text{E3} \times 0.05 \times 0.105}{2.873\text{E-5}}$$

$$\underline{\hat{\tau}_F = 5.48 \text{ MN/m}^2}$$

A plot of the vertical and horizontal shearing stresses is shown in Figure 14.18.

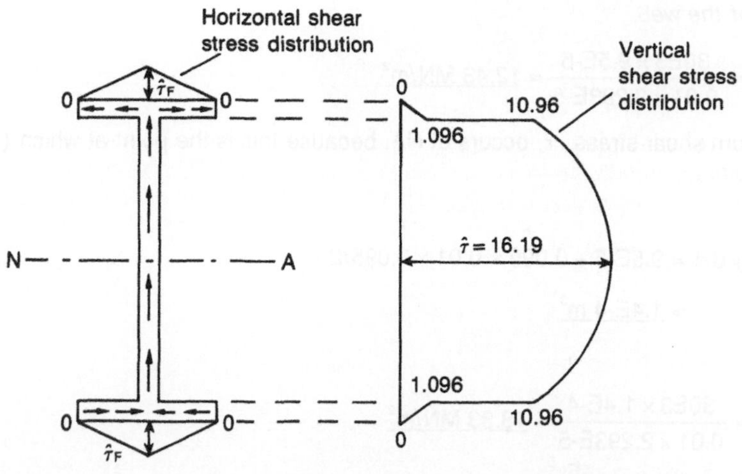

Figure 14.18 Vertical and horizontal shearing stress distributions (MN/m²)

EXAMPLE 14.6

Calculate and sketch the distribution of vertical and horizontal shear stress in the channel bar of Figure 14.19, when it is subjected to a vertical shearing force of 30 kN.

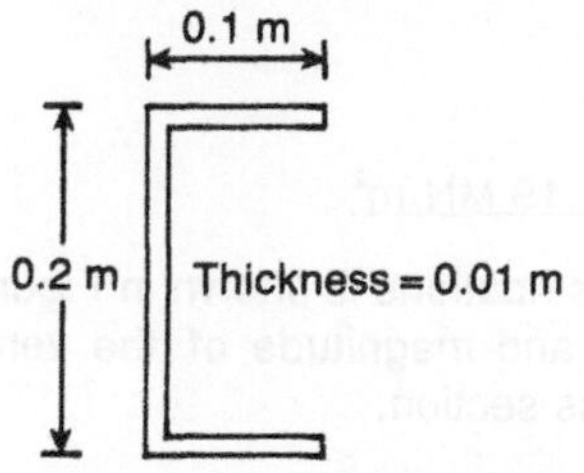

Figure 14.19 Channel section

SOLULTION

$$I = \frac{0.1 \times 0.2^3}{12} - \frac{0.09 \times 0.18^3}{12} = 2.293\text{E-5 m}^4$$

Vertical shearing stress

At the *bottom of the flange,*

$$\int y\,dA = 0.1 \times 0.01 \times 0.095 = \underline{9.5\text{E-5 m}^3}$$

and

$$\tau_1 = \frac{30\text{E3} \times 9.5\text{E-5}}{0.1 \times 2.293\text{E-5}} = \underline{1.24 \text{ MN/m}^2}$$

At the *top of the web,*

$$\tau_2 = \frac{30E3 \times 9.5E\text{-}5}{0.01 \times 2.293E\text{-}5} = \underline{12.43 \text{ MN/m}^2}$$

The maximum shear stress, $\hat{\tau}$, occurs at NA, because this is the point at which $(\int y\,dA)/b$ is a maximum.

@ NA,

$$\int y\,dA = 9.5E\text{-}5 + 0.095 \times 0.01 \times 0.095/2$$

$$= \underline{1.4E\text{-}4 \text{ m}^3}$$

and

$$\hat{\tau} = \frac{30E3 \times 1.4E\text{-}4}{0.01 \times 2.293E\text{-}5} = \underline{18.33 \text{ MN/m}^2}$$

Horizontal shearing stress

From equation (14.2), it can be seen that the horizontal shearing stress varies linearly along the flange width, from zero at the right edge to a maximum value of $\hat{\tau}_F$ at the intersection between the flange and the web.

From equation (14.2),

$$\tau = \frac{F(B-Z)\bar{y}}{I}$$

and

$$\hat{\tau} = FB\bar{y}/I = 30E3 \times 0.09 \times 0.095/2.293E\text{-}5 = \underline{11.19 \text{ MN/m}^2}$$

A plot of the vertical and horizontal shearing stress distributions is shown in Figure 14.20, where the arrows are used to indicate the direction and magnitude of the vertical and horizontal shearing stresses acting on the beam's cross-section.

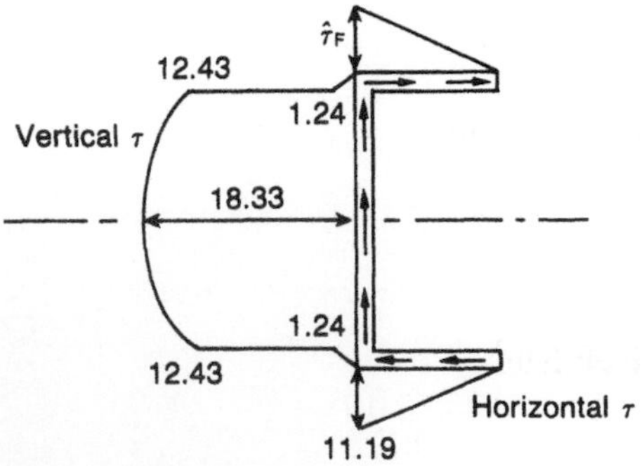

Figure 14.20 Plot of vertical and horizontal shear stress distributions

14.4 Shear centre

When slender symmetrical-section cantilevers are subjected to transverse loads in a laboratory, good agreement is usually found between the predictions of simple bending theory and experimental observations. When however, similar tests are carried out on cantilevers with unsymmetrical sections, such as channel bars, angle irons, etc., comparison between theoretical predictions and experimental observations is usually poor. The explanation for this can be obtained by considering the channel section of Figure 14.21, and assuming that the shearing force F is applied through its centroid.

From Figure 14.20, the shear stress distribution due to bending will be as shown in Figure 14.21(a), where the magnitude and direction of the arrows are used to indicate the magnitude and direction of the shearing stresses due to bending.

From Figure 14.21(a), it can be seen that horizontal equilibrium is achieved by the horizontal shearing force in the top flange being equal, but opposite, to the horizontal shearing force in the bottom flange.

Similarly, vertical equilibrium is achieved by the internal resisting vertical shearing force in the web being equal and opposite to the applied external vertical shearing force F. However, from Figure 14.21(a), it can be seen that rotational equilibrium is not achieved, so that if F is applied through the centroid of the section, the beam will twist, as shown in Figure 14.21(b), and as a result, shearing stresses due to torsion will occur in addition to shearing stresses due to bending.

In general, it is advisable to eliminate shearing stresses due to torsion, as these can be relatively large, and to achieve this, it is necessary to ensure that F acts at a point where rotational equilibrium is achieved. This point is called the shear centre, and some typical positions of shear centre are shown in Figure 14.22, together with distributions of shear stress due to bending caused by F.

NB The term *shear flow* has been introduced in this section, where *shear flow* q = *shear stress* × *wall thickness* = τt.

In certain cases, when the shear stress varies inversely with the wall thickness, so

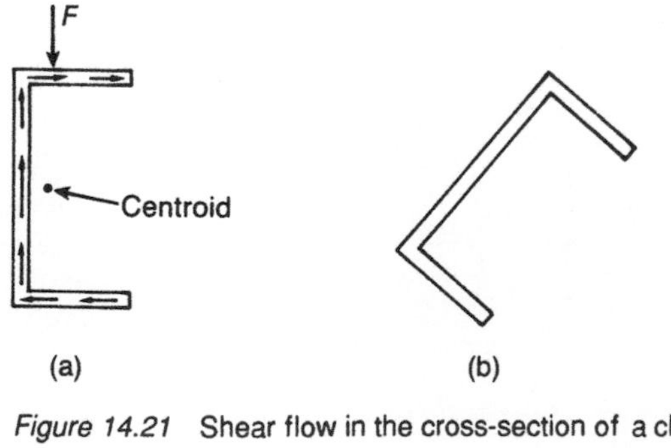

Figure 14.21 Shear flow in the cross-section of a channel bar: (a) shear flow; (b) twisting

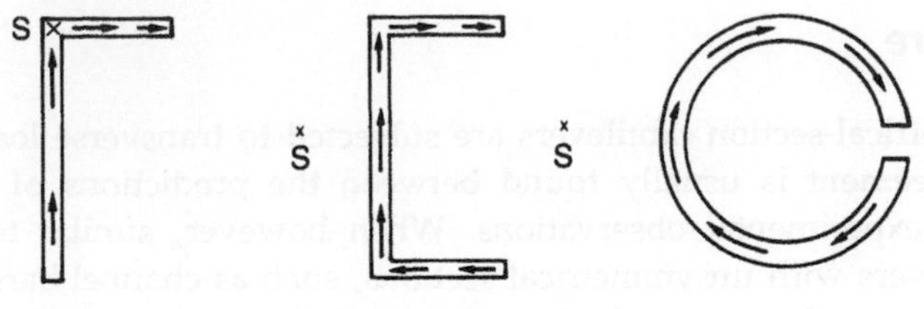

Figure 14.22 Some shear centre positions 'S'

that the shear flow q is constant, it may be found convenient to carry out the calculation using q instead of τ.

EXAMPLE 14.7

Determine the shear centre position for the channel-section in Figure 14.19.

SOLUTION

From Figure 14.20, it can be seen that the horizontal shearing stress distribution in the flange is linear, so that the average shear stress in the flange is

$$\tau_{av} = 11.19/2 = \underline{5.595\ \text{MN/m}^2}$$

Let F_F be the resisting shearing force in the flange. Then

$$F_F = 5.595 \times 0.01 \times 0.095 = 5.315\text{E-3 MN}$$
$$\underline{F_F = 5.315\ \text{kN}}$$

Now from vertical equilibrium considerations, the resisting shearing force in the web is

$$\underline{F = 30\ \text{kN}}$$

Let the shear centre position be at S in the beam section, as shown in Figure 14.23, and by taking moments about S,

$$F \times \Delta = F_F \times 0.19$$

or

$$30\Delta = 5.32 \times 0.19$$
$$\underline{\Delta = 0.0337\ \text{m} = 3.37\ \text{cm}}$$

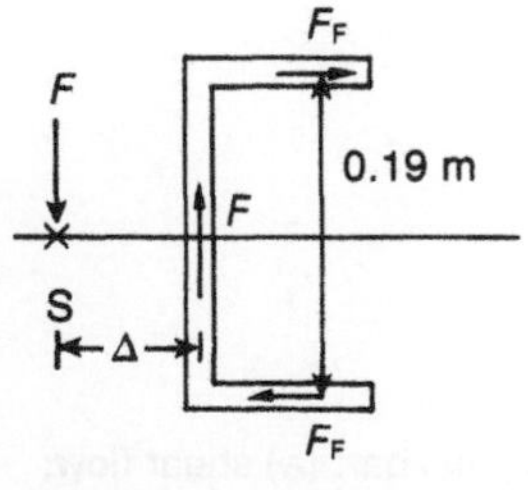

Figure 14.23 Shearing forces on channel section

EXAMPLE 14.8

Determine the shear centre position for the thin-walled curved section shown in Figure 14.24, which is of constant thickness t.

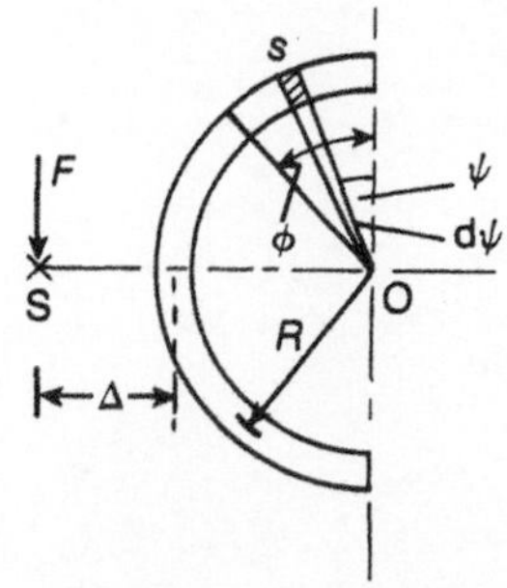

Figure 14.24 Thin-walled curved section

SOLUTION

At any distance s, the shear stress due to bending is

$$\tau_s = \frac{F}{tI}\int y\,\mathrm{d}A$$

where $\int y\,\mathrm{d}A$ is the first moment of area of the element, shaded in Figure 14.24, about NA, i.e.

$$\tau_s = \frac{F}{tI}\int_0^{\phi}(R\cos\psi)(tR\,\mathrm{d}\psi) = \frac{FR^2\sin\phi}{I} \tag{14.8}$$

The shear flow in the section is shown in Figure 14.25, where the magnitude and directions of the arrows are intended to give a measure of the magnitude and direction of the internal resisting shearing stresses.

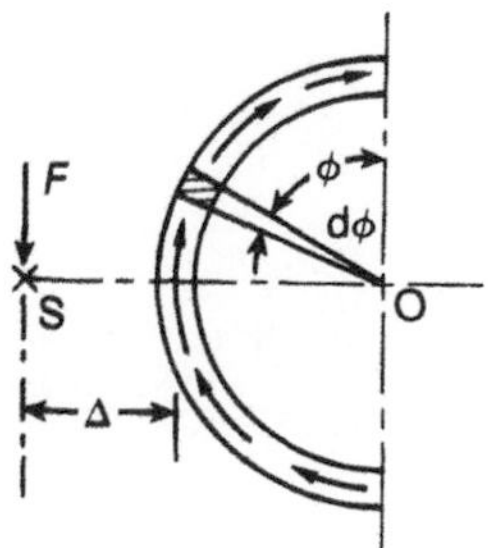

Figure 14.25 Shear flow in thin-walled curved section

To determine Δ, take moments about O, i.e.

$$F(\Delta + R) = \int_0^{\pi} \tau_s \times (R\,d\phi) \times t \times R$$

$$= \frac{2FR^4 t}{I}$$

or

$$\underline{\Delta = \frac{2R^4 t}{I} - R} \tag{14.9}$$

***To determine** I*

From Chapter 3,

$$I = \int y^2\,dA$$

$$= \int_0^{\pi} (R\cos\phi)^2 tR\,d\phi$$

$$= tR^3 \int_0^{\pi} \frac{(1+\cos 2\phi)}{2}\,d\phi$$

$$= \frac{tR^3}{2}\left[\phi + \frac{\sin 2\phi}{2}\right]_0^{\pi}$$

$$I = \pi R^3 t/2 \tag{14.10}$$

Substituting equation (14.10) into (14.9),

$$\underline{\Delta = R(4/\pi - 1) = 0.273R}$$

EXAMPLE 14.9

Determine the shear centre position for the thin-walled section of Figure 14.26, which is of uniform thickness t.

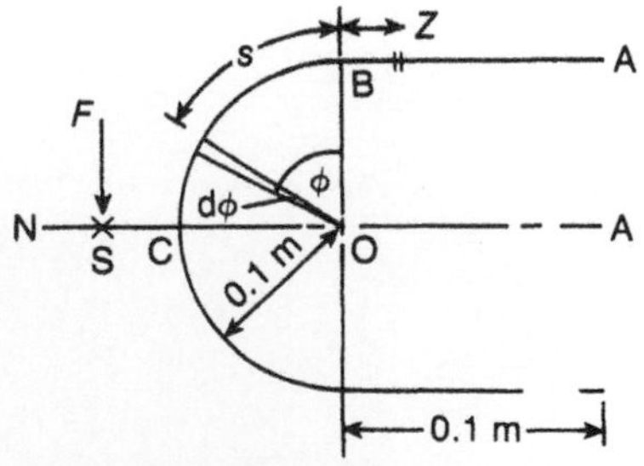

Figure 14.26 Curved section with flanges

SOLUTION

Let I be the second moment of area of section about NA, which can be obtained with the assistance of equation (14.10). Then

$$I = \frac{\pi R^3 t}{2} + 2\left(\frac{0.1t^3}{12} + t \times 0.1 \times 0.1^2\right)$$

but as the section is thin, higher-order terms involving t can be ignored; therefore

$$I = \frac{\pi \times 0.1^3 t}{2} + 2 \times 0.001t$$

$$\underline{I = 3.57\text{E-}3t}$$

Consider the top flange

τ_F = shear stress in the top flange at Z

$= F(B - Z)\bar{y}/I$

$$@\ Z = 0.1 \quad \tau_A = 0 \tag{14.11}$$

$$@\ Z = 0 \quad \tau_B = FB\bar{y}/I = F \times 0.1 \times 0.1/(3.57\text{E-}3t)$$

$$\underline{\tau_B = 2.8F/t} \tag{14.12}$$

Consider the curved section

$$@\ s \quad \tau_s = \frac{F}{tI}\int y\,\mathrm{d}A$$

$$= \frac{F}{t \times 3.57\text{E-}3t}\left(\int_0^{\phi} (R\cos\psi\ tR\,\mathrm{d}\psi) + 0.01 \times t \times R\right)$$

$$= \frac{F}{3.57\text{E-}3t}(R^2 \sin\phi + 0.1R)$$

$$\underline{\tau_s = \frac{F}{0.357t}(1 + \sin\phi)} \tag{14.13}$$

$@\ \phi = 0 \quad \tau_B = 2.8F/t$ as required (see equation (14.12))

$@\ \phi = 90° \quad \tau_C = 5.6F/t$ = maximum shear stress due to bending

To calculate the shear centre position

Let F_F be the resisting shearing force in the flange. Then

$$F_F = \frac{(\tau_A + \tau_B)}{2} \times 0.1t$$

$$\underline{F_F = 0.14F}$$

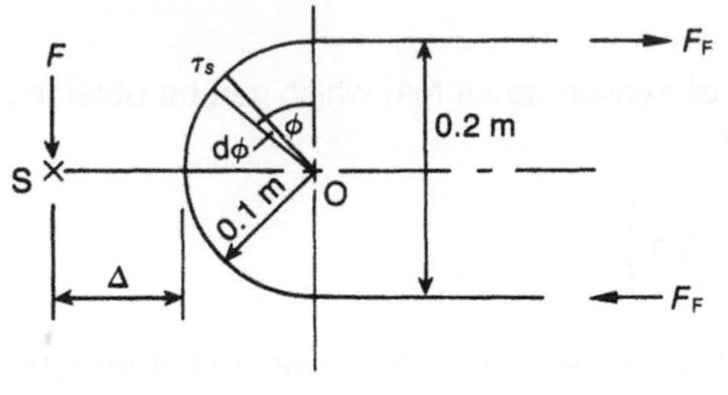

Figure 14.27 Thin-walled open section

To obtain Δ, consider rotational equilibrium about the point O of the section, as shown in Figure 14.27.

$$F(\Delta + 0.1) = F_F \times 0.2 + \int_0^\pi \tau_s \times R \times d\phi \times t \times R$$

$$= 0.2 \times 0.14F + \int_0^\pi \frac{F(1+\sin\phi)}{0.357t} \times 0.1^2 t\, d\phi$$

$$= 0.028F + 0.028F[\phi - \cos\phi]_0^\pi$$

$$= 0.028F + 0.028F\{[\pi + 1] - [0 - 1]\}$$

$$= 0.172F$$

Therefore

$\underline{\Delta = 0.072 \text{ m}}$

14.5 Shear centre positions for closed thin-walled tubes

The main problem of determining the shear stress due to bending in closed thin-walled tubes is that, initially, the shear stress is not known at any point. To overcome this difficulty, the assumption is made that the shear stress due to bending at a certain point in the section has an unknown value of τ_0, as shown in Figure 14.28. Megson [29] has shown that for a thin-walled closed tube, the relationship between

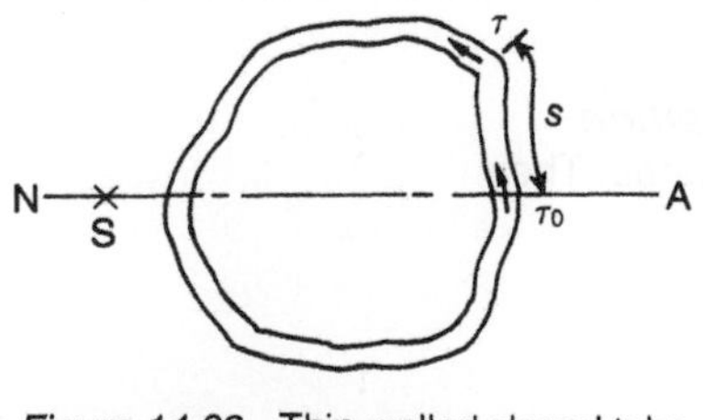

Figure 14.28 Thin-walled closed tube

the twist and the shear stress is given by

$$\oint \frac{\tau \, ds}{G} = 2A \frac{d\theta}{dz}$$

where

A = enclosed area of the cross-section of the tube
$d\theta/dz$ = twist/unit length
θ = angle of twist
z = distance along the axis of the tube
τ = shearing stress due to bending at any distance s from the 'starting' point.
$= \tau_0 + \tau_s$
τ_0 = shearing stress due to bending at the 'starting' point
τ_s = shearing stress due to bending at any distance s for an equivalent *open* tube
ds = elemental length
G = rigidity modulus

If F is the shearing force applied through the shear centre S, then there will be no twist, i.e.

$$\frac{d\theta}{dz} = 0 = \oint \frac{\tau \, ds}{G}$$

or

$$\oint (\tau_0 + \tau_s) \, ds = 0$$

but

$$\tau_0 = \text{constant}$$

Therefore

$$\tau_0 \oint ds = -\oint \tau_s \, ds$$

or

$$\tau_0 = -\frac{\oint \tau_s \, ds}{\oint ds} \tag{14.14}$$

Once τ_0 is determined from equation (14.14), τ can be found.

EXAMPLE 14.10

Determine the shear centre position for the thin-walled closed tube of Figure 14.29, which is of uniform thickness t.

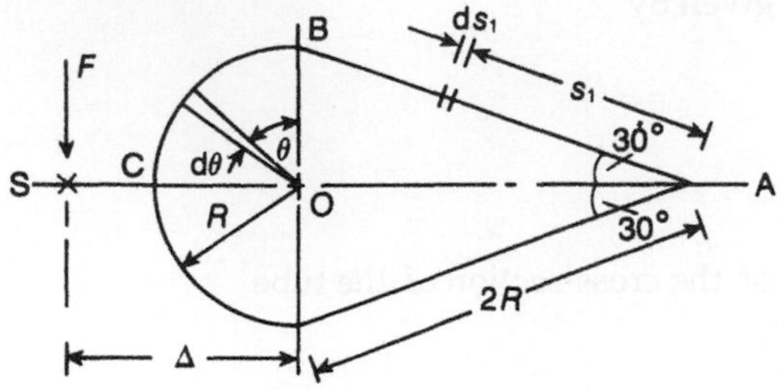

Figure 14.29 Thin-walled closed tube

SOLUTION

Let τ_0 be the shear stress due to bending at the point A. Then

$$I = 2\int_0^{2R} (s_1 \sin 30°)^2 \times (t\,ds_1) + \int_0^{\pi} (R\cos\phi)^2 \times (tR\,d\phi)$$

$$= 0.5t\int_0^{2R} s_1^2\,ds_1 + tR^3\int_0^{\pi} \cos^2\phi\,d\phi$$

$$= \frac{0.5t}{3}[s_1^3]_0^{2R} + tR^3\int_0^{\pi} \frac{(1+\cos 2\phi)}{2}\,d\phi$$

$$= \frac{4R^3t}{3} + \frac{tR^3}{2}\left[\phi + \frac{\sin 2\phi}{2}\right]_0^{\pi}$$

$$\underline{I = R^3t\left(\frac{4}{3} + \frac{\pi}{2}\right) = 2.904R^3t}$$

Consider AB

At any distance s_1,

$$\tau_{s1} = \frac{F}{tI}\int (s_1 \sin 30°) \times (t\,ds_1)$$

$$= \frac{0.5F}{I}\left[\frac{s_1^2}{2}\right]_0^{s_1}$$

$$\tau_{s1} = 0.25Fs_1^2/I \qquad (14.15)$$

Consider BC

At any distance ϕ,

$$\tau_{\phi} = \frac{F}{tI}\left(\int_0^{\phi} (R\cos\phi) \times (tR\,d\phi) + (2Rt) \times (R/2)\right)$$

$$= \frac{F}{I}R^2\{[\sin\phi]_0^{\phi} + 1\}$$

$$\underline{\tau_{\phi} = FR^2(1+\sin\phi)/I} \qquad (14.16)$$

Now from equation (14.14),

$$\tau_0 = -\frac{\oint \tau_s \, ds}{\oint ds}$$

but

$$\oint ds = 2R \times 2 + \pi R = \underline{7.14R} \qquad (14.17)$$

and

$$\oint \tau_s \, ds = \oint \tau_{s1} \, ds_1 + \oint \tau_\phi R \, d\phi$$

$$= 2\int_0^{2R} \frac{0.25Fs_1^2}{I} \, ds_1 + \int \frac{FR^2}{I}(1 + \sin\phi) R \, d\phi$$

$$= \frac{0.5F}{I}\left[\frac{s_1^3}{3}\right]_0^{2R} + \frac{FR^3}{I}[\phi - \cos\phi]_0^\pi$$

$$= \frac{F}{I}\left(\frac{4R^3}{3} + R^3\{[\pi + 1] - [0 - 1]\}\right)$$

$$= \frac{FR^3}{I}\left(\tfrac{4}{3} + \pi + 2\right)$$

$$= \frac{6.475FR^3}{2.904R^3t}$$

$$\oint \tau_s \, ds = 2.23F/t \qquad (14.18)$$

Hence, from equations (14.15) to (14.18),

$$\tau_0 = -\frac{2.23F}{t} \times \frac{1}{7.14R}$$

$$\underline{\tau_0 = -0.312F/Rt} \qquad (14.19)$$

so that if τ_{AB} is the shear stress at any point between A and B, then

$$\tau_{AB} = -\frac{0.312F}{Rt} + \frac{0.25Fs_1^2}{I}$$

$$= -\frac{0.312F}{Rt} + \frac{0.25Fs_1^2}{2.904R^3t}$$

$$\tau_{AB} = -\frac{0.312F}{Rt} + \frac{0.0861Fs_1^2}{R^3t} \qquad (14.20)$$

If τ_{BC} is the shear stress at any point between B and C, then

$$\tau_{BC} = -\frac{0.312F}{Rt} + \frac{FR^2(1+\sin\phi)}{2.904R^3t}$$

$$= -\frac{0.312F}{Rt} + \frac{F(1+\sin\phi)}{2.904Rt}$$

$$\tau_{BC} = \frac{F}{Rt}\left(0.0324 + \frac{\sin\phi}{2.904}\right) \quad (14.21)$$

Taking moments about 0,

$$F\Delta = \int_0^{\pi} \tau_{BC} \times (tR\,d\phi) \times R + 2\int_0^{2R} \tau_{AB} \times (t\,ds_1) \times R\cos 30°$$

$$= \frac{FR^2t}{2.904Rt}\int_0^{\pi}(0.0941 + \sin\phi)\,d\phi + 1.732Rt\int_0^{2R}\left(-\frac{0.312F}{Rt} + \frac{0.0861Fs_1^2}{R^3t}\right)ds_1$$

$$= \frac{FR}{2.904}[0.0941\phi - \cos\phi]_0^{\pi} + 1.732Rt\left[-\frac{0.312Fs_1}{Rt} + \frac{0.0861Fs_1^3}{3R^3t}\right]_0^{2R}$$

$$= \frac{FR}{2.904}\{[0.0941\pi + 1] - [-1]\} + 1.732FRt\left[-\frac{0.312\times 2R}{Rt} + \frac{8\times 0.0861R^3}{3R^3t}\right]$$

$$= FR(0.79 - 0.683)$$

Therefore

$$\Delta = 0.107R$$

14.6 Shear deflections

Deflections of beams usually consist of two components, namely deflections due to bending and deflections due to shear. If a beam is long and slender, then deflections due to shear are small compared with deflections due to bending. If, however, the beam is short and stout, then deflections due to shear cannot be neglected.

Deflections due to shear are caused by shearing action alone, where each element of the beam tends to change shape, as shown in Figure 14.30, where F is the shearing force acting on a typical element.

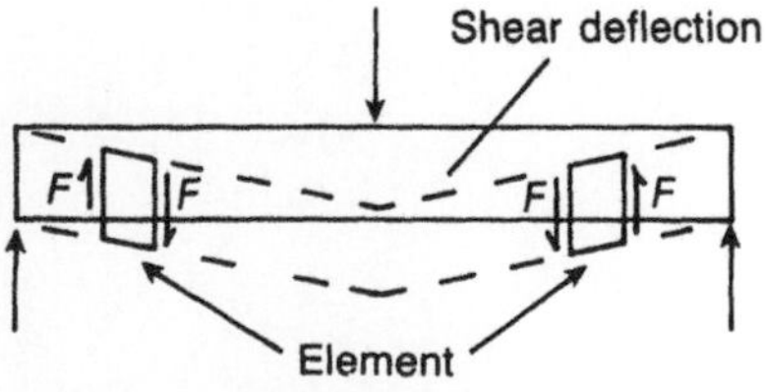

Figure 14.30 Shear deflection of a beam

EXAMPLE 14.11

Determine the value of the maximum deflection due to shear for the end-loaded cantilever of Figure 14.31. The cantilever is of uniform rectangular section, width b and depth d.

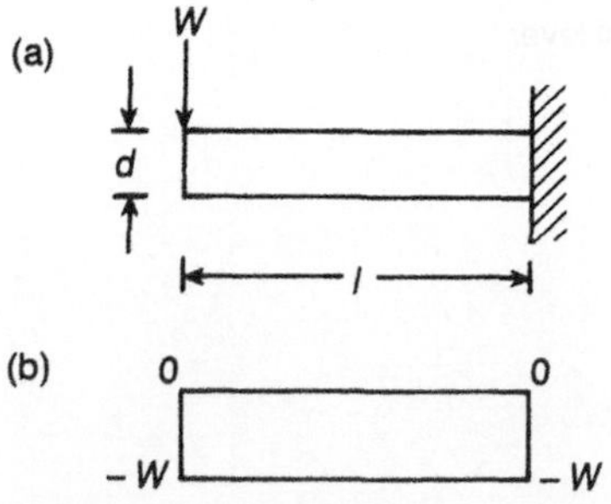

Figure 14.31 Shearing force distribution in an end-loaded cantilever: (a) cantilever; (b) shearing force diagram

SOLUTION

The shearing forces acting on the cantilever are of a constant value W, causing the shear deflected form shown in Figure 14.32.

From equation (14.3), the value of shear stress at any distance y from the neutral axis is

$$\tau = \frac{6W}{bd^3}\left(\frac{d^2}{4} - y^2\right)$$

From Chapter 9, the total shear strain energy of the cantilever is

$$\text{SSE} = \int \frac{\tau^2}{2G}\,\text{d(vol)}$$

$$= \frac{2}{2}\int_0^{d/2} \frac{36W^2}{Gb^2d^6}\left(\frac{d^2}{4} - y^2\right)^2 lb\,\text{d}y$$

$$= \frac{36W^2l}{Gbd^6}\int_0^{d/2}\left(\frac{d^4}{16} - \frac{d^2y^2}{2} + y^4\right)\text{d}y$$

$$\text{SSE} = \frac{3W^2l}{5Gbd} \qquad (14.22)$$

If δ_s is the maximum deflection due to shear, then the work done by the load W is

$$\text{WD} = \tfrac{1}{2}W\delta_s \qquad (14.23)$$

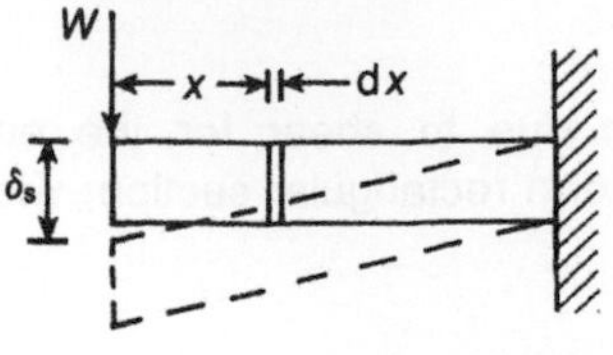

Figure 14.32 Shear deflected form of cantilever

Equating (14.22) and (14.23),

$$\delta_s = \frac{6Wl}{5Gbd} \tag{14.24}$$

EXAMPLE 14.12

Determine the value of the maximum deflection due to shear for a cantilever, assuming that it is subjected to a uniformly distributed load *w*, as shown in Figure 14.33.

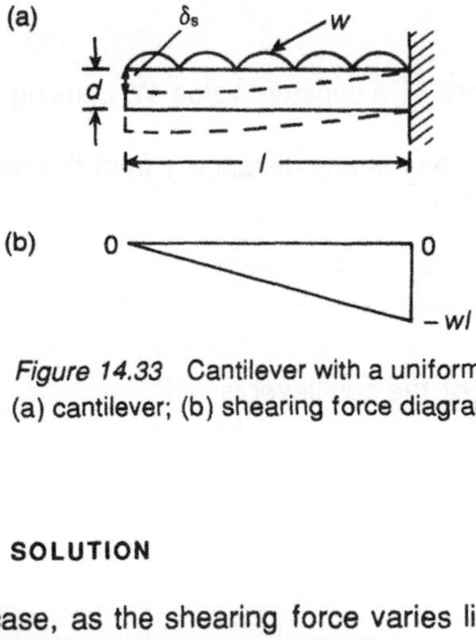

Figure 14.33 Cantilever with a uniformly distributed load: (a) cantilever; (b) shearing force diagram

SOLUTION

In this case, as the shearing force varies linearly along the length of the cantilever, the shear deflection will vary parabolically.

At any distance x from the free end, the shearing force on an element of the beam of length dx is

$$F = wx$$

From equation (14.24), the shear deflection of this element is

$$\frac{6wx\,dx}{5Gbd}$$

Therefore the total shear deflection of the cantilever is

$$\delta_s = \int_0^l \frac{6wx}{5Gbd}\,dx$$

$$\delta_s = \frac{3wl^2}{5Gbd}$$

14.6.1 *Total deflection of an end-loaded cantilever*

From chapter 5, the maximum deflection due to bending of an end-loaded cantilever is

$$\delta_b = \frac{Wl^3}{3EI}$$

which for a rectangular section of width b and depth d becomes

$$\delta_b = \frac{Wl^3}{3E} \times \frac{12}{bd^3} = \frac{4Wl^3}{Ebd^3}$$

Now, the total deflection is

$$\delta = \delta_b + \delta_s = \frac{4Wl^3}{Ebd^3} + \frac{6Wl}{5Gbd}$$

$$\delta = \frac{4Wl^3}{Ebd^3}\left[1 + \frac{3}{4}\left(\frac{d}{l}\right)^2\right] \qquad (14.25)$$

where the second term in the brackets represents the component of deflection due to shear, and it is assumed that $E = 2.5G$.

From equation (14.25), it can be seen that the deflection due to shear is important when (d/l) becomes relatively large.

14.7 Warping

The effects of warping have not been included in this chapter as the subject of the warping of thin-walled sections is beyond the scope of this book.

Megson [29] describes warping as out-of-plane deformation of a cross-section, particularly when an unsymmetrical section is not loaded through its shear centre, as shown in Figure 14.34.

The longitudinal direct stresses caused by warping are of particular importance when the beam is restrained from axial movement. Warping is not of importance in solid circular-section beams and in thin-walled tubes of circular and square cross-section.

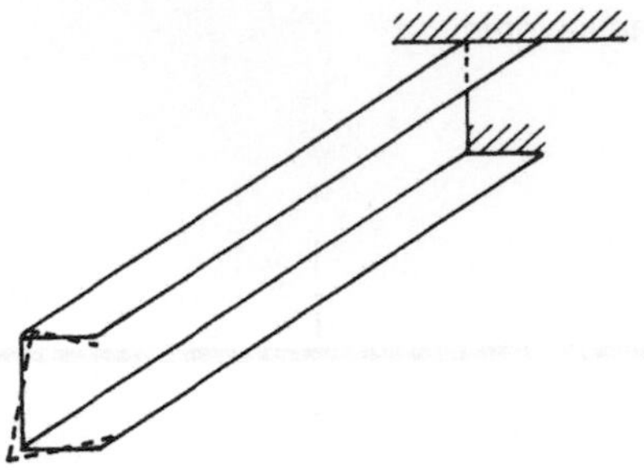

Figure 14.34 Warping of a cross-section

EXAMPLES FOR PRACTICE 14

1. A beam of length 3 m is simply supported at its ends and subjected to a uniformly distributed load of 200 kN/m, spread over its entire length. If the beam has a uniform cross-section of depth 0.2 m and width 0.1 m, determine the position and value of the maximum shearing stress due to bending. What will be the value of the maximum shear stress at mid-span?

{22.5 MPa @ NA @ the ends; 0}

2. Determine the maximum values of shear stress due to bending in the web and flanges of the sections of Figure 14.35 when they are subjected to vertical shearing forces of 100 kN.

{(a) 97.83 MPa, 35.87 MPa; (b) 127.4 MPa, 52.5 MPa}

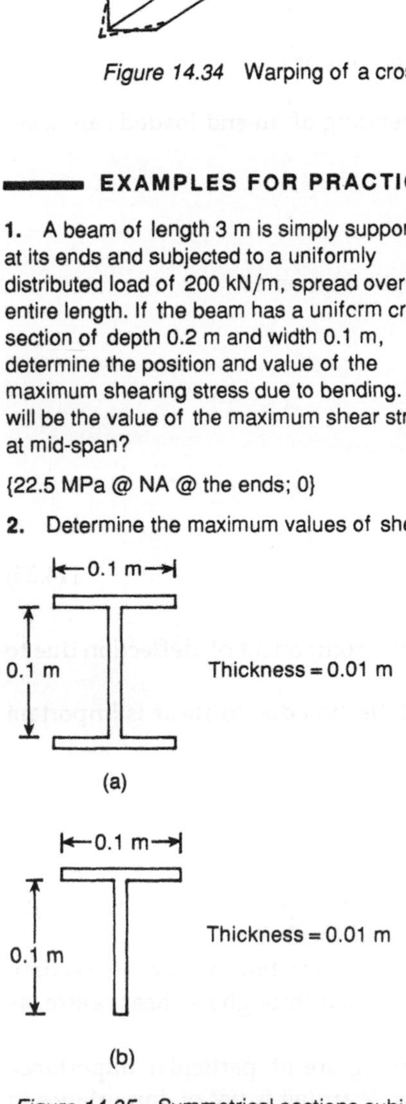

Figure 14.35 Symmetrical sections subjected to vertical shearing forces: (a) RSJ; (b) tee section

3. Determine an expression for the maximum shearing stress due to bending for the section of Figure 14.36, assuming that it is subjected to a shearing force of 0.5 MN acting through its centroid and in a perpendicular direction to NA.

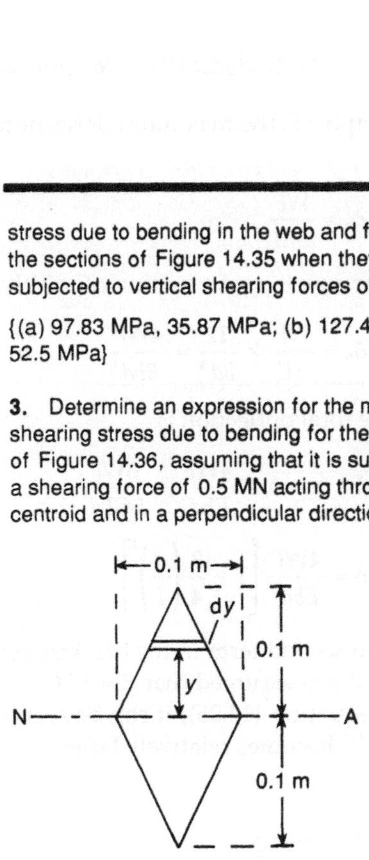

Figure 14.36

{$\tau = 5000[0.01 - (3y^2 - 20y^3)]/(1 - 10y)$, $\hat{\tau} = 56.25/\text{MPa}$ @ $y = \pm 0.025$ m}

4. Determine the value of the maximum shear stress for the cross-section of Figure 14.37, assuming that it is subjected to a shearing force of magnitude 0.5 MN acting through its centroid and in a perpendicular direction to NA.

{18.44 MPa}

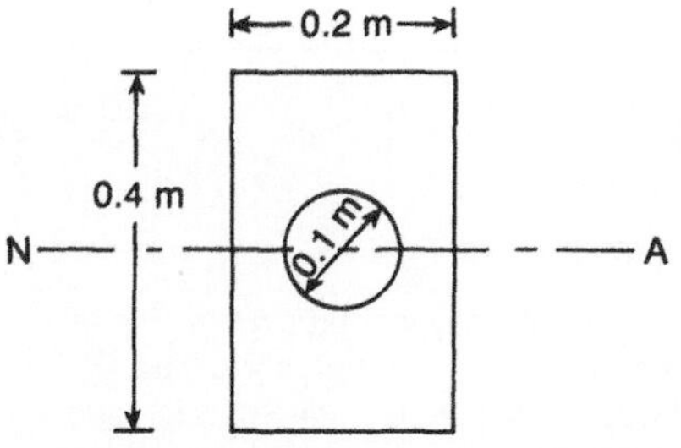

Figure 14.37 Rectangular section with a hole

5. A simply supported beam, with a cross-section as shown in Figure 14.38, is subjected to a centrally placed concentrated load of 100 MN, acting through its centroid and perpendicular to NA.

Determine the values of the vertical shearing stress at intervals of 0.1 m from NA.

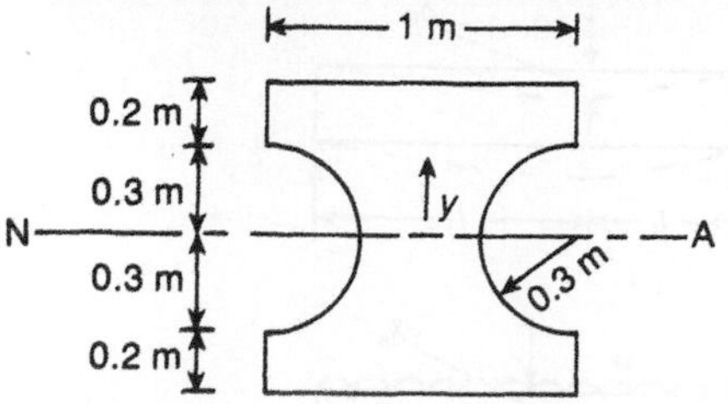

Figure 14.38 Complex cross-section

{@ $y=0$, $\tau_0=173.7$, $\tau_{0.1}=156.95$, $\tau_{0.2}=114.49$, $\tau_{0.3}=51.95$, $\tau_{0.4}=29.22$; @$y=0.5$, $\tau_{0.5}=0$ – all in MPa}

6. Determine the positions of the shear centres for the thin-walled sections of Figures 14.39(a) and (b).

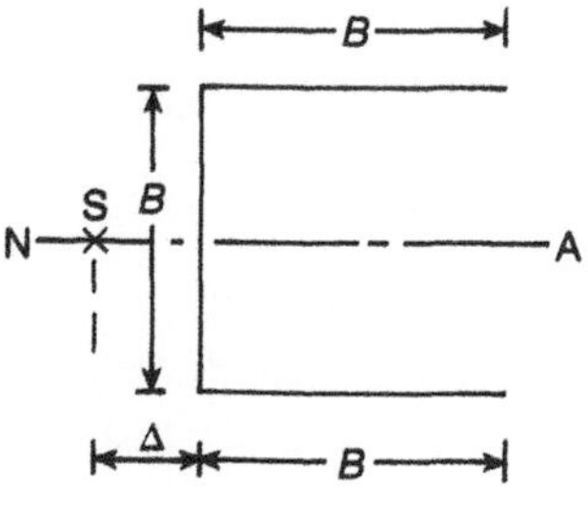

(a)

Figure 14.39 Thin-walled open sections: (a) channel section

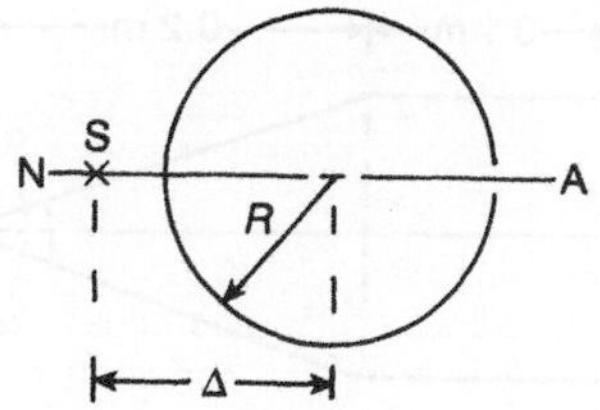

(b)

Figure 14.39 (*continued*) (b) split tube

{(a) $\Delta = 0.429B$; (b) $\Delta = 2R$}

7. Determine the shear centre positions for the thin-walled sections of Figures 14.40(a) and (b).

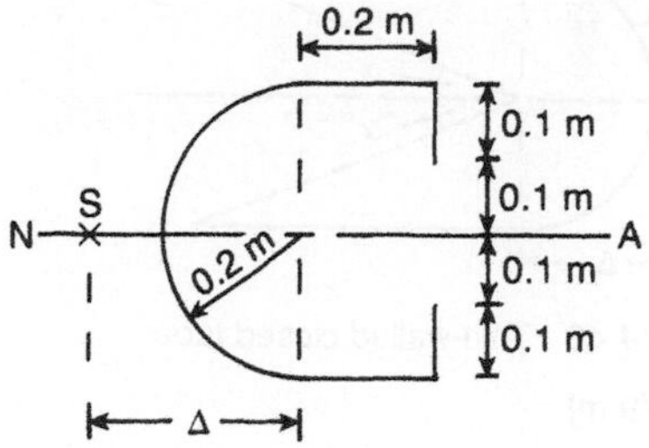

(a)

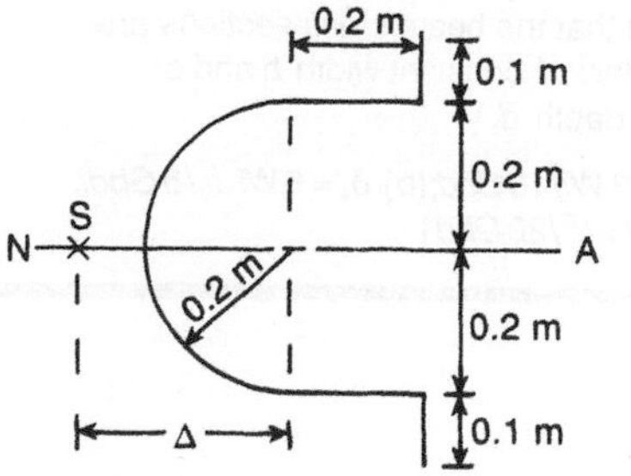

(b)

Figure 14.40 Thin-walled complex open sections

{(a) $\Delta = 0.396$ m; $\Delta = 0.35$ m}

8. Determine the shear centre position for the thin-walled closed tube of Figure 14.41 which is of uniform thickness. (Portsmouth, 1984).

{$\Delta = -0.032$ m}

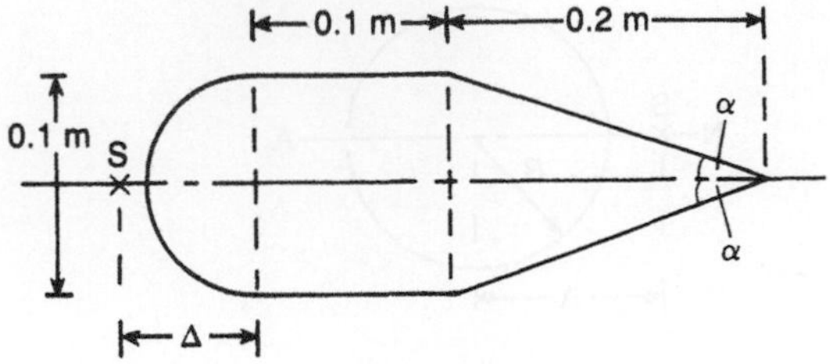

Figure 14.41 Thin-walled closed tube

9. Determine the shear centre position for the thin-walled closed tube of Figure 14.42, which is of uniform thickness. (Portsmouth, 1985).

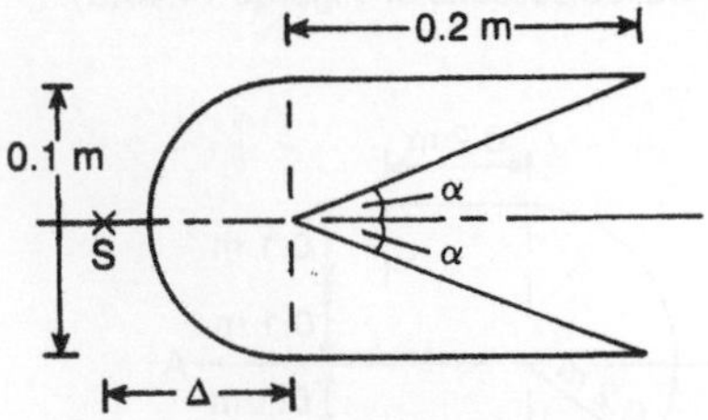

Figure 14.42 Thin-walled closed tube

{Δ = 0.079 m}

10. Determine the maximum deflections due to shear for the simply supported beams of Figures 14.43(a)–(c). In all cases, it may be assumed that the beam cross-sections are rectangular, of constant width *b* and of constant depth *d*.

{(a) $\delta_s = 3Wl/10Gbd$; (b) $\delta_s = 6Wl_1l_2/5Gbdl$; (c) $\delta_s = 3wl^2/20Gbd$}

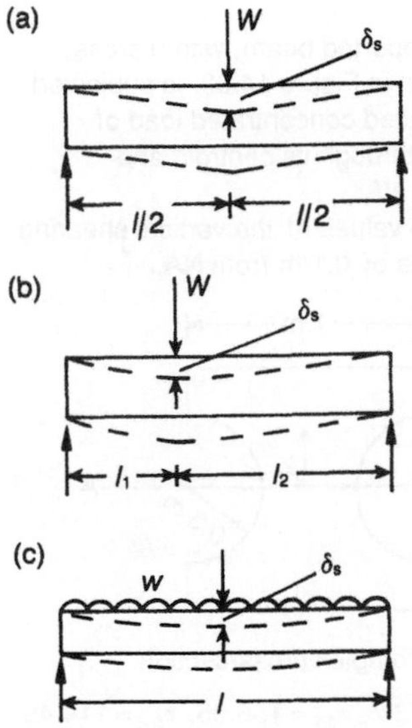

Figure 14.43 Shear deflections of simply supported beams: (a) centrally loaded; (b) off-centre load; (c) uniformly distributed load

15

The matrix displacement method

15.1 Introduction

The finite element method [5] is one of the most powerful methods of solving partial differential equations, particularly if these equations apply over complex shapes. The method consists of sub-dividing the complex shape into several elements of simpler shape, each of which is more suitable for mathematical analysis. The process then, as far as structural analysis is concerned, is to obtain the elemental stiffnesses of these simpler shapes and then, by considering equilibrium and compatibility at the inter-element boundaries, to assemble all the elements, so that a mathematical model of the entire structure is obtained.

Hence, owing to the application of loads on this mathematical model, the 'deflections' at various points of the structure can be obtained through the solution of the resulting simultaneous equations. Once these 'deflections' are known, the stresses in the structure can be determined through Hookean elasticity.

Each finite element is described by 'nodes' or 'nodal points', and the stiffnesses, displacements, loads, etc., are all related to those nodes. Finite elements vary in shape, depending on the systems they have to describe, and some typical finite elements are shown in Figure 15.1.

Now, the finite element method is a vast topic, covering problems in structural mechanics, fluid flow, heat transfer, acoustics, etc., and because of this, it is beyond the scope of the present book. However, because the finite element method is based on the matrix displacement method, a brief description of the latter will be given in the present chapter.

15.2 The matrix displacement method

The matrix displacement method is also known as the stiffness method. It is based on obtaining the stiffness of the entire structure by assembling together all the individual stiffnesses of each member of element of the structure. When this is

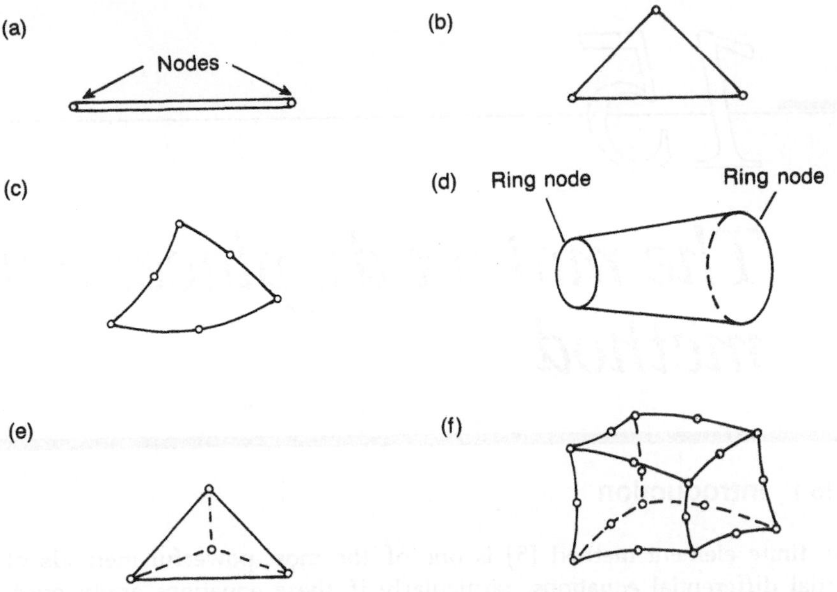

Figure 15.1 Some typical finite elements: (a) one-dimensional rod element, with end nodes; (b) two-dimensional triangular element with corner nodes; (c) curved triangular plate, with additional 'mid-side' nodes; (d) truncated conical element with ring nodes; (e) solid tetrahedral element with corner nodes; (f) 20-node curved brick element

done, the mathematical model of the structure is subjected to the externally applied loads, and by solving the resulting simultaneous equations, the nodal deflections are determined. Once the nodal deflections are known, the stresses in the structure can be obtained through Hookean elasticity.

Now, in the small-deflection theory of elasticity, a structure can be said to behave like a complex spring, where each member or element of the structure can be regarded as an individual spring, with a different type and value of stiffness.

Thus to introduce the method, let us consider the single elemental spring of Figure 15.2.

Let

k = stiffness of spring
= slope of load–deflection relationship for the spring

X_1 = axial force at node 1

X_2 = axial force at node 2

u_1 = nodal displacement at node 1 in the direction of X_1

u_2 = nodal displacement at node 2 in the direction of X_2

In Figure 15.2, 1 and 2 are known as the nodes or nodal points.

Figure 15.2 Spring element

From Hooke's law,

$$X_1 = k(u_1 - u_2) \tag{15.1}$$

and from considerations of equilibrium,

$$X_2 = -X_1$$

$$= k(u_2 - u_1) \tag{15.2}$$

If equations (15.1) and (15.2) are put into matrix form, they appear as follows:

$$\begin{Bmatrix} X_1 \\ X_2 \end{Bmatrix} = \begin{bmatrix} k & -k \\ -k & k \end{bmatrix} \begin{Bmatrix} u_1 \\ u_2 \end{Bmatrix} \tag{15.3}$$

or

$$\{P_i\} = [k]\{u_i\}$$

where

$\{P_i\}$ = a vector of elemental nodal forces

$\{u_i\}$ = a vector of elemental nodal displacements

15.3 The structural stiffness matrix [*K*]

Consider a simple structure composed of two elemental springs, as shown in Figure 15.3, where

k_a = stiffness of spring 1–2

k_b = stiffness of spring 2–3

15.3.1 To determine [K]

$[K]$ will be of order 3×3 because there are three degrees of freedom, namely u_1, u_2 and u_3, and it can be obtained as follows.

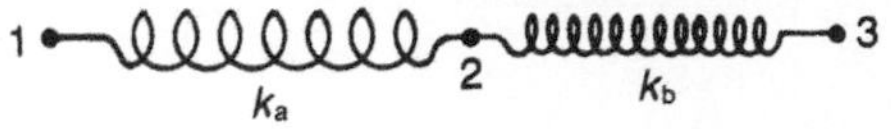

Figure 15.3 Simple structure

For *element 1–2*, from equation (15.3) the elemental stiffness matrix for spring 1–2 is given by (15.4), where the components of stiffness are related to the nodal displacements, u_1 and u_2:

$$[k_{1\text{-}2}] = \begin{array}{c} \begin{matrix} u_1 & \quad u_2 \end{matrix} \\ \begin{bmatrix} k_a & -k_a \\ -k_a & k_a \end{bmatrix} \begin{matrix} u_1 \\ u_2 \end{matrix} \end{array} \tag{15.4}$$

Similarly for *element* 2–3,

$$[k_{2\text{-}3}] = \begin{array}{c} \begin{matrix} u_2 & \quad u_3 \end{matrix} \\ \begin{bmatrix} k_b & -k_b \\ -k_b & k_b \end{bmatrix} \begin{matrix} u_2 \\ u_3 \end{matrix} \end{array} \tag{15.5}$$

Superimposing the components of stiffness corresponding to the displacements, u_1, u_2 and u_3, from (15.4) and (15.5), the stiffness matrix for the entire structure is obtained as follows:

$$[K] = \begin{array}{c} \begin{matrix} u_1 & \quad u_2 & \quad u_3 \end{matrix} \\ \begin{bmatrix} k_a & -k_a & 0 \\ k_a & (k_a + k_b) & -k_b \\ 0 & -k_b & k_b \end{bmatrix} \begin{matrix} u_1 \\ u_2 \\ u_3 \end{matrix} \end{array} \tag{15.6}$$

15.3.2 *Method of solution*

The above matrix is singular because free-body displacements have been allowed, i.e. it is necessary to apply boundary conditions. If the system is fixed at node 3, such that the deflection of 3 is zero, then the equations become

$$\begin{Bmatrix} Q_1 \\ Q_2 \\ R \end{Bmatrix} = \left[\begin{array}{cc:c} k_a & -k_a & 0 \\ -k_a & (k_a + k_b) & -k_b \\ \hdashline 0 & -k_b & k_b \end{array}\right] \begin{Bmatrix} u_1 \\ u_2 \\ 0 \end{Bmatrix}$$

or

$$\left\{\begin{array}{c} q_F \\ \hdashline R \end{array}\right\} = \left[\begin{array}{c:c} K_{11} & K_{12} \\ \hdashline K_{21} & K_{22} \end{array}\right] \left\{\begin{array}{c} u_F \\ \hdashline 0 \end{array}\right\}$$

That is, the nodal displacements $\{u_F\}$ are given by

$$\{u_F\} = [K_{11}]^{-1}[q_F\} \tag{15.7}$$

and the reactions $\{R\}$ are given by

$$\{R\} = [K_{21}]\{u_F\} \tag{15.8}$$

where Q_1 and Q_2 are loads applied to nodes 1 and 2, respectively, and

$\{q_F\}$ = a vector of externally applied loads corresponding to the free displacements

$[K_{11}]$ = that part of the structural stiffness matrix corresponding to the free displacements

$$= \begin{bmatrix} k_a & -k_a \\ -k_a & (k_a + k_b) \end{bmatrix}$$

$\{u_F\}$ = a vector of free displacements

$$= \begin{Bmatrix} u_1 \\ u_2 \end{Bmatrix}$$

For large stiffness matrices of a banded form, $[K_{11}]$ is, in general, not inverted, and solution is carried out by Gaussian elimination or by Choleski's method.

15.4 Elemental stiffness matrix for a plane rod

(A rod is defined as a member of a framework which resists its load axially, e.g. a member of a pin-jointed truss.)

Under an axial load X, a rod of length l and uniform cross-sectional area A will deflect a distance

$$u = Xl/AE$$

Consider the one-dimensional rod shown in Figure 15.4.

$$X_1 = \frac{AE}{l}(u_1 - u_2) = \text{axial force at node 1}$$

$$X_2 = \frac{AE}{l}(u_2 - u_1) = \text{axial force at node 2}$$

or in matrix form:

$$\begin{Bmatrix} X_1 \\ X_2 \end{Bmatrix} = \frac{AE}{l}\begin{bmatrix} 1 & -1 \\ -1 & 1 \end{bmatrix}\begin{Bmatrix} u_1 \\ u_2 \end{Bmatrix}$$

The above can be seen to be of similar form as (15.3). Hence, the elemental stiffness

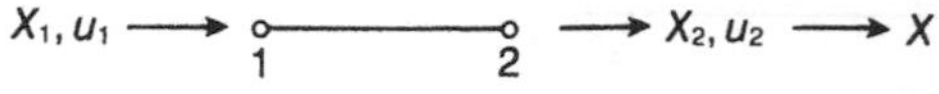

Figure 15.4

matrix for a plane rod is given by

$$[k] = \frac{AE}{l}\begin{bmatrix} 1 & -1 \\ -1 & 1 \end{bmatrix} \tag{15.9}$$

Equation (15.9) is the *elemental stiffness matrix for a rod in local co-ordinates,* but in practice it is more useful to obtain the elemental stiffness matrix in global co-ordinates.

Let, Ox° and Oy° be the global axes and Ox and Oy the local axes, as shown in Figure 15.5

In global co-ordinates, both u and v displacements are important; hence, (15.9) must be written as follows.

$$[k] = \frac{AE}{l}\begin{array}{c} \begin{matrix} u_1 & v_1 & u_2 & v_2 \end{matrix} \\ \begin{bmatrix} 1 & 0 & -1 & 0 \\ 0 & 0 & 0 & 0 \\ -1 & 0 & 1 & 0 \\ 0 & 0 & 0 & 0 \end{bmatrix} \end{array} \begin{matrix} u_1 \\ v_1 \\ u_2 \\ v_2 \end{matrix} \tag{15.10}$$

From Figure 15.5, it can be seen that at node 2

$$\begin{aligned} X_2 &= X_2^0 \cos\theta + Y_2^0 \sin\theta \\ Y_2 &= -X_2^0 \sin\theta + Y_2^0 \cos\theta \end{aligned} \tag{15.11}$$

Similar expressions apply to node 1. Hence, in matrix form

$$\begin{Bmatrix} X_1 \\ Y_1 \\ X_2 \\ Y_2 \end{Bmatrix} = \left[\begin{array}{cc|cc} \cos\theta & \sin\theta & 0 & 0 \\ -\sin\theta & \cos\theta & 0 & 0 \\ \hline 0 & 0 & \cos\theta & \sin\theta \\ 0 & 0 & -\sin\theta & \cos\theta \end{array}\right] \begin{Bmatrix} X_1^0 \\ Y_1^0 \\ X_2^0 \\ Y_2^0 \end{Bmatrix}$$

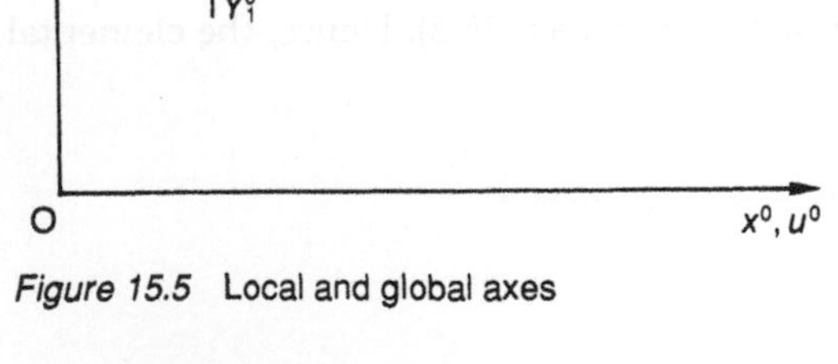

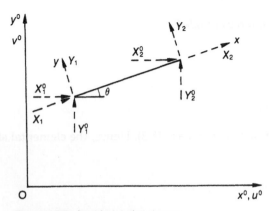

Figure 15.5 Local and global axes

i.e.

$$\{P\} = [\Xi]\{P^0\}$$

where

$[\Xi]$ = a matrix of directional cosines

$\{P^0\}$ = a vector of elemental nodal forces in global co-ordinates

Now, $[\Xi]$ can be seen to be orthogonal, as $\cos^2\theta + \sin^2\theta = 1$, $\cos\theta \times -\sin\theta + \sin\theta \times \cos\theta = 0$, etc.

Hence,

$$[\Xi]^{-1} = [\Xi]^T$$

Therefore

$$\{P^0\} = [\Xi]^T\{P^0\}$$

Similarly,

$$\{u\} = [\Xi]\{u^0\} \tag{15.12}$$

Now,

$$\{P\} = [k]\{u\}$$

Therefore

$$[\Xi]\{P^0\} = [k][\Xi]\{u^0\}$$

$$\{P^0\} = [\Xi]^T[k][\Xi]\{u^0\}$$

and hence,

$$[k^0] = [\Xi]^T[k][\Xi] \tag{15.13}$$

Similarly,

$$\{K^0\} = [\Xi]^T[K][\Xi] \tag{15.14}$$

Hence, from equations (15.10) and (15.13), the *elemental stiffness matrix for a rod in global co-ordinates* is as follows:

$$[k^0] = \frac{AE}{l}\begin{array}{c} \begin{matrix} u_1^0 & v_1^0 & u_2^0 & v_2^0 \end{matrix} \\ \begin{bmatrix} C^2 & CS & -C^2 & -CS \\ CS & S^2 & -CS & -S^2 \\ -C^2 & -CS & C^2 & CS \\ -CS & -S^2 & CS & S^2 \end{bmatrix} \end{array}\begin{matrix} u_1^0 \\ v_1^0 \\ u_2^0 \\ v_2^0 \end{matrix} \tag{15.15}$$

where

$C = \cos\theta$

$S = \sin\theta$

EXAMPLE 15.1

Determine the forces in the members of the plane pin-jointed truss of Figure 15.6. It may be assumed that AE is constant for all members.

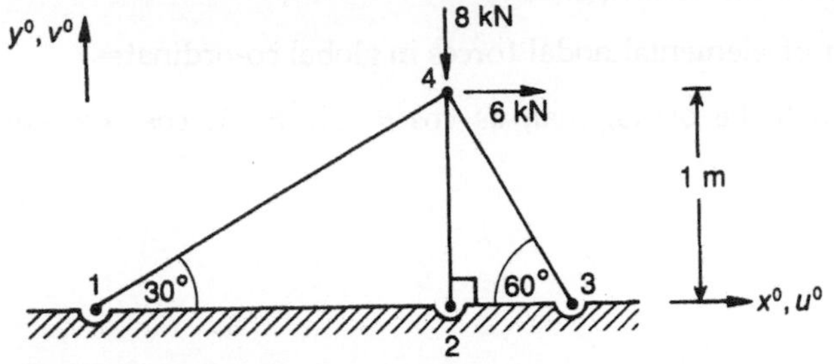

Figure 15.6 Plane pin-jointed truss

SOLUTION

Member 1–4

$\theta = 30°$ C = 0.866 S = 0.5 $l = 2$ m

The member points in the direction from node 1 to node 4 are as shown in Figure 15.7.

Now as the member is firmly pinned at node 1, it is only necessary to consider the components of the stiffness matrix corresponding to the free displacements u_4^0 and v_4^0. Hence, from equation (15.15),

$$[k_{1-4}^0] = \frac{AE}{2}\begin{bmatrix} 0.75 & 0.433 \\ 0.433 & 0.25 \end{bmatrix} \begin{matrix} u_4^0 \\ v_4^0 \end{matrix} \quad (\text{columns } u_4^0,\ v_4^0)$$

$$= AE\begin{bmatrix} 0.375 & 0.216 \\ 0.216 & 0.125 \end{bmatrix} \tag{15.16}$$

Member 2–4

$\theta = 90°$ C = 0 S = 1 $l = 1$ m

The member points in the direction from node 2 to node 4 are as shown in Figure 15.8.

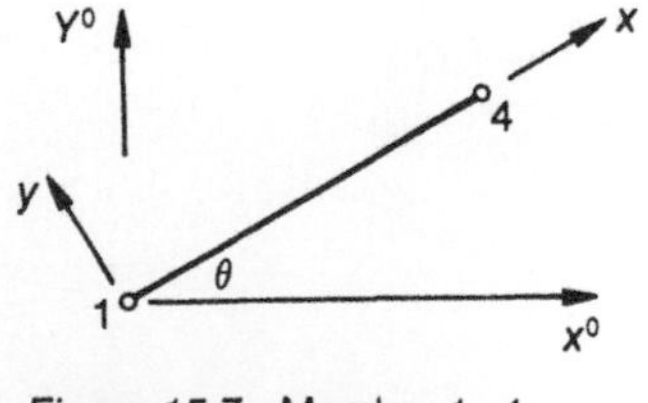

Figure 15.7 Member 1–4

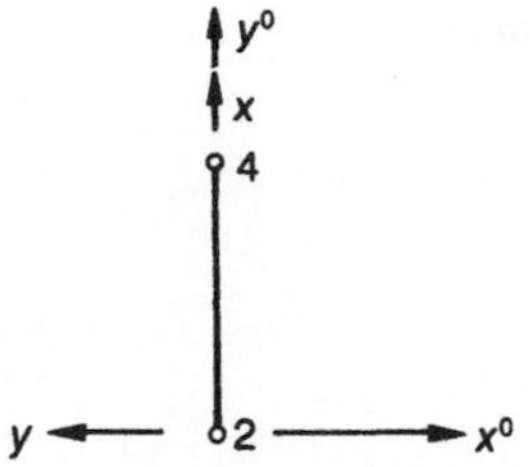

Figure 15.8 Member 2–4

$$[K^0_{2-4}] = \frac{AE}{1}\begin{bmatrix} 0 & 0 \\ 0 & 1 \end{bmatrix} \begin{matrix} u^0_4 \\ v^0_4 \end{matrix} \quad \text{(columns } u^0_4,\ v^0_4\text{)} \tag{15.17}$$

Member 4–3

$\theta = -60°$ C = +0.5 S = −0.866 $l = 1.155$ m

The member points in the direction from node 4 to node 3, as shown in Figure 15.9.

$$[K^0_{4-3}] = \frac{AE}{1.155}\begin{bmatrix} 0.25 & -0.433 \\ -0.433 & 0.75 \end{bmatrix} \begin{matrix} u^0_4 \\ v^0_4 \end{matrix}$$

$$= AE\begin{bmatrix} 0.216 & -0.375 \\ -0.375 & 0.649 \end{bmatrix} \tag{15.18}$$

From (15.16) to (15.18), the *structural stiffness matrix* $[K_{11}]$ corresponding to the *free displacements* is

$$[K_{11}] = AE\begin{bmatrix} (0.375 + 0 + 0.216) & (0.216 + 0 - 0.375) \\ (0.216 + 0 - 0.375) & (0.125 + 1 + 0.649) \end{bmatrix} \begin{matrix} u^0_4 \\ v^0_4 \end{matrix}$$

$$= AE\begin{bmatrix} 0.591 & -0.159 \\ -0.159 & 1.744 \end{bmatrix} \begin{matrix} u^0_4 \\ v^0_4 \end{matrix} \tag{15.19}$$

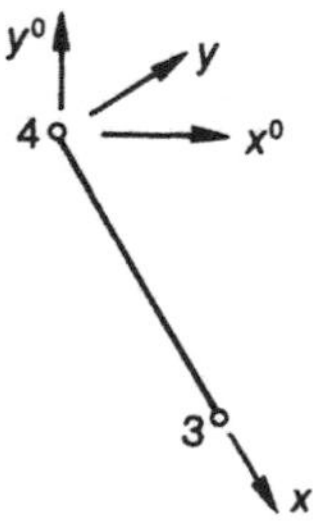

Figure 15.9 Member 4–3

The *vector of loads* corresponding to the *free displacements* is

$$\{q_F\} = \begin{Bmatrix} 6 \\ -8 \end{Bmatrix} \begin{matrix} u_4^0 \\ v_4^0 \end{matrix} \qquad (15.20)$$

Substituting (15.19) and (15.20) into (15.7)

$$\{u_F\} = \begin{Bmatrix} u_4^0 \\ v_4^0 \end{Bmatrix} = \frac{1}{AE}\begin{bmatrix} 0.591 & -0.159 \\ -0.159 & 1.774 \end{bmatrix}^{-1} \begin{Bmatrix} 6 \\ -8 \end{Bmatrix}$$

$$= \frac{1}{AE}\begin{bmatrix} 1.774 & 0.159 \\ 0.159 & 0.591 \end{bmatrix} \begin{Bmatrix} 6 \\ -8 \end{Bmatrix} \Big/ (0.591 \times 1.774) - 0.159^2)$$

$$= \frac{1}{AE}\begin{bmatrix} 1.734 & 0.155 \\ 0.155 & 0.578 \end{bmatrix} \begin{Bmatrix} 6 \\ -8 \end{Bmatrix}$$

or

$$\underline{u_4^0 = 9.16/AE} \qquad (15.21)$$

and

$$\underline{v_4^0 = -3.69/AE} \qquad (15.22)$$

The forces in the members of the framework can be obtained from the theory of Hookean elasticity, if the axial extension or contraction of each element is known. This can be achieved by resolving (15.21) and (15.22) along the local axis of each element, as follows.

Member 1–4

$u_1 = 0$

and u_4 can be obtained from (15.12):

$$u_4 = \lfloor C \quad S \rfloor \begin{Bmatrix} u_4^0 \\ v_4^0 \end{Bmatrix}$$

$$= \lfloor 0.866 \quad 0.5 \rfloor \frac{1}{AE} \begin{Bmatrix} 9.16 \\ -3.69 \end{Bmatrix}$$

$$= 6.088/AE$$

From Hooke's law,

$$F_{1-4} = \text{axial force in member 1–4}$$

$$= \frac{AE}{l} \times (u_4 - u_1)$$

$$= \frac{AE}{2} \times \frac{6.088}{AE} = \underline{3.04 \text{ kN (tensile)}}$$

Member 2–4

$u_2 = 0$

and

$$u_4 = \lfloor C \quad S \rfloor \begin{Bmatrix} u_4^0 \\ v_4^0 \end{Bmatrix}$$

$$= \lfloor 0 \quad 1 \rfloor \frac{1}{AE} \begin{Bmatrix} 9.16 \\ -3.69 \end{Bmatrix}$$

$$= -3.69/AE$$

$F_{2\text{-}4}$ = axial force in member 2–4

$$= \frac{AE}{1} \times (u_4 - u_2)$$

$$= AE \times (-3.69/AE) = \underline{-3.69 \text{ kN (compressive)}}$$

Member 4–3

$u_3 = 0$

and

$$u_4 = \lfloor +0.5 \quad -0.866 \rfloor \frac{1}{AE} \begin{Bmatrix} 9.16 \\ -3.69 \end{Bmatrix}$$

$$u_4 = +7.776/AE$$

$$F_{4\text{-}3} = \frac{AE}{1.155}(u_3 - u_4)$$

$$= \frac{AE}{1.155} \times \left(\frac{-7.776}{AE} \right) = \underline{-6.73 \text{ kN (compressive)}}$$

The method can be applied to numerous other problems, which are beyond the scope of this book, but if the reader requires a greater depth of coverage, he or she should consult references 16–20, 29.

EXAMPLE FOR PRACTICE 15

1. Determine the nodal displacements and member forces in the plane pin-jointed trusses of Figure 15.10(a) and (b). For all members, AE is constant.

{(a) $u_4^0 = -11.673/AE$, $v_4^0 = 5.54/AE$, $F_{1\text{-}4} = -0.88$ kN, $F_{2\text{-}4} = 5.54$ kN, $F_{3\text{-}4} = 6.42$ kN; (b) $u_4^0 = 2.585/AE$, $v_4^0 = -6.545/AE$, $F_{1\text{-}4} = 2.756$ kN, $F_{2\text{-}4} = 6.029$ kN, $F_{3\text{-}4} = 1.98$ kN}

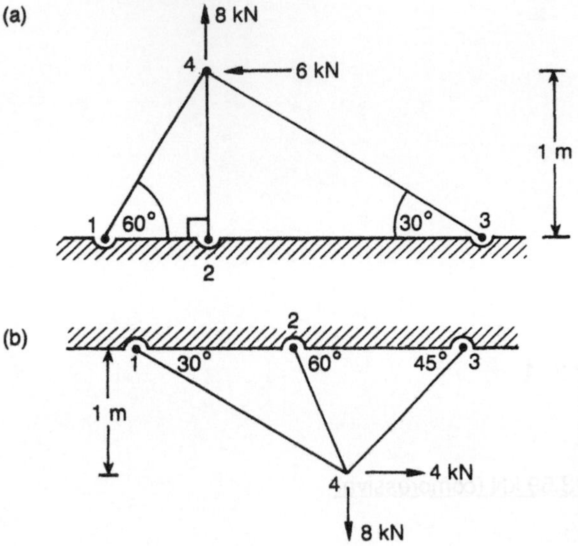

Figure 15.10

16

Experimental strain analysis

16.1 Introduction

In this chapter, a brief description will be given of some of the major methods in experimental strain analysis, and in particular, to the use of electrical resistance strain gauges and photoelasticity.

The aim in this chapter is to expose the reader to various methods of experimental strain analysis, and to encourage him or her to consult other publications which cover this topic in a more comprehensive manner [23–25].

16.2 Electrical resistance strain gauges

The elastic strain in most structures, constructed from steel, aluminium alloy, etc., seldom exceeds 0.1%, and it is evident that such small magnitudes of strain will need considerable magnification to record them precisely. This feature presented a major problem to structural engineers in the past, and it was not until the 1930s that this problem was resolved, when the electrical resistance strain gauge was invented. This gauge was invented as a direct result of requiring lighter aircraft structures, although the principal that the electrical resistance strain gauge is based on was discovered as early as 1856, by Lord Kelvin, when he observed that the electrical resistance of copper and iron wires varied with strain. The discovery of Lord Kelvin was that, when a length of wire is strained, its electrical resistance changes and, within certain limits, this relationship is linear, and can be expressed as follows:

$$\text{Strain} \propto \text{change of electrical resistance; or strain } (\varepsilon) = K\,\frac{\Delta R}{R} \tag{16.1}$$

where

K is known as the gauge factor, and is dependent on the material of construction (i.e. it is a material constant). For an ordinary Cu/Ni gauge, K is about 2.

R is the electrical resistance of the strain gauge (ohms).
ΔR is the change of electrical resistance of the strain gauge (ohms) due to ε.

As K is known, and R and ΔR can be measured, ε can be readily obtained from equation (16.1).

16.2.1 Temperature compensation

If a strain gauge is subjected to a change of temperature, it very often suffers a larger change in length than that caused by external loadings. To overcome this deficiency, it is necessary to attach another strain gauge (called a 'dummy' gauge) to a piece of material with the same properties as the structure itself, and to subject this piece of material to the same temperature changes as the structure, but not to 'constrain' the 'dummy' gauge, or allow it to undergo any external loading.

The dummy gauge should have identical properties to the active gauge, and for static analysis, the two gauges should be connected together in the form of a Wheatstone bridge, as shown in Figure 16.1.

16.2.2 Pressure compensation

If an electrical resistance strain gauge is subjected to fluid pressure, its electrical resistance changes owing to the Poisson effect of the gauge material. To overcome this deficiency, it is necessary to subject the strain gauge to the same pressure medium conditions as the active gauge.

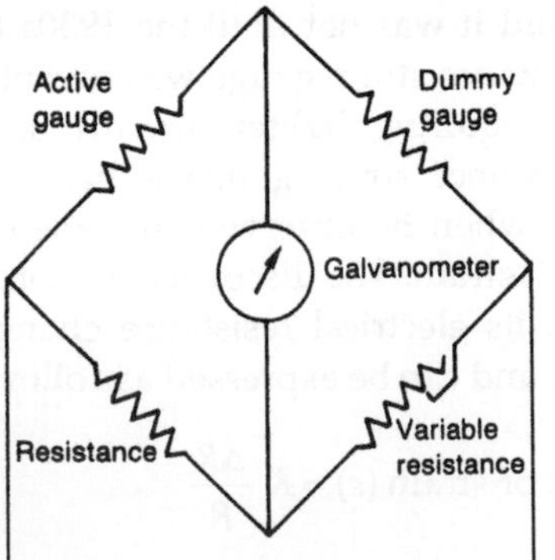

Figure 16.1 The 'null' method of measuring strains

16.2.3 The 'null' method of measuring strains

Figure 16.1 shows the circuitry for this method of strain measurement, which is suitable for static analysis.

Although the active gauge and the dummy gauge may have the same initial electrical resistance, after attaching these gauges to their respective surfaces, there may be small differences in their electrical resistances, so that the galvanometer will become unbalanced. Thus, prior to loading the structure, it will be necessary to balance the galvanometer, by suitably changing the electrical resistance on the variable resistance arm. Once the structure is loaded, and the gauge experiences strain, the galvanometer will once again become unbalanced, and by balancing the galvanometer, the change of electrical resistance due to this load can be measured. Normally, the strain gauge equipment has facilities which allow it to directly record strain, providing the gauge factor is pre-set for the particular batch of gauges.

Further loading of the structure will cause the galvanometer to become unbalanced, once again, and by rebalancing the galvanometer, the strain can be recorded for this particular loading condition.

16.2.4 The 'deflection' method of measuring strain

It is evident from Section 16.2.3 that the 'null' method is only suitable for static analysis, where there is a sufficient time available to record each individual strain. Thus, for dynamic analysis, or where many measurements of static strain are required in quick succession, the 'null' method is unsuitable.

One method of overcoming this problem is to use the 'deflection' method of strain analysis, where the strain gauge circuit of Figure 16.2 is used.

In this case, as the measured strains are very small, they need to be amplified prior to being recorded. The strain recorder can take many forms, varying from chart recorders and storage oscilloscopes to computers. Thus, it is evident that this method

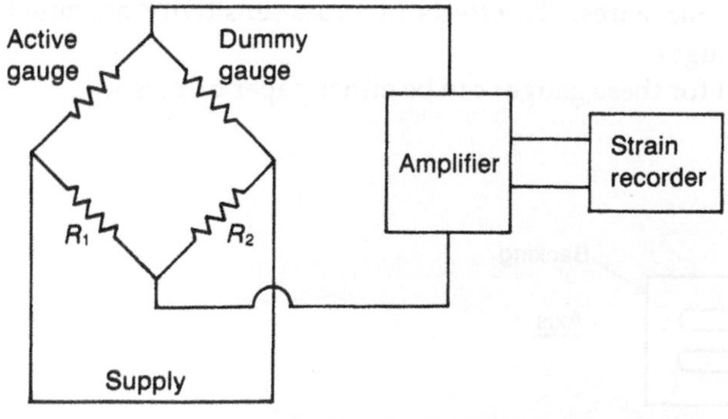

Figure 16.2 Strain gauge circuit for the 'deflection' method of strain measurement

of measuring strain is much more expensive than the null method, because of its requirements for large-scale amplification and sensitive strain recorders, but, nevertheless, it has been successfully employed for dynamic strain recording. The electrical supply shown in Figure 16.2 can be either AC or DC.

16.3 Types of electrical resistance strain gauge

The two most common types of electrical resistance strain gauge are the wire gauge and the foil gauge, and these gauges are now described.

16.3.1 Wire gauges

One of the simplest forms of the wire gauge is the *zigzag* gauge, where the diameter of its wire can be as small as 0.025 mm, as shown in Figure 16.3. The reason why this gauge wire is of zigzag form is because a minimum length of wire is required, so that the supplied power can be sufficiently small to prevent heating of the gauge itself. The gauge is constructed by sandwiching a length of high-resistance wire in zigzag form, between two pieces of paper.

It should be noted that the gauge is attached to the structure via its backing, and that the latter is required as an electrical insulation. The problem with the zigzag gauge is that it is prone to cross-sensitivity.

Cross-sensitivy is an undesirable property of a strain gauge, as the high-resistance wire, which is perpendicular to the axis of the gauge, measures erroneous strains in this direction, in addition to the required strains which lie along the axis of the gauge.

To some extent, cross-sensitivity can be reduced by employing the strain gauge of Figure 16.4. This gauge is constructed by placing thin strands of high-resistance wire parallel to each other, and connecting them together by low-resistance welded cross-bars. Thus, as the electrical resistance of the cross-bars is small compared with the electrical resistance of the wires, the effects of cross-sensitivity are much smaller than for zigzag wire gauges.

The backing material for these gauges can be either paper or plastic.

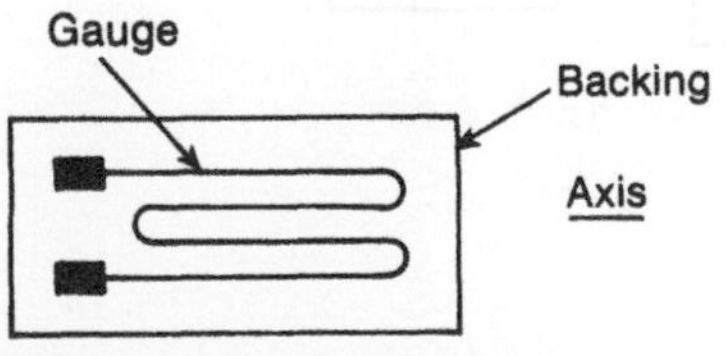

Figure 16.3 A linear zigzag wire gauge

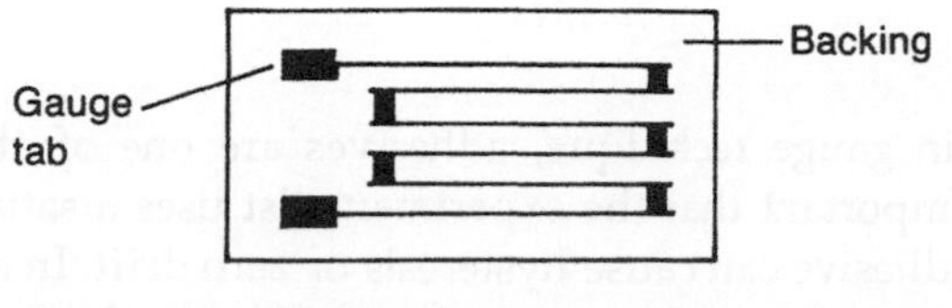

Figure 16.4 A linear wire gauge, with welded cross-bars

16.3.2 *Foil gauges*

Another popular strain gauge is the foil gauge of Figure 16.5, which was invented by Saunders-Roe in 1952. This gauge is etched out of flat metal foil, so that its electrical resistance along its axis can be very large compared with its electrical resistance perpendicular to its axis. This property allows successful construction of such gauges to be as small as 0.2 mm (1/128 in) or as long as 2.54 m (100 in), with negligible cross-sensitivity effects. In the cases of these extreme lengths of strain gauge, considerable expertise is required to attach these gauges successfully to the structure. Small strain gauges are required for strain investigations in regions of stress concentrations, and strain gauges of length 2.54 m are used to investigate the longitudinal strength characteristics of ship structures.

In general, the most popular lengths of linear gauges vary between about 5 and 10 mm. The main advantage of the foil gauge is that because it is etched out of flat metal foil, it and its backing can be very thin, so that the gauge can be in a more intimate contact with the structure than can a wire gauge.

Disadvantages of foil gauges are that they are generally more expensive than wire gauges, and that for very thin foil gauges, they can easily fracture, if roughly handled. Foil gauges are usually plastic backed.

16.4 Gauge material

Most electrical resistance strain gauges of normal resistance (about 100 ohms) are constructed from a copper/nickel alloy, where the proportion of Cu:Ni usually varies from about 55:45 to 60:40. For such gauges, the gauge factor is about 2, and depends on the Cu:Ni ratio.

A popular alloy for high-resistance gauges (about 1000 ohms) is called *Nichrome*, which consists of 75% Ni, 12% Fe, 11% Cr and 2% Mn.

Figure 16.5 A linear foil gauge

16.5 Gauge adhesives

In the electrical resistance strain gauge technique, adhesives are one of the most important considerations. It is important that the experimentalist uses a satisfactory adhesive, as an unsatisfactory adhesive can cause hysteresis or zero drift. In addition to this, another important consideration is the preparation of the surface to which the gauge is to be attached.

Prior to attaching the gauge, it should be ensured that the surface is clean and free from rust, paint, dirt, grease, etc. This surface should not be highly polished; if it is, it should be roughened a little with a suitable abrasive, and then it should be degreased with a suitable solvent, such as acetone, alcohol, etc. After application of the strain gauge cement to the appropriate part of the structure's surface, and in some cases to the back of the gauge itself, the gauge should be gently placed onto its required position.

To ensure that no air bubbles become trapped between the gauge and the surface of the structure, the thumb, or finger, should be pressed firmly onto the surface of the gauge, and gently rolled to and fro, until any excess strain gauge cement is squeezed out.

The adhesion time of the gauge varies from a few minutes to several days, depending on the type of adhesive used and the environment that the adhesive is to be exposed to.

Gauge adhesives are generally either organic based or ceramic based, the former being satisfactory for temperatures below 260 °C and the latter for temperatures in excess of this. A brief description of some of the different types of gauge adhesive will now be given.

16.5.1 *Cellulose acetates*

These are among the most common forms of adhesive (e.g. Durofix) that can be bought from high street shops. They adhere by evaporation, and take from 24 hours to 3 days to gain full strength, depending on the surrounding temperature. They are usually used for paper-backed gauges.

16.5.2 *Epoxy resins*

These require the mixing of a resin with a hardener. A popular combination is to use Araldite strain gauge cement, with either Araldite hardener HY951 or Araldite hardener HY956, in the following proportions by weight:

Araldite strain gauge cement	100
with either HY951 hardener	4 to 4.5
or HY956 hardener	8 to 10

Another epoxy resin adhesive is the M-Bond epoxy supplied by Welwyn Strain measurements.

These cements adhere by curing, where the time taken to gain full strength can take 24 hours, but in the case of Araldite, the curing time can be accelerated by exposing the cement to ultra-violet radiation. These adhesives are suitable for either plastic-backed or paper-backed gauges.

16.5.3 *Cyanoacrylates*

These are pressure adhesives, which adhere by applying pressure to the cement, via the gauge. One of the earliest of these adhesives is called Eastman-Kodak 910, which allows a strain gauge to be used within a few minutes of attaching it to the structure.

Another more recent cyanoacrylate is M-Bond 200, which is supplied by Welwyn Strain Measurements.

16.5.4 *Norton-Rokide*

This adhesive is suitable for high-temperature work, and because of this, the strain gauge does not normally have any backing. Instead, the cement is first sprayed onto the surface, and then the gauge is firmly pressed down, but care has to be taken to ensure that some cement lies between the gauge and the structure, so that its resistance to earth does not break down. After the wires have been attached to the gauge tabs, further cement is sprayed over the gauge and the surrounding surface, so that it is encapsulated.

16.6 Water-proofing

If a strain gauge is exposed to water or damp conditions, its electrical resistance to earth will break down, and render the gauge useless. Thus, if a gauge is likely to be exposed to such an environment, it is advisable to water-proof the gauge and its wiring.

Some methods of water-proofing strain gauges are discussed below, but prior to using any of these methods, it should be assumed that the gauges and their surrounding surfaces are free from water.

Di-Jell is a micro-crystalline wax which has the appearance of a jelly-like substance. In general, it is only suitable for damp-proofing or for water-proofing when the water is stationary. To water-proof the gauge, the Di-Jell is simply applied to the gauge and its surrounding surface.

Silicone greases and petroleum jelly (Vaseline) can also be used for damp-proofing, but care should be taken to ensure that these substances are not subjected to temperatures which will melt them.

A more robust and permanent method of water-proofing strain gauges is that recommended by Welwyn Strain Measurements Limited. This consists of painting M-Coat A or D over the gauge and its adhesive, and then covering the M-Coat with aluminium foil. Finally, the whole surface is covered with M-Coat G, the electrical leads being first covered with M-Coat B, as shown in Figure 16.6.

Prior to the development of the Welwyn method of water-proofing strain gauges, the Saunders–Roe technique proved popular. This method consisted of applying successive coats of expoxy resin and glass cloth, and covering the whole with an impervious rubber-based solution.

Other methods of water-proofing consist simply of covering the gauge and its surrounding surface with various types of sealant, including automobile underbody sealants.

Line [26] carried out an investigation on the water-proofing qualities of a number of sealants, as shown in Table 16.1. Line used foil gauges of dimensions 2.54 cm length, 1.02 cm width. He measured the electrical resistance of these strain gauges in air, before immersion into water, and then took these measurements again, 1 hour and also 3 weeks later, after continuous immersion in water. He also measured the electrical resistances of these gauges at water pressures of 3.45 and 6.9 MPa. He

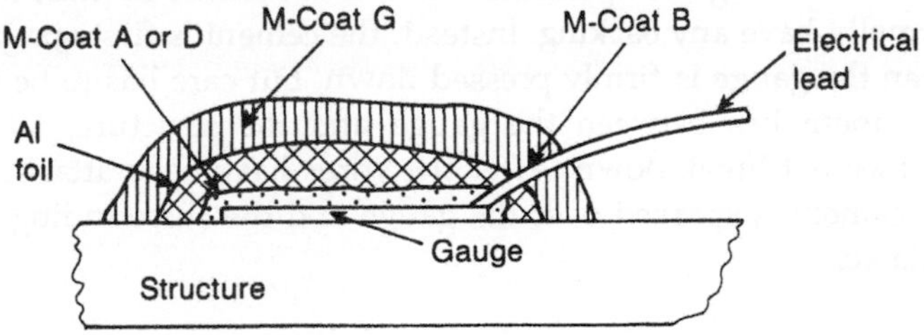

Figure 16.6 Welwyn method of water-proofing

Table 16.1 **Effect of pressure and water immersion on the electrical resistance of strain gauges**

Sealant	Electrical resistance (ohms)				
	Before	1 hour in water	3 weeks in water	3.45 MPa	6.9 MPa
Bostik 6	55	54	55	55	55
Underbody seal	56	56	55	56	56
M-Coat A	56	56	55	56	56
M-Coat D	56	55	55	55	55
Di-Jell	52	52	52	–	–

found that all the sealants were satisfactory under test, and he recorded the following observations:

(a) The underbody seal appeared to have softened and, although it was still waterproof, it could easily be chipped off by hand.
(b) Under the M-Coat A, it was possible to see that the metal was completely rust-free.
(c) The pressures were too low and the instruments were too insensitive to record a change of resistance due to the effects of pressure.

NB Dally and Riley [25] report on tests by Milligan [27], and by Brace [28], who found that the effect of pressure caused a strain of about 0.58×10^{-6} per MPa of pressure, and they concluded that for most problems, the effects of pressure on strain gauges can be ignored at pressures below 20.7 MPa (3000 lbf/in^2).

16.7 Other strain gauges

Other forms of gauge include *shear pairs* and *strain rosettes,* as described in Chapter 7, together with *crack measuring* and *diaphragm* gauges. Diaphragm gauges consist of a combination of radial and circumferential gauges, and crack measuring gauges consist of several parallel strands of wire and tabs.

16.8 Gauge circuits

Skilful use of strain gauge circuits can eliminate the use of dummy gauges and provide increased sensitivity.

16.8.1 Combined bending and torsion of circular section shafts

For circular-section shafts, under the effects of combined bending and torsion, it is convenient to use two pairs of 'shear pairs' as shown in Figure 16.7. By fitting two pairs of shear pairs, it is possible to record either the bending moment M or the torque T, depending on the circuit used.

The shear pairs must be fitted at 45° to the axis of the shaft, and a typical shear pair of strain gauges is shown in Figure 16.8. Shear pairs are so called because, under

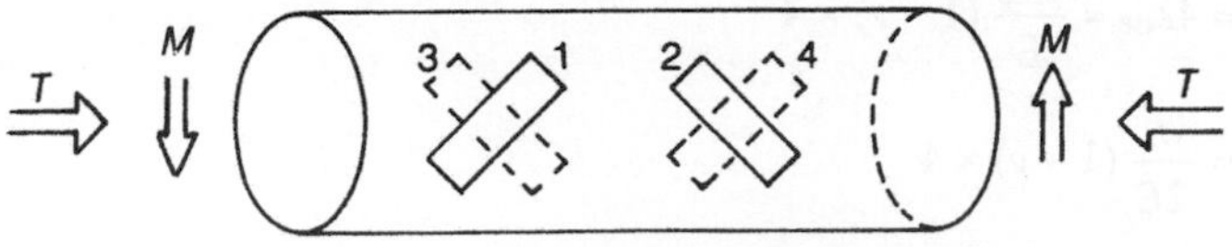

Figure 16.7 Shaft under combined bending and torque

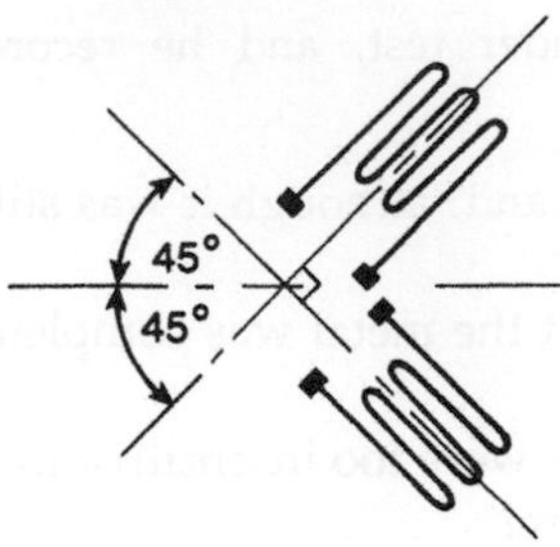

Figure 16.8 A shear pair of strain gauges

pure torque, the maximum principal stresses in a circular-section shaft, which are numerically equal to the maximum shear stress, lie at 45° to the axis of the shaft (see Chapter 7).

16.8.2 To determine M

Let

ε_{45B} = direct strain due to M, which lies at 45° to the axis of the shaft

ε_{45T} = direct strain due to T, which lies at 45° to the axis of the shaft

= $\gamma/2$ (see Chapter 7)

γ = maximum shear strain due to T

If the gauges are connected together in the form of a full Wheatstone bridge, as shown in Figure 16.9, γ will be eliminated. From Figure 16.9, the *output* will be

$$(\text{gauge 1} - \text{gauge 3}) - (\text{gauge 4} - \text{gauge 2})$$

$$= \left(\varepsilon_{45B} + \frac{\gamma}{2} + \varepsilon_{45B} - \frac{\gamma}{2}\right) - \left(-\varepsilon_{45B} - \frac{\gamma}{2} - \varepsilon_{45B} + \frac{\gamma}{2}\right)$$

$$= 4\varepsilon_{45B}$$

or

$$\text{Output} = 4\varepsilon_{45B} = \frac{\sigma_{45B}}{E}(1 - \nu) \times 4$$

$$= \frac{\sigma_B}{2E}(1 - \nu) \times 4$$

$$= 2\sigma_B(1 - \nu)/E$$

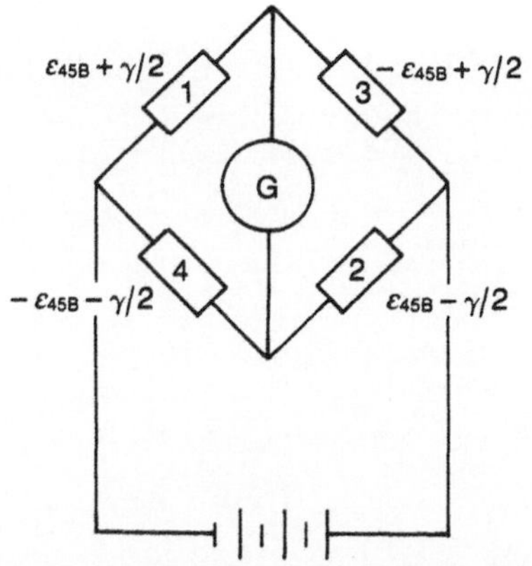

Figure 16.9 Circuit for measuring *M*

Therefore

$$\sigma_B = \frac{\text{Output}}{2(1-\nu)} \times E = \text{bending stress}$$

but

$$M = \sigma_B \times \frac{\pi d^4}{64} \times \frac{2}{d}$$

$$\underline{M = \frac{\pi d^3 \sigma_B}{32}} \quad (16.2)$$

16.8.3 *To determine* T

By adopting the circuit of Figure 16.10, ε_{45B} can be eliminated. The *output* from the circuit of Figure 16.10 will be

(gauge 1 – gauge 2) – (gauge 4 – gauge 3)

$$= \left(\varepsilon_{45B} + \frac{\gamma}{2} - \varepsilon_{45B} + \frac{\gamma}{2}\right) - \left(-\varepsilon_{45B} - \frac{\gamma}{2} + \varepsilon_{45B} - \frac{\gamma}{2}\right)$$

Output = 2γ

or

$\underline{\gamma = \text{output}/2}$

but

$$T = \tau \times \frac{\pi d^4}{32} \times \frac{2}{d}$$

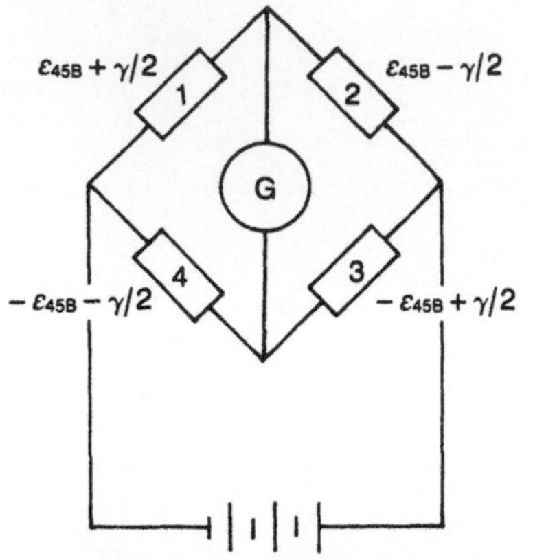

Figure 16.10 Circuit for measuring T

or

$$T = \frac{\pi G \gamma d^3}{16} \tag{16.3}$$

NB Strains due to axial loads, including *thermal effects,* will automatically be eliminated when the circuits of Figures 16.9 and 16.10 are adopted.

16.8.4 *Combined bending and axial strains*

Consider a length of beam under combined bending and axial load, as shown in Figure 16.11.

Let

ε_B = bending strain due to M

ε_D = direct strain due to P, plus thermal effects

The bending moment M can be obtained by adopting the circuit of Figure 16.12.

The output from the circuit of Figure 16.12 is

gauge 1 – gauge 2 = $\underline{2\varepsilon_B}$

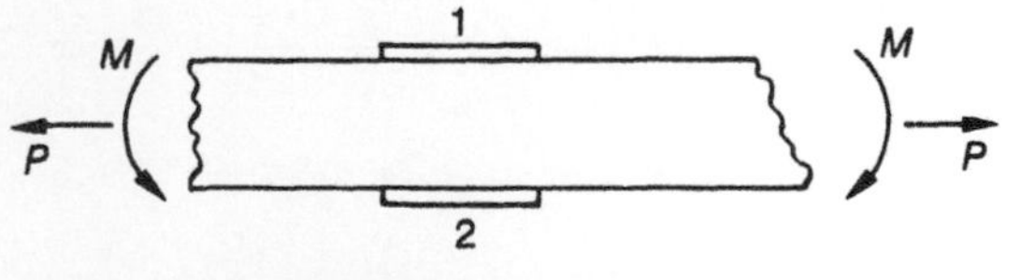

Figure 16.11 Combined M and P

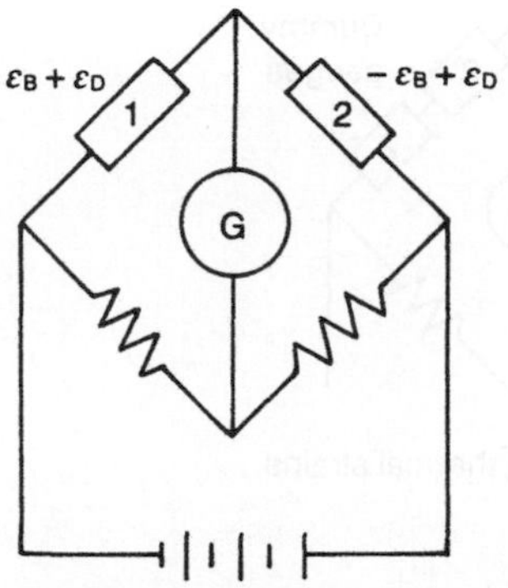

Figure 16.12 Circuit to determine *M*

or

$\underline{\varepsilon_B = \text{output}/2 = \sigma_B/E}$

Hence, M can be obtained.

16.8.5 Full bridge for measuring M

$$\begin{aligned}\text{Output} &= (\text{gauge 1} - \text{gauge 2}) - (\text{gauge 4} - \text{gauge 3})\\ &= 4\varepsilon_B\end{aligned}$$

i.e. the circuit of Figure 16.13 will give four times the sensitivity of a single strain gauge, and twice the sensitivity of the circuit of Figure 16.12. It will also automatically eliminate thermal strains.

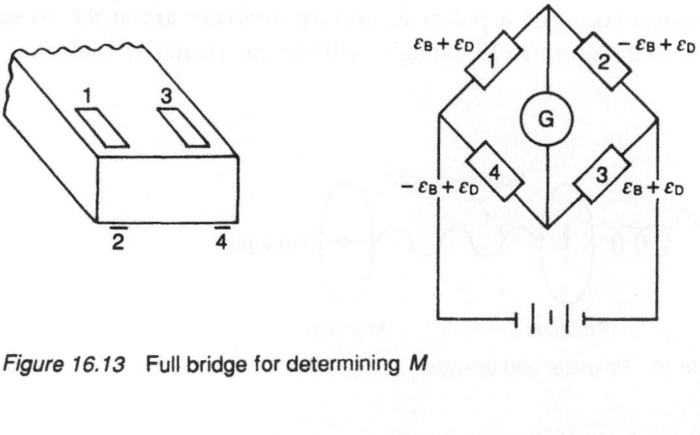

Figure 16.13 Full bridge for determining *M*

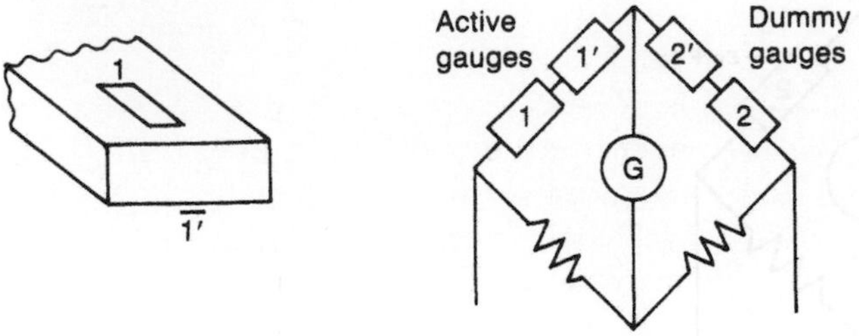

Figure 16.14 Circuit for eliminating bending and thermal strains

16.8.6 *Half-bridge for measuring axial strains (see Figure 16.14)*

Output $= 2\varepsilon_D = 2 \times \sigma_D/E$

Hence, P can be obtained.

16.9 Photoelasticity

The practice of photoelasticity is dependent on the shining of light through transparent or translucent materials, but prior to discussing the method, it will be necessary to make some definitions.

A beam of ordinary light vibrates in many planes, transverse to its axis of propagation. It is evident that, as light vibrates in so many different planes, it will be necessary for a successful photoelastic analysis to allow only those components of a beam of light to vibrate in a known plane.

This is achieved by the use of a *polarizer*, which, in general, only allows those components of light to pass through it that vibrate in a vertical plane. Polarizers are made from Polaroid sheet, and apart from photoelastic analysis, they are used for sunglasses ('shades' in the USA).

A polarizer which has a horizontal axis of transmission is called an *analyzer*. Thus, if the axes of transmission of a polarizer and an analyzer are at 90° to each other (crossed), as shown in Figure 16.15, no light will emerge from the analyzer.

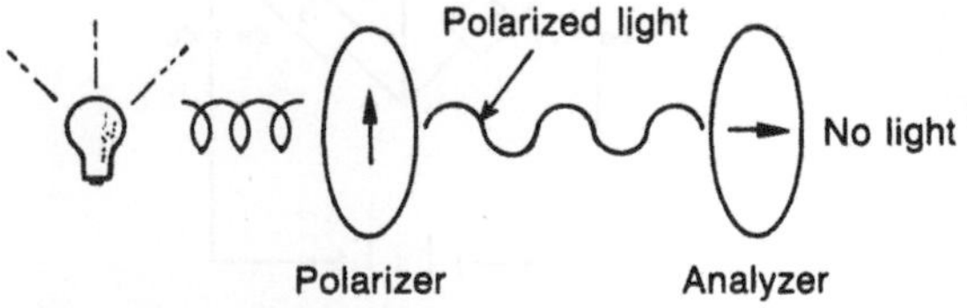

Figure 16.15 Polarizer and analyzer

16.9.1 Birefringence

The practice of photoelasticity is dependent on birefringence or double refraction. Birefringence can be described as the property possessed by some transparent and translucent materials, whereby a single ray of polarized light is split into two rays on emerging from the birefringent material, where the two rays are perpendicular to each other. In general, the two emerging rays are out of phase, and this is called the relative retardation. Some materials are permanently birefringent; others, such as that from which the models are constructed are only birefringent under stress. In the case of the latter, a single ray of polarized light is split into two rays on emerging from the model, where the directions of the two rays lie along the planes of the principal stresses. The relative retardation of the two rays, namely R_t, is proportional to the magnitude of the principal stresses and the thickness of the model, as follows:

$$R_t + C(\sigma_1 - \sigma_2)h \tag{16.4}$$

where

C = the stress optical coefficient (i.e. it is a material constant)

σ_1 = maximum principal stress

σ_2 = minimum principal stress

h = model thickness

Typical materials used for photoelastic models include epoxy resin (e.g. Araldite), polycarbonate (e.g. Lexan or Makrolan), urethane rubber, etc.

A plane polariscope is one of the simplest pieces of equipment used for photoelasticity. It consists of a light source, a polarizer, a model, an analyzer and a screen, as shown in Figure 16.16.

When the model is unstrained, no light will emerge, but when the model is strained, strain patterns will appear on the screen. This can be explained as follows.

The polarized light emerges from the model along the directions of the planes of the principal stresses, so that these rays will vibrate at angles to the vertical plane. On emerging from the analyzer, only those components of light which vibrate in a horizontal plane will be displayed on the screen, to give a measure of the stress distribution in the model. If daylight or white light is used in a plane polariscope, the stress patterns will involve all the colours of the rainbow. The reason for this is that

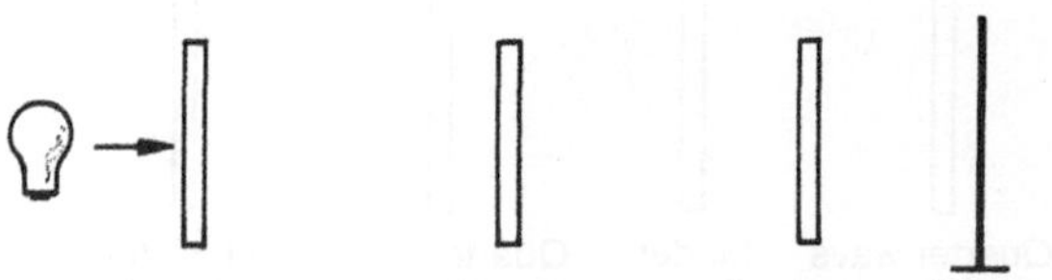

Figure 16.16 A plane polariscope

daylight or white light is composed of all the colours of the rainbow, each colour vibrating at a different frequency. The approximate frequency of deep red light is about 390×10^{12} Hz, and the approximate frequency of deep violet light is 770×10^{12} Hz, the speed of light in a vacuum being about 2.998×10^8 m/s (186 282 miles/s). The difference in frequencies between different coloured lights is one of the reasons why violet is on the inside of the rainbow and red is on the outside, and these differences probably account for why we can distinguish between different colours.

However, whereas the stress patterns are quite spectacular when daylight is used, it is difficult to analyze the model. For this reason, it is preferable to use *monochromatic light*.

Monochromatic light is light of one wavelength only, and when it is used, the stress patterns consist of dark lines against a light background. Typical lamps used to produce monochromatic light include mercury vapour and sodium.

The use of monochromatic light in a plane polariscope will produce *isoclinics* and *isochromatics*, which are defined as follows:

> *Isoclinic fringe patterns* are lines of constant stress direction, and occur when monochromatic light is used. They are useful for determining the principal stress directions.
>
> *Isochromatic fringe patterns* are lines of constant maximum shear stress, i.e. lines of constant $(\sigma_1 - \sigma_2)$, and occur when monochromatic light is used.
>
> *Isotropic points* are points on the model where $\sigma_1 = \sigma_2$, and *singular* points are points on the model where $\sigma_1 = \sigma_2 = 0$.

16.9.2 Circularly polarized light

One of the problems with using a plane polariscope is that both isochromatic and isoclinic fringes appear together, and much difficulty is experienced in distinguishing between two. The problem can be overcome by using a circular polariscope, which can extinguish the isoclinics. A plane polariscope can be converted to a circular polariscope by inserting *quarter-wave plates*, as shown in Figure 16.17.

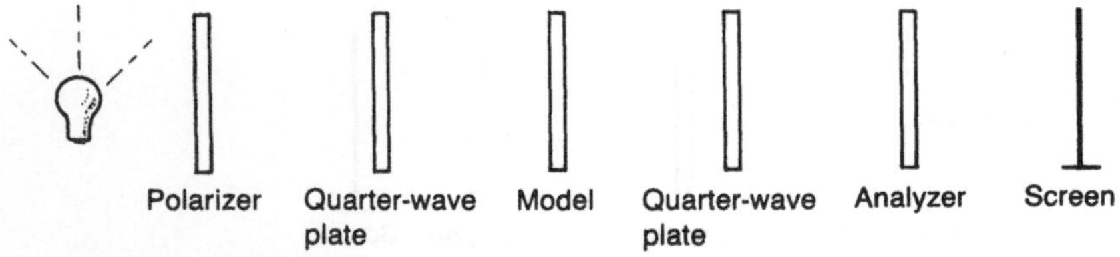

Figure 16.17 Circular polariscope

Quarter-wave plates are constructed from permanently birefringent materials, which have a relative retardation of $\lambda/4$, and to extinguish the isoclinics, they are placed with their fast axes at 45° to the planes of polarization of the polarizer and the analyzer, and at 90° to each other, where λ is the wavelength of the selected light.

16.9.3 Some notes on experimental photoelasticity

To obtain the isoclinics, remove the quarter-wave plates and rotate the polarizer and the analyzer by the same angle and in the same direction. This process will make the isoclinic fringes move with the change of angle, but will not affect the isochromatic fringes.

To obtain *isochromatic half-fringes* in a circular polariscope, place the axes of polarization of the polarizer in the same plane as the axis of polarization of the analyzer.

The *isochromatic fractional fringes* can be obtained by appropriately placing the plane of the polarizer at various angles, other than 0° and 90°, to the plane of polarization of the analyzer.

16.9.4 Material fringe value f

The material fringe value for any given photoelastic material is given by

$$f = \frac{\lambda}{2C} \text{ N/(m. fringe)} \tag{16.5}$$

where

C = stress optical coefficient

λ = wavelength of the selected light

If there are *n isochromatic fringes*, then from equations (10.4) and (10.5),

$$\sigma_1 - \sigma_2 = 2\hat{\tau} = nf/h \tag{16.6}$$

where

h = model thickness

$\hat{\tau}$ = maximum shear stress at the nth fringe

Thus, at the ith fringe, the maximum shear stress τ_i is given by

$$\underline{\hat{\tau}_i = i \times f/2h}$$

Typical values of f, under mercury green light, together with other material constants for some photoelastic materials, are given in Table 16.2.

Table 16.2 **Some material constants for photoelastic materials**

Material	*f* (kN/m. fringe)	*E* (MPa)	ν	UTS* (MPa)
Araldite B	10.8	2760	0.38	70
Makralon	7.0	2930	0.38	62
Urethane rubber	0.2	3	0.46	0.15*

* This quoted value is not the ultimate tensile (UTS) of the material, but its elastic limit.

16.9.5 Stress trajectories

These are a family of lines which lie orthogonally to each other. They are lines of constant principal stress, and they can be constructed from isoclinics.

16.9.6 Three-dimensional photoelasticity

One popular method in three-dimensional photoelasticity is to load the model whilst it is being subjected to a temperature between 120 and 180 °C. The model is then allowed to cool slowly to room temperature, but the load is not removed during this process. On removing the load, when the model is at room temperature, the stress system will be 'frozen' into the model, and then by taking thin two-dimensional slices from the model, a photoelastic investigation can be carried out – sometimes by immersing the slices into a liquid of the same refractive index as the model's material.

There are many other methods in three-dimensional photoelasticity, but descriptions of these methods are beyond the scope of this book.

16.10 Moiré fringes

These have no connection whatsoever with the photoelastic method. They are in fact interference patterns whereby a suitable pattern is placed or shone onto the structure. The pattern is then noted or photographed before and after deformation, and the two patterns are superimposed to produce 'fringes'. Examination of these fringes by the use of a comparator or a microscope can be made to determine the experimental strains.

The patterns can consist of lines, grids or dots, etc., which can be parallel radial or concentric, depending on the shape of the structure to be analyzed.

16.11 Brittle lacquer techniques

This is another method of experimental strain analysis, where a thin coating of a brittle lacquer is sprayed or painted onto the surface of the structure, before it is loaded. After loading the structure, the lacquer will be found to have cracked patterns, and these patterns will be related to the direction and magnitude of the maximum tensile stresses. One of the most popular brittle lacquers is called 'stress-coat', which is manufactured by the Magnaflux Corporation of the USA.

One of the problems with using brittle lacquer is that the cracks only occur under tension, so that if regions of compressive stresses are required to be examined, the lacquer must be applied to the surface of the structure while it is under load. On removing the load, the brittle lacquer will crack in those zones where the structure was in compression when loaded.

16.12 Semi-conductor strain gauges

These are constructed from a single crystal of silicon, and each gauge takes the form of a short rectangular filament.

Semi-conductor strain gauges are very sensitive to strain, where their change of resistance for a given strain can be about 100 times the value of the change of resistance of an electrical resistance strain gauge. Semi-conductor strain gauges are usually small (e.g. 0.5 mm), and because of this and their high sensitivity, they are particularly useful for experimental strain analysis in regions of high stress concentration. Their main disadvantage is that they are much more expensive than electrical resistance strain gauges. They are usually manufactured from either a P-type material or an N-type material. The gauges can be made to be self-temperature compensating, by constructing the gauge from a P-type element together with an N-type element, the two being connected in one half of a Wheatstone bridge.

16.13 Acoustical gauges

These gauges are based on measuring the magnitude of the resonating frequency of a piece of wire stretched between two knife edges. Initially, the wire must be in tension, so that any compressive strains it is likely to receive under load will not cause it to lose its bending stiffness. The wire is 'plucked' by an electromagnet, and another electromagnet, called the 'pick-up', receives the signals from the resonating wire. The signal received by the pick-up magnet is then amplified, and apart from the signal being sent to an oscilloscope, it is used further to excite the 'plucking' magnet, so that the amplitude of the resonating wire will be maximized.

One knife edge of the acoustical gauge is fixed, and the other knife edge is movable, so that the latter can transmit strain to the resonating wire. Change of length of the wire will cause its resonant frequency to change, which can then be compared with a reference acoustical gauge. This gauge can then be adjusted with the aid of a micrometer screw gauge attached to it, so that its resonant frequency is the same as the gauge under test, and, hence, the strain recorded.

Acoustical gauges vary in length from 2.54 cm (1 in) to 15.24 cm (6 in), and because of their bulk, they are generally only preferred to electrical resistance strain gauges in special circumstances, where their inherent robustness and long-term stability characteristics are considered to be of prime importance.

References

1 Ross, C. T. F. *Computational Methods in Structural and continuum Mechanics*, Ellis Horwood, Chichester, 1982.
2 Williams, J. G. *Stress Analysis of Polymers*, Longman, Harlow, 1973.
3 Macaulay, W. H. Note on the Deflection of Beams, *Messenger Math.*, **48**, 129–130, 1919.
4 Stephens, R. C. *Strength of Materials*, Arnold, London, 1970.
5 Ross, C. T. F. *Finite Element Methods in Engineering Science*, Ellis Horwood, Chichester, 1993.
6 Johnson, D. *Advanced Structural Mechanics*, Collins, London, 1986.
7 Rockey, K. C., Evans, H. R., Griffiths, D. W. and Nethercot, D. A. *The Finite Element Method*, 2nd Edition, Collins, London, 1983.
8 Saada, A. S. *Elasticity – Theory and Applications*, Pergamon, New York, 1974.
9 Den Hartog, J. P. *Advanced Strength of Materials*, McGraw-Hill, New York, 1952.
10 Timoshenko, S. P. and Goodier, J. N. *Theory of Elasticity*, McGraw-Hill-Kogakusha, New York, 1970.
11 Ford, Hugh and Alexander, J. M. *Advanced Strength of Materials*, Ellis Horwood, Chichester, 1977.
12 Ross, C. T. F. *Finite Element Programs for Axisymmetric Problems in Engineering*, Ellis Horwood, Chichester, 1984.
13 Ross, C. T. F. The Collapse of Ring-reinforced Circular Cylinders under Uniform External Pressure, *Trans. RINA*, **107**, 375–94, 1965.
14 Richards, T. H. *Energy Methods in Stress Analysis*, Ellis Horwood, Chichester, 1977.
15 Zienkiewicz, O. C. *The Finite Element Method*, 3rd Edition, McGraw-Hill, Maidenhead, 1977.
16 Cook, R. D. *Concepts and Applications of Finite Element Analysis*, 2nd Edition, Wiley, New York, 1981.
17 Segerlind, L. J. *Applied Finite Element Analysis*, 2nd Edition, Wiley, New York, 1984.
18 Irons, B. and Ahmad, S. *Techniques of Finite Elements*, Ellis Horwood, Chichester, 1980.
19 Fenner, R. T. *Finite Element Methods for Engineers*, Macmillan, London, 1975.
20 Davies, G. A. O. *Virtual Work in Structural Analysis*, Wiley, Chichester, 1982.
21 Smith, I. M. *Programming the Finite Element Method*, Wiley, Chichester, 1982.
22 Owen, D. R. J. and Hinton, E. *A Simple Guide to Finite Elements*, Pineridge, Swansea, 1980.

23 Coker, E. G. and Filon, L. N. G. *A Treatise on Photoelasticity*, Cambridge University Press, 1931.
24 Holister, G. S. *Experimental Stress Analysis – Principles and Methods*, Cambridge University Press, 1967.
25 Dally, J. W. and Riley, W. F. *Experimental Stress Analysis*, 2nd Edition, McGraw-Hill-Kogakusha, New York, 1978.
26 Line, D. R. Investigation into the Use of Some Commercial and Industrial Adhesives and Sealants for Electrical Resistance Strain Gauge Application, Final Year Project, Dept. of Mech. Eng., Portsmouth Polytechnic, 1977–1978.
27 Milligan, R. V. The Effects of High Pressure on Foil Strain Gages, *Exp. Mech.*, **4**, 25–36, 1964.
28 Brace, W. F. Effect of Pressure on Electrical Resistance Strain Gages, *Exp. Mech.*, **4**, 212–16, 1964.
29 Megson, T. H. G. *Aircraft Structures*, Edward Arnold, London, 1972.
30 Bolotin, V. V. *The Dynamic Stability of Elastic System*, Holden-Day, San Francisco, 1964.
31 Stroud, K. A. *Further Engineering Mathematics*, Macmillan, Basingstoke, 1986.

Index

ADVANCED APPLIED FINITE ELEMENT METHODS
CARL T.F. ROSS, Professor of Structural Dynamics, Department of Mechanical and Manufacturing Engineering, University of Portsmouth
ISBN 1-898563-51-9 450 pages 1998

A pragmatic and comprehensive coverage of applied finite elements in engineering science. It is a companion volume to the author's *Finite Element Techniques in Structural Mechanics* (Horwood Publishing, 1997) which begins at a lower level and is less interdisciplinary and rigorous. An abundance of worked examples appropriate to the techniques appear throughout the text. Examples for practice appear as problems in a section at the end of most chapters. ***Computer programs available on internet from*** **www.mech.port.ac.uk/sdelby/mbm/CTFRprog.html**

Contents: Matrix algebra; Basic structural concepts and energy theorems; The discrete system; Static analysis of pin-jointed trusses; Static analysis of rigid-jointed frames; Finite element analysis; In-plane quadrilateral elements; Vibrations of structures; Grillages; Non-linear structural mechanics; Steady-state field problems; Axisymmetric problems; Transient field problems; The modal analysis method; Mathematical modelling.

"An excellent introduction ... very readable, clear and easy to follow."
Chartered Mechanical Engineer

FINITE ELEMENT PROGRAMS IN STRUCTURAL ENGINEERING AND CONTINUUM MECHANICS
CARL T.F. ROSS, Professor of Structural Dynamics, Department of Mechanical and Manufacturing Engineering, University of Portsmouth
ISBN 1-898563-28-4 650 pages 1996

This undergraduate and postgraduate book will serve for courses in mechanical, civil, structural and aeronautical engineering; and naval architecture, written in a step-by-step methodological approach.

"Computer programs for finite element analysis ... students and lecturers find them of value"
Structural Engineer (Professor I.A. Macleod, Strathclyde University, Glasgow)

"These programs, written in Quick Basic, utilize the finite element method to solve a variety of engineering problems ranging from static and dynamic analysis in two and three dimensions to two-dimensional field ... recommended to readers with a strong theoretical background, and upper-division undergraduates through to professionals"
Choice, American Library Association (Dr.D.A. Pape, Alfred University, USA)

DYNAMICS OF MECHANICAL SYSTEMS

CARL T.F. ROSS, Professor of Structural Dynamics, Department of Mechanical Engineering and Manufacture, University of Portsmouth

ISBN: 1-898563-34-9 260 pages 1996

This fundamental introduction for first and second year undergraduates reading mechanical, civil, structural and aeronautical engineering, and naval architecture will also appeal to BTEC students. The step-by-step and methodical approach is aimed to help students who find difficulty with mathematics and Newtonian Physics.

Contents: Introduction to statics and dynamics; Kinematics of particles; Kinetics of particles: Force, mass, acceleration; Kinetics of particles: Work, Energy, power; Kinetics of particles: Momentum and impulse; Kinematics of rigid bodies; Kinetics of rigid godies; Gyroscopic theory and application; Free and forced vibrations; Appendices: Vector algebra, and mass moments of inertia.

Figures are all clear and quite amusing, especially Fig. 3.11, which shows a man sitting in a cart pulling himself up an incline using 'frictionless' pulleys and a 'massless' rope. The text is also clear ... achieves its target of providing help for students who have difficulties with this topic."

The Structural Engineer (Dr Brian Ellis)

FINITE ELEMENT TECHNIQUES IN STRUCTURAL MECHANICS

CARL T.F. ROSS, Professor of Structural Dynamics, Department of Mechanical and Manufacturing Engineering, University of Portsmouth

ISBN: 1-898563-25-X 224 pages 1997

Contents: Introduction to matrix algebra; The matrix displacement method; Finite element method; Dynamics of structures; Elastic buckling and the non-linear behaviour of structures; Modal method of analysis.

"All Carl Ross' previous books are very clear and well written. This thoroughly interesting text is no exception. The worked examples and practice problems are particularly useful. I will continue to recommend Professor Ross' books to my students. For anyone requiring an introduction to finite element analysis, this text is excellent."

Journal of Strain Analysis (Dr. S.J. Hardy, Swansea University College of Wales)

"Dr Ross has succeeded in introducing an elementary text which should definitely help any beginner. It is easy to read and the reader may enjoy solving examples for practice given at the end of most chapters."

Journal of Mechanical Engineering Science (Dr El Zafrani, Cranfield University)

"Very readable, clear and easy to follow. It should be especially useful to those new to the finite element method."

Chartered Mechanical Engineer (Dr Peter Hartley)

ENGINEERING MATHEMATICS

N. CHALLIS & H. GRETTON, Department of Engineering, Sheffield Hallam University

ISBN: 1-898563-65-9 *ca.* 250 pages 1999

This book aims to help students to take a modern approach to practical engineering mathematical techniques, fully embracing modern technology, i.e, hand-held machines, spreadsheets, symbol manipulators. With so much technological power available, the emphasis broadens from mechanics of solution to include specifying the problem, asking if the answer is appropriate and convincng oneself and others of this. The book encourages a range of solution approaches using SONG (an acronym of Symbolic, Oral, Numerical and Graphical), reflecting the Harvard Reform Calculus movement. This addresses the richness of the ideas, deepens understanding, and allows confirmation of solutions. Development of key skills is integrated: communication both written and oral, IT use, problem solving and modelling. The structure should help students become more conscious of and responsible for their own learning, for example using self-assessments.

Contents: Numbers, algebra and functions; Differential and integral calculus; Linear simultanious equations and matrices; Ordinary differential equations; Sequences and series; Basic statistics and probability.

MACRO-ENGINEERING AND THE EARTH: World Projects for Year 2000
Festchrift in Honour of Frank Davidson

UWE KITZINGER, President, International Association of Macro-Engineering Societies, *and* First President of Templeton College, Oxford, *and* ERNST G. FRANKEL, Professor of Ocean Engineering, Massachusetts Institute of Technology, Cambridge, Mass, USA

ISBN: 1-898563-59-4 350 pages 1998

This book delivers a wealth of ideas, comments and projects on energy, communications, transport, management, human resources, and financial and legal issues which macro-engineering can contribute towards the solution of the Earth's environmental problems. Some twenty engineers and scholars identify problems in the next century whose solutions call for international political planning and a more collaborative, peaceful and prosperous world order.

Contents: Turbines with a twist; Power from space; Electricity development; Saving the Earth - the whole Earth; Large climate-changing projects; A sustainable global city; A new urban model; World water supply; A Rhone-Algeria aqueduct; Ocean farming; World-scale seamless transportation; An Asia-Europe railway and Middle-East island; Swiss Alpine tunnel projects; Satellites in world communications; Military and civilian telemedicine; Starting a macro-project on the right foot; Collaborative leadership; High-performance partnering practices; Military-civilian corps; Prime contracting: Profit with peril? Knowledge networking; Concepts of eminent domain; Macro-concepts in China; Challenges for the twenty-first century.

"A cornucopia of brilliant innovative ideas."

Interdisciplinary Science Reviews (Professor J.E. Harris, MBE, FRS)

ANALYSIS OF ENGINEERING STRUCTURES

B. BEDENIK, Faculty of Civil Engineering, University of Maribor, Slovenia, *and*
C. BESANT, Head of Computer-Aided Systems Engineering, Imperial College of Science, Technology and Medicine, University of London

ISBN: 1-898563-55-1 450 pages 1998

This text is a fundamental coverage for advanced undergraduates and postgraduates of structural engineering, and professionals working in industrial and academic research. The methods for structural analysis are explained in great detail, being based on basic static, kinematics and energy methods previously discussed in the text.

Considerable attention is given to practical applications whereby each theoretical analysis is reinforced with worked examples. An innovative approach enables influence lines calculations (using *ψ-functions* developed by one of the authors) in a far simpler manner than by any previously known method. Basic algebra given in the appendices provides the necessary mathematical tools to understand the text.

Contents: Introduction; Definitions and basic concepts; Statically determinate structures; Kinematics of structures; Basic concepts of structural analysis; Deformations; Stiffness and flexibility; The force method; The displacement method; The finite element method; Bridge analysis; Computer applications.

MACRO-ENGINEERING: MIT Brunel Lectures on Global Infrastructure

Editors: FRANK P. DAVIDSON, ERNST G. FRANKEL, & C. LAWRENCE MEADOR Macro-Engineering Research Group, School of Engineering, Massachusetts Institute of Technology, Cambridge, USA

ISBN 1-898563-33-0 208 pages 1997

International authorities from engineering, oceanography, academia, public service, finance and law make available the latest reflections on large-scale engineering for building a better world for tomorrow. Their book describes how great and imaginative concepts may be refined, tested, adapted, financed, implemented and put to use.

Contents: The Suez canal re-visited: 19th century global infrastructure; Operation Mulberry: A floating transportable harbour for World-War II Normandy invasion; Financial engineering of the Channel Tunnel; Old cities and new towns for tomorrow's infrastructure; High speed rail transport in Europe; Lessons learnt from major projects; Fail-safe transport and road travel: Low-impact, high volume transportation systems; Prefabricated and relocatable artificial island technology; Military software and the clean-up of the Alaskan oil spill; Progress for the next century: Survey and suggestions.

"The readership is vast. Some of the world's most significant engineering achievements are recorded. Detailed historical research, combined with the often anecdotal style of the lectures, makes for a very interesting read."

The Structural Engineer

Printed and bound by CPI Group (UK) Ltd, Croydon, CR0 4YY
03/10/2024
01040343-0019